H. alper

SO-AXV-423

Advances in

ORGANOMETALLIC CHEMISTRY

VOLUME 25

Advances in Organometallic Chemistry

EDITED BY

F. G. A. STONE

DEPARTMENT OF INORGANIC CHEMISTRY
THE UNIVERSITY
BRISTOL, ENGLAND

ROBERT WEST

DEPARTMENT OF CHEMISTRY
UNIVERSITY OF WISCONSIN
MADISON, WISCONSIN

VOLUME 25

1986

ACADEMIC PRESS, INC.
Harcourt Brace Jovanovich, Publishers
Orlando   San Diego   New York   Austin
London   Montreal   Sydney   Tokyo   Toronto

COPYRIGHT © 1986 BY ACADEMIC PRESS, INC.
ALL RIGHTS RESERVED.
NO PART OF THIS PUBLICATION MAY BE REPRODUCED OR
TRANSMITTED IN ANY FORM OR BY ANY MEANS, ELECTRONIC
OR MECHANICAL, INCLUDING PHOTOCOPY, RECORDING, OR
ANY INFORMATION STORAGE AND RETRIEVAL SYSTEM, WITHOUT
PERMISSION IN WRITING FROM THE PUBLISHER.

ACADEMIC PRESS, INC.
Orlando, Florida 32887

United Kingdom Edition published by
ACADEMIC PRESS INC. (LONDON) LTD.
24–28 Oval Road, London NW1 7DX

LIBRARY OF CONGRESS CATALOG CARD NUMBER: 64-16030

ISBN 0–12–031125–9

PRINTED IN THE UNITED STATES OF AMERICA

86 87 88 89   9 8 7 6 5 4 3 2 1

Contents

Silenes

A. G. BROOK and KIM M. BAINES

Metalla-Derivatives of β-Diketones

CHARLES M. LUKEHART

Organometallic Sonochemistry

KENNETH S. SUSLICK

Carbene and Carbyne Complexes of Ruthenium, Osmium, and Iridium

MARK A. GALLOP and WARREN R. ROPER

v

Borabenzene Metal Complexes

GERHARD E. HERBERICH and HOLGER OHST

The Synthesis of Organometallics by Decarboxylation Reactions

GLEN B. DEACON, SUELLEN J. FAULKS, and GEOFFREY N. PAIN

Detection of Transient Organometallic Species by Fast Time-Resolved IR Spectroscopy

MARTYN POLIAKOFF and ERIC WEITZ

Carbonyl Derivatives of Titanium, Zirconium, and Hafnium

DAVID J. SIKORA, DAVID W. MACOMBER, and MARVIN D. RAUSCH

ADVANCES IN ORGANOMETALLIC CHEMISTRY, VOL. 25

Silenes

A. G. BROOK and KIM M. BAINES

Lash Miller Chemical Laboratories
University of Toronto
Toronto M5S 1A1, Canada

I

INTRODUCTION AND HISTORY

This review describes the current status of *silenes* (silaethylenes, silaethenes), molecules which contain a silicon–carbon double bond. The heart of the material is derived from a computer-based search of the literature which we believe reports all silenes that have been described to date, either as isolated species, chemically trapped species, proposed intermediates (in reactions where some experimental evidence has been provided), or as the result of molecular orbital calculations. Ionized species

1

Copyright © 1986 by Academic Press, Inc.
All rights of reproduction in any form reserved.

are not included nor are a few uncharacterized high molecular weight species claimed to contain silicon–carbon double bonds. While all silenes are reported, not all references to each silene are included (e.g., $Me_2Si=CH_2$ has more than 50 references), and thus only some of the more recent references are given in some cases. The computer search is believed to be complete to December 20, 1984. The reported 153 silenes, listed by empirical and structural formula and including appropriate references, are given in Table I.

TABLE I

KNOWN SILENES WITH REFERENCES

Empirical formula	Structural formula	Refs.
$CSiF_4$	$F_2Si=CF_2$	1,2
CH_2SiD_2	$D_2Si=CH_2$	4–8
CH_2SiD_2	$H_2Si=CD_2$	9
CH_2SiF_2	$H_2Si=CF_2$	1–3
CH_2SiF_2	$F_2Si=CH_2$	1–3,10,11
CH_2SiFCl	$FClSi=CH_2$	10
CH_2SiCl_2	$Cl_2Si=CH_2$	10
CH_2SiF_2	$FHSi=CHF$	3
CH_3SiCl	$ClHSi=CH_2$	7,8,12
CH_3SiF	$H_2Si=CHF$	3,13
CH_3SiF	$FHSi=CH_2$	3,13
CH_3SiD	$HDSi=CH_2$	7,8
CH_3SiNO_2	$H_2Si=CHNO_2$	13
CH_3SiNO_2	$NO_2(H)Si=CH_2$	13
CH_4SiO	$H_2Si=CHOH$	13,14
CH_4SiO	$HOSiH=CH_2$	13,14
CH_4Si	$H_2Si=CH_2$	4,7,13,15–17
CH_6Si_2	$H_2Si=CHSiH_3$	13
CH_6Si_2	$H_3Si(H)Si=CH_2$	13
CH_6Si_2O	$H_3SiO(H)Si=CH_2$	13
CH_6Si_2O	$H_2Si=CHOSiH_3$	13
C_2H_3SiN	$NC(H)Si=CH_2$	13
C_2H_3SiN	$H_2Si=CHCN$	13
C_2H_4Si	$H_2Si=C=CH_2$	19,20–23
C_2H_4Si	$H_2C=Si=CH_2$	19,21,22,24
C_2H_4Si	$(-HSi=CHCH_2-)$	21,22
$C_2H_4Si_2$	$(-SiH=CH-SiH=CH-)$	25
$C_2H_4Si_2$	$(-HSi=CH-CH=SiH-)$	25
C_2H_5SiCl	$Me(Cl)Si=CH_2$	11,26
C_2H_5SiF	$Me(F)Si=CH_2$	11,26
C_2H_6Si	$H_2Si=CHMe$	3,27
C_2H_6Si	$MeHSi=CH_2$	13,28–32

TABLE I (*continued*)

Empirical formula	Structural formula	Refs.
$C_2H_6Si_2$	$H_2Si{=}C{=}CHSiH_3$	20
$C_2H_6Si_2O_3$	$H_2C{=}Si(OH){-}O{-}Si(OH){=}CH_2$	33
C_3SiD_8	$(CD_3)_2Si{=}CD_2$	34,35
$C_3H_2SiD_6$	$(CD_3)_2Si{=}CH_2$	35–38
C_3H_4Si	$({-}SiH{=}CH{-}CH{=}CH{-})$	39,40
$C_3H_5SiD_3$	$(CD_3)MeSi{=}CH_2$	36,41
$C_3H_6SiD_2$	$Me_2Si{=}CD_2$	34,35
C_3H_6Si	$(CH_2{=}CH)HSi{=}CH_2$	42
C_3H_6Si	$H_2Si{=}CH(CH{=}CH_2)$	42
C_3H_8Si	$Me_2Si{=}CH_2$	43–48
C_4H_8Si	$(CH_2{=}CH)MeSi{=}CH_2$	49,50
C_4H_8Si	$({-}CH_2CH_2CH_2{-})Si{=}CH_2$	51
C_4H_8Si	$Me(H)Si{=}CHCH{=}CH_2$	52,53
C_4H_8Si	$[{-}(Me)Si{=}CH{-}CH_2CH_2{-}]$	50
C_4H_8Si	$MeHSi{=}C({-}CH_2CH_2{-})$	52
$C_4H_{10}Si$	$Me_2Si{=}CHMe$	54–58
$C_4H_{12}Si_2$	$Me_3Si(H)Si{=}CH_2$	59
C_5H_6Si	$({-}SiH{=}CH{-}CH{=}CH{-}CH{=}CH{-})$	39,60–64
C_5H_8Si	$({-}MeSi{=}CHCH_2CH{=}CH{-})$	144
$C_5H_{10}Si$	$(CH_2{=}CHCH_2)MeSi{=}CH_2$	65
$C_5H_{12}Si_2$	$[{-}(Me_3Si)Si{=}CHCH_2{-}]$	145
$C_5H_{14}Si_2$	$Me_3SiSiMe{=}CH_2$	65,66
C_6H_8Si	$[{-}Si(Me){=}CH{-}CH{=}CH{-}CH{=}CH{-}]$	67–71
$C_6H_{14}Si_2$	$[{-}(Me_3Si)Si{=}CHCH_2CH_2{-}]$	53
$C_6H_{14}Si_2$	$[{-}CH_2CH(SiMe_3){-}Si({=}CH_2){-}]$	53
$C_6H_{14}Si_2$	$Me_3Si(H)Si{=}CHCH{=}CH_2$	53
$C_6H_{16}Si_2$	$Me_2Si{=}CH(SiMe_3)$	72
C_7H_8Si	$PhHSi{=}CH_2$	73
$C_7H_{10}Si$	$Me_2Si{=}C({-}CH{=}CH{-}CH{=}CH{-})$	74–76
$C_7H_{10}Si$	$[{-}Si(Me){=}CH{-}CH{=}C(Me){-}CH{=}CH{-}]$	77
$C_7H_{18}SiGe$	$Me_2Si{=}C(Me_3Ge)Me$	78
$C_7H_{18}Si_2$	$Me_2Si{=}C(Me_3Si)Me$	78
$C_7H_{18}Si_2$	$Me_2Si{=}CHCH_2SiMe_3$	133
$C_7H_{20}Si_3$	$Me_3SiSiMe_2Si(Me){=}CH_2$	79
$C_8H_{18}Si$	$Me_2Si{=}CHCH_2CMe_3$	80
$C_8H_{18}Si$	$Me_2Si{=}CHCH_2CHMeCH_2CH_3$	141
$C_8H_{18}Si$	$Me_2Si{=}CH(CH_2)_4CH_3$	141
$C_8H_{18}Si_2$	$Me_2Si{=}CHCH_2SiMe_2CH{=}CH_2$	81,82,133
$C_8H_{18}Si_2$	$Me_2Si{=}CHCH{=}CHSiMe_3$	140
$C_9H_{10}Si$	$Ph(CH_2{=}CH)Si{=}CH_2$	73
$C_9H_{10}Si$	$[{-}(Me)Si{=}CH{-}(o{\text-}C_6H_4){-}CH_2{-}]$	83
$C_9H_{12}Si$	$PhEtSi{=}CH_2$	73
$C_9H_{12}Si$	$Me_2Si{=}CHPh$	84
$C_9H_{12}Si$	$[{-}C({=}SiMe_2){-}C({=}CH_2)CH{=}CHCH{=}CH{-}]$	149

(*continued*)

TABLE I (continued)

Empirical formula	Structural formula	Refs.
$C_9H_{14}Si$	$[-SiH=CH-CH=C(t\text{-Bu})-CH=CH-]$	77
$C_9H_{20}Si_2O_2$	$Me_2Si=C(SiMe_3)COOEt$	85,86,146
$C_9H_{24}Si_3$	$Me_2Si=C(SiMe_3)_2$	72,87–90,147
$C_{10}H_{12}Si_2$	$H_2Si=C[-CH=CH-(o\text{-}C_6H_4)-CH(SiH_3)-]$	91
$C_{10}H_{12}Si_2$	$H_2Si=C[-CH(SiH_3)-CH=CH-(o\text{-}C_6H_4)-]$	91
$C_{10}H_{12}Si_2$	$H_2Si=C[-CH(SiH_3)-CH=(o\text{-}C_6H_4)=CH-]$	91
$C_{10}H_{14}SiO_2W$	$(Me_2Si=CH_2)WH(CO)_2(C_5H_5)$	92
$C_{10}H_{14}SiO$	$Me_2Si=CHCH_2CH[-C(=O)-CH=CH-CH=CH-]$	93
$C_{10}H_{16}Si$	$[-Si(Me)=CH-CH=C(t\text{-Bu})-CH=CH-]$	77
$C_{10}H_{19}SiCl$	$n\text{-}BuClSi=CH(c\text{-}C_5H_9)$	94
$C_{10}H_{22}Si_2$	$Me_2Si=CMeCH_2SiMe_2(CMe=CH_2)$	133
$C_{10}H_{28}Si_4$	$(Me_3Si)_2Si=CHSiMe_3$	95
$C_{11}H_{14}Si$	$Me_2Si=CPh(CH=CH_2)$	96,148
$C_{11}H_{16}Si$	$Ph(t\text{-}Bu)Si=CH_2$	73
$C_{11}H_{20}Si_2$	$Me_2Si=C[-CH(SiMe_3)-CH=CH-CH=CH-]$	97–102
$C_{11}H_{30}Si_4O$	$(Me_3Si)_2Si=CMe(OSiMe_3)$	103,104
$C_{12}H_{10}SiO$	$[-(Si=CHCH=CHCH=C)-(o\text{-}C_6H_4)CH_2O-]$	105
$C_{12}H_{20}Si_2$	$PhMeSi=CHCH_2SiMe_3$	81,82
$C_{12}H_{22}Si_2$	$Me_2Si=C[-CH(SiMe_3)CH=CMeCH=CH-]$	97,99,106–108
$C_{12}H_{22}Si_2$	$Me_2Si=C[-CH(SiMe_3)CH=CHCH=CMe-]$	91
$C_{12}H_{22}Si_2O$	$Me_2Si=C[-CH(SiMe_3)CH=CHCH=C(OMe)-]$	91
$C_{13}H_8SiCl_2$	$Cl_2Si=(C_{13}H_8)(9\text{-fluorene})$	109
$C_{13}H_{10}Si$	$[-SiH=(o\text{-}C_6H_4)=CH-(o\text{-}C_6H_4)-]$	110
$C_{13}H_{12}Si$	$Ph_2Si=CH_2$	111–114
$C_{13}H_{20}Si$	$PhMeSi=CHCH_2CMe_3$ (Z,E)	115–118
$C_{13}H_{20}Si_2$	$Me_2Si=C=CPh(SiMe_3)$	119
$C_{13}H_{22}Si$	$[-Si(t\text{-}Bu)=CH-CH=C(t\text{-}Bu)-CH=CH-]$	120
$C_{13}H_{22}Si_2$	$Me_2Si=CPhCH_2SiMe_3$	82,100,133
$C_{14}H_{36}Si_4O$	$(Me_3Si)_2Si=C(CMe_3)(OSiMe_3)$	104,121–124
$C_{15}H_{22}Si_2$	$Me_2Si=C[-CHSiMe_3CH=CH-(o\text{-}C_6H_4)-]$	91
$C_{15}H_{22}Si_2$	$Me_2Si=C[-CHSiMe_3CH=(o\text{-}C_6H_4)=CH-]$	91
$C_{15}H_{24}SiO_2W$	$(Me_2Si=CH_2)WH(CO)_2(C_5Me_5)$	92
$C_{15}H_{28}Si_2$	$Me_2Si=C[-CHSiMe_3CH=C(CMe_3)CH=CH-]$	99
$C_{15}H_{36}Si_3$	$Me_2Si=C(SiMe_3)[SiMe(t\text{-}Bu)_2]$	90,125
$C_{15}H_{36}Si_3$	$(t\text{-}Bu)_2Si=C(SiMe_3)_2$	90,142
$C_{16}H_{20}Si_2$	$[-SiMe=(C-CHSiMe_3CH=CHCH=C)-(o\text{-}C_6H_4)-]$	138,139
$C_{16}H_{32}Si_4O$	$(Me_3Si)_2Si=CPh(OSiMe_3)$	104,121,123,124
$C_{17}H_{30}OSi_2O$	$Me_2Si=C(SiMe_3)C(=O)C_{10}H_{15}$	126
$C_{17}H_{40}Si_4O$	$(Me_3Si)_2Si=C(OSiMe_3)[C(Me)-(CH_2)_5-]$	123
$C_{17}H_{42}Si_4O$	$(Me_3Si)_2Si=C(OSiMe_3)C(Et)_3$	122–124
$C_{19}H_{20}Si$	$Me_2Si=CPhCH=CHCPh=CH_2$	135
$C_{19}H_{28}Si_3$	$Ph_2Si=C(SiMe_3)_2$	136,143
$C_{19}H_{28}Si_3$	$Me_2Si=C(SiMe_3)(SiMePh_2)$	136
$C_{19}H_{28}Si_3$	$MePhSi=CSiMe_3(SiMe_2Ph)$	136

TABLE I (*continued*)

Empirical formula	Structural formula	Refs.
$C_{19}H_{28}Si_3$	$Me_2Si{=}C(SiMe_2Ph)_2$	*136*
$C_{19}H_{44}Si_3O$	$Me_2Si{=}C(SiMe_3)[Me(t\text{-}Bu)_2Si]\cdot C_4H_8O$	*127*
$C_{20}H_{24}Si_2$	$[{-}(Me)Si{=}CPhCH{=}CHCPh(SiMe_3){-}]$	*128*
$C_{20}H_{24}Si_2$	$PhMeSi{=}C[{-}CHSiMe_3CH{=}CH{-}(o\text{-}C_6H_4){-}]$	*91*
$C_{20}H_{24}Si_2$	$Me_2Si{=}C[{-}CHSiMe_2PhCH{=}CH{-}(o\text{-}C_6H_4){-}]$	*91*
$C_{20}H_{24}Si_2$	$Me_2Si{=}CPhCH{=}CHCPh{=}SiMe_2$	*133,134*
$C_{20}H_{26}Si_2$	$Me_2Si{=}CPhCH_2SiMe_2(CPh{=}CH_2)$	*133*
$C_{20}H_{28}Si_3$	$Ph_2Si{=}C{=}C(SiMe_3)_2$	*136*
$C_{20}H_{42}Si_4O$	$(Me_3Si)_2Si{=}C(OSiMe_3)C_{10}H_{15}$	*122–124,129*
$C_{21}H_{22}Si_2$	$[{-}SiPh{=}(\overline{C{-}CHSiMe_3CH{=}CHCH{=}C}){-}(o\text{-}C_6H_4){-}]$	*138,139*
$C_{21}H_{26}Si$	$(t\text{-}Bu)_2Si{=}(C_{13}H_8)(9\text{-fluorene})$	*130*
$C_{21}H_{28}Si_3$	$Ph_2Si{=}C{=}C{=}C(SiMe_3)_2$	*136*
$C_{21}H_{30}Si_3$	$(Me_3SiCH_2)_2Si{=}(C_{13}H_8)(9\text{-fluorene})$	*131*
$C_{21}H_{32}Si_3$	$(p\text{-}MeC_6H_4)_2Si{=}C(SiMe_3)_2$	*137*
$C_{21}H_{32}Si_3$	$(m\text{-}MeC_6H_4)_2Si{=}C(SiMe_3)_2$	*137*
$C_{21}H_{32}Si_3$	$(o\text{-}MeC_6H_4)_2Si{=}C(SiMe_3)_2$	*137*
$C_{21}H_{32}Si_3$	$Me_2Si{=}C(SiMe_3)[SiMe(p\text{-}MeC_6H_4)_2]$	*137*
$C_{21}H_{32}Si_3$	$Me_2Si{=}C(SiMe_3)[SiMe(m\text{-}MeC_6H_4)_2]$	*137*
$C_{21}H_{32}Si_3$	$p\text{-}MeC_6H_4MeSi{=}CSiMe_3(p\text{-}MeC_6H_4SiMe_2)$	*137*
$C_{21}H_{32}Si_3$	$m\text{-}MeC_6H_4MeSi{=}CSiMe_3(m\text{-}MeC_6H_4SiMe_2)$	*137*
$C_{21}H_{32}Si_3$	$Me_2Si{=}C[SiMe_2(p\text{-}MeC_6H_4)]_2$	*137*
$C_{21}H_{32}Si_3$	$Me_2Si{=}C[SiMe_2(m\text{-}MeC_6H_4)]_2$	*137*
$C_{22}H_{32}Si_3$	$(p\text{-}MeC_6H_4)_2Si{=}C{=}C(SiMe_3)_2$	*137*
$C_{22}H_{32}Si_3$	$(m\text{-}MeC_6H_4)_2Si{=}C{=}C(SiMe_3)_2$	*137*
$C_{22}H_{32}Si_3$	$(o\text{-}MeC_6H_4)_2Si{=}C{=}C(SiMe_3)_2$	*137*
$C_{23}H_{32}Si_3$	$(p\text{-}MeC_6H_4)_2Si{=}C{=}C{=}C(SiMe_3)_2$	*137*
$C_{23}H_{32}Si_3$	$(m\text{-}MeC_6H_4)_2Si{=}C{=}C{=}C(SiMe_3)_2$	*137*
$C_{23}H_{32}Si_3$	$(o\text{-}MeC_6H_4)_2Si{=}C{=}C{=}C(SiMe_3)_2$	*137*
$C_{28}H_{46}Si_5$	$[{-}(Me_3Si)Si{=}CPhCSiMe_3{=}CPhC(SiMe_3)_2{-}]$	*138,139*
$C_{31}H_{26}Si$	$Me_2Si{=}C({-}CPh{=}CPh{-}CPh{=}CPh{-})$	*132*
$C_{31}H_{28}Si$	$Me_2Si{=}CPhCPh{=}CPhCPh{=}CH_2$	*135*
$C_{32}H_{32}Si_2$	$Me_2Si{=}CPhCPh{=}CPhCPh{=}SiMe_2$	*133,134*

Prior to 1966–1967, based on numerous attempts to synthesize molecules containing a silicon–carbon double bond (*150–152*) the firm position had been reached that, if they could be formed at all, silenes were (would be) too unstable to survive. It was believed that the π bond would be very weak because of poor overlap, due to the differences in energy and the longer bond distances between the adjacent $2p$ and $3p$ orbitals of carbon and silicon, respectively (*153,154*). In 1967 Gusel'nikov and Flowers (*155*) reported experiments which reopened the entire issue of π bonding between silicon and carbon. In the pyrolysis of 1,1-dimethyl-1-silacyclobutane there was clear evidence for the formation of ethylene and indirect evidence for the formation of 1,1-dimethylsilene, based on trapping experiments and on the isolation of its self-dimerization product, 1,1,3,3-tetramethyl-1,3-disilacyclobutane [Eq. (1)]. These experiments

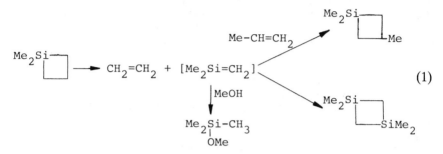

$$(1)$$

prompted a variety of investigations on double bonding between elements of the second and third row of the periodic table and culminated, after a decade of intense research activity, in papers describing the isolation of stable silenes and disilenes (as well as other species such as those containing the P=C or As=C bonds). During the 17 years since the Gusel'nikov work, evidence for the intermediacy of (or calculations concerning) more than 150 silenes has been published, and what follows is an attempt to survey and summarize the current state of knowledge of this class of compound.

A number of earlier reviews are available which survey various topics in detail. These include works by Gusel'nikov and Nametkin (*156*), Barton (*157*), Coleman and Jones (*158*), and Schaefer (*159*) as well as brief sections in "Comprehensive Organic Chemistry" (*160*) and "Comprehensive Organometallic Chemistry" (*161*).

As a result of the intense activity in the area of silenes in recent years, there are now clear expectations of what the "normal" behavior of a "simple" silene (i.e., one not substituted with very large or very polarizing groups) will be. Such a silene, usually formed in a high temperature reaction, will be a highly reactive transient species, readily trapped by alcohols to form alkoxysilanes, or by dienes to form silacyclohexenes. In the absence of a trapping agent, head-to-tail dimerization to a stable 1,3-disilacyclobutane is a major reaction pathway, but rearrangements to isomeric silylenes, particularly in the presence of uv radiation or high temperatures, are also a very common fate of these reactive species.

As can be seen from Table I, the silenes that have been investigated range from the very simplest ($H_2Si=CH_2$) to complex and heavily substituted species which are indefinitely stable at room temperature, and whose chemical and physical properties, including crystal structures, have been thoroughly documented. What follows is a digest of the available information.

II

SYNTHESIS OF SILENES

The silenes that have been synthesized to date have ranged in stability all the way from those which are isolated only if trapped in argon matrices at very low temperatures (e.g., $Me_2Si=CH_2$, $MeSiH=CH_2$) to solids which survive for years if kept under argon, away from oxygen and moisture. However, many silenes have been recognized or identified as reaction intermediates only on the basis of trapping experiments (often with alcohols or dienes), or as the products of their self-dimerization, or as products of intramolecular isomerization. A wide variety of methods has been used for silene synthesis, most of which can be assigned to one of three general categories: thermal, photochemical, and 1,2-elimination reactions.

A. *Thermal and Photochemical Routes*

A much explored pathway to simple silenes involves the thermolysis of silacyclobutanes at 400–700°C, the original Gusel'nikov–Flowers (*155*) route. Such temperatures are not readily conducive to the isolation and study of reactive species such as silenes except under special conditions, and flash thermolysis, or low pressure thermolysis, coupled with use of liquid nitrogen or argon traps has frequently been employed if study of the physical properties is desired. Under these high temperature conditions rearrangements of simple silenes to the isomeric silylenes have been observed which can lead to complications in the interpretation of results (*53,65*). Occasionally phenyl-substituted silacyclobutanes have been photolyzed at 254 nm to yield silenes (*113*) as has dimethylsilacyclobutane in the vapor phase (147 nm) (*162*).

A somewhat milder route which appears to be devoid of the complications of isomerization is the retro-Diels–Alder reaction of bicyclo [2.2.2] octadienes, frequently substituted with aryl groups (*5,30,53,65*), [Eq. (2)], and recently Wiberg (*88,90*) described a very mild route involving both [2 + 2] and [2 + 4] cycloreversions which occur at 60°C to generate $Me_2Si=C(SiMe_3)_2$. However, the generality of this latter source of silenes has not been established yet [Eq. (3)].

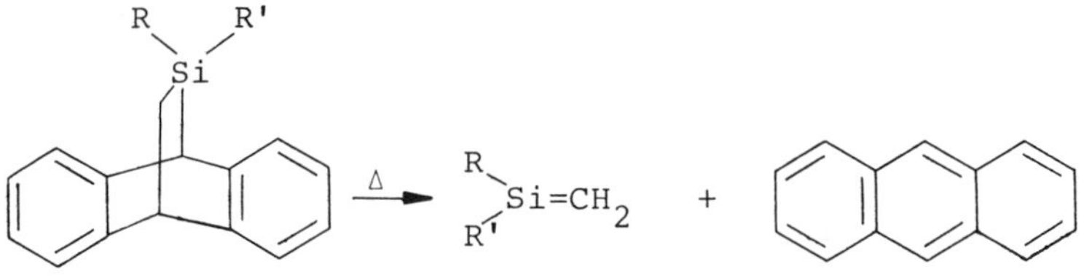

$$R_2Si \rightarrow \overset{\Delta}{\longrightarrow} \quad \underset{R'}{\overset{R}{\diagdown}}Si=CH_2 \quad + \quad \text{(anthracene)} \tag{2}$$

$$\underset{Ph}{\overset{Me_2Si-C(SiMe_3)_2}{\underset{Me_3Si-N}{\diagdown}}} \underset{\overset{60°}{\rightleftharpoons}}{} \quad \begin{array}{c} Me_2Si=C(SiMe_3)_2 \\ + \\ Me_3SiN=CPh_2 \end{array} \quad \underset{\overset{60°}{\rightleftharpoons}}{} \quad \underset{Me_3Si}{\overset{Me_2Si \!\!-\!\! C(SiMe_3)_2}{\underset{N \!\!-\!\! CPh_2}{|\quad\quad|}}} \tag{3}$$

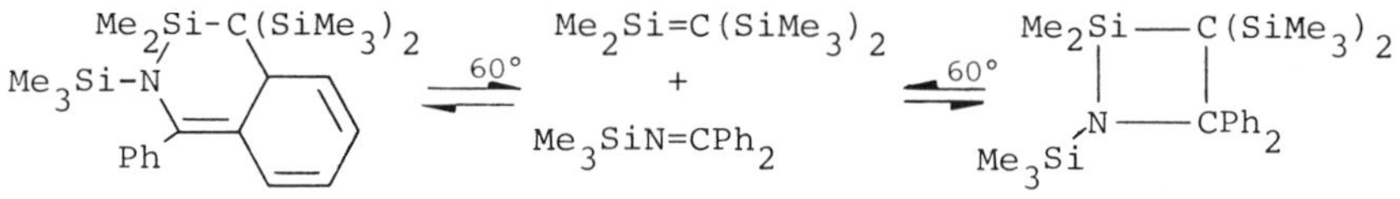

Another thermal route which has been particularly useful for the generation of silaaromatics (silabenzene, silatoluene, 4-substituted silaaromatics, etc.) and silafulvenes is the "ene" reaction of allylsilanes, as illustrated [Eq. (4)].

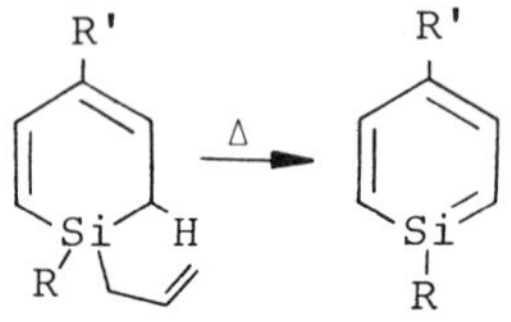

$$\underset{R'}{\overset{R'}{\diagup}}\overset{H}{\underset{Si}{\diagdown}} \overset{\Delta}{\longrightarrow} \quad \underset{R}{\overset{R'}{\diagup}}\overset{}{\underset{Si}{\diagdown}} \tag{4}$$

$$R = R' = H \ (\text{ref. } 71)$$
$$R = R' = \underline{t}\text{-Bu} \ (\text{ref. } 120)$$
$$R = Me, \ R' = H \ (\text{ref. } 69)$$

Two indirect routes to silenes, one derived from silylenes and the other from silylcarbenes, are of some generality and importance. Silylenes (e.g., $Me_3Si\text{---}\dot{\ddot{S}}i\text{---}\triangleleft$) (53) have been derived from the thermolysis of either methoxy or chloro polysilyl compounds. Thermolysis resulted in the elimination of trimethylmethoxy- or trimethylchlorosilane and yielded the silylene, which, based on products of trapping, clearly had rearranged in part to the isomeric silene [Eq. (5)]. Alternatively the silylene Me_2Si: has

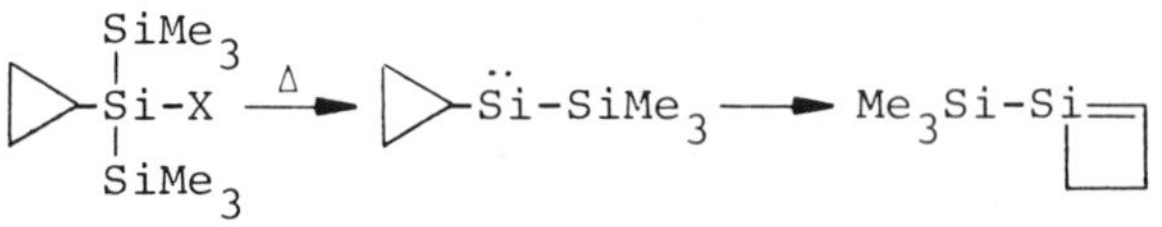

$$\overset{SiMe_3}{\underset{SiMe_3}{\triangleright\!\!-\!\!Si\!\!-\!\!X}} \overset{\Delta}{\longrightarrow} \triangleright\!\!-\!\!\ddot{S}i\!\!-\!\!SiMe_3 \longrightarrow Me_3Si\!\!-\!\!Si{=}\square \tag{5}$$

been generated photochemically at 254 nm from dodecamethylcyclohexasilane, $(Me_2Si)_6$, with subsequent isomerization effected by 450 nm radiation at less than 35 K to give methylsilene [Eq. (6)] (28,163). In another

$$(Me_2Si)_6 \xrightarrow{h\nu} (Me_2Si)_5 + Me_2Si: \xrightarrow{h\nu} MeSiH=CH_2 \tag{6}$$

approach, Ando's group has photolyzed or thermolyzed a variety of silyldiazoalkanes yielding α-silylcarbenes which smoothly rearrange to give silenes having a variety of functional groups attached to carbon [Eq. (7)] (*85,86,126*).

$$\text{Me}_3\text{SiSiMe}_2\text{CN}_2\text{R} \xrightarrow[\text{or } \Delta]{h\nu} \text{Me}_3\text{SiSiMe}_2\ddot{\text{C}}\text{R} \longrightarrow \text{Me}_2\text{Si}=\text{CR}(\text{SiMe}_3)$$

$$\text{R}=\text{COOEt, } \text{COC}_{10}\text{H}_{15}$$

$$(7)$$

Several di- or polysilyl systems have been found to be useful precursors for the photochemical generation of silenes. Vinyldisilanes cleanly yield silylmethylsilenes (*133*), while alkynyldisilanes yield mixtures of silylated silaallenes and silacyclopropenes [Eq. (8)] (*119,136*). Aryldisilanes when photolyzed form species presumed to be silenes, but showing unusual chemical behavior (see below) [Eq. (9)] (*97–102*).

$$\text{R}_3\text{SiSiR}_2\text{CH}=\text{CH}_2 \xrightarrow{h\nu} \text{R}_2\text{Si}=\text{CHCH}_2\text{SiR}_3$$

$$(8)$$

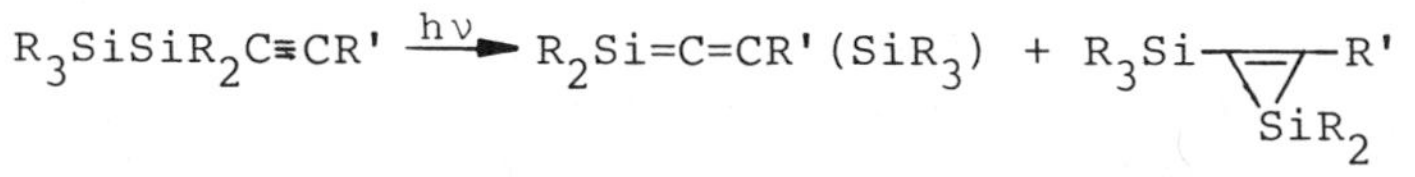

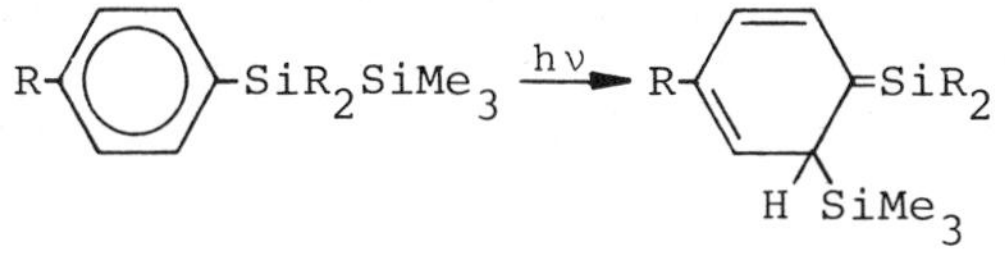

$$\text{R}_3\text{SiSiR}_2\text{C}\equiv\text{CR'} \xrightarrow{h\nu} \text{R}_2\text{Si}=\text{C}=\text{CR'}(\text{SiR}_3) \ + \ \text{R}_3\text{Si}\underset{\text{SiR}_2}{\overset{}{\triangle}}\text{R'}$$

$$(9)$$

Finally, research in our group has shown that a wide variety of polysilylacylsilanes consistently undergo very clean photochemical 1,3-rearrangements of silyl groups from silicon to oxygen and yield silenes, some of which are remarkably long-lived, and two of which have been crystallized [Eq. (10)] (*104,122–124*).

$$\text{R}_3\text{SiSiR}_2\overset{\overset{\text{O}}{\|}}{\text{C}}\text{R'} \longrightarrow \text{R}_2\text{Si}=\text{C}\overset{\text{OSiR}_3}{\underset{\text{R'}}{\diagup}}$$

$$(10)$$

$$(\text{Me}_3\text{Si})_3\text{Si}\overset{\overset{\text{O}}{\|}}{\text{C}}\text{R} \longrightarrow (\text{Me}_3\text{Si})_2\text{Si}=\text{C}\overset{\text{OSiMe}_3}{\underset{\text{R}}{\diagup}}$$

B. *1,2-Eliminations*

The other important general route to silenes involves 1,2-elimination reactions of MX or MY from adjacent silicon and carbon atoms. Usually the metal, normally lithium, is attached to carbon (by metallation), and the halogen (usually Cl, F) or Y group is attached to silicon [Y = PhS, Ph_2PO_2, $(PhO)PhPO_2$, $(PhO)_2PO_2$, p-tolylSO$_3$]. Wiberg has studied these systems in depth, particularly as applied to synthesis of the silene $Me_2Si{=}C(SiMe_3)_2$, and has provided much quantitative data concerning rates and temperatures required for elimination to occur (usually in the range of +50 to −90°C) (*87*). A useful review article is available (*90*).

The other important route to silenes via eliminations has been studied by Jones *et al.* through addition of *t*-butyllithium to vinylchloro- or fluorosilanes followed by 1,2-elimination of the LiX [Eq. (11)]. While silenes

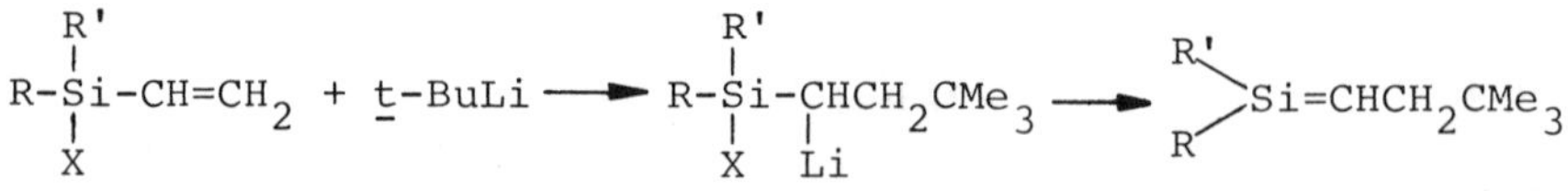

$$
\underset{\underset{X}{|}}{\overset{\overset{R'}{|}}{R-Si}}-CH{=}CH_2 \;+\; \underline{t}\text{-BuLi} \longrightarrow \underset{\underset{X}{|}\;\underset{Li}{|}}{\overset{\overset{R'}{|}}{R-Si}}-CHCH_2CMe_3 \longrightarrow \overset{R'}{\underset{R}{\diagup}}Si{=}CHCH_2CMe_3
$$

$$(11)$$

appear to be formed directly in hydrocarbon solvents, in more basic solvents like tetrahydrofuran it seems probable that coupling reactions of the α-lithiosilanes occur, leading to dimeric 1,3-disilacyclobutanes and other products, without the actual intervention of the silene (*141,164*).

In summary, it is clear that in the relatively few years since the synthesis of the first silene was reported, a wide variety of reliable routes has been developed which can lead cleanly to silenes that vary widely in structure and stability.

III

PHYSICAL PROPERTIES OF SILENES

A. *Infrared Spectra*

While infrared characterization of several silenes has been carried out, it has not always been possible to assign the observed bands unequivocally. Table II summarizes the available infrared data. The infrared spectra of reactive silenes, generated by the pyrolysis of suitable precursors, have most often been recorded in an argon matrix at 10 K, whereas the infrared

TABLE II

INFRARED SPECTRA OF SILENES

Silene	IR method temperature (K)	IR bands observed[a] (cm^{-1})	Calculated Si=C stretching frequency (cm^{-1})	Reference for observed spectrum	Reference[b] for calculated spectrum
$H_2C=SiH_2$	Ar matrix, 10	2239(m), 2219(m), 985(w), 927(w), 817(s), 741(s)	992	5,6	4
$H_2C=SiD_2$	Ar matrix, 10	1635(m), 1600(m), 952(m), 759(s), 719(s), 396(w)	980	5,6	4
$H_2C=SiCl_2$	Ar matrix, 10	1008(m), 732(s), 593(m)	1020	5	4
$MeHSi=CH_2$	Ar matrix, 30	2188(s), 1412(m), 1397(m), 1300(m), 1254(s), 988(s), 830(s), 880(s), 812(vs), 714(m), 732(m), 688(w), 615(m)	1049	28,6	28

(continued)

TABLE II (continued)

Silene	IR method temperature (K)	IR bands observed[a] (cm^{-1})	Calculated Si=C stretching frequency (cm^{-1})	Reference for observed spectrum	Reference[b] for calculated spectrum
Me$_2$Si=CH$_2$	Ar matrix, 10	1259, 1251, 1003, 825, 817, 643	1014	165,37	38
(CD$_3$)$_2$Si=CH$_2$	Ar matrix, 10	1016, 1001, 769, 687, 579	977, 1016	165	38
(CD$_3$)MeSi=CH$_2$	Ar matrix, 10	1259, 1025, 1016, 975, 817, 755, 606	—	36	—
D$_2$C=SiMe$_2$	Ar matrix, 12	1259, 1251, 1117, 895, 820, 776, 617, 535, 508	—	34	—
D$_2$C=Si(CD$_3$)$_2$	Ar matrix, 12	1112, 1028, 1002, 866, 732, 657, 501	—	34	—
Me$_2$Si=CHMe	Ar matrix, 8	3020, 2980, 2965, 2940, 2900, 2870, 1450, 1410, 1370, 1315, 1255, 1120, 978, 883, 808, 795, 712, 708, 645, 608, 358	—	56	—

Compound	Conditions	Frequencies (cm⁻¹)		Ref.	
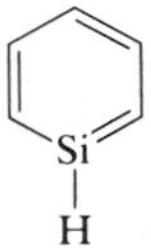 benzene ring–Si=, H	Ar matrix, 10	2217, 1526, 1500, 1353, 1259, 886, 598, 566, 418	—	*172*	—
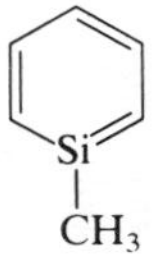 benzene ring–Si=, CH_3	Ar matrix, 23	1530, 1500, 1410, 1360, 1268, 980, 965, 900, 890, 883, 842, 770, 697, 655, 563	—	*68*	—
$(Me_3Si)_2Si=C(OSiMe_3)(CMe_3)$	Thin film / C_6D_6 solution	1136, 1042 / 1130	—	*104*	—
$(Me_3Si)_2Si=C(OSiMe_3)(CEt_3)$	C_6D_6 solution	1133	—	*122*	—
$(Me_3Si)_2Si=C(OSiMe_3)(C_{10}H_{15})$	Nujol	1263(m), 1255(m), 1247(m), 1195(w), 1135(s), 1101(w), 1007(w), 933(w)	—	*122*	—
	Solid	1310(w), 1257(m), 1194(m), 1170(s), 1132(s), 1097(w), 1002(m), 970(w), 932(w)			

[a] Tentative assignment/assignment of the Si=C stretching band is underlined.
[b] Remaining calculated frequencies and assignments are also given in these references.

13

TABLE III Chemical Shifts

Silene	$\delta(Si^{a,a'})$	$\delta(Si^b)$	$\delta(Si^c)$
I Me$_3$Si^a, Me$_3$Si$_{a'}$ — Si^b=C(OSi^{c}Me$_3$)(CMe$_3$)	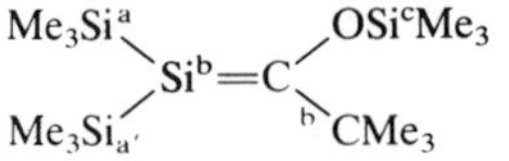 −12.1 −12.6	41.5	13.4
II Me$_3$Si^a, Me$_3$Si$_{a'}$ — Si^b=C(OSi^{c}Me$_3$)(CEt$_3$)	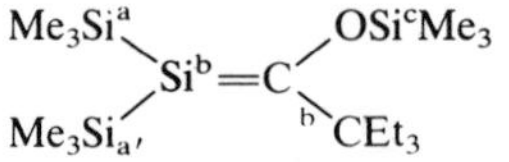 −12.4 −12.9	54.3	12.7
III Me$_3$Si^a, Me$_3$Si$_{a'}$ — Si^b=C(OSi^{c}Me$_3$)(Me, cyclohexyl)	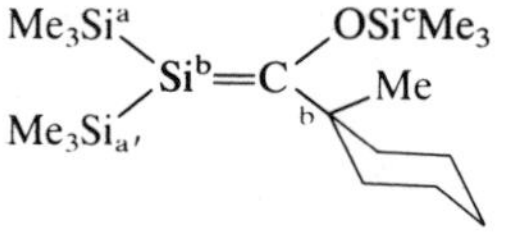 −12.3 −12.6	43.5	13.6
IV Me$_3$Si^a, Me$_3$Si$_{a'}$ — Si^b=C(OSi^{c}Me$_3$)(cyclohexyl)	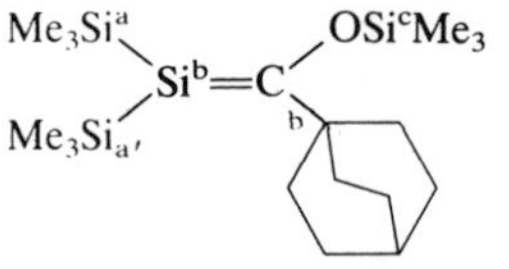 −12.3 −12.8	42.4	13.3
V Me$_3$Si^a, Me$_3$Si$^{a'}$ — Si^b=C(OSi^{c}Me$_3$)(adamantyl)	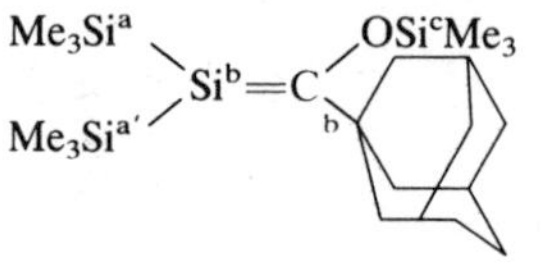 −12.8 −13.4	41.4	12.9
VI Me$_3$Si^{c}O, Me — Si^b=C(Si^{a}Me$_2$t-Bu)(adamantyl)	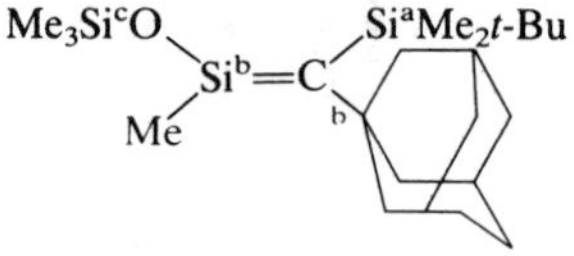 8.6	126.5	6.4
VII Me, Me — Si^b=C(Si^{a}Me$_3$)(Si$^{a'}$me-t-Bu$_2$)	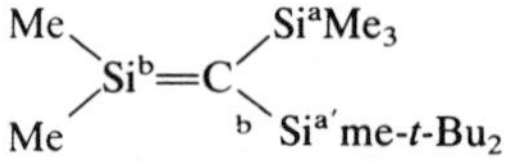 −4.6^d 6.0	144.2	—

[a] ^{29}Si δ in ppm relative to external Me$_4$Si, in C$_6$D$_6$. ^{13}C δ in ppm relative to
[b] J in Hz.
[c] Not measured.
[d] ^{13}C and ^{29}Si in C$_6$D$_5$CD$_3$ at −70° C.

spectra of the more stable, highly substituted silenes have been obtained by more conventional methods. The data can be summarized as follows: (i) the Si=C stretching band of deuterated or deuteriomethylated silenes appears to be in the range of 950–1150 wavenumbers, whereas for the parent silene or simple methylated derivatives, is falls in a much narrower range of 985–1005 wavenumbers (*37,165*), (ii) silenes of the general structure (Me$_3$Si)$_2$Si=C(OSiMe$_3$)(alkyl) appear to have a characteristic

AND COUPLING CONSTANTS FOR SILENES[a,b]

$\delta(C^b)$	$^1J(Si{=}C)$	$^1J(Si^a{-}Si^b)$	$^1J(Si^aMe)$	Ref.
212.7	83.5	70.8	48.0	123
207.3	83.9	73.2	48.0	123
212.9	85.0	72.1	47.7	123
212.7	—[c]	—[c]	—[c]	207
214.2	84.4	72.0	47.6	123
118.1	—[c]	—	—[c]	200
77.2	—	—	—	167

C_6D_6, unless otherwise noted.

band at approximately 1130–1136 cm^{-1}; however, the origin of this band is still uncertain (122), (iii) the Si$=$C stretching band of the silaaromatics remains unassigned, (iv) the Si—H stretching modes in the parent silene, 1-methylsilene, and silabenzene are found approximately 100 cm^{-1} higher than the Si—H stretching mode in silanes. This is convincing evidence that the hydrogen atom is indeed attached to an sp^2-hybridized silicon atom since the same trend occurs in carbon chemistry.

B. Nuclear Magnetic Resonance Spectra

The bulky, stable silenes of Brook *et al.* (*104,122–124,168*) and Wiberg *et al.* (*166,167*) have been the only systems capable of being studied by nuclear magnetic resonance (NMR) spectroscopy to date. Table III lists the ^{13}C and ^{29}Si chemical shifts and the relevant coupling constants of these compounds.

The ^{29}Si chemical shifts of the alkylsilyl and trimethylsiloxy groups fall within the normal range observed for this type of silicon. However, it is interesting to note that the two trimethylsilyl groups in silenes I–V are not chemically equivalent, thus indicating no free rotation about the silicon–carbon double bond even at 60°C. The same holds true for the two methyl groups attached to the sp^2-hybridized silicon in silene VII (^{13}C, −2.10, 10.81 ppm). The ^{29}Si shifts of the sp^2-hybridized silicon atoms fall at unusually low field. Again, this is consistent with a deshielded sp^2-hybridized silicon in a double bond. The range for silenes I to V is upfield from the ^{29}Si shift observed for an aryl-substituted disilene [(mesityl)$_2$Si=Si(mesityl)$_2$, 63.6 ppm] (*169*), reflecting the strong shielding character of trimethylsilyl groups. When these groups are replaced with a deshielding trimethylsiloxy group (and an alkyl group, as in silene VI) or simple alkyl groups (as in silene VII), the signal of the sp^2-hybridized silicon moves a considerable distance downfield (values of 126.5 and 144.2 ppm, respectively, have been observed). The deshielding effect of the siloxy group [and possibly the $(Me_3Si)_2Si$=group] also affects the chemical shift of the sp^2-hybridized carbon in silenes I–V. These shifts are at considerably lower field than the corresponding ^{13}C shift in alkenes (80–150 ppm) (*170*). The sp^2-hybridized carbon in silene VI, substituted with a trialkylsilyl rather than a trimethylsiloxy group, is considerably more shielded (118.1 ppm), and when substituted with two trialkylsilyl groups as in silene VII the shielding is even greater (77.2 ppm) (*167*).

The larger $^1J(Si=C)$ and $^1J(Si^a—Si^b)$ coupling constants of silenes I to V compared to systems having only sp^3-hybridized atoms (*123*) is another indication of sp^2-hybridization at silicon b.

C. Ultraviolet Spectra

The available ultraviolet (uv) data on silenes can be divided into three categories: (1) simple silenes, (2) silaaromatics, and (3) highly substituted silenes. The uv spectra of the silene $H_2Si=CH_2$, 1,1-dideuteriosilene, and 1,1-dichlorosilene (category 1) were recorded in an argon matrix at 10 K. They all show a characteristic absorption at approximately 245–260 nm

(5). The uv spectra of silabenzene and silatoluene (category 2), recorded in a frozen argon matrix, are characteristic of a π-perturbed aromatic system. The uv absorptions are shifted to longer wavelengths, compared to benzene, and possess considerable fine structure; $\lambda = 212, 272, 320$ nm for silabenzene (172), and $\lambda = 307, 314, 322$ nm for silatoluene (68). This series of bathochromic shifts has been observed with other donor-substituted heterobenzenes (171). Silenes of the general structure $(Me_3Si)_2Si{=}C(OSiMe_3)R$, where $R = CMe_3$, CEt_3 and 1-adamantyl (category 3), exhibit absorption maxima near 340 nm (104,122). Their extinction coefficients have been estimated to be 5200, 7060, and 7400 for $R = CMe_3$, CEt_3, and 1-adamantyl, respectively (104,122), consistent with the π character of a double bond.

D. *The Silicon–Carbon Double Bond Length*

There have now been four experimental determinations of a silicon–carbon double bond length. The first of these was a gas phase electron diffraction study of 1,1-dimethylsilene (173). This study was the subject of much controversy since the experimentally determined bond length, 1.83 Å, was much longer than the one predicted by *ab initio* calculations (1.69–1.71 Å, see below) (159). Since the calculations were carried out at a relatively high level of theory and the effects of electron correlation on determining the Si$=$C bond length were considered, the validity of the data extracted from the electron diffraction study is in serious doubt.

The second determination of a silicon–carbon double bond length came from the X-ray crystal structure of 1,1-bis(trimethylsilyl)-2-(trimethyl-siloxy)-2-(1-adamantyl)-1-silaethene (**1**) (122). Again, the experimentally

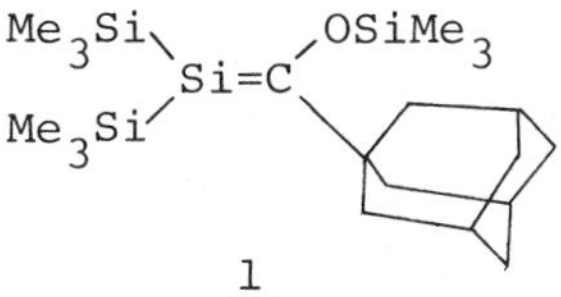

$$Me_3Si\diagdown_{Si=C}\diagup^{OSiMe_3}$$
$$Me_3Si\diagup \qquad \diagdown$$

1

determined Si$=$C bond length, 1.764 Å, is longer than the theoretical predictions. However, the difference is not so great in this case, and the lengthening of the bond can be attributed to a combination of steric and electronic effects (13,122,123) (see below).

Recently, Wiberg has reported the X-ray crystal structure of the tetrahydrofuran (THF) adduct of another stable silene, **2** (166). The silicon–carbon bond length in this less sterically hindered silene is 1.747 Å. Owing to the coordination of the THF molecule, the silicon atom has considerable tetrahedral character and the attached methyl groups do not lie in the plane containing the double bond and the sp^2-hybridized carbon atom with its attached silyl groups (*166*). The tetrahydrofuran molecule, acting as an electron donor, would be expected to increase the length of the Si=C bond. Wiberg has recently reported the crystal structure of the THF-free silene $Me_2Si=C(SiMe_3)(SiMe(t\text{-}Bu)_2)$(*167*) and, as expected, the silicon–carbon bond length is considerably shorter, 1.702 Å, and is in good agreement with *ab initio* calculations. Since this Si=C bond does not contain any highly electronegative substituents, it seems reasonable to conclude that its length is representative of silenes with small or only moderately bulky substituents.

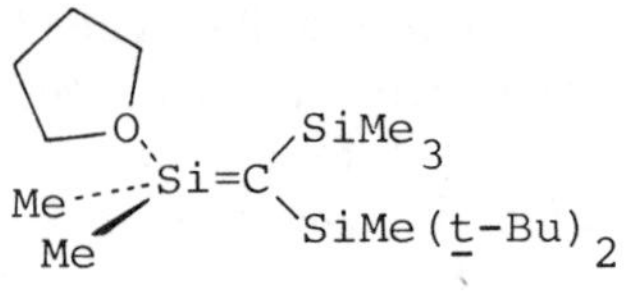

E. *Photoelectron Spectra*

The ionization potential of silaethene has been found to be 8.8 eV, in close agreement with *ab initio* calculations (8.95 eV)(*174*). Two experimental determinations of the ionization potential of 1,1-dimethylsilene have been made, one by transient photoelectron spectroscopy (8.3 eV)(*175*) and one by high-temperature PE spectroscopy (7.98 eV)(*176*). These values agree reasonably well with the calculated value of 8.14 eV (*176*). The low value of the ionization potential of 1,1-bis(trimethylsilyl)-2-(trimethylsiloxy)-2-(1-adamantyl)-1-silaethene, 7.7 eV, indicates a very electron rich species (due to the oxygen atom) that loses an electron readily (*122*). The first (IE_1 = 8.0 eV), second (IE_2 = 9.3 eV), and fourth (IE_4 = 11.3 eV) ionization potentials of silabenzene have been measured and are in satisfactory agreement with the calculated values (*70*). The ionization potentials obtained from the PE spectrum of silatoluene are consistent with the values obtained for silabenzene (*69*). As expected, the attachment of a methyl group lowered the first ionization potential of silatoluene to 7.7 eV, whereas it had no noticeable effect on the second ionization potential (IE_2 = 9.1 eV).

IV

CALCULATIONS CONCERNING
THE PROPERTIES OF SILENES

The silicon–carbon double bond has attracted the attention of theoretical chemists who have now performed a large number of high quality calculations on various properties of silenes. It is gratifying to observe that there is close agreement between experiment and calculation in most properties investigated, as summarized below.

A. The π Bond Strength of Silenes and Their Singlet–Triplet Energy Levels

The silicon–carbon π bond strength has been estimated by two different methods. The first method approximates the strength of the π bond by computation of the rotational barrier of a silene; the second method involves estimations based on kinetic and thermochemical arguments. Several calculations have been done to determine the rotational barrier of silenes (17,76,177–179): the most accurate ones include the effects of electron correlation. Thus, Ahlrichs and Heinzmann (180) have determined the π bond strength, ΔE_{π}, for silaethene to be 46 kcal/mol, and Hanamura et al. (181) have determined the ΔE_{π} for 1,1-dimethylsilene to be 47 kcal/mol. Walsh (182,183), on the other hand, has estimated the π bond strength of 1-methylsilene to be 39 ± 5 kcal/mol based on bond strengths and kinetic arguments, and Potzinger et al. (184) obtained the value of 37 kcal/mol for the π bond strength in 1,1-dimethylsilene based on similar reasoning. The silicon–carbon bond strength has also been estimated indirectly using the heats of formation of 1,1-dimethylsilene and 1-methylsilene determined experimentally by ion cyclotron resonance spectroscopy. By this method, Hehre and co-workers (185,186) have obtained π bond strengths of 34 and 42 kcal/mol for 1,1-dimethylsilene and 1-methylsilene, respectively, in reasonable agreement with the calculated values.

Hood and Schaefer (187), using state-of-the-art *ab initio* calculations have determined the singlet–triplet energy level separation of the parent silene to be 38.5 kcal/mol, and Hanamura et al. (181) determined the singlet–triplet energy splitting in 1,1-dimethylsilene to be 36 kcal/mol. It is believed that earlier calculations tended to underestimate the singlet–triplet energy separations in silenes (180,188–190).

B. *Vibrational Frequencies*

A large number of calculations have been done to determine the vibrational frequencies of some simple silenes. The parent silene and its deuterated analogs have been investigated (*4,9,188*) as well as 1,1-dimethylsilene, its deuterated derivatives (*38,181*), and 1,1-dichlorosilene (*4*). For the undeuterated compounds a Si=C stretching vibration of approximately 1000 cm^{-1} was calculated. Deuteration usually results in a decrease of this value.

C. *The Silicon–Carbon Double Bond Length*

Numerous calculations have been done to predict the length of the silicon–carbon double bond. A summary of the early work on this subject can be found in an excellent account by Schaefer (*159*), and therefore only more recent work will be reviewed here. In his review, Schaefer, using state-of-the-art calculations, determined the double bond length of the parent silene to be 1.705 Å. Kohler and Lischka (*188*) using similar calculations, obtained a value of 1.714 Å, very close to Schaefer's value. The double bond length in 1,1-dimethylsilene has been determined using a generalized valence bond approach by Hanamura, Nagase, and Morokuma (*181*), and their prediction of 1.728 Å is consistent with the value Schaefer obtained by using a double zeta quality basis set (*159*). Thus, it can be concluded that the length of the Si=C bond in the parent or methyl-substituted silenes is approximately 1.71 Å.

The effect of substituents on the calculated length of the silicon–carbon double bond has been studied (*2,3,13,14*). Apeloig and Karni (*13*) have noted the following general trends: (i) substitution at the silicon end generally shortens the Si=C bond relative to the parent silene and (ii) substitution at the carbon end generally lengthens the Si=C bond relative to the parent silene. These general observations are believed to be mainly a result of the electronic effects of the substituents on the polarization of the silicon–carbon double bond $Si^{\delta+}=C^{\delta-}$. Substituents that increase the polarization (the degree of ionicity) are expected to shorten the bond and vice versa. Other factors that Apeloig and Karni believe may contribute to the lengthening and shortening of the Si=C bond are the changes that occur in the sizes of the valence orbitals of carbon and silicon as a result of substitution. π conjugation between the substituent and the double bond is believed to be of little or no consequence.

An interesting conclusion that Apeloig and Karni have made, based on additional calculations, is that the effects of substituents on the Si=C bond length are approximately additive. Thus, they are able to predict the double bond length of highly substituted silenes which are normally inaccessible because the cost of a reliable self-consistent field calculation is prohibitive.

D. Reactivity of Silenes

The dimerization of silenes to 1,3-disilacyclobutanes is a common fate of these reactive species, and thus the energetics of this pathway are of interest. Ahlrichs and Heinzmann (180) have predicted the head-to-tail dimerization of the parent silene to be highly exothermic by 76 kcal/mol. The energy of activation was also calculated and predicted to be very small (less than 14 kcal/mol). The dimerization energies of various monosubstituted silenes have also been investigated (13) and, again, are expected to be highly exothermic (from 70 to 84 kcal/mol). The polar nature of the silicon–carbon double bond is believed to be the main contributing factor to the high reactivity. (2,180,183).

Apeloig and Karni (13) have also studied the effects of substituents on the reactivity of silenes by the frontier molecular orbital (FMO) approach. They have concluded that, concerning electronic factors, the polarity of the carbon–silicon double bond, and thus the coefficients of the frontier orbitals, play a more important role than the energies of these orbitals in controlling the reactivity of silenes.

E. Miscellaneous Properties

The first ionization potentials of several silenes have been predicted using MINDO/3 calculations (177). The ionization potentials found for the parent silene, 8.95 ± 0.1 eV, (174) and 1,1-dimethylsilene, 7.7 eV (184), 7.5 ± 0.3 eV (191), and 8.14 eV (176), have also been estimated from ab initio (174,176) or thermochemical (184,191) calculations.

Heats of formation of some silenes have been calculated using heats of atomization, obtained by various computational methods (17,177,178,192). $H_2Si{=}CH_2$ was predicted in this way to have a ΔH_f^0 of 33.3 kcal/mol which agrees reasonably well with the value obtained by Walsh (182) based on thermochemical and bond energy arguments. Potzinger et al. (184) and Gusel'nikov and Nametkin (191) have estimated the heat of formation of 1,1-dimethylsilene to be 18.2 ± 6 kcal/mol and 15.5 ± 6.2 kcal/mol, respectively, in reasonable agreement with the experimental value obtained by ion cyclotron resonance spectroscopy (20.5 kcal/mol) (185,186). In both cases, their calculations are based on data obtained from a kinetic study of the pyrolysis of dimethylsilacyclobutane.

F. Calculations on Silabenzene, Silaallenes, and Related Molecules

1. Silabenzene and Its Isomers

The geometries of silabenzene and four of its isomers were optimized using *ab initio* calculations at the STO-3G level (isomers **3–5** and **7**)(*60,193*) or using semi-empirical MNDO calculations (isomer **6**)(*194*).

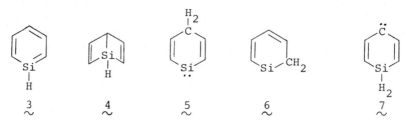

The relative energies of these geometries were obtained at the 3-21G^* level (*60*). The silicon–carbon bond length of 1.72 Å in silabenzene (isomer **3**), lies between the values of a localized double bond (1.64 Å) and a single bond (1.86 Å)(*60*) calculated at the same level of theory. This suggests that the molecule possesses some delocalized aromatic character. Silabenzene is predicted to be the most stable isomer of the five structures; however, the silylene isomers are predicted to lie only 20–25 kcal/mol higher in energy (*60*). In simpler systems (i.e., $H_2C=SiH_2 \rightarrow H_3C—\ddot{S}iH$) the silene and silylene isomers are predicted to be thermoneutral (*197*). Thus, the authors concluded that the stability of isomer **3** relative to isomers **5** and **6** is due to aromaticity in silabenzene (*60*). A similar measure of the delocalization energy in silabenzene was obtained by Schlegel and co-workers (*193*), using heats of isomerization data. Thus, it was concluded that the delocalization energy of silabenzene is about two-thirds that of benzene. Aromatic character in other planar, cyclic silicon–containing ring systems has been sought; however, calculations of Gordon *et al.* (*61*) reveal that neither the silacyclopentadienyl anion nor the silacyclopropenyl cation possess any degree of aromaticity.

The heat of formation (13.7 kcal/mol) and the first ionization potential (8.6 eV) of silabenzene have also been calculated using the MINDO/3 method (*177*).

2. Silaallene and Its Isomers

Three different research groups have examined the geometry and relative stabilities of silaallene and its various isomers (*21,22,24*). Silaallenes have been proposed as reaction intermediates by Kumada (*119*).

Despite the differing levels of calculations, the same general conclusions were reached. The silicon–carbon double bonds in 1-silaallene (1.69 Å) and 2-silaallene (1.70 Å) are shorter than in isolated silenes at the same level of theory. This trend is also observed in the analogous carbon series. 1-Silaallene is thermodynamically more stable than 2-silaallene by ~21 kcal/mol (22). Intuitively, this is what would have been expected, realizing the low ability of silicon to participate in multiple bonds. As may be expected from simpler systems (i.e., $H_2Si=CH_2$)(197), silylene isomers (for example, structures 8 and 9) are considerably more stable (approximately 15 kcal/mol) than their silaallene counterparts.

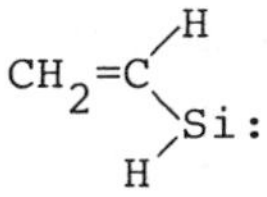

$$CH_2=C\overset{H}{\underset{H}{\diagdown}}Si:$$

$$\overset{\cdot\cdot}{Si}$$
$$H\diagup\diagdown H$$
$$H \quad\quad H$$

$$\underset{\sim}{8} \quad\quad\quad\quad\quad\quad \underset{\sim}{9}$$

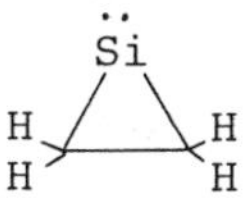

The rotational barriers of both silaallene isomers have been estimated (24) and the transition states for rotation studied in detail (19). Thus, 1-silaallene is predicted to have a rotational barrier of 35.9 kcal/mol and 2-silaallene a rotational barrier of 20.1 kcal/mol. 2-Silaallene is predicted to have a heat of formation of 31.3 kcal/mol and a first ionization potential of 9.17 eV by MINDO/3 calculations (177).

3. Miscellaneous Calculations

A theoretical study has been done on disilacyclobutadienes and their various isomers (25). As has been noted before, structures containing silylenes (R_2Si:) are more stable than those containing silicon–carbon multiple bonds. Silacyclobutadiene structures have also been investigated theoretically (40). Silacyclobutadiene has been found to possess anti-aromatic character, though only a fraction of the amount found for the analogous cyclobutadiene. Trinquier and Malrieu (42) have studied linear isomers of silabutadienes by ab initio "valence only" calculations. They conclude that conjugation is important in stabilizing these isomers.

V

DIMERIZATION OF SILENES

Dimerization is one of the most common reactions of silenes, particularly in the absence of an effective trapping agent.

$$R_2Si{=}CR'_2 \longrightarrow \begin{array}{c} R_2Si{-}CR'_2 \\ | \qquad | \\ R'_2C{-}SiR_2 \end{array} \qquad (12)$$

Indeed this reaction of silenes was one of the earliest recognized by Gusel'nikov and Flowers (155) in their pioneering work. With most silenes, i.e., those without substituents which create unusual polarization of the double bond, head-to-tail dimerization is both expected, based on the significant polarization of the bond as $Si^{\delta+}{=}C^{\delta-}$, and normally observed, although in a few cases some head-to-head dimerization occurs and has been interpreted as the result of a competing radical pathway (78). The mechanism by which dimerization occurs, nominally a concerted thermal [2 + 2] cycloaddition which is forbidden on the basis of non-conservation of orbital symmetry (198), has not been investigated in depth. The calculated activation energy for dimerization has been estimated at 14 kcal/mol (180), much lower than expected for the "forbidden" [2 + 2] dimerization of alkenes, which may reflect a relaxation of the symmetry restriction because of the heteronuclear bonds, but other stepwise processes, either hetero- or homopolar, cannot be excluded at present. Once formed, the resulting 1,3-disilacyclobutanes are stable species which do not readily revert thermally back to the silene precursor.

As inferred above, head-to-head dimerization of silenes has been observed in a number of cases (104.122–124). Members of the family of silenes $(Me_3Si)_2Si{=}C(OSiMe_3)R$ (10), derived from photolysis of $(Me_3Si)_3SiCOR$, undergo only head-to-head dimerization, if at all, the exceptions being those cases where no dimerization occurs because the R group is so bulky (CEt_3, 1-adamantyl, 1-methylcyclohexyl) that formation of the carbon–carbon bond of the 1,2-disilacyclobutane ring cannot occur because of steric hindrance. In these cases no dimer is observed and the silene monomers are stable and in principle isolable. The siloxysilenes (10, Scheme 1) actually undergo two types of head-to-head dimerization, leading in each case to formation of a silicon–silicon bond between two silene molecules. One result is the formation of 1,2-disilacyclobutanes (11), referred to above. The other result is a linear dimer (12), unsaturated at one end and saturated at the other, which arises from those silenes which have allylic hydrogen within the R group attached to the carbon of the double bond (i.e., R = —CH⌐, as shown). This type of dimer could arise through two molecules of silene undergoing an "ene" type cycloaddition reaction, forming 12 as shown. A few cases of "ene" type reaction with other types of silene have been observed (see below), but another mechanism seems more likely, as will be explained. In terms of the actual experimental results, linear dimers are formed exclusively when R = Me

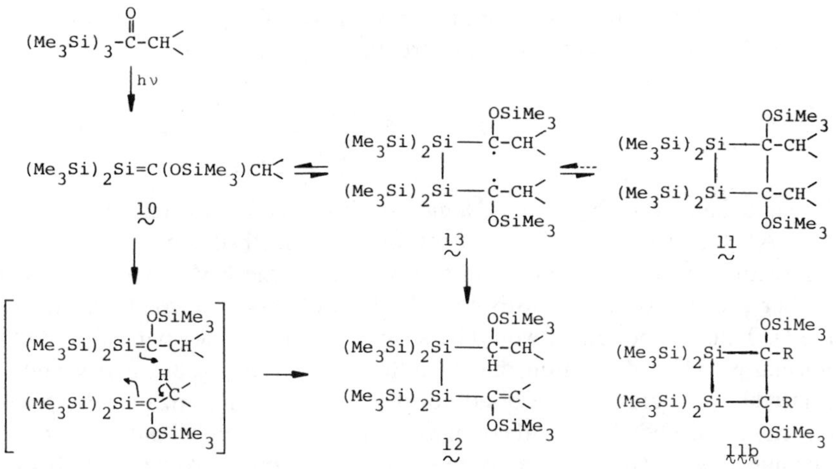

SCHEME 1

or Et, the latter in two geoisomeric forms, while both linear and cyclic dimers are observed with R = isopropyl or benzyl, and only cyclic dimers are observed with R = Ar, bicyclo[2.2.2]octyl and CMe_3.

While the dimers **11b** with R = i-Pr and Ph are stable solids, when R = t-Bu the solid dimer (mp 86°C) on dissolution in any inert solvent immediately dissociates to form a monomer–dimer mixture where the monomeric silene predominates at room temperature in a ratio of about 4:1 (*104*). However, on cooling, the proportions shift to form a 3:1 excess of dimer over monomer before the dimer precipitates from the solution. This reversible behavior can be explained, based on the crystal structure of the dimer. The C—C bond in the ring has been shown to be 1.66 Å long (*124*), reflecting severe steric hindrance between the adjacent t-butyl and siloxy groups and suggesting that the bond can barely form and is very weak. Essentially the same behavior is observed when R = bicyclo[2.2.2]octyl (*207*). This explains why, with bulkier R groups (CEt_3, 1-adamantyl, 1-methylcyclohexyl) which would create even greater steric hindrance, no dimers are formed and why photolysis solutions of these acylsilanes show only the presence of long-lived silenes which, in solution, slowly revert over a period of a week or two back to the parent acylsilane (*122*).

The reason for the observed head-to-head dimerization is of some interest. Unlike "simple" silenes, where the double bond is significantly polarized, $Si^{\delta+}=C^{\delta-}$ it is believed that the siloxysilenes are not significantly polarized overall, because of the effects of the substituents, which can be described, at least partly, in terms of contributions of the canonical forms shown below to the overall structure.

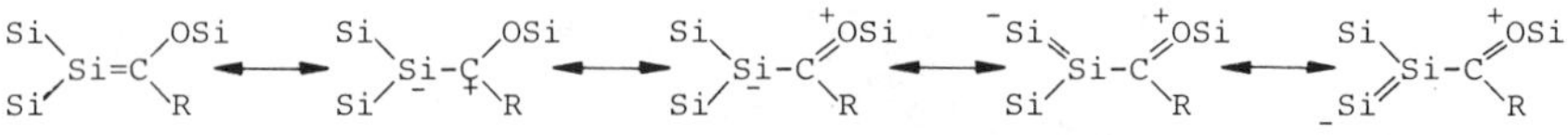

The unusually long $Si=C$ bond length observed (1.76 Å versus the normal 1.70 Å), the unusual shielding of the sp^2-hybridized silicon, and deshielding of the sp^2-hybridized carbon as revealed by the NMR data discussed earlier are all consistent with contributions of this type to the structure (*123*). Thus the net effect would be a cancellation (at least partially) of the normal $Si=C$ polarization due to differences in electronegativity and a relatively nonpolar double bond not driven toward dimerization by polar effects. Hence head-to-head coupling at silicon by a radical pathway might become the most favorable route for dimerization. In keeping with this interpretation is the presence, in each case where a "stable" monomeric silene is observed, of a persistent ESR signal consistent with a carbon-

$$\overset{\displaystyle O—Si}{\underset{}{|}}$$

centered radical of the structure Si—Si—$\overset{.}{C}$—R (*208*). A similar ESR signal is also formed instantaneously along with the monomeric silene when the *t*-butyl dimer **11b** is dissolved in inert solvent. This observation is strongly suggestive that both the formation from monomer and the dissociation of dimer back to monomer involves a common (di-)radical intermediate, such as **13**. Applying this mechanism to other systems, the diradical **13** can be formed from two monomers, **10**, and can reversibly dissociate (hence the radical itself may not be long-lived, although the ESR signal is), or may go forward forming either linear dimer **12** by disproportionation or cyclic dimer **11**, provided the R groups do not provide too much steric hindrance.

VI

REACTIONS OF SILENES: ADDITIONS TO THE DOUBLE BOND

As silenes represent a much more reactive higher energy π system than alkenes, it is not surprising that they show many of the reactions of alkenes as well as some behavior not generally exhibited by alkenes. However, the relative instability of silenes has meant that in most cases little mechanistic information on the observed reactions is available, and it is by no means certain in all cases that the same mechanism prevails for reactions of silenes and alkenes with the same reagent.

Two reactions have come to be extensively used with silenes, arising from the need to trap the short-lived species cleanly and in high yield, as evidence either of their formation or of the extent of their formation. These are the addition of alcohols, usually methanol, across the double bond to yield an alkoxysilane, and the Diels–Alder reaction with a diene, often 2,3-dimethylbutadiene. Each is an example of the two different types of addition to the Si=C double bond.

A. Addition of σ-Bonded Reagents

Many reagents, while undergoing a σ bond cleavage, add readily across the ends of a carbon–silicon double bond: these include acids (e.g., HCl, HOAc), H_2O, PhSH, halogens (especially Cl_2 and Br_2), alkyl halides (CCl_4, $CHCl_3$, $PhCH_2Cl$), a few cases of organometallic reagents [RMgX and RLi (100,199)], and a variety of silicon, germanium, and tin compounds of the type R_3MY (Y = OMe, NR_2, OPh, etc.) (90). Concerning the addition of alcohols, Wiberg (90) has suggested that a complex forms initially (in which the oxygen of the alcohol acting as a Lewis base interacts with the silicon of the silene acting as a Lewis acid), followed by proton migration to the carbon of the silene, in effect a two-step mechanism [Eq. (13)].

$$R_2Si=CR'_2 + ROH \longrightarrow R_2Si=CR'_2 \longrightarrow R_2Si-CR'_2 \qquad (13)$$

If this mechanism is correct, the rate of reaction through this nucleophilic addition mechanism with various alcohols should be related to the nucleophilicity of the alcohol, and Wiberg has observed the following relative rates consistent with this proposal: MeOH (96), EtOH (62), i-PrOH (48), t-BuOH (32), n-pentylOH (8), cyclohexanol (4), phenol (1) (90). This order obviously contains both basicity (nucleophilicity) and steric factors and is in accord with the proposed mechanism. Similarly, a series of amines were found to react with the relative rates i-PrNH$_2$ (97), t-BuNH$_2$ (48), PhNH$_2$ (2.2). Note that of the related and essentially equally sterically hindered Lewis bases, i-PrNH$_2$ (97) reacts more rapidly than the weaker base i-PrOH (48), as does t-BuNH$_2$ (48) relative to t-BuOH (32), entirely consistent with this addition being dependent on the strength of the nucleophile.

The stereochemistry of alcohol (and related reagents) addition to a silene is not known with certainty. The present authors have recently synthesized stable silenes where two geoisomers are possible (200) (also see Section VII). While NMR (^{1}H, ^{13}C, ^{29}Si) data indicated the presence of only one isomer in solution (although a small amount, perhaps 5% or less, of the geometric isomer could have been present), addition of methanol gave a mixture of diastereomers in proportions ranging from approximately 2:1 to 4:1 depending on the structure of the silene and the exact conditions. This suggests, consistent with the Wiberg mechanism, that the addition is not a one-step concerted two-bond formation, although kinetic effects could perhaps explain the above observation.

A variety of Me_3SiY compounds, such as Me_3SiOMe and Me_3SiOPh, have been observed to add easily to several silenes as do various Me_3SiNR_2 compounds (88). Jones (115) has reported that this reaction is a stereo-specific one, even at 300°C (the temperature at which the silene was generated), since different precursors to geoisomeric silenes yielded di-astereomeric silene adducts with high stereoselectivity. This remarkable result is the only study to date which provides strong evidence that additions to silenes may be stereospecific: a related study on geoisomeric disilenes by West (202) using alcohols, HCl (gas), or Cl_2 in pentane gave various mixtures of 1,2 adducts indicating clearly the absence of stereo-selectivity. In this case, not necessarily applicable to silenes, a radical mechanism has been implicated.

Many halogen compounds react fairly readily with silenes. Thus our highly substituted silenes get hot in CCl_4 or $CHCl_3$ while undergoing addition, and less hindered ones react with benzyl chloride (but not MeI or Me_3SiCl). Wiberg (88,90) has found that his stable silene $Me_2Si{=}C(SiMe_3)_2$ reacts only slowly with Me_3SiCl but rapidly with its Ge or Sn analogs.

Thus it is clear that a wide variety of polar σ-bonded reagents will react with the silicon–carbon double bond. In general, the reactions can be explained on the basis of a two-step addition reaction initiated by nucleophilic attack on silicon.

B. Addition of π-Bonded Reagents

A second category of silene reactions involves interactions with π-bonded reagents which may include homonuclear species such as 1,3-dienes, alkynes, alkenes, and azo compounds as well as heteronuclear reagents such as carbonyl compounds, imines, and nitriles. Four modes of reaction have been observed: nominal [2 + 2] cycloaddition (thermally forbidden on the basis of orbital symmetry considerations), [2 + 4] cycloadditions accompanied in some cases by the products of apparent "ene" reactions (both thermally allowed), and some cases of (allowed) 1,3-dipolar cycloadditions.

Many silenes have been found to undergo Diels–Alder reactions with 2,3-dimethylbutadiene in high yield, and hence this has been used as an important routine way of characterizing silenes. The reaction has been studied in some depth by Wiberg (*88,90*) for the silene $Me_2Si{=}C(SiMe_3)_2$, where the effect of temperature on yield of the Diels–Alder product versus "ene" product versus "silene dimer" has been studied (increasing the temperature from $-80°$ to $-40°C$ substantially increases the yield of Diels–Alder product mainly at the expense of silene dimer). In addition, the effect of modifying the structure of the diene has been quantitatively established, electron-rich (more nucleophilic) dienes reacting more rapidly with the (electrophilic) silenes. Typical relative rates are

These data, and particularly the inhibitory effect of cis methyl groups on the 1 and 4 carbon atoms on rate of reaction, are consistent with this reaction occurring by a synchronous process which may show considerable substrate selectivity.

[2 + 2] reactions with alkynes (e.g., phenylpropyne) are cleaner than [2 + 4] reactions with 2,3-dimethylbutadiene with highly crowded silenes, e.g., $(Me_3Si)_2Si{=}C(OSiMe_3)R$ ($R{=}t$-Bu), presumably because of steric effects, but neither reaction occurs cleanly in more crowded situations (R = 1-adamantyl, CEt_3, etc.). In cases of [2 + 2] reactions where azo compounds were involved, the resulting four-membered diazasilacyclobutane was found to decompose, leading to a silaimine ($R_2Si{=}NR$) and an imine ($R_2C{=}NR$) (*88*). Other [2 + 2] reactions with carbon–carbon double bonds are described later.

Some remarkable chemistry is observed when silenes react with heteroatom systems, in particular carbonyl compounds ($>C{=}O$) and imines ($>C{=}N{-}R$). The reaction with ketones was first described by Sommer (*203*), who postulated formation of an intermediate siloxetane which could not be observed and hence was considered to be unstable even at room temperature, decomposing spontaneously to a silanone (normally isolated as its trimer and other oligomers) and the observed alkene [Eq. (14)]. Many efforts have been made to demonstrate the existence of the siloxetane, but it is only very recently that claims have been advanced for the isolation of this species. In one case (*86*) an alternative formulation for the product obtained has been advanced (*204*). In a second case (*121*) involving reaction of a highly hindered silene with cyclopentadienones, only NMR data are available to date in support of the proposed structure. However, Wiberg has recently described some remarkably stable siloxetanes (*88,90*) (see below).

$$R_2Si=CR'_2 + O=CR''_2 \longrightarrow \begin{array}{c} R_2Si\text{---}CR'_2 \\ | \qquad\quad | \\ O\text{-----}CR''_2 \end{array} \longrightarrow \begin{array}{c} [R_2Si=O] \\ + \\ R'_2C=CR''_2 \end{array} \quad (14)$$

An interesting related reaction involves the creation of an α-ketosilene which spontaneously cyclized to give a siloxetene in quantitative yield (*126*) [Eq. (15)]. Recently Wiberg (*88,90*) has described several coupling

$$Me_3Si\text{-}SiMe_2\overset{O}{\underset{N_2}{\overset{\|}{\underset{\|}{C}}}}\text{-}C\text{-}C_{10}H_{15} \xrightarrow{h\nu} Me_3Si\text{-}SiMe_2\overset{O}{\overset{\|}{\ddot{C}}}\text{-}\overset{\|}{C}\text{-}C_{10}H_{15}$$

$$(15)$$

$$\begin{array}{cc} Me_2Si=C\text{-}SiMe_3 \\ | \\ O=C\text{-}C_{10}H_{15} \end{array} \longrightarrow \begin{array}{cc} Me_2Si\text{---}C\text{-}SiMe_3 \\ | \qquad\quad \| \\ O\text{-----}C\text{-}C_{10}H_{15} \end{array}$$

reactions of his stable silene **14** with benzophenone (or the related silyl imine) in which competing [2 + 2] and [2 + 4] reactions are observed (the latter involving destruction of the aromatic π system of a benzene ring) (see Scheme 2).

This extraordinary reaction to yield the six-membered ring adduct **15** appears to be the kinetically favored route, but heating (which might have been expected to facilitate a 1,3-prototropic shift to restore aromatic conjugation and stability) leads instead to the four-membered ring isomer **16**, evidently the thermodynamically favored species. That this compound appears to be so stable and does not rapidly collapse to $(Me_2SiO)_n$ and the alkene **17** is extraordinary, given the number of attempts by various groups over the years to form siloxetanes. However, at 118°C over 20 hours the alkene **17** is formed in high yield by a process which may involve radicals (*90*). In contrast, the corresponding nitrogen analog **18**, when warmed to 60°C, does not yield Si=N— and alkene, but gives the silene **14** and imine **20**; hence this compound is used as a useful source ("store") of $Me_2Si=C(SiMe_3)_2$. Wiberg has described a large amount of careful detailed work on this silene, some of which involves very unusual chemistry as noted above. (One cannot but look forward to crystal

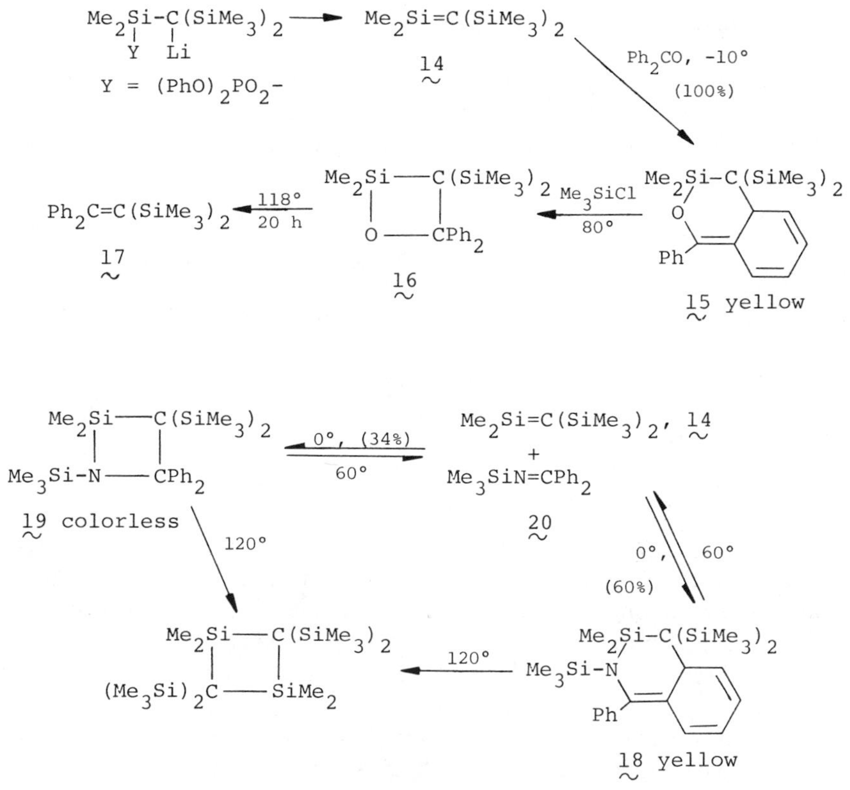

$Me_2Si-C(SiMe_3)_2 \longrightarrow Me_2Si=C(SiMe_3)_2$
    | |
    Y   Li   14

Y = $(PhO)_2PO_2^-$ $Ph_2CO, -10°$ (100%)

$Ph_2C=C(SiMe_3)_2 \xleftarrow[20\ h]{118°}$ $Me_2Si\!-\!-\!C(SiMe_3)_2$ $\xleftarrow[80°]{2\ Me_3SiCl}$ $Me_2Si-C(SiMe_3)_2$

17   O$-\!-$CPh$_2$ O

               16   Ph

                                       15 yellow

$Me_2Si\!-\!-\!C(SiMe_3)_2$ $\xrightarrow[60°]{0°,\ (34\%)}$ $Me_2Si=C(SiMe_3)_2,\ 14$

$Me_3Si-N\!-\!-\!CPh_2$ +

19 colorless $Me_3SiN=CPh_2$

           120°   20   0°,   60°

                                           (60%)

$Me_2Si\!-\!-\!C(SiMe_3)_2$ $\xleftarrow{120°}$ Me_3Si-N $Me_2Si-C(SiMe_3)_2$

$(Me_3Si)_2C\!-\!-\!SiMe_2$ Ph

                                       18 yellow

SCHEME 2

structure, and ^{13}C- and ^{29}Si-NMR data, to unambiguously confirm the structure of some of the compounds described, e.g., **15, 16, 18, 19**).

In general "ene" reactions have not been well investigated with silenes. They have been studied in depth by Wiberg using the silene $Me_2Si=C(SiMe_3)_2$ (*88,90*), and the relative reactivity with different alkenes has been established. They have been found by Jones (*116*) to be major pathways among other competing reactions (e.g., dimerization) of the silene PhMeSi=CHCH$_2$-*t*-Bu and they have also been observed occasionally as minor products in reactions with some members of the silene family $(Me_3Si)_2Si=C(OSiMe_3)R$. Ene-like reactions were also found to be important processes with the Kumada family of compounds derived from the photolysis of aryl disilanes (see Section VI,C), but radical pathways may be involved in the formation of the products.

C. Reactions of Silenes: Unusual Behavior of Some Families

Two families of silenes warrant special comment. The first arises from the photolysis of a number of aryldisilanes as investigated by Kumada et al. (97–102), who reported that a rearrangement occurs to give products formulated as silenes, as shown in Eq. (16).

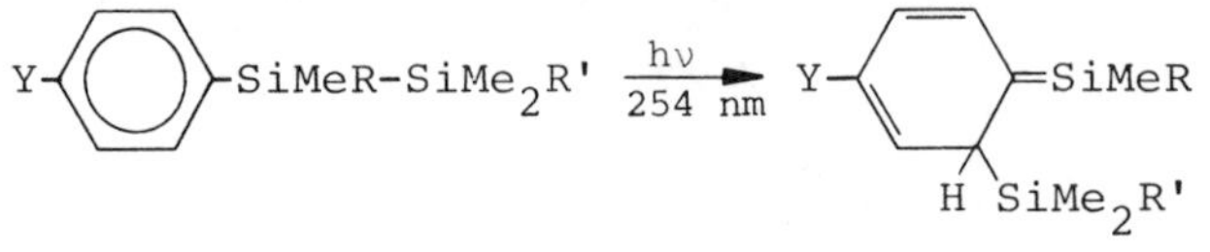

$$Y-\langle\bigcirc\rangle-\text{SiMeR-SiMe}_2\text{R}' \xrightarrow[254 \text{ nm}]{h\nu} Y-\langle\bigcirc\rangle=\text{SiMeR}$$
$$\text{H} \quad \text{SiMe}_2\text{R}'$$

$$(16)$$

$$Y = \text{H, Me, } \underline{t}\text{-Bu}$$
$$R = \text{Me, Ph}$$
$$R' = \text{Me, Ph}$$

This family of products shows behavior rather different from what is usually observed for simple silenes. Thus, there is no report of dimers being formed although high molecular weight material is normally present. No 1,3 H shifts to restore aromaticity are observed in phenyl (as opposed to naphthyl) systems. While alcohols like MeOH add, none of the expected 1,2-addition products were found; instead 1,4- and 1,6-adducts were obtained. Reactions with ketones do not occur by the usual pathways, and, unlike normal silenes, reactions with dienes do not exhibit the normal Diels–Alder behavior but react with the involvement of only one double bond. Alkenes also add abnormally. The reactions of the Kumada compounds, summarized in Scheme 3, may be those of silenes exhibiting unusual behavior due to the presence of an extended π system. However, as both Kumada (102) and Sakurai (205) have suggested, these products are more consistent with radical-type behavior involving species such as **21** which on H-abstraction lead to the observed aromatized products. There is ESR evidence for radicals in systems of this type (205). Because of the almost complete difference in behavior of these species from that of "normal" silenes, whether lightly or heavily substituted, they cannot be regarded as typical silenes.

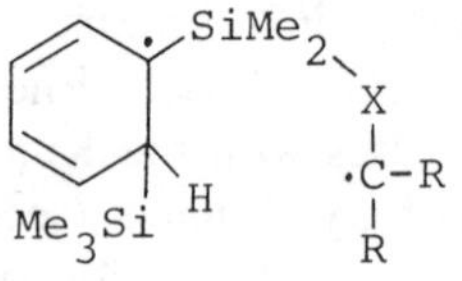

21

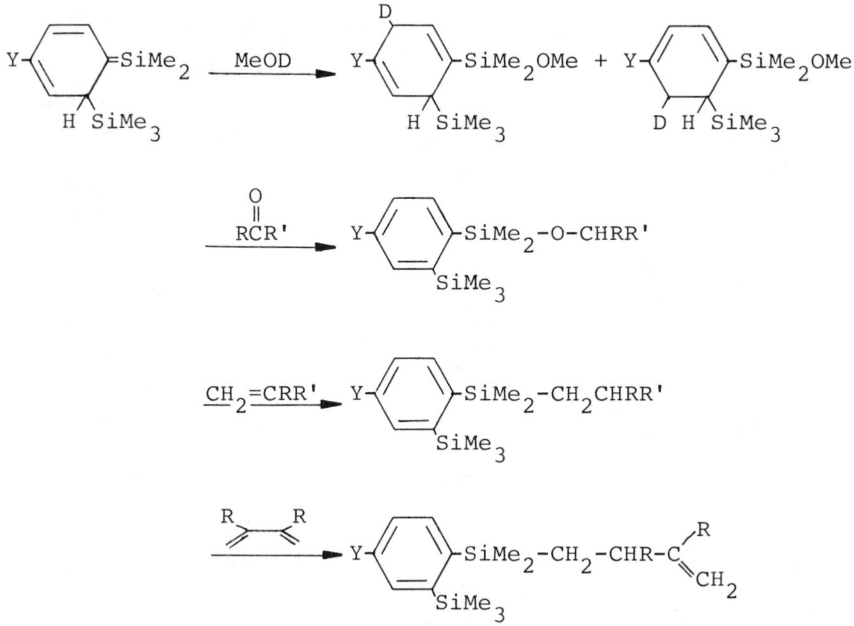

Y⟨ ⟩–SiMe$_2$ $\xrightarrow{\text{MeOD}}$ Y⟨ ⟩–SiMe$_2$OMe + Y⟨ ⟩–SiMe$_2$OMe

H SiMe$_3$ H SiMe$_3$ D H SiMe$_3$

$$\xrightarrow[\text{RCR'}]{\overset{O}{\|}}$$ Y⟨ ⟩–SiMe$_2$–O–CHRR'

SiMe$_3$

$\xrightarrow{CH_2=CRR'}$ Y⟨ ⟩–SiMe$_2$–CH$_2$CHRR'

SiMe$_3$

$\xrightarrow{}$ Y⟨ ⟩–SiMe$_2$–CH$_2$–CHR–C$\underset{CH_2}{\overset{R}{\diagup}}$

SiMe$_3$

SCHEME 3

Another family of silenes, those derived from the photolysis of acyldi- or polysilanes at $\lambda > 360$ nm, also show somewhat unusual behavior compared to simpler silenes. These silenes exhibit great stability which in some cases has allowed isolation of solid silenes, and which has allowed acquisition of much physical data relating to silicon–carbon double bonds as mentioned earlier.

Photolysis of acyldisilanes at $\lambda > 360$ nm (103,104) was shown, based on trapping experiments, to yield both silenes **22** and the isomeric siloxycarbenes **23**, but with polysilylacylsilanes only silenes **24** are formed, as shown by trapping experiments and NMR spectroscopy (104,122–124) (see Scheme 4). These silenes react conventionally with alcohols, 2,3-dimethylbutadiene (with one or two giving some evidence of minor amounts of ene-like products), and in a [2 + 2] manner with phenylpropyne. Ketones, however, do not react cleanly. Perhaps the most unusual behavior of this family of silenes is their exclusive head-to-head dimerization as described in Section V. More recently it has been found that these silenes undergo thermal [2 + 2] reactions with butadiene itself (with minor amounts of the [2 + 4] adduct) and with styrene and vinylnaphthalene. Also, it has been found that a dimethylsilylene precursor will

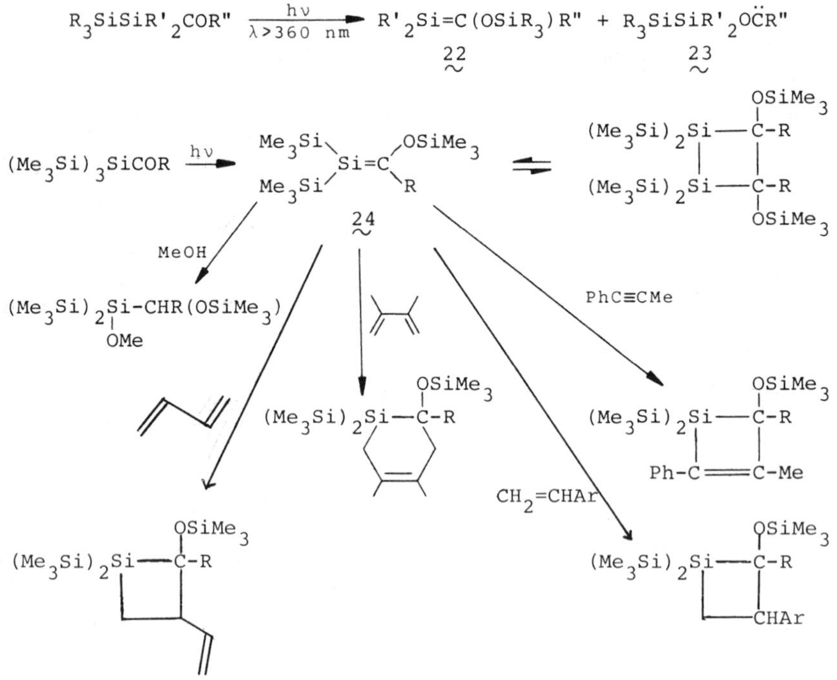

$$R_3SiSiR'_2COR'' \xrightarrow[\lambda>360\ nm]{h\nu} R'_2Si=C(OSiR_3)R'' + R_3SiSiR'_2O\ddot{C}R''$$

$$\underset{22}{} \qquad\qquad \underset{23}{}$$

SCHEME 4

add dimethylsilylene (Me_2Si:) across a silene double bond to give a disilacyclopropane (209).

Photolysis of a mesitylsilene (24, R=mesityl) led to insertion into one of the methyl C—H bonds yielding a benzocyclobutene (209): insertion of silenes into C—H bonds is not a very common reaction of silenes.

During an investigation of the role of the trimethylsilyl group attached to silicon in stabilizing silenes such as 24 we have recently observed a new and remarkable rearrangement (200)(Scheme 5). Photolysis of the bis-trimethylsilyl-t-butylacylsilane 25 in the conventional manner gave only one isomeric silene 26 (stereochemistry unknown), as judged by ^{13}C- or ^{29}Si-NMR spectroscopy.

Treatment of the photolysate with methanol gave two diastereomeric methoxysilanes in about 3:1 proportions, consistent with methanol addition being a nonstereospecific process. During extended photolysis (8 hours) to convert all of 25 to 26 it was observed that the concentration of 26 decreased, and a set of signals characteristic of a different silene grew in (no change of relative proportions occurred in the dark). The chemical

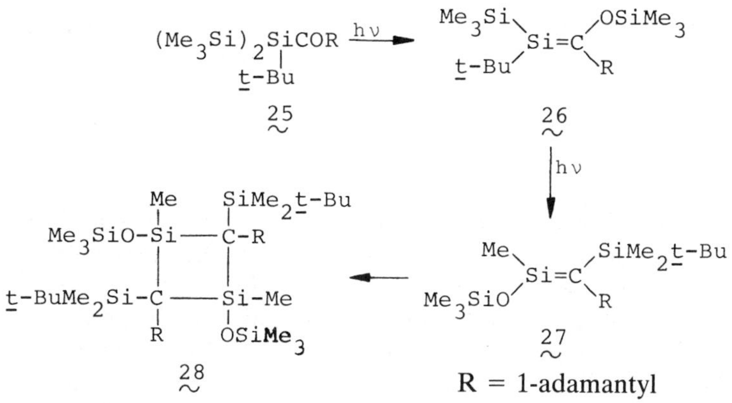

$$(Me_3Si)_2SiCOR \xrightarrow{h\nu}$$

25

$$\underset{t\text{-Bu}}{\overset{Me_3Si}{\diagdown}} Si=C \underset{R}{\overset{OSiMe_3}{\diagup}}$$

26

$$\downarrow h\nu$$

28

$$\underset{Me_3SiO}{\overset{Me}{\diagdown}} Si=C \underset{R}{\overset{SiMe_2t\text{-Bu}}{\diagup}}$$

27

R = 1-adamantyl

SCHEME 5

shifts of these signals suggested the absence of silyl groups attached directly to the sp^2-hybridized silicon and the absence of a siloxy group attached to sp^2-hybridized carbon, inferring that considerable rearrangement of groups had occurred. The disappearance of **25** followed clean first order kinetics, and the overall process **25** → **26** → **27** clearly was a consecutive sequence kinetically. The NMR signals of the new silene **27** survived for periods of time varying from days to weeks, at which time a head-to-tail dimer **28** precipitated from the solution (a second minor dimer was also formed). A crystal structure confirmed the structure of **28** as the cis head-to-tail dimer of the precursor **27**, and the structure of **27** was confirmed using ^{13}C DEPT and ^{29}Si INEPT NMR spectra. This is a remarkable rearrangement in which three of the four groups present in the original silene **26** apparently have migrated in forming **27**. Exactly how this occurs is not yet known: a possible pathway has been proposed (*200*). This remarkable behavior is not unique to the acylsilane **25**: its analog in which phenyl replaces *t*-butyl gives related rearrangements. It is thus evident that silenes possess a rich and complex photochemistry.

VII

MOLECULAR REARRANGEMENTS OF SILENES

A number of molecular rearrangements of silenes have been reported, some of which are particularly interesting either because of the energetics involved, or for the complexity of the processes involved.

A rearrangement of some interest involving a very simple silene is that of the interconversion of 1-methylsilene and dimethylsilylene [Eq. (17)].

$$\text{MeSiH=CH}_2 \rightleftharpoons \text{Me-}\overset{..}{\text{Si}}\text{-Me} \qquad (17)$$

Numerous calculations had suggested that while methylsilene and dimethylsilylene were essentially equi-energetic, the interconversion barrier was of the order of at least 40 kcal/mol (159). Conlin (206) reported that in high temperature thermolyses of 1-methyl-1-silacyclobutane the products from trapping experiments indicated that both methylsilene and its isomer dimethylsilylene were present, the latter presumably arising from the former because of the high temperatures involved, a suggestion which was subsequently convincingly confirmed (30). Reisenauer et al. (12) had shown that the interconversion in either direction could be effected with radiation of the appropriate wavelength (λ = 254 nm to go from silene to silylene and λ > 400 nm to go from silylene to silene), the latter result also having been observed by Drahnak, Michl, and West (163). An earlier surprising observation by these latter authors (given the >40 kcal/mol energy barrier) that methylsilene rapidly rearranged to dimethylsilylene at 100 K has now been explained as a rearrangement "promoted by the act of trapping" of the reaction intermediate (29). Thus, what was once an apparent controversial anomaly now seems to have been resolved, the results being consistent with theoretical calculations. Comparable rearrangements do not seem to occur with 1,1-dimethylsilene.

Barton has reported a wide variety of elegant studies in which various silenes or silylenes have been created, usually thermally, and their subsequent rearrangements investigated in terms of the observed products of trapping (51,53,65,145). It has been clearly established that interconversion between silenes and silylenes, especially where H atoms or Me_3Si groups migrate, are facile processes. In some cases, radicals can be the precursors to silenes (65).

At the other end of the scale of molecular complexity Wiberg has reported the rapid equilibration of the methyl groups in $(CD_3)_2Si{=}C(SiMe_3)_2$ when liberated thermally from a "silene store" (90), the half-life of exchange being about 0.5 hour at 120°C and complete statistical redistribution being observed after 5 hours at 120°C [Eq. (18)]. However, what is not entirely clear is why, if the silene $Me_2Si{=}C(SiMe_3)_2$ dimerizes at −100°C, it does not dimerize when liberated from the "store" at 120°C—the explanation that the "concentration remains so small that dimerization does not occur" is not entirely convincing.

$$(CD_3)_2Si=C\begin{array}{l}SiMe_3\\SiMe_3\end{array} \;\rightleftharpoons\; (CD_3)_2MeSi-C\begin{array}{l}SiMe_2\\SiMe_3\end{array} \;\rightleftharpoons\; (CD_3)MeSi=C\begin{array}{l}SiMe_2(CD_3)\\SiMe_3\end{array}$$

$$\quad\;\; Me_2Si=C\begin{array}{l}SiMe_2(CD_3)\\SiMe_2(CD_3)\end{array} \;\rightleftharpoons\; (CD_3)Me_2Si-C\begin{array}{l}SiMe_2(CD_3)\\SiMe_2\end{array}$$

$$\tag{18}$$

A related isomerization was observed with the silene t-$Bu_2Si=C(SiMe_3)_2$ [or its salt precursor t-Bu_2SiF-—$CLi(SiMe_3)_2$ in ether at room temperature (166)] which isomerizes to $Me_2Si=C(SiMe_3)(SiMe$-t-$Bu_2)$, a relatively stable solid silene which was isolated [or to the related salt Me_2SiF—$CLi(SiMe_3)(SiMe$-t-$Bu_2)$]. This latter silene apparently exchanges Me groups between silicon atoms so rapidly that only a single broad ^{1}H-NMR resonance signal is observed at 30°C for the Me groups, whether as Me_2Si, Me_3Si, or MeSi. In contrast, however, at −70°C well-separated ^{13}C signals were seen in the NMR spectrum for each type of Me group: Me_2Si (at −2.10 and 10.81 ppm), t-Bu_2MeSi (at 5.85 ppm), and Me_3Si (at 7.06 ppm). Separate resonances were also observed in the ^{29}Si spectrum for each type of Si-Me present. Presumably the lower temperature greatly slowed the rate of the exchange process.

A related "scrambling" of groups in a silene has also been reported by Eaborn (143) to explain the structure of compounds isolated from the thermolysis of tris(trimethylsilyl)fluorodiphenylsilylmethane at 450°C, where Me and Ph groups freely interchange between silicon atoms [Eq. (19)]. A related rearrangement is probably also involved in the photochemical silene-to-silene isomerizations derived from acylpolysilanes described earlier.

$$(Me_3Si)_3C-SiPh_2F \longrightarrow \begin{array}{l}Me_3Si\\Me_3Si\end{array}\!\!\!C=SiPh_2 \;\rightleftharpoons\; \begin{array}{l}Me_2Si\\Me_3Si\end{array}\!\!\!C-SiMePh_2$$

$$\begin{array}{l}PhMe_2Si\\Me_2Si\end{array}\!\!\!C-SiMe_2Ph \;\rightleftharpoons\; \begin{array}{l}PhMe_2Si\\Me_3Si\end{array}\!\!\!C=SiMePh$$

$$\tag{19}$$

It seems obvious that groups attached to the ends of a silicon-carbon double can scramble under a variety of conditions, but the limited observations to date vary so widely in experimental conditions required that it is not at all clear what the specific requirements are for this process to occur.

VIII

SUMMARY

From the data given above it is obvious that in less than two decades great strides have been made in our knowledge of silenes. A wide variety of species have been synthesized, and some silenes have been found sufficiently stable that their crystal structures and physical properties have been accurately determined. Experimental results are in close accord with theoretical calculations (for example, Si=C bond length, infrared stretching frequency, ionization potential): calculations of the properties of a silicon–carbon double bond have found great appeal with theoretical chemists, and highly sophisticated calculations have led to very reliable results. Much has been learned about the chemical behavior of silenes, although more work remains to be done to understand fully some reaction mechanisms and to explain the apparently anomalous behavior of some species. The stereospecificities of many reactions with silenes are not yet fully known, nor the mechanisms understood, and recent results from our laboratory, and Wiberg's (90), indicate that remarkably complex intra-molecular rearrangements can take place under very mild conditions (much milder than with alkenes, for example). Thus, while most of the "fundamentals" of silene chemistry can be said to have been studied and are now understood, one can anticipate further exciting developments, particularly in the area of molecular rearrangements, and, it is to be hoped, in applications to synthetic or applied chemistry.

REFERENCES

1. Gowenlock, B. G., and Hunter, J. A., *J. Organometal. Chem.* **140**, 265 (1977).
2. Damrauer, R., and Williams, D. R., *J. Organometal. Chem.* **66**, 241 (1974).
3. Gordon, M. S., *J. Am. Chem. Soc.* **104**, 4352 (1982).
4. Baskir, E. G., Mal'tsev, A. K., and Nefedov, O. M., *Izv. Akad. Nauk. SSSR, Ser. Khim.* 1314 (1983).
5. Maier, G., Mihm, G., and Reisenauer, H. P., *Angew. Chem.* **93**, 615 (1981).
6. Auner, N., and Grobe, J., *Z. Anorg. Allg. Chem.* **459**, 15 (1979).
7. Maier, G., Mihm, G., Reisenauer, H. P., and Littmann, D., *Chem. Ber.* **117**, 2369 (1984).
8. Maier, G., Mihm, G., and Reisenauer, H. P., *Chem. Ber.* **117**, 2351 (1984).
9. Schlegel, H. B., Wolfe, S., and Mislow, K., *Chem. Commun.* 246 (1975).
10. Auner, N., and Grobe, J., *J. Organometal. Chem* **222**, 33 (1981).
11. Auner, N., and Grobe, J., *Z. Anorg. Allg. Chem.* **485**, 53 (1982).
12. Reisenauer, H. P., Mihm, G., and Maier, G., *Angew. Chem.* **94**, 864 (1982).
13. Apeloig, Y., and Karni, M., *J. Am. Chem. Soc.* **106**, 6676 (1984).
14. Gordon, M. S., and George, C., *J. Am. Chem. Soc.* **106**, 609 (1984).
15. Nagase, S., and Kudo, T., *THEOCHEM* **12**, 35 (1983).

16. Nagase, S., and Kudo, T., *J. Chem. Soc. Chem. Commun.* 141 (1984).
17. Bell, T. N., Kieran, A. F., Perkins, K. A., and Perkins, P. G., *J. Phys. Chem.* **88**, 1334 (1984).
18. Gusel'nikov, L. E., Sokolova, V. M., Volnina, E. A., Kerzina, Z. A., Nametkin, N. S., Komalenkova, N. G., Baskirova, S. A., and Chernyshev, E. A., *Dokl. Akad. Nauk. SSSR* **260**, 348 (1981).
19. Krogh-Jespersen, K., *J. Comput. Chem.* **3**, 571 (1982).
20. Ishikawa, M., Sugisawa, H., Fuchikami, T., Kumada, M., Yamabe, T., Kawakami, H., Fukui, K., Veki, Y., and Shizuka, H., *J. Am. Chem. Soc.* **104**, 2872 (1982).
21. Lien, M. H., and Hopkinson, A. C., *Chem. Phys. Lett.* **80**, 114 (1981).
22. Gordon, M. S., and Koob, R. D., *J. Am. Chem. Soc.* **103**, 2939 (1981).
23. Hopkinson, A. C., and Lien, M. H., *J. Organometal. Chem.* **206**, 287 (1981).
24. Barthelat, J. C., Trinquier, G., and Bertrand, G., *J. Am. Chem. Soc.* **101**, 3785 (1979).
25. Holme, T. A., Gordon, M. S., Yabushita, S., and Schmidt, M. W., *Organometallics* **3**, 583 (1984).
26. Auner, N., and Grobe, J., *J. Organometal. Chem.* **222**, 33 (1981).
27. Gordon, M. S., *Chem. Phys. Lett.* **76**, 163 (1980).
28. Arrington, C. A., Klingensmith, K. A., West, R., and Michl, J., *J. Am. Chem. Soc.* **106**, 525 (1984).
29. Arrington, C. A., West, R., and Michl, J., *J. Am. Chem. Soc.* **105**, 6176 (1983).
30. Conlin, R. T., and Kwak, Y. W., *Organometallics* **3**, 918 (1984).
31. Davidson, I. M. T., Ijadi-Maghsadi, S., Barton, T. J., and Tillman, N., *J. Chem. Soc. Chem. Commun.* 478 (1984).
32. Alnaimi, I. S., Weber, W. P., Nazran, A. S., and Griller, D., *J. Organometal. Chem.* **272**, C10 (1984).
33. Li, W. -H., Wang, C. -H., and Hu, C. -Y., *K'o Hsueh T'ung Pao* **26**, 191 (1981).
34. Khabashesku, V. N., Baskir, E. G., Maltsev, A. K., and Nefedov, O. M., *Izv. Akad. Nauk. SSSR, Ser. Khim.* 238 (1983).
35. Mal'tsev, A. K., Khabashesku, V. N., and Nefedov, O. M., *J. Organometal. Chem.* **271**, 55 (1984).
36. Mal'tsev, A. K., Khabashesku, V. N., and Nefedov, O. M., *J. Organometal. Chem.* **226**, 11 (1982).
37. Gusel'nikov, L. E., Volkova, V. V., Avakyan, V. G., and Nametkin, N. S., *J. Organometal. Chem.* **201**, 137 (1980).
38. Avakyan, V. G., Gusel'nikov, L. E., Volkova, V. V., and Nametkin, N. S., *Dokl. Akad. Nauk. SSSR* **254**, 657 (1980).
39. Gentle, T. M., and Muetterties, E. L., *J. Am. Chem. Soc.* **105**, 304 (1983).
40. Gordon, M. S., *J. Chem. Soc. Chem. Commun.* 1131 (1980).
41. Mal'tsev, A. K., Khabashesku, V. N., Baskir, E. G., and Nefedov, O. M., *Izv. Akad. Nauk. SSSR, Ser. Khim.* 222 (1980).
42. Trinquier, G., and Malrieu, J. P., *J. Am. Chem. Soc.* **103**, 6313 (1981).
43. DuPuy, C. H., and Damrauer, R., *Organometallics* **3**, 362 (1984).
44. Gusel'nikov, L. E., Volkova, V. V., Avakyan, V. G., Nametkin, N. S., Voronkov, M. G., Kirpichenko, S. V., Suslova, E. N., *Dokl. Akad. Nauk. SSSR* **272**, 892 (1983).
45. Gusel'nikov, L. E., Volkova, V. V., Avakyan, V. G., Nametkin, N. S., Voronkov, M. G., Kirpichenko, S. V., and Suslova, E. N., *J. Organometal. Chem.* **254**, 173 (1983).
46. Davidson, I. M. T., Wood, I. T., *J. Chem. Soc. Chem. Commun.* 550 (1982).
47. Gusel'nikov, L. E., Polyakov, Y., Zairkin, V. G., and Nametkin, N. S., *Dokl. Akad. Nauk. SSSR* **274**, 598 (1984).

48. Avakyan, V. G., Volkova, V. V., Gusel'nikov, L. E., and Nametkin, N. S., *Dokl. Akad. Nauk. SSSR* **274**, 1112 (1984).
49. Davidson, I. M. T., Fenton, A. M., Jackson, P., and Lawrence, F. T., *J. Chem. Soc. Chem. Commun.* 806 (1982).
50. Barton, T. J., Burns, G. T., Goure, W. F., and Wulff, W. D., *J. Am. Chem. Soc.* **104**, 1149 (1982).
51. Barton, T. J., Burns, G. T., and Gschneidner, D., *Organometallics* **2**, 8 (1983).
52. Burns, G. T., and Barton, T. J., *J. Am. Chem. Soc.* **105**, 2006 (1983).
53. Burns, S. A., Burns, G. T., and Barton, T. J., *J. Am. Chem. Soc.* **104**, 6140 (1982).
54. Mal'tsev, A. K., Korolev, V. A., Khabashesku, V. N., and Nefedov, O. M., *Dokl. Akad. Nauk. SSSR* **251**, 1166 (1980).
55. Barton, T. J., and Hoekman, S. K., *Synth. React. Inorg. Met.-Org. Chem.* **9**, 297 (1979).
56. Chapman, O. L., Chang, C. C., Kolc, J., Jung, M. E., Lowe, J. A., Barton, T. J., and Tumey, M. L., *J. Am. Chem. Soc.* **98**, 7844 (1976).
57. Chedekel, M. R., Koglund, M., Kreeger, R. L., and Shechter, H., *J. Am. Chem. Soc.* **98**, 7846 (1976).
58. Barton, T. J., and Kline, E. A., *J. Organometal. Chem.* **42**, C21 (1972).
59. Davidson, I. M. T., and Scrampton, R. J., *J. Organometal. Chem.* **271**, 249 (1984).
60. Chandrasekhar, J., Schleyer, P. v. R., Baumgaertner, R. O. W., and Reetz, M. T., *J. Org. Chem.* **48**, 3453 (1983).
61. Gordon, M. S., Boudjouk, P., and Anwari, F., *J. Am. Chem. Soc.* **105**, 4972 (1983).
62. Reyes, L. M., and Canuto, S., *THEOCHEM* **6**, 77 (1982).
63. Bock, H., Rosmus, P., Solouki, B., and Maier, G., *J. Organometal. Chem.* **271**, 145 (1984).
64. Maier, G., Mihm, G., Baumgaertner, R. O. W., and Reisenauer, H. P., *Chem. Ber.* **117**, 2337 (1984).
65. Barton, T. J., Burns, S. A., and Burns, G. T., *Organometallics* **1**, 210 (1982).
66. Barton, T. J., and Jacobi, S. A., *J. Am. Chem. Soc.* **102**, 7979 (1980).
67. Barton, T. J., and Vuper, M., *J. Am. Chem. Soc.* **103**, 6788 (1981).
68. Kreil, C. L., Chapman, O. L., Barton, T. J., and Burns, G. T., *J. Am. Chem. Soc.* **102**, 841 (1980).
69. Bock, H., Bowling, R. A., Solouki, B., Barton, T. J., and Burns, G. T., *J. Am. Chem. Soc.* **102**, 429 (1980).
70. Solouki, B., Rosmus, P., Bock, H., and Maier, G., *Angew. Chem.* **92**, 56 (1980).
71. Barton, T. J., and Burns, G. T., *J. Am. Chem. Soc.* **100**, 5246 (1978).
72. Sekiguchi, A., and Ando, W., *Chem. Lett.* 871 (1983).
73. Bertrand, G., Dubac, J., Mazerolles, P., and Ancelle, J., *J. Chem. Soc. Chem. Commun.* 382 (1980).
74. Barton, T. J., Burns, G. T., Arnold, E. V., and Clardy, J., *Tetrahedron Lett.* **22**, 7 (1981).
75. Nakadaira, Y., Sakaba, H., and Sakurai, H., *Chem. Lett.* 1071 (1980).
76. Ustynyuk, Y. A., Zakharov, P. I., Azizov, A. A., Shchembelov, G. A., and Gloriozov, I., *J. Organometal. Chem.* **96**, 195 (1975).
77. Maerkl, G., Rudnick, D., Schulz, R., and Schweig, A., *Angew. Chem.* **94**, 211 (1982).
78. Barton, T. J., and Hoekman, S. K., *J. Am. Chem. Soc.* **102**, 1584 (1980).
79. Barton, T. J., Burns, S. A., and Burns, G. T., *Organometallics* **1**, 210 (1982).
80. Conlin, R. T., Bessellieu, M. P., Jones, P. R., and Pierce, R. A., *Organometallics* **1**, 396 (1982).
81. Ishikawa, M., Fuchikami, T., and Kumada, M., *J. Organometal. Chem.* **149**, 37 (1978).

82. Ishikawa, M., Fuchikami, T., and Kumada, M., *J. Organometal. Chem.* **117**, C58 (1976).
83. Barton, T. J., and Burns, G. T., *Organometallics* **1**, 1455 (1982).
84. Golino, C. M., Bush, R. D., Sommer, L. H., and On, P., *J. Am. Chem. Soc.* **97**, 1957 (1975).
85. Ando, W., Sekiguchi, A., and Sato, T., *J. Am. Chem. Soc.* **103**, 5573 (1981).
86. Ando, W., Sekiguchi, A., and Sato, T., *J. Am. Chem. Soc.* **104**, 6830 (1982).
87. Wiberg, N., Preiner, G., Schieda, O., and Fischer, G., *Chem. Ber.* **114**, 3505 (1981).
88. Wiberg, N., Preiner, G., and Schieda, O., *Chem. Ber.* **114**, 3518 (1981).
89. Wiberg, N., and Preiner, G., *Angew. Chem.* **90**, 393 (1978): *Angew. Chem.* **89**, 343 (1977).
90. Wiberg, N., *J. Organometal. Chem.* **273**, 141 (1984).
91. Ishikawa, M., Oda, M., Miyoshi, N., Fabry, L., Kumada, M., Yamabe, T., Akagi, K., and Fukui, K., *J. Am. Chem. Soc.* **101**, 4612 (1979).
92. Lewis, C., and Wrighton, M. S., *J. Am. Chem. Soc.* **105**, 7768 (1983).
93. Bertrand, G., Mazerolles, P., and Ancelle, J., *Tetrahedron* **37**, 2459 (1981).
94. Damon, R. E., and Barcza, S., US Pat. Appl. 74002, CA 95, 115739b.
95. Sekiguchi, A., and Ando, W., *Tetrahedron Lett.* **24**, 2791 (1983).
96. Tzeng, D., Fong, R. H., Soysa, H. S. D., and Weber, W. P., *J. Organometal. Chem.* **219**, 153 (1981).
97. Ishikawa, M., Fuchikami, T., and Kumada, M., *J. Organometal. Chem.* **127**, 261 (1977).
98. Ishikawa, M., Fuchikami, T., and Kumada, M., *J. Organometal. Chem.* **162**, 223 (1978).
99. Ishikawa, M., Fuchikami, T., and Kumada, M., *J. Organometal. Chem.* **133**, 19 (1977).
100. Ishikawa, M., Fuchikami, T., and Kumada, M., *J. Organometal. Chem.* **173**, 117 (1979).
101. Ishikawa, M., *Pure Appl. Chem.* **50**, 11 (1978).
102. Ishikawa, M., Fuchikami, T., Sugaya, T., and Kumada, M., *J. Am. Chem. Soc.* **97**, 5923 (1975).
103. Brook, A. G., and Harris, J. W., *J. Am. Chem. Soc.* **98**, 3381 (1976).
104. Brook, A. G., Harris, J. W., Lennon, J., and El Sheikh, M., *J. Am. Chem. Soc.* **101**, 83 (1979).
105. Khovrin, G. V., Zhitlova, I. V., and Gribov, L. A., *Izv. Timiryazev. Sel'skokhoz. Akad.* (5), 172 (1971).
106. Ishikawa, M., Oda, M., Nishimura, N., and Kumada, M., *Bull. Chem. Soc. Jpn.* **56**, 2795 (1983).
107. Ishikawa, M., Fuchikami, T., and Kumada, M., *J. Organometal. Chem.* **118**, 155 (1976).
108. Ishikawa, M., Fuchikami, T., and Kumada, M., *J. Organometal. Chem.* **118**, 139 (1976).
109. Shuvaev, A. T., Zemlyanov, A. P., Kolodyazhnyi, Yu. V., Osipov, O. A., Tatevosyan, M. M., Eliseev, V. N., and Morgunova, M. M., *Izv. Akad. Nauk. SSSR. Ser. Fiz.* **38**, 541 (1974).
110. McCarthy, W. Z., *Diss. Abstr. Int B* **45**, 868 (1984).
111. Sommer, L. H., and McLick, J., *J. Organometal. Chem.* **101**, 171 (1975).
112. Roark, D. N., and Sommer, L. H., *J. Chem. Soc. Chem. Commun.* 167 (1973).
113. Boudjouk, P., and Sommer, L. H., *J. Chem. Soc. Chem. Commun.* 54 (1973).
114. Boudjouk, P., Roberts, J. R., Golino, C. M., and Sommer, L. H., *J. Am. Chem. Soc.* **94**, 7926 (1972).
115. Jones, P. R., and Lee, M. E., *J. Am. Chem. Soc.* **105**, 6725 (1983).
116. Jones, P. R., Lee, M. E., and Lin, L. T., *Organometallics* **2**, 1039 (1983).
117. Jones, P. R., and Lee, M. E., *J. Organometal. Chem.* **232**, 33 (1982).
118. Jones, P. R., and Cheng, A. H.-B., VII International Symposium on Organosilicon Chemistry, Kyoto, 1984, Abstract 3A1010.

119. Ishikawa, M., Fuchikami, T., and Kumada, M., *J. Am. Chem. Soc.* **99**, 245 (1977).
120. Maerkl, G., and Hofmeister, P., *Angew. Chem.*, **91**, 863 (1979).
121. Maerkl, G., and Horn, M., *Tetrahedron Lett.* **24**, 1477 (1983).
122. Brook, A. G., Nyburg, S. C., Abdesaken, F., Gutekunst, B., Gutekunst, G., Kallury, R. K. M. R., Poon, Y. C., Chang, Y. -M., and Wong-Ng, W., *J. Am. Chem. Soc.* **104**, 5667 (1982).
123. Brook, A. G., Abdesaken, F., Gutekunst, G., and Plavac, N., *Organometallics* **1**, 994, (1982).
124. Brook, A. G., Nyburg, S. C., Reynolds, W. F., Poon, Y. C., Chang, Y. -M., Lee, J. -S., and Picard, J. -P., *J. Am. Chem. Soc.* **101**, 6750 (1979).
125. Wiberg, N., and Wagner, G., *Angew. Chem.* **95**, 1027 (1983).
126. Sekiguchi, A., and Ando, W., VII International Symposium on Organosilicon Chemistry, Kyoto, 1984, Abstract 3A1030.
127. Wiberg, N., Wagner, G., Muller, G., and Riede, J., *J. Organometal. Chem.* **271**, 381 (1984).
128. Barton, T. J., and Wulff, W. D., *J. Am. Chem. Soc.* **101**, 2735 (1979).
129. Brook, A. G., Nyburg, S. C., Kallury, R. K. M. R., and Poon, Y. C., *Organometallics* **1**, 987 (1982).
130. Barton, T. J., and Tully, C. R., *J. Organometal. Chem.* **172**, 11 (1979).
131. Zemlyanskii, N. N., Borisova, I. V., Bel'skii, V. K., Ustynyuk, Yu. A., and Beletskaya, I. P., *Izv. Akad. Nauk. SSSR, Ser. Khim.* **32**, 956 (1983).
132. Sekiguchi, A., and Ando, W., *J. Am. Chem. Soc.* **103**, 3579 (1981).
133. Sakurai, H., Kamiyama, Y., and Nakadaira, Y., *J. Am. Chem. Soc.* **98**, 7424 (1976).
134. Nakadaira, Y., Kanouchi, S., and Sakurai, H., *J. Am. Chem. Soc.* **96**, 5623 (1974).
135. Nakadaira, Y., Kanouchi, S., and Sakurai, H., *J. Am. Chem. Soc.* **96**, 5621 (1974).
136. Ishikawa, M., Kovar, D., Fuchikami, T., Nishimura, K., and Kumada, M., *J. Am. Chem. Soc.* **103**, 2324 (1981).
137. Ishikawa, M., Nishimura, K., Ochiai, H., and Kumada, M., *J. Organometal. Chem.* **236**, 7 (1982).
138. Sugisawa, S., Ph. D. thesis, Kyoto University, 1983.
139. Ishikawa, M., Tabohashi, T., Kumada, M., and Iyoda, J., *J. Organometal. Chem.* **264**, 79 (1984).
140. Burns, G. T., and Barton, T. J., *J. Organometal. Chem.* **216**, C5 (1981).
141. Jones, P. R., Cheng, A. H. -B., and Albanesi, T. E., *Organometallics* **3**, 78 (1984).
142. Wiberg, N., and Wagner, G., unpublished results.
143. Eaborn, C., Happer, D. A. R., Hitchcock, P. B., Hopper, S. P., Safa, K. D., Washburne, S. S., and Walton, D. M., *J. Organometal. Chem.* **186**, 309 (1980).
144. Burns, G. T., and Barton, T. J., *J. Organometal. Chem.* **209**, C25 (1981).
145. Barton, T. J., and Burns, G. T., *Tetrahedron Lett.* **24**, 159 (1983).
146. Barton, T. J., and Hussman, G. P., *Organometallics* **2**, 692 (1983).
147. Wiberg, N., Preiner, G., and Schieda, O., *Chem. Ber.* **114**, 2087 (1981).
148. Valkovich, P. B., and Weber, W. P., *Tetrahedron Lett.* 2153 (1975).
149. Okazaki, R., Kang, K. T., and Inamoto, N., *Tetrahedron Lett.* **22**, 235 (1981).
150. Schlenk, W., *Ann.* **394**, 221 (1912).
151. Kipping, F. S., *J. Chem. Soc.* 104 (1927).
152. Kipping, F. S., Murray, A. G., and Maltby, J. G., *J. Chem. Soc.* 1108 (1929).
153. Pitzer, K. S., *J. Am. Chem. Soc.* **70**, 2140 (1948).
154. Mullikin, R. S., *J. Am. Chem. Soc.* **72**, 4493 (1950).
155. Gusel'nikov, L. E., and Flowers, M., *Chem. Commun.* 864 (1967).
156. Gusel'nikov, L. E., and Nametkin, N. S., *Chem. Rev.* **79**, 529 (1979).

157. Barton, T. J., *Pure Appl. Chem.* **52**, 615 (1980).
158. Coleman, B., and Jones, M., *Rev. Chem. Intermed.* **4**, 297 (1981).
159. Schaefer, H. F., *Acc. Chem. Res.* **15**, 283 (1982).
160. Fleming, I., *in* "Comprehensive Organic Chemistry," Vol. 3, pp. 671-675 Pergamon, Toronto, 1979.
161. Armitage, D. A., *in* "Comprehensive Organometallic Chemistry," Vol. 2, pp. 80-86 Pergamon, Toronto, 1982.
162. Tokach, S., Boudjouk, P., and Koob, R. D., *J. Phys. Chem.* **82**, 1203 (1978).
163. Drahnak, T. J., Michl, J., and West, R., *J. Am. Chem. Soc.* **103**, 1845 (1981).
164. Jones, P. R., Lim, T. F., and Pierce, R. A., *J. Am. Chem. Soc.* **102**, 4970 (1980).
165. Nefedov, O. M., Mal'tsev, A. K., Khabashesku, V. N., and Korolev, V. A., *J. Organometal. Chem.* **201**, 123 (1980).
166. Wiberg, N., Wagner, G., Muller, G., and Riede, J., *J. Organometal. Chem.* **271**, 381 (1984).
167. Wiberg, N., Wagner, G., and Muller, G., *Angew. Chem. Int. Ed. Engl.* **24**, 229 (1985).
168. Brook, A. G., Abdesaken, F., Gutekunst, B., Gutekunst, G., and Kallury, R. K., *J. Chem. Soc. Chem. Commun.* 191 (1981).
169. West, R., Fink, M. J., and Michl, J., *Science* **214**, 1343 (1981).
170. Levy, G. C., and Nelson, G. L., "Carbon-13 Nuclear Magnetic Resonance for Organic Chemists," p. 59 Wiley (Interscience), New York, 1972.
171. Jutzi, P., *Angew. Chem. Int. Ed. Engl.* **14**, 232 (1975).
172. Maier, G., Mihm, G., and Reisenauer, H. P., *Angew. Chem. Int. Ed. Engl.* **19**, 52 (1980).
173. Mahaffy, P. G., Gutowsky, R., and Montgomery, L. K., *J. Am. Chem. Soc.* **102**, 2854 (1980).
174. Rosmus, P., Bock, H., Solouki, B., Maier, G., and Mihm, G., *Angew. Chem. Int. Ed. Engl.* **20**, 598 (1981).
175. Koenig, T., and McKenna, W., *J. Am. Chem. Soc.* **103**, 1212 (1981).
176. Dyke, J. M., Josland, G. D., Lewis, R. S., and Morris, A., *J. Phys. Chem.* **86**, 2913 (1982).
177. Dewar, M. J. S., Lo, D. H., and Ramsden, D. A., *J. Am. Chem. Soc.* **97**, 1311 (1975).
178. Blustin, P. H., *J. Organometal. Chem.* **105**, 161 (1976).
179. Curtis, M. D., *J. Organometal. Chem.* **60**, 63 (1973).
180. Ahlrichs, R., and Heinzmann, R., *J. Am. Chem. Soc.* **99**, 7452 (1977).
181. Hanamura, M., Nagase, S., and Morokuma, K., *TetrahedronLett.* **22**, 1813 (1981).
182. Walsh, R., *Acc. Chem. Res.* **14**, 246 (1981).
183. Walsh, R., *J. Organometal. Chem.* **38**, 245 (1972).
184. Basu, S., Davidson, I. M. T., Laupert, R., and Potzinger, P., *Ber. Bunsenges. Phys. Chem.* **83**, 1282 (1979).
185. Pietro, W. J., Pollack, S. K., and Hehre, W. J., *J. Am. Chem. Soc.* **101**, 7126 (1979).
186. Pau, C. F., Pietro, W. J., and Hehre, W. J., *J. Am. Chem. Soc.* **105**, 16 (1983).
187. Hood, D. M., and Schaefer, H. F., *J. Chem. Phys.* **68**, 2985 (1978).
188. Kohler, H. J., and Lischka, H., *J. Am. Chem. Soc.*, **104**, 5884 (1982).
189. Strausz, O. P., Gammie, L., Theodorakopoulos, G., Mezey, P. G., and Csizmadia, I. G., *J. Am. Chem. Soc.* **98**, 1622 (1976).
190. Strausz, O. P., Robb, M. A., Theodorakopoulos, G., Mezey, P. G., and Csizmadia, I. G., *Chem. Phys. Lett.* **48**, 162 (1977).
191. Gusel'nikov, L. E., and Nametkin, N. S., *J. Organometal. Chem.* **169**, 155 (1979).
192. Verwoerd, W. S., *J. Comp. Chem.* **3**, 445 (1982).
193. Schlegel, H. B., Coleman, B., and Jones, M., *J. Am. Chem. Soc.* **100**, 6499 (1978).
194. Dewar, M. J. S., and Thiel, W., *J. Am. Chem. Soc.* **99**, 4907 (1977).

195. For other calculations involving silabenzene, see Refs. 177 and 196.
196. Blustin, P. H., *J. Organometal. Chem.* **166,** 21 (1979).
197. Goddard, J. D., Yoshioka, Y., and Schaefer, H. F., *J. Am. Chem. Soc.* **102,** 7644 (1980).
198. Woodward, R. B., and Hoffmann, R., *Angew. Chem. Int. Ed. Engl.* **8,** 781 (1969).
199. Brook, A. G., and McClenaghan, J., unpublished studies.
200. Brook, A. G., Safa, K. D., Lickiss, P. D., and Baines, K. M., *J. Am. Chem. Soc.* **107,** 4338 (1985).
201. Wiberg, N., and Kopf, H., unpublished studies summarized in ref. 90.
202. West, R., Fink, M. J., Michalczyk, M. J., De Young, D. J., and Michl, J., VII International Symposium on Organosilicon Chemistry, Kyoto 1984, Abstract 1B1030.
203. Golino, C. M., Bush, R. D., Roark, D. N., and Sommer, L. H., *J. Organometal. Chem.* **66,** 29 (1974).
204. Barton, T. J., and Hussman, G. P., *Organometallics* **2,** 692 (1983).
205. Sakurai, H., Nakadaira, Y., Kira, M., Sugiyama, H., Yoshida, K., and Takiguchi, T., *J. Organometal. Chem.* **184,** C36 (1980).
206. Conlin, R. T., and Wood, D. L., *J. Am. Chem. Soc.* **103,** 1843 (1981).
207. Brook, A. G., Behnam, B., and Ford, R. R., unpublished studies.
208. Cooper, J., Hudson, A., and Jackson, R. A., *J. Chem. Soc., Perkin II* 1933 (1973).
209. Brook, A. G., and Wessely, H. -J., *Organometallics* **4,** 1487 (1985).

ADVANCES IN ORGANOMETALLIC CHEMISTRY, VOL. 25

Metalla-Derivatives of
β-Diketones

CHARLES M. LUKEHART

Department of Chemistry
Vanderbilt University
Nashville, Tennessee 37235

I

INTRODUCTION

A complex containing metalla-β-diketonato ligands was first reported by the author and co-workers in 1975. This discovery led to a general investigation of this chemistry and to the synthesis of several related transition metal organometallic compounds. A review of the early developments of this research has been published (*1*). In this article, more recent results pertaining to metalla-β-diketonate chemistry are reviewed. The relationship of these results to other areas of organometallic chemistry is also mentioned.

The term "metalla-β-diketonate" refers to an anionic diacyl complex, such as $L_nM(RCO)(R'CO)^-$, in which the two acyl ligands occupy adjacent coordination sites of the metal atom. Resonance stabilization of the negative charge by π-delocalization onto the two acyl ligands gives Lewis structures **1** and **2**. The two acyl ligands acquire an acyl/carbene

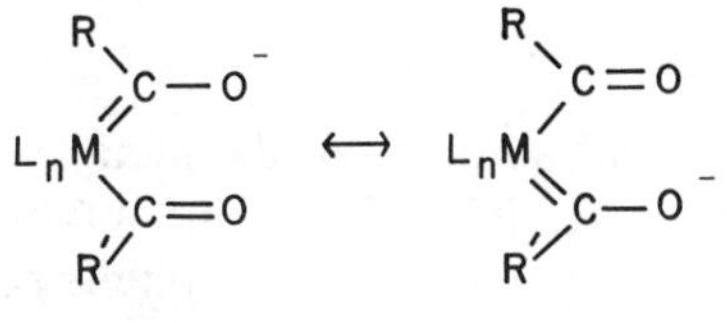

1   2

45

Copyright © 1986 by Academic Press, Inc.
All rights of reproduction in any form reserved

hybrid electronic structure. When the metal moiety is cis-$(OC)_4Re$ or $(\eta\text{-}C_5H_5)(OC)Fe$, for example, the metal atom obeys the EAN rule. Structures **1** and **2** are formally related to organic β-diketonate anions where the sp^2 CH methine group has been replaced by the metal fragment, L_nM. Because of this analogy, these organometallic anions are referred to as metalla-β-diketonates. The metal moieties listed above are now regarded as being isolobal to a methine group (2,3).

The first chemical analogy between β-diketonate and metalla-β-diketonate anions to be established was the formation of metal tris-chelate complexes containing metalla-β-diketonate anions as bidentate, chelating ligands (3). In these complexes, the two acyl ligands bridge the two metal atoms, M and M'. In recent years, bridging acyl ligands have become an important type of ligand in both dinuclear (eg., **4** and **5**) and polynuclear (eg., **6** and **7**) complexes (4–6). Complex **7** is an example of a complex in which an acyl ligand bridges two metals connected by an M—M single bond.

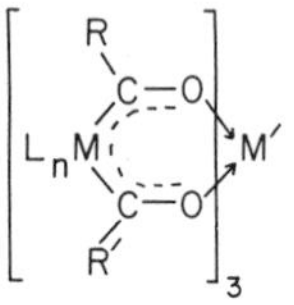

3

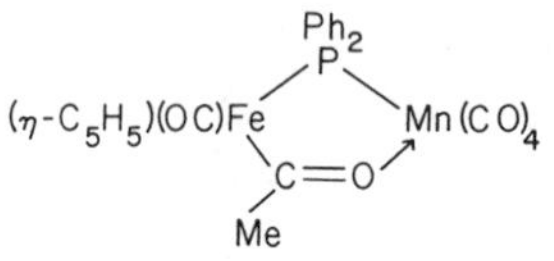

4

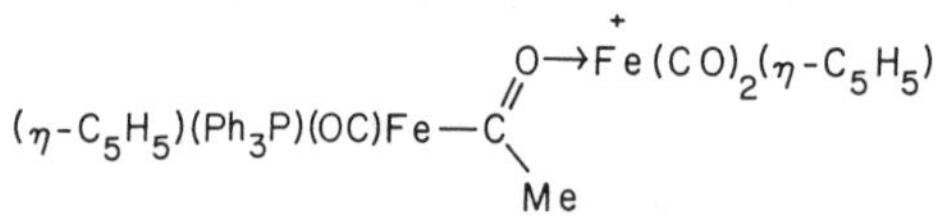

5

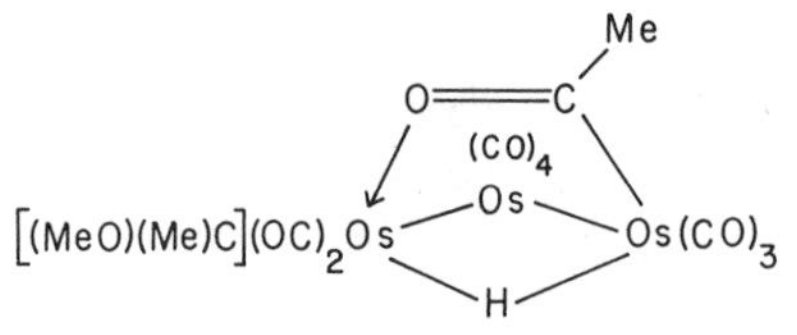

6

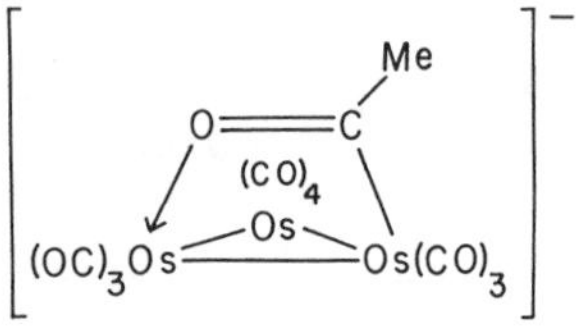

7

Another class of molecules related to the metalla-β-diketonates are the metalla-β-diphosphinites, $L_nM(R_2PO)_2^-$ (7–10). These anionic complexes can be protonated to give neutral enolic tautomers or coordinated to main group elements or metal ions to give polynuclear complexes. The metalla-β-diketonate anions form the same types of complexes.

II

GENERAL CLASSES OF COMPOUNDS

A. Metalla-β-diketonates

Metalla-β-diketonate complexes, such as **1**, are conveniently prepared by reacting acylmetal carbonyl complexes with strong bases that can also react as nucleophiles, such as organolithium, Grignard, or boron hydride reagents [Eq. (1)]. These reactions can be followed by IR spectroscopy.

$$RC(O)M(CO)_xL_y + R'Li \rightarrow Li[(OC)_{x-1}L_yM(RCO)(R'CO)] \tag{1}$$

8 **9**

The carbonyl and acyl C—O stretching vibrations of **8** shift to lower frequencies as **9** is formed. In addition, if **8** is an octahedral complex then the nucleophilic attack of R' occurs at a carbonyl ligand cis to the acyl ligand to give the required cis relative orientation of the acyl ligands in the metalla-β-diketonate anion, **9**. The carbonyl ligand C—O stretching *pattern* of **9** in the IR spectrum confirms this structural change.

The first *cis*-diacylmetalate complex, *cis,cis*-[MgBr]$_2$[Os(MeCO)$_2$Br$_2$-(CO)$_2$]·5THF, was reported by L'Eplattenier (*11*). This complex was prepared from MeMgBr and OsBr$_2$(CO)$_4$. Casey and Bunnell established that methyllithium adds to a cis carbonyl ligand of PhC(O)Mn(CO)$_5$ by determining the X-ray structure of [Me$_4$N][*cis*-(OC)$_4$Mn(PhCO)(MeCO)] (*12,13*). Darensbourg *et al.* similarly prepared [MgCl][*cis*-(OC)$_4$Mn-(PhCH$_2$CO)$_2$] from benzoylpentacarbonylmanganese and PhCH$_2$MgCl (*14*).

Our recognition of these *cis*-diacylmetalate complexes as metalla-β-diketonate species led to an extension of this chemistry (*1*). Nucleophilic addition to carbonyl ligands of acyl complexes gave metalla-β-diketonates of the following types: [*cis*-(OC)$_4$Mn(RCO)(R'CO)]$^-$, in which R is alkyl or benzyl and R' is alkyl, alkoxide, or NMe$_2$; [*fac*-(OC)$_3$(RNC)Mn(MeCO)$_2$]$^-$, where R is Me, C$_6$H$_{11}$, or *t*-butyl; [*cis*-(OC)$_4$Re(RCO)(R'CO)]$^-$, in which R is alkyl or benzyl and R' is alkyl or H; and [(η-C$_5$H$_5$)(OC)Fe(RCO)(R'CO)]$^-$, where R is alkyl and R' is alkyl or benzyl.

We recently reported the X-ray structure of the benzamidinium salt of a rhenaacetylacetonate anion, [PhC(NH$_2$)$_2$][*cis*-(OC)$_4$Re(MeCO)$_2$] (*15*). This structure is an important reference for other structural data of metalla-β-diketonate derivatives. The Re—C(acyl) and C(acyl)—O average distances of 2.182(6) and 1.239(9) Å indicate some degree of π-delocalization between the Re atom and the acetyl ligands. Structural comparisons indicate that μ-η^2-acetyl ligands will have C—O distances in the range of 1.25–1.42 Å, although most of these ligands will have

C—O distances in the range of 1.25–1.29 Å. For example, in complexes **4** and **6**, the bridging acetyl C—O distances are 1.262(9) and 1.28(2) Å, respectively.

Casey *et al.* have studied the decarbonylation reactions of [*cis*-$(OC)_4M(MeCO)(PhCO)]^-$, in which M is Mn or Re *(16,17)*. These complexes lose a carbonyl ligand to form five-coordinate intermediates of the type $[(OC)_3M(MeCO)(PhCO)]^-$. Reversible methyl migration proceeds much more rapidly than does phenyl migration. In the course of these studies, a phosphine substituted rhena-β-diketonate complex, [*fac*-$(OC)_3(Et_3P)Re(MeCO)(PhCO)]^-$, was prepared.

When acetylpentacarbonylrhenium is treated with Li[BEt$_3$H] in THF solution at $-78°C$ the rhenaformylacetonate complex [*cis*-$(OC)_4$Re-$(MeCO)(HCO)]^-$ is formed *(18)*. This species has a half-life of 8.1 minutes in THF solution at 32°C. Related manganese formyl complexes have been reported by Gladysz *et al.* *(19,20)*. More recently Gladysz *et al.* have reported the preparation of the following types of metalla-formyl-acylate anions: [*cis*-$(OC)_4Mn(HCO)(RCO)]^-$, in which R is Ph, CH_2OMe, C(O)OMe, or CF_3; $[(\eta-C_5H_5)(OC)Fe(HCO)(RCO)]^-$, where R is Me, Ph, or *p*-C_6H_4OMe; and $[(\eta-C_5H_5)(OC)_2Mo(HCO)(RCO)]^-$, in which R is C(O)OMe *(21)*. All of these species are kinetically unstable relative to various decomposition products. Sweet and Graham have prepared the diformylrhenate complex $[(\eta-C_5H_5)(ON)Re(HCO)_2]^-$. This complex is formed by treating $(\eta-C_5H_5)Re(NO)(CO)(HCO)$ with Li[BEt$_3$H] *(22)*.

Neutral *bis*(acyl)Fe(CO)$_4$ complexes have been reported by Stewart *et al.* *(23)*. The acyl ligands have perfluoroalkyl substituents. These complexes are isoelectronic to the anionic diacyltetracarbonylmanganate species mentioned above.

B. *Metalla-β-diketones*

Metalla-β-diketones are readily prepared by protonation of metalla-β-diketonate anions. The first reported metalla-β-diketone was the rhenaacetylacetone molecule [*cis*-$(OC)_4Re(MeCO)_2]H$ (**10**) *(1,24)*. Diagnostic indications of the formation of this type of complex are (1) the appearance of a *cis*-$L_2M(CO)_4$ pattern in the carbonyl C—O stretching

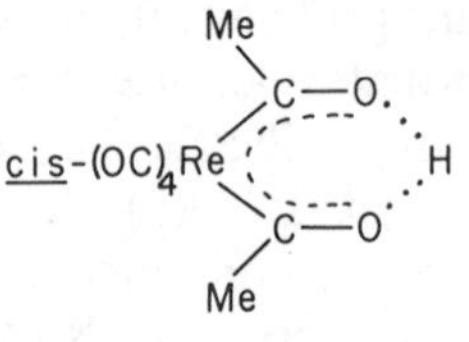

10

region of the IR spectrum at frequencies expected for a neutral complex, (2) the presence in the IR spectrum of an acyl (C—O) stretching band at ~ 1520 cm^{-1} and an O $\cdots$ H $\cdots$ O vibrational stretching band at ~ 1650 cm^{-1}, and (3) the appearance of a very low resonance ($\sim \delta$ 21 in CS$_2$ solution) in the ^{1}H-NMR spectrum for the enolic proton. Other known metalla-β-diketones include [cis-(OC)$_4$Re(MeCO)(RCO)]H, in which R is i-Pr or PhCH$_2$ (1); [cis-(OC)$_4$Re(MeCO)(PhCO)]H (17); [cis-(OC)$_4$Re(Me$_3$CCO)$_2$]H (25); and [(η-C$_5$H$_5$)(OC)Fe(MeCO)(RCO)]H, where R is Me, i-Pr, or Ph$_2$HC (1). Enolic proton resonances for ferra-β-diketones appear in the range of δ 19.1–20.1.

The most well-characterized metalla-β-diketone is **10** in that both X-ray and neutron diffraction structures have been determined (**11** and **12**) (26).

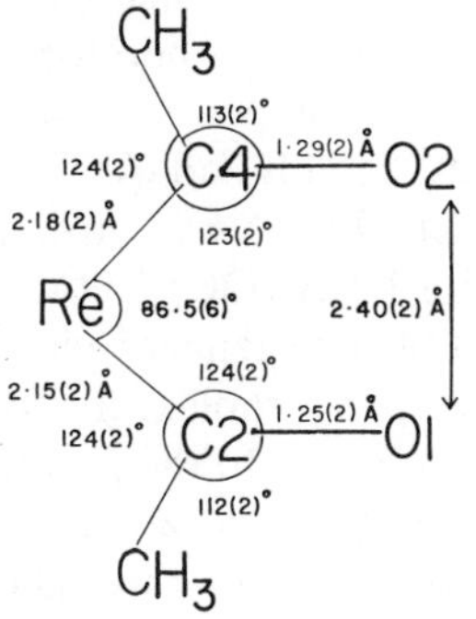

I I	I2
(x-ray)	(neutron)

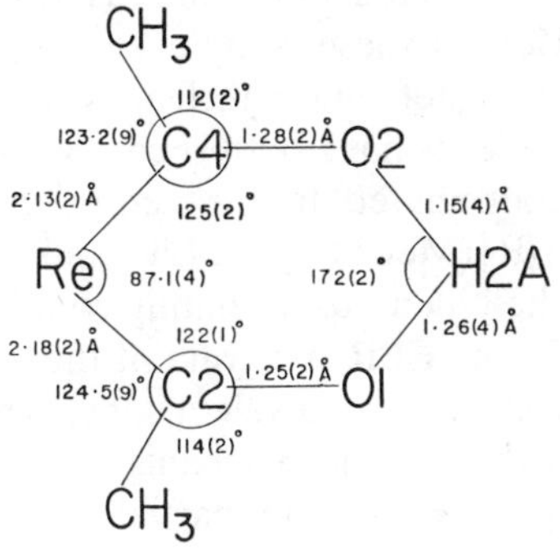

Both structural determinations give an O-1 $\cdots$ O-2 bite distance of 2.40(2) Å. The corresponding bite distances in tetraacetylethane (which exists as the dienolic tautomer) and acetylacetone are 2.42 and 2.535 Å, respectively, (27,28). The intra-chelate ring bond distances indicate slight asymmetry; however, this ring is symmetrical within the ± 1.5 σ limit. The neutron structure clearly establishes the presence of an enolic hydrogen atom. Although the O $\cdots$ H $\cdots$ O hydrogen bond appears to be slightly asymmetrical, the large thermal motion of the hydrogen atom prevents a more precise definition of this structural unit. The O-1—H-2A—O-2 angle of 172(2)° is consistent with a nearly linear O $\cdots$ H $\cdots$ O hydrogen bond.

C. Metalla-β-diketonato Complexes

A large number of complexes have been prepared in which metalla-β-diketonate anions coordinate to either metal ions or B(X)(Y) moieties. The chemistry of such metal complexes has been reviewed previously (1). To summarize briefly, tris-chelate complexes (**3**) of Al(III) or Ga(III)

containing mangana-, rhena-, and ferra-β-diketonato ligands can be prepared from the corresponding metalla-β-diketonate anions and either
$AlCl_3$ or $GaCl_3$ via direct chloride displacement. The X-ray structure of
$[cis\text{-}(OC)_4Mn(MeCO)_2]_3Al$ confirms the bidentate, chelating coordination
of the manganaacetylacetonato ligands. Seventeen such complexes have
been prepared. In solution, the tris-chelate complexes containing unsym-
metrically substituted metalla-β-diketonato ligands exhibit the expected
geometrical isomerism.

Rhena-β-diketonato complexes of Mg(II), Zn(II), Cu(II), Fe(III),
Cr(III) and U(IV) have been isolated also (1). These complexes are
prepared only by treating a rhena-β-diketone with a metal complex
containing basic ligands, e.g., $Cu(OMe)_2$, $Fe(OEt)_3$, and $Cr[N(i\text{-}Pr)_2]_3$.
Apparently, the reaction proceeds by loss of the conjugate acid of the basic
ligand. Most of these complexes are paramagnetic and, therefore, represent paramagnetic polynuclear, organometallic compounds. Several of
these complexes have both rhena-β-diketonato ligands and other ancillary
ligands coordinated to the central metal ion. The X-ray structures of
$[cis\text{-}(OC)_4Re(MeCO)_2]_2Cu$ (29) and $(\eta\text{-}C_5H_5)_2U[cis\text{-}(OC)_4Re(MeCO)_2]_2$
confirm the bidentate, chelating coordination of the rhenaacetylacetonato
ligands. In the EPR spectrum of the above copper complex, the unpaired
Cu electron shows equivalent superhyperfine coupling to the two rhenium
nuclei. Although the mechanism of this superhyperfine coupling is not
known, it may indicate partial electron delocalization throughout the
rhenaacetylacetonato ligands.

In related work, an unusual ytterbium(III) complex, $[(\eta\text{-}C_5Me_5)_2\text{-}$
$Yb]_2[Fe_3(CO)_7(\mu\text{-}CO)_4]$, was shown by Tilley and Andersen to contain
four isocarbonyl ligands (30). The iron portion of this complex acts
formally as a 1,3,5-trimetalla-β-diketonato ligand coordinated to the two
Yb(III) ions.

Metalla-β-diketones or metalla-β-diketonate anions react with trigonal
boron compounds, BX_2Y, in which $X = Y = $ halogen or $X = Cl$ and
$Y = Ph$, to afford neutral (metalla-β-diketonato)$B(X)(Y)$ complexes.
Complexes of this type are known with mangana-, rhena-, and ferra-β-
diketonato ligands. The earlier chemistry of these complexes has also been
previously reviewed (1).

Recent results have now clarified the geometrical isomerism present in
(ferra-β-diketonato)BF_2 complexes. The X-ray structure of $[(\eta\text{-}C_5H_5)(OC)\text{-}$
$Fe(MeCO)(i\text{-}PrCO)]BF_2$ (13), where R is Me and R' is i-Pr, reveals a slightly
boat-shaped ferra-chelate ring in which the CO ligand occupies an "axial" site
and the C_5H_5 ligand occupies an "equatorial" site. An IR spectrum of this
complex in KBr shows a strong C—O stretching band at 2005 cm^{-1}. In
solution, however, the IR spectrum of this compound shows two C—O

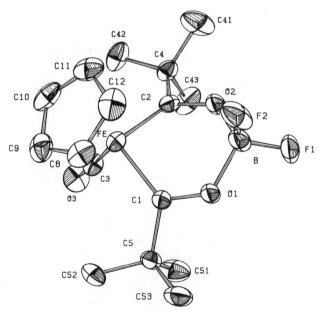

stretching bands—one at ~2010 cm^{-1} and one at ~1970 cm^{-1}. The relative intensities of these bands varies with solvent polarity, so the presence of another isomer is indicated (1).

The structure of the isomer having a carbonyl C—O band at ~1970 cm^{-1} has been determined for a related complex. Infrared spectra of $[(\eta\text{-}C_5H_5)(OC)Fe(Me_3CCO)_2]BF_2$ in KBr (as well as in various solvents) show a single carbonyl C—O stretching band at ~1960 cm^{-1}. The X-ray structure of this complex reveals a boat-shaped structure like **14** (in which R = R' = Me$_3$C) where the C$_5$H$_5$ ligand now occupies an "axial" site in closest proximity to the BF$_2$ "axial" fluorine atom (25). An ORTEP view of this complex is shown in Fig. 1. These results establish the structures of both geometrical isomers (**13** and **14**) for this type of complex.

FIG. 1. An ORTEP view of $[(\eta\text{-}C_5H_5)(OC)Fe(Me_3CCO)_2]BF_2$ with hydrogen atoms omitted for clarity (ellipsoids at 30% probability).

Presumably, the relative stability of each isomer depends on intramolecular steric repulsions, such as R ... R', C_5H_5 ... F (axial), and C_5H_5 ... R, R' interactions.

In other recent work, (ferra-β-diketonato)BF_2 complexes containing various vinyl substituents have been prepared [Eq. (2)] (32). When the

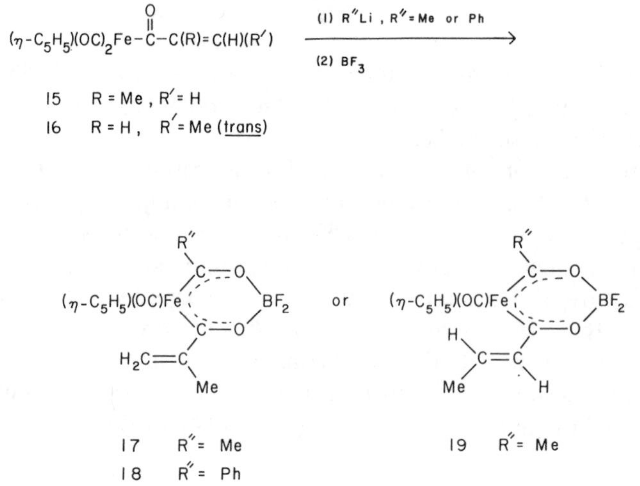

$$(\eta\text{-}C_5H_5)(OC)_2Fe\overset{\overset{O}{\parallel}}{-}C{-}C(R){=}C(H)(R') \xrightarrow[\text{(2) } BF_3]{\text{(1) } R''Li, \ R''=Me \ \text{or} \ Ph}$$

15　R = Me , R' = H
16　R = H , R' = Me (trans)

17　R'' = Me
18　R'' = Ph

19　R'' = Me

(2)

methacryloyl and crotonyl iron complexes **15** and **16** are treated with methyl- or phenyllithium followed by BF_3, the substituted vinyl (ferra-β-diketonato)BF_2 complexes **17–19** are formed. The X-ray structure of **17** has been determined, and an ORTEP view of this molecule is shown in Fig. 2. The C-5—C-6 distance of 1.317(3) Å represents a normal C=C double bond length, and the ferra-chelate ring has a boat structure of type **13**.

Complexes **17–19** can be written in one valence structure as α, β-unsaturated carbonyl compounds in which the carbonyl oxygen atom is coordinated to a $BF_2(OR)$ Lewis acid. The C=C double bonds of such organic systems are activated toward certain reactions, like Diels–Alder additions, and complexes **17-19** show similar chemistry. Complexes **17** and **18** undergo Diels–Alder additions with isoprene, 2,3-dimethyl-1,3-butadiene, *trans*-2-methyl-1,3-pentadiene, and cyclopentadiene to give Diels–Alder products **20–23** as shown in Scheme 1 for complex **17** (32). Compounds **20–23** are prepared in crude product yields of 75–98% and are isolated as analytically pure solids in yields of 16–66%. The X-ray structure of the isoprene product **20** has been determined and the ORTEP diagram (shown in Fig. 3) reveals the regiochemistry of the Diels–Alder addition. The C-14=C-15 double bond distance is 1.327(4) Å, and the

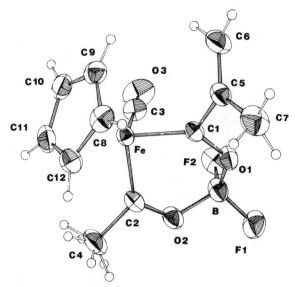

FIG. 2. An ORTEP view of complex **17** with thermal ellipsoids at 30% probability. Six half-hydrogen atoms are located about C-4.

ferra chelate ring has a boat structure of type **14**. Proton-NMR spectra indicate that the addition product of opposite stereochemistry is present in less than 5% relative abundance.

The addition of 2,3-dimethyl-1,3-butadiene to **17** gives only one structural isomer, **21**. The pseudo-first order half-life of this Diels–Alder addition reaction is 47.5 minutes at 22.2°C in neat diene solution. The E_a of this reaction is estimated to be 9.2 kcal/mol. This reaction rate is ∼ 50 times faster than the rate of addition of this diene to methyl methacrylate (*33*).

An ORTEP diagram of product **22** (Fig. 4) shows that the Diels–Alder addition of *trans*-2-methyl-1,3-pentadiene to **17** occurs with the same regiochemistry as does the isoprene addition. The relative stereochemistry at C-5 and C-16 is shown also, and the ferra-chelate ring has a boat structure of type **14**. The C-14=C-15 double bond distance is 1.318(4) Å. In going from **17** to **20** and **22** and the ferra-chelate ring boat structure inverts from type **13** to type **14**, presumably because of the significant increase in the steric size of the chelate-ring substituents. Cyclopentadiene adds to **17** to give **23** as an essentially 1:1 mixture of endo/exo isomers. The ferra-chelate ring boat structure of each of these isomers is also of type **14**.

Complex **19** does not appear to undergo Diels–Alder addition with isoprene; additions with 2,3-dimethyl-1,3-butadiene and cyclopentadiene do take place, however. The cyclopentadiene product exists solely as the endo isomer.

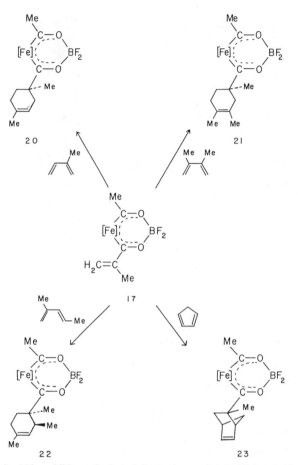

SCHEME 1. Diels–Alder additions of selected dienes to complex **17**. [Fe] = $(\eta\text{-}C_5H_5)(OC)Fe$.

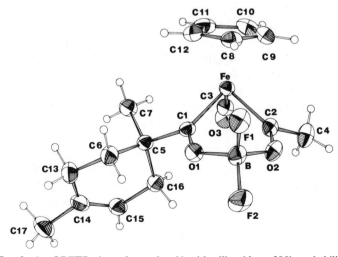

FIG. 3. An ORTEP view of complex **20** with ellipsoids at 30% probability.

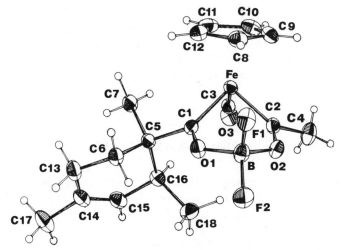

Fig. 4. An ORTEP view of complex 22 with ellipsoids at 30% probability.

Diels–Alder additions to 17–19 parallel the known Diels–Alder additions to vinyl substituted Fischer carbene complexes (34). It appears that both $M(CO)_5$ carbene moieties, where M is Cr or W, and (ferra-β-diketonato)BF_2 substituents can activate C=C double bonds toward Diels–Alder additions.

D. Triacylrhenato Complexes

Acylpentacarbonylrhenium complexes react with 2 molar equivalents of methyllithium to give triacylrhenate dianions of the type $Li_2[fac$-$(OC)_3Re(RCO)(MeCO)_2]$ (1). A similar manganese dianion, $[fac$-$(OC)_3Mn(PhCH_2CO)_3]^{2-}$, has been prepared in situ by Darensbourg et al., using a large excess of $PhCH_2MgCl$ as the nucleophilic reagent (14). These triacylrhenate dianions form bis-chelate complexes with Al(III), Ga(III), Zr(IV), and Hf(IV). Reaction with the boron trihalides gives neutral complexes of the type $[fac$-$(OC)_3Re(MeCO)_3]BX$, in which X is F, Cl, Br, or I. When the boron bromide complex is treated with $AgBF_4$ in alcohol solution, the corresponding boron alkoxide complex $[fac$-$(OC)_3Re(MeCO)_3]B(OR)$ is formed.

The X-ray structures of the boron bromide and boron chloride complexes confirm the tridentate coordination of the triacetylrhenato ligand to the boron atom. Both molecules have crystallographically imposed mirror symmetry. Although both compounds decompose appreciably during data collection, a relatively larger data set could be measured for the boron bromide complex. Within a $\pm 2\sigma$ limit, this complex exhibits idealized C_{3v} symmetry with average values for the B—O and acetyl C—O distances of 1.50(2) and 1.36(2) Å, respectively (35). The chemistry of these (triacyltri-carbonylrhenato)BX complexes is discussed below.

III

METALLA-β-KETOIMINES

Earlier work demonstrated that rhena-β-diketones readily undergo Schiff-base condensation with ammonia or primary amines to give rhena-β-ketoimines **24** and **25** [Eq. (3)] (1). When R is i-Pr, condensation occurs

$$[cis\text{-}(OC)_4Re(RCO)(MeCO)]H + H_2NR' \xrightarrow{\ -H_2O\ }$$

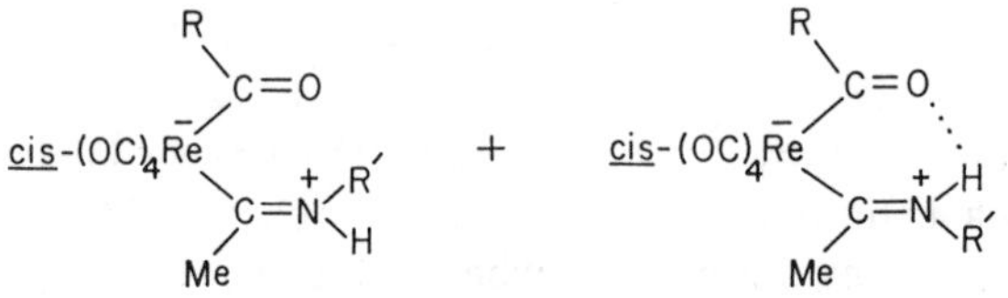

(3)

24   25

almost exclusively at the acetyl ligand. The X-ray structure of the N-phenylrhenaacetylacetonimine complex reveals the zwitterionic structure **24** (where R is Me and R' is phenyl). The C=N double bond distance is 1.26(2) Å. Adjacent molecules are held together by a chain network of N—H $\cdots$ O intermolecular hydrogen bonds formed between the acetyl oxygen atoms and the imine N—H groups. This isomer is referred to as the "interisomer" for this reason. In solution, interisomers isomerize to and equilibrate with the other geometrical isomer, **25**, via slow rotation about the carbon–nitrogen multiple bond. Intramolecular N—H $\cdots$ O hydrogen bonding is presumed to be present in these "intraisomers" based on proton-NMR data. The iminium N—H proton resonances for inter- and intraisomers appear at about δ 11.4 and δ 15.3, respectively, when R' is aryl, and in the ranges δ 8.91–9.82, and δ 13.06–13.58, respectively, when R' is H or alkyl. The low field N—H resonances for the intraisomers are consistent with the presence of intramolecular hydrogen bonding (as is observed for the enolic protons of metalla-β-diketones).

In recent years, the formation of rhena-β-ketoimine complexes has been extended to more complex amines that have biological importance. These amine derivatives have been prepared to extend the Schiff-base condensation [Eq. (3)] to amines having a wide range of solubilities and to demonstrate the formation of heavy atom labeled derivatives of biologically important molecules. Such derivatization should alter the distribution and transport properties of these amines, and the presence of the Re moiety might facilitate the detection and isolation of these or related species. The possibility of preparing the analogous Tc derivatives also exists.

Rhena-β-ketoimine derivatives of several selected 2-ethylamino compounds have been prepared (36). These amines include 2-chloroethylamine (a DNA-alkylating reagent), cystamine (a heparin antagonist), histamine (a potent vasodilator), tryptamine and O-methylserotonin (two indole alkaloids), and O,O-dimethyldopamine (an adrenergic drug derivative).

Twenty one rhena-β-ketoimine derivatives of 14 different amino acids and one dipeptide have also been synthesized (37,38). The amino acid derivatives are prepared according to Eq. (3) in which the primary amine is an amino acid ester. A dipeptide derivative is formed via normal peptide coupling reactions, as shown in Eq. (4). In this reaction sequence, the ethyl

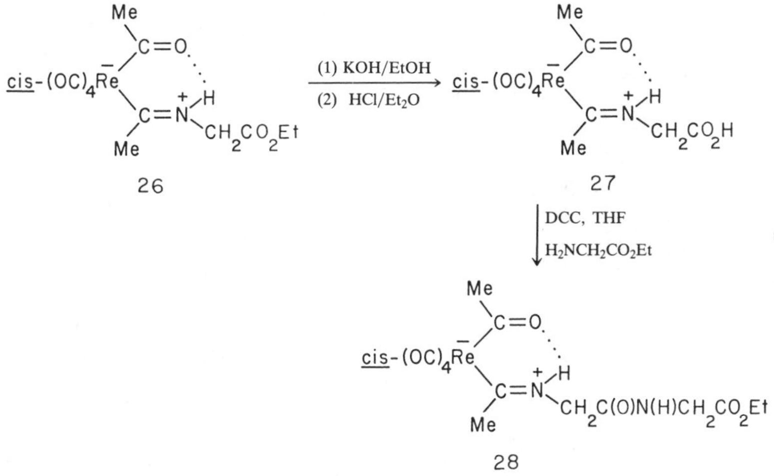

$$(4)$$

rhenaglycinate complex 26 is hydrolyzed to the free acid 27. Coupling this rhenaglycine complex to ethyl glycinate using dicyclohexylcarbodiimide (DCC) gives the ethyl rhenaglycylglycinate complex 28. In this sequence the rhena moiety acts as an N-terminus protecting group. In slightly related work, Ioganson et al. have prepared $Re(CO)_3$ complexes which contain glycylglycine as a ligand (39).

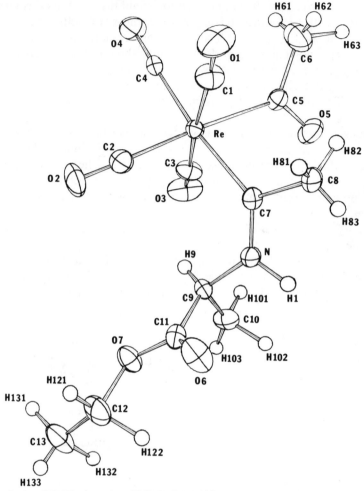

FIG. 5. An ORTEP view of *cis*-(OC)₄Re(MeCO)[MeCN(L-CHMeCO₂Et)(H)] with ellipsoids at 20% probability.

An X-ray structure of the analogous ethyl L-rhenaalaninate complex revealed the expected intermolecular hydrogen bonding between rhena groups with no significant structural distortions within the amino acid moiety (37). The ORTEP view of this complex is shown in Fig. 5. The C-7—N distance of 1.29(2) Å represents considerable carbon–nitrogen multiple bonding. When this complex is treated with 6 N HCl/acetone, the free amino acid is obtained.

Rhenaacetylacetonimine derivatives of the following amino acid esters have been prepared: L-Val(OMe), L-Leu(OEt), L-Phe(OEt), L-Ser(OEt), L-Tyr(OEt), L-Cys(OEt), L-Cys—Cys(OMe)$_2$, L-Met(OEt), D,L-Asp(OMe)$_2$, L-His(OMe), L-Trp(OEt), and L-Lys(OEt). Unprotected functional groups such as aliphatic or phenolic hydroxyl groups, sulfhydryl or mercaptyl substituents, and basic heterocyclic rings do not prevent Schiff-base condensation. Selective rhena labeling of both the N^α and N^ϵ amino groups of ethyl lysinate has been accomplished. These amino acid derivatives might be useful reagents for peptide synthesis.

More potentially useful rhena-β-ketoimine complexes have been prepared in which the rhena-β-ketoimine group is attached to a molecular fragment that reacts with specific functional groups. In Eq. (5), the

$$[\textit{cis}\text{-}(OC)_4Re(MeCO)_2]H \xrightarrow[\text{(2) ClCH}_2\text{C(O)Cl, Et}_3\text{N}]{\text{(1) H}_2\text{N(CH}_2)_2\text{NHMe}} \textit{cis}\text{-}(OC)_4\overset{-}{Re}\begin{array}{c} \text{Me} \\ \diagdown \\ \diagup C{=}O \\ \diagup C{=}\overset{+}{N}{-}H \\ \text{Me} \end{array}{}^{CH_2CH_2N(Me)\overset{O}{\overset{\|}{C}}CH_2X} \tag{5}$$

$$\textbf{(29)} \ X = Cl; \ \textbf{(30)} \ X = I$$

synthesis of a rhenaacetylacetonimine complex containing a chloroacetamide group, **29**, is shown (*40*). The rhena-β-ketoimine product of the initial condensation with *N*-methyl ethylenediamine has been isolated and characterized, also.

When complex **29** is treated with NaI in acetone solution, the corresponding iodoacetamide complex **30** is formed *in situ*. This complex reacts with sulfhydryl groups to form the rhena-labeled products **31** [Eq. (6)].

$$\textbf{30} + HSR \xrightarrow[-[\text{Et}_3\text{NH}] \ I]{+\text{Et}_3\text{N}} \textit{cis}\text{-}(OC)_4\overset{-}{Re}\begin{array}{c} \text{Me} \\ \diagdown \\ \diagup C{=}O \\ \diagup C{=}\overset{+}{N}{-}H \\ \text{Me} \end{array}{}^{CH_2CH_2N(Me)\overset{O}{\overset{\|}{C}}CH_2SR} \tag{6}$$

$$31$$

Using this procedure, the S-labeled rhena-β-ketoimine derivatives of benzyl mercaptan, ethanethiol, Me$_3$COC(O)NHCH$_2$CH$_2$SH, *N*-acetyl-L-cysteine, and *N*-acetyl-D,L-penicillamine have been prepared.

In related work, complexes **32** and **33** have been synthesized (*41*). The precursor complex in each preparation is the rhenaacetylacetonimine complex **25**, where $R' = CH_2CH_2OH$. This complex is formed by direct Schiff-base condensation with 2-aminoethanol. Complex **32** is formed when this precursor compound is treated with MeP(O)F$_2$, and complex **33** is formed similarly by treatment with *N,N'*-carbonyldiimidazole. The fluoromethylphosphonate and carbonylimidazolate ester moieties in **32** and **33** are functional groups that are known to react with hydroxyl substituents. Our preliminary results have shown that **32** reacts with ethanol to give the ethyl methylphosphonate derivative. Complexes **32** and **33** might be useful reagents for the heavy atom labeling of OH groups.

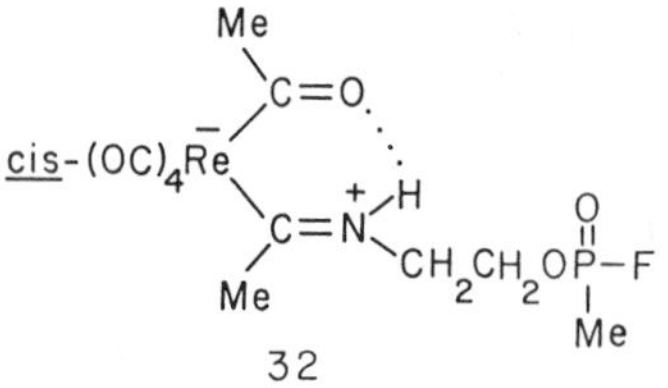

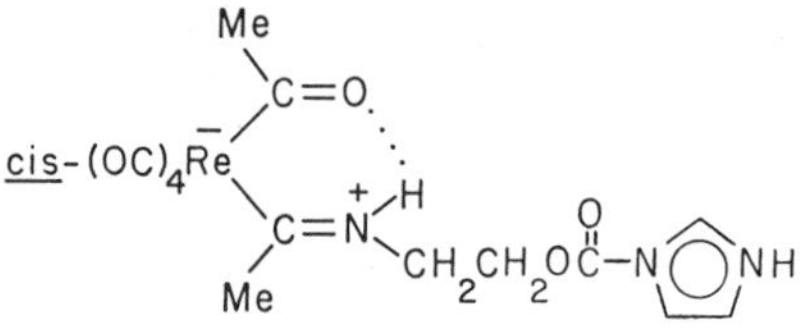

In other reaction chemistry, deprotonation of the rhenaacetylacetonimine complex **24** (R = Me and R' = H), with NaH gives the corresponding rhenaacetylacetoniminato anion. This anion reacts with BCl_3, BBr_3, BI_3, and BCl_2Ph to give (rhena-β-ketoiminato)B(X)(Y) complexes (**42**). The X-ray structure of [cis-(OC)$_4$Re(MeCO)(MeCNH)]B(Cl)(Ph) confirms that the rhenaacetylacetoniminato anion coordinates to the boron atom as a bidentate, chelating ligand (**43**). The rhena chelate ring is slightly boat-shaped like that found in (metalla-β-diketonato)BF_2 complexes as discussed above. A delocalized π-electronic structure is evident from the molecular structure, in contrast to the localized zwitterionic structure of the interisomers of the rhena-β-ketoimines **24**. The coordination of the rhena-β-ketoiminato ligand to the boron atom models the presumed structure of the intraisomers **25**.

IV

INTERLIGAND CARBON–CARBON COUPLING REACTIONS

The most interesting and unique reaction yet discovered for a metalla-β-diketonate molecule is the interligand C—C coupling of the two acyl carbon donor atoms. This very general reaction was discovered serendipitiously while attempting to explore the reaction chemistry of (ferra-β-diketonato)BF_2 complexes. Because the acyl ligands in these compounds have hybrid acyl/carbene character, the methyl substituents of the ferra chelate ring should presumably be relatively acidic. If treatment with a base could produce an α-enolate anion, then subsequent alkylation or acylation might permit a convenient route for introducing additional functionality on the ferra chelate ring.

When complex **34** is treated with KH in THF solution, rapid evolution of hydrogen gas occurs and an anionic complex is formed [Eq. (7)]. Metathetical exchange of K^+ for Me_4N^+ gives **35** as a product. An ORTEP view of

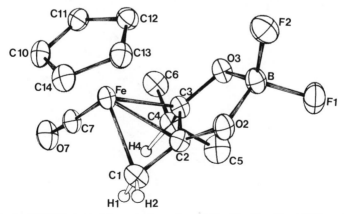

the anion of **35** is shown in Fig. 6 (*44*). This X-ray structure reveals that deprotonation of the methyl substituent of **34** did occur; however, the α-enolate anion rearranges to give an η-allyl complex shown as **35** in the "all sigma" Lewis structure. Of course, the more conventional π-bonded representation would be an alternative description of the η-allyl ligand. In forming **35**, the acyl carbon–acyl carbon nonbonded distance of 2.633(2) Å in **34** shortens to 1.410(6) Å when the C-2—C-3 bond of **35** is formed.

FIG. 6. An ORTEP view of the anionic portion of **35** with ellipsoids at 30% probability. Some hydrogen atoms are deleted for clarity.

Also, the Fe $\cdots$ methyl contact distance of 3.005(2) Å in **34** shortens to 2.085(5) Å when the Fe—C-1 bond is formed. The most dramatic feature of this rearrangement is that the two acyl carbon donor atoms of **34** have undergone a transannular, interligand C—C coupling reaction.

This rearrangement has been generalized to also include mangana- and rhena-β-diketonato boron difluoride complexes [Eq. (8)] (*45*). A diagnostic indication of the formation of allyl products **36** is the observation of the

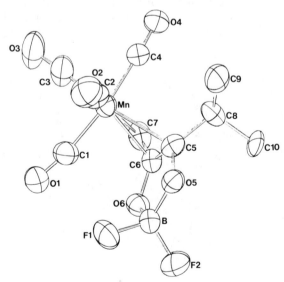

$$L_nM = (\eta\text{-}C_5H_5)(OC)Fe,$$
$$cis\text{-}(OC)_4Mn, \text{ or}$$
$$cis\text{-}(OC)_4Re$$
$$R = Me, i\text{-}Pr$$

36

(8)

syn and anti allyl protons in the ^{1}H-NMR spectra. Because of the boron ring, the R substituents of complexes **36** are required to occupy anit sites of the allyl ligand. Deprotonation can be effected with KH, tetramethyl-piperidine, or even pyridine. For the rhenium complex where R is *i*-Pr, deprotonation of the methine carbon of the isopropyl group also occurs,

FIG. 7. An ORTEP view of the anionic portion of complex **36** (in which $L_nM = cis\text{-}(OC)_4Mn$ and R = *i*-Pr) with ellipsoids at 30% probability. Hydrogen atoms are deleted for clarity. The cation is PPN$^+$.

although deprotonation of the methyl substituent is preferred kinetically.

Figure 7 shows an ORTEP diagram for the tetracarbonylmanganese product **36** in which R is *i*-Pr. The presence of an η-allyl ligand is evident. Atoms C-5 and C-6 are the original acyl ligand donor atoms prior to deprotonation. After deprotonation, these atoms have undergone interligand C—C coupling to give a normal C—C allyl distance of 1.39(2) Å. In contrast to other *cis*-$M(CO)_4$ metalla-β-diketonate derivatives which have the axial carbonyl ligands titled slightly *toward* the metalla-β-diketonate chelate ring, this complex has a slight square pyramidal distortion in which the two axial carbonyl ligands are titled *away* from the η-allyl ligand. The possible significance of this structural feature is discussed below.

It is evident that this interligand C—C coupling of "acyl" carbon donor atoms is both a general and a facile reaction. Although several groups have reported ligand coupling reactions, the coupling of carbonyl and isocyanide ligands is perhaps most closely related to these coupling reactions. Bercaw *et al.* have reported that a Zr(II) complex affords coupling of the two terminal CO ligands of $[(C_5H_4R)Fe(CO)_2]_2$ to give **37** and **38** [Eq. (9)]

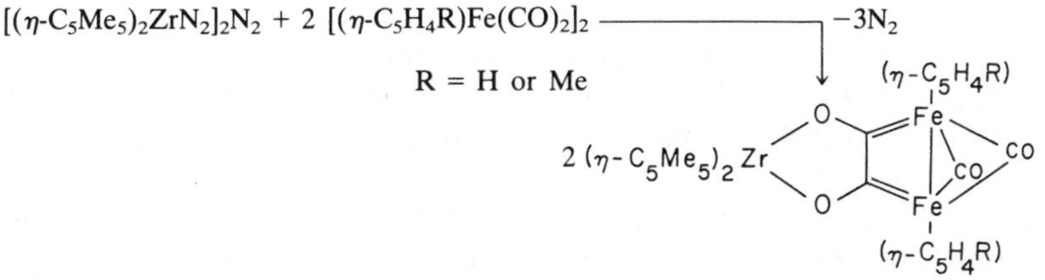
[(η-C5Me5)2ZrN2]2N2 + 2 [(η-C5H4R)Fe(CO)2]2
−3N2
R = H or Me
(η-C5H4R)
2 (η-C5Me5)2 Zr
O
O
Fe
Fe
CO
CO
(η-C5H4R)

(37) R = H
(38) R = Me

(9)

(*46*). The X-ray structure of **37** confirms the interligand C—C coupling. Cramer *et al.* have reported a coupling reaction that is believed to be quite similar to Eq. (8) using an organouranium complex [Eq. (10)] (*47*). The

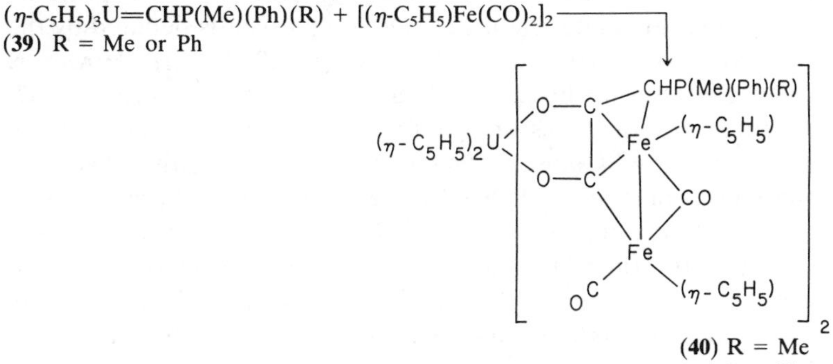
(η-C5H5)3U=CHP(Me)(Ph)(R) + [(η-C5H5)Fe(CO)2]2
(39) R = Me or Ph
CHP(Me)(Ph)(R)
(η-C5H5)2U
O—C
O—C
Fe
(η-C5H5)
CO
Fe
OC
(η-C5H5)
2

(10)

(40) R = Me
(41) R = Ph

uranium ylide complex **39** adds across the C—O bond of a CO ligand in $[(C_5H_5)Fe(CO)_2]_2$ with concomitant interligand C—C coupling between the carbon atom of this CO group and the carbon atom of a bridging CO ligand to give the products **40** and **41**. The X-ray structure of **40** has been determined. The newly formed ligand can be described as an η-allyl ligand containing an iron moiety as a substituent on one of the terminal allyl carbon atoms.

Lippard *et al.* have demonstrated a reductive C—C coupling between the carbon donor atoms of terminal isocyanide ligands [Eq. (11)] (*48,49*).

$$[\text{Mo(CNR)}_6\text{I}]\text{I} \xrightarrow[\text{H}_2\text{O (trace)}]{\text{Zn/THF}} \left[\begin{array}{c} \text{R} \\ | \\ \text{H}-\text{N} \\ \diagdown \\ \text{C} \\ \| \diagdown \\ \diagdown \quad \text{Mo(CNR)}_4\text{I} \\ \text{C} \\ \diagup \\ \text{H}-\text{N} \\ | \\ \text{R} \end{array} \right] \text{I} \qquad (11)$$

(42) $\text{R} = \text{Me}_3\text{C}$

(43) $\text{R} = \text{Me}_3\text{C}$

The X-ray structure of **43** confirms the interligand C—C bond formation. In going from **42** to **43**, the interligand, nonbonded carbon–carbon distance of 2.36 Å in **42** decreases to a carbon–carbon multiple bond length of 1.38 Å in **43**.

Because the interligand C—C coupling of the acyl carbon donor atoms in metalla-β-diketonate complexes [Eq. (8)] is such a general (though unusual) reaction that occurs very facilely, we have proposed (without proof) a formal description of how this type of coupling might take place (*44,50*). This formalism has been adopted by others to explain the C—C coupling shown in Eq. (10) (*47*).

Scheme 2 shows a representative description for this coupling reaction. A general (metalla-β-diketonato)BF$_2$ complex can be described by Lewis structure **44** or its zwitterionic equivalent Lewis structure **45**. Deprotonation generates the α-enolate anion **46**. Electron pair shifts denoted by the broken curved arrows gives the η-allyl product **48** directly. However, electron pair shifts shown by the solid curved arrows gives structure **47**, which can then rearrange as shown to form **48**. Stabilization of the conjugate base **46** as **47** seems reasonable because the negative charge on the exomethylene substituent of **46** is now delocalized onto the chelate ring oxygen atoms. Furthermore, complex **47** has conventional vinyl and carbene ligands. In going from **44** to **48** the formal oxidation state and electron count on M does not change. In the rearrangement of **46** to **48**, the metal atom is formally reduced by one electron from the electron pair on

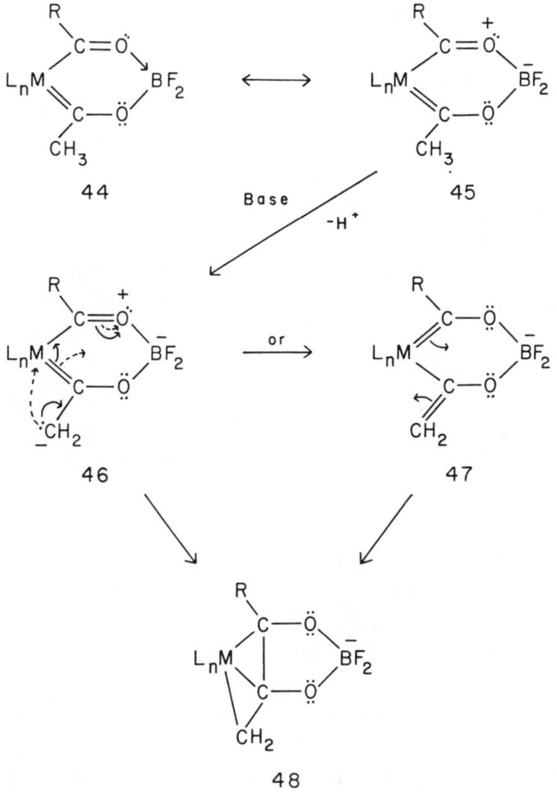

SCHEME 2. A representative description for the interligand C—C coupling reaction.

the exomethylene carbon, while it formally loses one electron upon C—C formation.

Within the isolobal formalism, the conversion of **47** to **48** is a symmetry-allowed process, if it were to proceed as a concerted reaction (*50*). Structure **47** represents a *transoid*-2-metalla-1,3-butadiene. In the σ-bonding description, complex **48** represents formally a 1-metalla-bicyclo[1.1.0]butane. Therefore, the conversion of **47** to **48** represents a thermally allowed, concerted $[_\pi 2_a + _\pi 2_s]$ ring closure, in analogy to the pericyclic ring opening of bicyclo[1.1.0]butanes to give *trans,trans*-1,3-butadienes.

In addition, the structures of the 15-electron, d^7 $M(CO)_4$ metalla moieties of complexes of type **44** and **48** are quite different, and this structural difference might reflect a change in the symmetry of the fragment orbital set as **44** is converted to **48**. The structures of the *cis*-$(OC)_4M$ groups in [*cis*-$(OC)_4Mn(MeCO)_2]_3Al$ and [*cis*-$(OC)_4$-

Re(MeCO)$_2$]H, which are analogous to **44**, reveal a slight tilting of the two axial carbonyl ligands toward the metalla-chelate ring giving a slight tetrahedral distortion. These complexes have metalla moieties that are isolobal to sp^2 methine groups. Hoffmann's elegant work on the fragment orbitals of transition metal carbonyl moieties indicates that d^7 M(CO)$_4$ moieties have an optimum C_{2v} geometry with a slight tetrahedral distortion (*51*). However, if d^7 M(CO)$_4$ moieties have a square-pyramidal distortion such that the L(apical)—M—L(basal) angle is ~105°, then the fragment orbitals form a threefold set much like the three hybrid acceptor orbitals of an sp^3 CH group. The structure of anion **48**, in which R = *i*-Pr and L$_n$M = *cis*-(OC)$_4$Mn (see Fig. 7), shows a slight square-pyramidal distortion of the Mn(CO)$_4$ group with a L(apical)—M—L(basal) angle of 96°. If the coordination geometry of the metalla moieties in these metalla-β-diketonate complexes is determined predominantly by electronic factors, then a distinction between metalla groups acting as either sp^2 CH or sp^3 CH isolobal analogs might be possible.

From a more practical view, transannular, interligand C—C coupling might occur when (1) the two carbon donor atoms have some degree of unsaturation or acyl character, (2) the two carbon atoms are in close contact, and (3) the metal atom to which these carbon atoms are bonded undergoes an initial reduction (as by the exomethylene group in Scheme 2). Internal stabilization of the product of reductive coupling would be desirable also (*44*). A similar set of requirements was proposed recently for the reductive coupling of CO and CNR ligands (*52*).

The general interligand C—C coupling reaction shown in Eq. (8) for (metalla-β-diketonato)BF$_2$ compounds also occurs directly with metalla-β-diketonate anions, thereby precluding the need to prepare the neutral difluoroboron complexes (*53*). As a "one-pot" synthesis, metal carbonyl acetyl compounds can be converted to neutral η-allyl complexes [Eq. (12)].

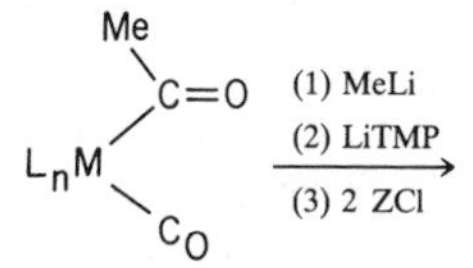

$$L_nM = (\eta\text{-}C_5H_5)(OC)Fe \text{ or } cis\text{-}(OC)_4Mn$$

(12)

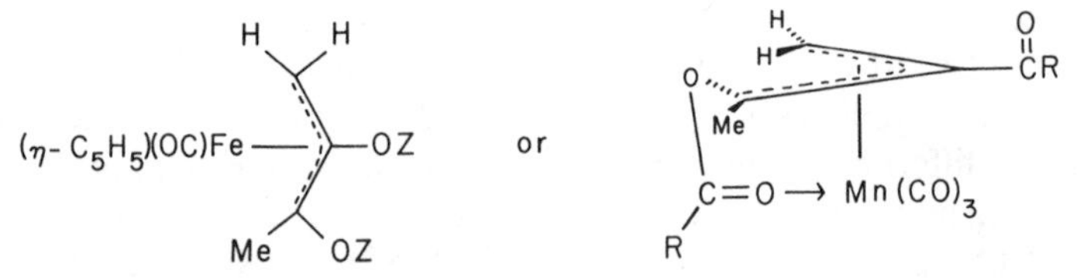

Z = MeC(O), PhC(O), or Me$_3$Si   R = Me or Ph

In this procedure, methyllithium adds to a CO ligand adjacent to the acetyl ligand giving a metalla-β-diketonate anion [cf. Eq. (1)]. The addition of lithium tetramethylpiperidide (LiTMP) removes a proton from one of the acetyl methyl groups and leads to the acyl C—C coupling. The η-allyl products formed are dianionic because two of the allyl carbon atoms bear O⁻ substituents. Diacylation or disilylation affords the neutral products shown in Eq. (12). One of the allyl ester substituents in the manganese complexes displaces a CO ligand and coordinates to the manganese atom. This type of chelated allyl ligand is observed in the isoelectronic complex {(OC)₃Fe[η-CH₂C(H)C(H)(CH₂C=ŌMe)]}PF₆ (54) and related compounds.

This interligand C—C coupling has been extended to (triacyltricarbonyl-rhenato)BX complexes to give (η-allyl)(boroxycarbene) complexes **49** [Eq. (13)] (55). In this reaction, the (triacylrhenato)BX complexes can be

R = Me or i-Pr
X = F or Cl

$$(13)$$

represented formally as having rhena-β-diketonato and boroxycarbene ligands. Deprotonation of a methyl substituent of the rhena-β-diketonato portion would lead to the interligand C—C coupling as described in Scheme 2. When R = i-Pr, a minor structural isomer is observed where the C—C coupling has occurred between the acyl carbon atoms of the two acetyl ligands.

Very recent work has extended the interligand C—C coupling reaction into several areas. The Diels–Alder addition products **20** and **21** undergo deprotonation at the chelate ring methyl substituent upon treatment with base to give the concomitant formation of an η-allyl product via an interligand C—C coupling (56). These results indicate that a two-carbon fragment can be coupled to the relatively complex ligand already obtained from the Diels–Alder addition. This procedure might be applicable to the synthesis of purely organic compounds.

To probe the possibility that *transoid*-2-metalla-1,3-butadiene species (like **47**) would spontaneously rearrange to η-allyl products (like **48**) via interligand C—C coupling, the Fe–vinyl complex **50** was treated first with methyllithium and then with benzoyl chloride [Eq. (14)] (57). Attack by

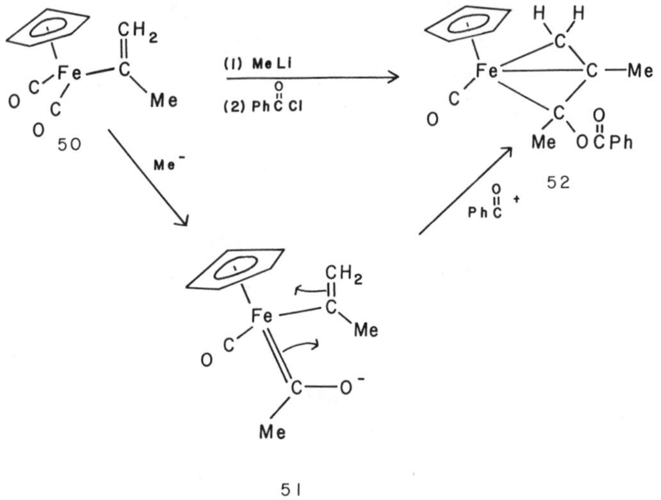

$$(14)$$

methyllithium at a carbonyl ligand should give **51**, which is a 2-metalla-1,3-butadiene. Interligand C—C coupling and acylation should afford the neutral η-allyl complex **52**. Complex **52** is isolated from this reaction as a thermally unstable oil. It has been characterized unambiguously by its exact mass spectrum. However, this reaction is more complex than that shown because reproducing this reaction under presumably identical conditions gives isolated yields of **52** in the range of only 0 to 8%. Of course, **51** might not be a suitable analog to **47** or to the intermediate in Eq. (*12*) because the vinyl ligand should bear an oxy substituent rather than a methyl group, and because lithium ion stabilization of the transoid structure of **51** would probably not occur.

Finally, one of the above criteria proposed as a requirement for the initiation of interligand C—C coupling is that the metal atom needs to undergo a formal one-electron reduction. In Scheme 2, the exomethylene group serves this purpose in going from **46** (or **47**) to **48**. Interesting results have been obtained when the reducing agent is an external chemical reagent or when reduction is effected electrochemically, as shown in Scheme 3 (*25*). The (ferra-β-diketonato)BF$_2$ complex **53** contains *t*-butyl substituents on the chelate ring, so α-enolate carbanion formation with concomitant C—C coupling cannot occur by the deprotonation route shown in Scheme 2. When this complex is treated with excess Na/Hg, sodium naphthalenide, sodium benzophenone ketyl, or electrochemical reduction over a mercury pool, the reaction solution rapidly darkens in color to give eventually a black residue. After basic hydrolysis and then neutralization with acid, pivaloin is isolated in yields of 44, 45, 56, and 26%, from the respective reducing agents mentioned above.

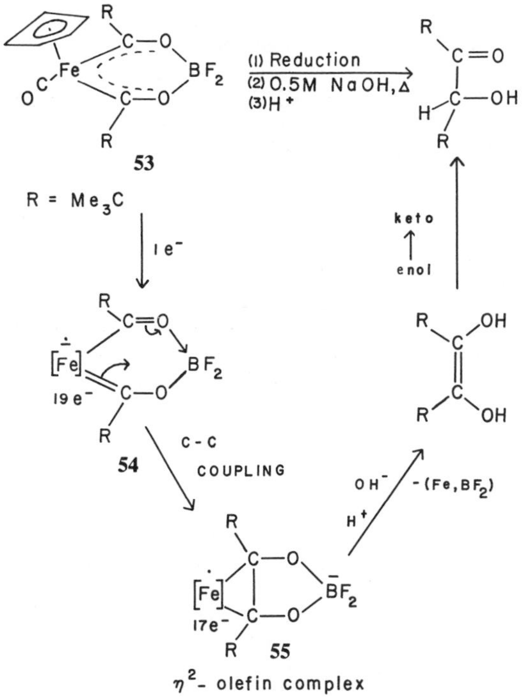

SCHEME 3. Proposed mechanism for the formation of pivaloin by an interligand C—C coupling of two carbon donor atoms.

We propose that one-electron reduction of **53** gives the 19-electron complex **54**, which then undergoes spontaneous interligand C—C coupling to give **55**. This complex is a 17-electron, η-alkene compound. Thermal dissociation of the iron moiety and standard hydrolytic removal of the BF$_2$ group would give the enol tautomer of pivaloin after neutralization. This enol tautomer would then rearrange to the more stable keto tautomer. Further reduction of the dissociated 17-electron iron moiety would give an *overall* two-electron reduction for the C—C coupling. Regardless of the actual mechanism of this reaction, the direct isolation of pivaloin from this interligand C—C coupling and subsequent "one-pot" workup procedure is an interesting observation. In separate control reactions, **53** yields pivaloin only when a reduction step is included. It is hoped that this coupling reaction might enhance the synthetic utilization of metalla-β-diketonate complexes. Possible applications might lead to new methods of synthesizing acyloins or *cis*-glycols as well as to a better understanding of carbonyl coupling reactions, such as those used in the preparation of pinacol.

Note Added in Proof. The stereochemistry of the Diels–Alder addition reactions shown in Scheme 1 to give products **20–22** has been examined by 400 MHz ^{1}H-NMR. These data indicate that the cycloadditions occur with a high degree of stereoselectivity (having diastereoselectivities of greater than 94:6). The chiral Fe environment of the dienophile **17** strongly influences the direction of the facial attack of the diene reactants.

Acknowledgments

The author thanks the National Science Foundation, the Donors of the Petroleum Research Fund, administered by the American Chemical Society, the Research Fellowship program of the Alfred P. Sloan Foundation, and the University Research Council of Vanderbilt University for grants that supported the above research. The author is very grateful for the dedication and enthusiasm shown by those who have chosen to work with him at Vanderbilt University. The names of these persons are prominently displayed in the listed references.

References

1. C. M. Lukehart, *Acc. Chem. Res.* **14**, 109 (1981).
2. R. Hoffmann, *Angew. Chem., Int. Ed. Engl.* **21**, 711 (1982).
3. F. G. A. Stone, *Angew. Chem., Int. Ed. Engl.* **23**, 89 (1984).
4. R. P. Rosen, J. B. Hoke, R. R. Whittle, G. L. Geoffroy, J. P. Hutchinson, and J. A. Zubieta, *Organometallics* **3**, 846 (1984), and references therein.
5. S. J. LaCroce and A. R. Cutler, *J. Am. Chem. Soc.* **104**, 2312 (1982), and references therein.
6. C. M. Jensen, C. B. Knobler, and H. D. Kaesz, *J. Am. Chem. Soc.* **106**, 5926 (1984).
7. D. M. Roundhill, R. P. Sperline, and W. B. Beaulieu, *Coord. Chem. Rev.* **26**, 263 (1978).
8. K. R. Dixon and A. D. Rattray, *Inorg. Chem.* **16**, 209 (1977).
9. R. T. Paine, E. N. Duesler, and D. C. Moody, *Organometallics* **1**, 1097 (1982), and references therein.
10. U. Schubert, R. Werner, L. Zinner, and H. Werner, *J. Organometal. Chem.* **253**, 363 (1983).
11. F. L'Eplattenier, *Inorg. Chem.* **8**, 965 (1969).
12. C. P. Casey and C. A. Bunnell, *J. Chem. Soc. Chem. Commun.* 733 (1974).
13. C. P. Casey and C. A. Bunnell, *J. Am. Chem. Soc.* **98**, 436 (1976).
14. D. Drew, M. Y. Darensbourg, and D. J. Darensbourg, *J. Organometal. Chem.* **85**, 73 (1975).
15. P. G. Lenhert, C. M. Lukehart, P. D. Sotiropoulos, and K. Srinivasan, *Inorg. Chem.* **23**, 1807 (1984).
16. C. P. Casey and D. M. Scheck, *J. Am. Chem. Soc.* **102**, 2728 (1980).
17. C. P. Casey and L. M. Baltusis, *J. Am. Chem. Soc.* **104**, 6347 (1982), and references therein.
18. K. P. Darst and C. M. Lukehart, *J. Organometal. Chem.* **171**, 65 (1979).
19. J. A. Gladysz, *Adv. Organometal. Chem.* **20**, 1 (1982), and references therein.
20. J. A. Gladysz and J. C. Selover, *Tetrahedron Lett.* 319 (1978).
21. J. C. Selover, M. Marsi, D. W. Parker, and J. A. Gladysz, *J. Organometal. Chem.* **206**, 317 (1981).
22. J. R. Sweet and W. A. G. Graham, *J. Am. Chem. Soc.* **104**, 2811 (1982).

23. D. W. Hensley, W. L. Wurster, and R. P. Stewart, Jr., *Inorg. Chem.* **20,** 645 (1981).
24. C. M. Lukehart and J. V. Zeile, *J. Am. Chem. Soc.* **98,** 2365 (1976).
25. R. Srinivasan, M. S. thesis, Vanderbilt University, 1984. See also L. C. Hall, C. M. Lukehart, and R. Srinivasan, *Organometallics* **4,** 2071 (1985).
26. A. J. Schultz, K. Srinivasan, R. G. Teller, J. M. Williams, and C. M. Lukehart, *J. Am. Chem. Soc.* **106,** 999 (1984).
27. J. P. Schaefer and P. J. Wheatley, *J. Chem. Soc. A* 528 (1966).
28. A. Camerman, D. Mastropaolo, and N. Camerman, *J. Am. Chem. Soc.* **105,** 1584 (1983).
29. P. G. Lenhert, C. M. Lukehart, and L. T. Warfield, *Inorg. Chem.* **19,** 311 (1980).
30. T. D. Tilley and R. A. Andersen, *J. Am. Chem. Soc.* **104,** 1772 (1982).
31. P. G. Lenhert, C. M. Lukehart, and L. T. Warfield, *Inorg. Chem.* **19,** 2343 (1980).
32. P. G. Lenhert, C. M. Lukehart, and L. A. Sacksteder, *J. Am. Chem. Soc.* **108,** in press (1986).
33. A. I. Konovalov, *Dok. Akad, Nauk SSSR* **149,** 1334 (1963).
34. W. D. Wulff and D. C. Yang, *J. Am. Chem. Soc.* **105,** 6726 (1983).
35. A. J. Baskar and C. M. Lukehart, *J. Organometal. Chem.* **254,** 149 (1983).
36. C. M. Lukehart and M. Raja, *Inorg. Chem.* **21,** 1278 (1982).
37. A. J. Baskar, C. M. Lukehart, and K. Srinivasan, *J. Am. Chem. Soc.* **103,** 1467 (1981).
38. D. Afzal and C. M. Lukehart, *Inorg. Chem.* **22,** 3954 (1983).
39. A. A. Ioganson, Yu G. Kovalev, and E. D. Korniets, *Bull. Acad. Sci. USSR* **31,** 1466 (1982).
40. M. Raja, Ph.D. thesis, Vanderbilt University, August, 1984.
41. J. H. Davis, Jr., and C. M. Lukehart, unpublished results.
42. C. M. Lukehart and M. Raja, *Inorg. Chem.* **21,** 2100 (1982).
43. P. G. Lenhert, C. M. Lukehart, and K. Srinivasan, *Inorg. Chem.* **32,** 438 (1984).
44. C. M. Lukehart and K. Srinivasan, *J. Am. Chem. Soc.* **103,** 4166 (1981).
45. P. G. Lenhert, C. M. Lukehart, and K. Srinivasan, *J. Am. Chem. Soc.* **106,** 124 (1984).
46. D. H. Berry, J. E. Bercaw, A. J. Jircitano, and K. B. Mertes, *J. Am. Chem. Soc.* **104,** 4712 (1982).
47. R. E. Cramer, K. T. Higa, S. L. Pruskin, and J. W. Gilje, *J. Am. Chem. Soc.* **105,** 6749 (1983).
48. C. M. Giadomenico, C. T. Lam, and S. J. Lippard, *J. Am. Chem. Soc.* **104,** 1263 (1982), and references therein.
49. C. T. Lam, P. W. R. Corfield, and S. J. Lippard, *J. Am. Chem. Soc.* **99,** 617 (1977).
50. C. M. Lukehart and K. Srinivasan, *Organometallics* **1,** 1247 (1982).
51. M. Elian and R. Hoffmann, *Inorg. Chem.* **14,** 1058 (1975).
52. R. Hoffmann, C. N. Wilker, S. J. Lippard, J. L. Templeton, and D. C. Brower, *J. Am. Chem. Soc.* **105,** 146 (1983).
53. C. M. Lukehart and K. Srinivasan, *Organometallics* **2,** 1640 (1983).
54. A. D. U. Hardy, and G. A. Sim, *J. Chem. Soc. Dalton Trans.* 2305 (1972).
55. C. M. Lukehart and W. L. Magnuson, *J. Am. Chem. Soc.* **106,** 1333 (1984).
56. C. M. Lukehart and L. A. Sacksteder, in preparation (1986).
57. C. M. Lukehart and J. B. Myers, Jr., in preparation (1986).

Organometallic Sonochemistry

KENNETH S. SUSLICK

School of Chemical Sciences
University of Illinois at Urbana-Champaign
Urbana, Illinois 61801

I

INTRODUCTION AND INTENT

In 1894, the destroyer H.M.S. Daring failed to meet specifications: its speed and efficiency were inexplicably low. Sir John Thornycroft and Sidney Barnaby observed severe vibration and excessive slippage of the ship's screw propeller. After replacing four sets of blades, a solution to the problem was found by simply increasing the surface area of the propeller and decreasing its angular velocity. Their description of the observation of associated bubble formation on the moving propeller was the first report of the phenomenon known as cavitation (1), which occurs both during turbulent flow and during ultrasonic irradiation of liquids. In attempting to explain such observations, Lord Rayleigh described (2) in 1917 the first mathematical model for the collapse of cavities in incompressible liquids and predicted *enormous* local temperatures (10,000 K) and pressures (10,000 atm) during such collapse.

The chemical (3) and biological (4) effects of ultrasound were first reported by Loomis more than 50 years ago. In spite of early work in the area of sonochemistry, interest within the chemical community remained

73

Copyright © 1986 by Academic Press. Inc.
All rights of reproduction in any form reserved.

exceedingly modest until the past few years. With the advent of inexpen-
sive and reliable sources of ultrasound, however, increasing use of
sonochemistry in a variety of reactions is being reported. The purpose of
this review is to act as a critical introduction for those interested in the
chemical effects of ultrasound on organometallic systems. An overview of
the physics of acoustic cavitation is required to explain the origin of
sonochemical reactivity, and a brief summary of the general reactivity
patterns observed for all sonochemical reactions is essential for a sense of
perspective. The primary thrust, however, will be on organometallic
sonochemistry, nearly all of which has been reported during the past 10
years. Interested readers are referred to earlier reviews of general
sonochemical phenomenon (5–7).

A number of terms in this area will be unfamiliar to most chemists.
Cavitation is the formation of gas bubbles (or cavities) in a liquid and
occurs when the pressure within the liquid drops sufficiently lower than the
vapor pressure of the liquid. Cavitation can occur from a variety of causes:
turbulent flow, laser heating, electrical discharge, boiling, radiolysis, or
acoustic irradiation. We shall be concerned exclusively with *acoustic
cavitation*. When sound passes through a liquid, it consists of expansion
(negative-pressure) waves and compression (positive-pressure) waves.
These cause preexisting bubbles to grow and recompress. Acoustic cavitation can lead, as discussed later, to an implosive collapse of such cavities
with associated high-energy chemistry. The importance of acoustic cavitation extends well beyond its chemical effects, since it is relevant to studies of
heat transport, liquid tensile strengths, and superheating and boiling
phenomena (8,9). Furthermore, because ultrasound is heavily used both for
medical treatment (hyperthermia for soft tissue traumas) and diagnosis
(sonography of fetal development), the biological and chemical effects of
ultrasound are of immediate importance to the health services community
(10–12). We use the symbol -)-)-)→ in this article to indicate ultrasonic
irradiation or "sonication" of a solution leading to a sonochemical
reaction. *Sonocatalysis* will be restricted in its use only to the creation of a
catalytically competent intermediate by ultrasonic irradiation; we shall not
refer to a simple sonochemical rate enhancement of an already ongoing
reaction by this term.

II

MECHANISMS OF THE CHEMICAL EFFECTS OF ULTRASOUND

The velocity of sound in water is ~1500 m/second; ultrasound spans the
frequencies of 20 KHz to 10 MHz, with associated acoustic wavelengths of
7.6 to 0.015 cm. Clearly no direct coupling of the acoustic field with

chemical species on a molecular level can account for sonochemistry. Instead, the chemical effects of ultrasound derive from several different physical mechanisms, depending on the nature of the system. All represent "nonlinear" acoustic phenomena: the propagation of high amplitude sound waves results in effects which can be described only with the inclusion of terms not linear with the acoustic waves' displacement amplitude. An extensive literature dealing with nonlinear propagation of sound exists (9,13,14), but is beyond the scope of this review.

Acoustic cavitation can be considered to involve at least three discrete stages: nucleation, bubble growth, and, under proper conditions, implosive collapse. The dynamics of cavity growth and collapse are strikingly dependent on local environment: we therefore will consider separately cavitation in a homogeneous liquid and cavitation near a liquid–solid interface.

A. Nucleation of Cavitation

The tensile strength of a *pure* liquid is determined by the attractive intermolecular forces which maintain its liquid state. On that basis, the calculated tensile strength of water, for example, is in excess of -1000 atmospheres (15). In practice however, the measured threshold for initiation of cavitation is never more than a small fraction of that: tap water will cavitate at a negative acoustic pressure of a few atmospheres. The tensile strength increases upon purification, but even after exhaustive purification and submicrometer filtering, water will withstand only -200 atmospheres for a few seconds (16). One also needs to rationalize two other methods which increase the cavitation threshold: vacuum degassing (17) and initial hydrostatic pressurization (18). Indeed, if the observed tensile strengths of liquids did approach their theoretical limits, the acoustic intensities required to initiate cavitation would be well beyond that generally available, and no sonochemistry would be observed in homogeneous media!

These observations demonstrate that cavitation is initiated at a nucleation site where the tensile strength is dramatically lowered. An obvious site would be small gas bubbles present in the liquid. Free gass bubbles, however, are caught in a double bind: small ones of the size needed for acoustic cavitation (a few micrometers in radius) will redissolve in a few seconds, whereas larger ones will rapidly rise to the surface (19). The nucleation mechanism generally accepted at this time involves gas entrapped in small-angle crevices of particulate contaminants (20–22) as shown schematically in Fig. 1. As the crevice-stabilized nucleus is subjected to large, negative acoustic pressures, the bubble volume grows, releasing small free bubbles into solution or undergoing violent collapse itself. Those actions which remove such nucleation sites (e.g., ultrafiltration to remove

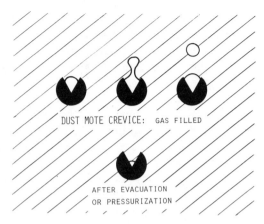

FIG. 1. Nucleation of acoustic cavitation.

particulates, evacuation or pressurization to flood the crevices, etc.) will thus increase the cavitation threshold. In liquids undergoing cavitation, one should note that after the initial cycle of cavitation, the implosive collapse of bubbles generates microcavities which can then serve as nucleation sites for the next cycle (*23*).

B. *Cavitation in Homogeneous Media*

Flynn proposed the generally accepted division of cavitation phenomenon in homogeneous liquids into (*1*) transient cavitation, in which a short-lived bubble undergoes large excursions of size in a few acoustic cycles and may terminate in a violent collapse, and (*2*) stable cavitation, in which a bubble oscillates many times with limited change about its equilibrium radius (*24*). Both stable and transient cavitation may occur simultaneously in a solution, and a bubble undergoing stable cavitation may change to transient cavitation if the radius becomes suitable for efficient collapse. It is transient cavitation which gives rise to sonochemistry. An idealized pictorial representation of this scheme is shown in Fig. 2. Several exhaustive reviews of acoustic cavitation dynamics have been published (*8,25–27*) so this discussion will be limited to a qualitative overview.

The oscillatory behavior of cavities in an acoustic field has been well-described by a variety of mathematical models derived from Rayleigh's original approach with the inclusion of various nonideal liquid properties (*28–33*). Let us examine, as an example, one such equation of

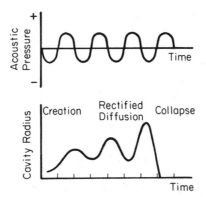

FIG. 2. Idealized representation of Bubble growth and collapse during transient cavitation.

motion, developed by Plesset (28):

$$\rho R \ddot{R} + 1.5(\rho \dot{R}^2) - (P_0 + 2\sigma/R_0)(R_0/R)^{3\gamma}$$
$$+ P_0[1 - (P_A/P_0)\cos(\omega t)] + 2\sigma/R + 4\mu \dot{R}/R = 0 \qquad (1)$$

where R is the instantaneous radius of the cavity, $\dot{R}$ and $\ddot{R}$ the velocity and acceleration, respectively, of the bubble's surface, R_0 the equilibrium radius, P_0 the ambient pressure, P_A the acoustic pressure amplitude, $P_0 + 2\sigma/R_0$ the effective pressure inside the cavity at the equilibrium radius, ω the acoustic frequency, ρ the liquid density, σ the surface tension, μ the effective liquid viscosity, and γ the polytropic exponent of the gas within the bubble. The inertial effects are contained in the first two terms, the internal pressure in the third, the external applied pressure in the fourth, the surface tension in the fifth, and viscosity damping in the sixth. Such equations have been solved numerically for varying degrees of approximation. All have difficulty, however, in accurately calculating the dynamics of bubble motion during the latter stages of implosive collapse where sonochemical events are expected to originate.

One can easily calculate from such equations, however, what size cavity would undergo maximum expansion when subjected to a given acoustic field. Minnaert, for example, derives (34) (from a simplified model which assumed a noncondensable gas and neglected viscosity) this resonant size of a transient cavity as

$$R_r = (2\pi\omega)^{-1} (3\eta P_0/\rho)^{1/2} \qquad (2)$$

where R_r is the resonant size. More complete determinations (8) do not lead to significant differences for frequencies less than 300 KHz at $P_0 = 1$

atmosphere. At 20 KHz, a typical frequency of laboratory ultrasonic irradiations, R_r is calculated to be 170 μm and at 1 MHz, 3.3 μm.

Bubbles which are well below this optimal resonant size will still undergo transient cavitation if the acoustic field is sufficiently large. Given a well-defined acoustic field, one would wish to know which size cavities will undergo transient cavitation, which will undergo stable cavitation, and which will simply redissolve. The first class involves Blake's mechanism for transient cavitation, in which the bubble grows rapidly under the instigation of the expansion wave of a single acoustic cycle (35–38). The minimum acoustic pressure at which such growth can still occur (the "Blake threshold") is derived from equations of motion similar to the one already discussed, in terms of the ambient pressure, the liquid surface tension, and the initial radius of the bubble. Bubbles much larger than this resonant size will not be capable of undergoing transient cavitation due to the nonnegligible inertial term: they would be unable to respond to the imposed pressure changes within the time frame of the acoustic frequency.

Cavities below this resonant size are still capable of growth, however, through the process known as rectified diffusion (39,40). Even when far from resonance with the sound field, a bubble will undergo small oscillations. Since the surface area of such a bubble is slightly larger during the negative-pressure portion of an acoustic cycle than during the positive-

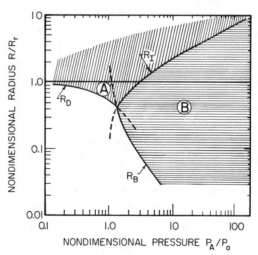

FIG. 3. Thresholds of cavitation. Region A: Bubble growth through rectified diffusion only. Region B: Bubble growth through transient cavitation. R_D, Threshold for rectified diffusion; R_I, threshold for predomination of inertial effects; R_B, Blake threshold for transient cavitation. [After R. E. Apfel (8).]

pressure portion, more gas will diffuse into the bubble during expansion than will diffuse out again during recompression. Thus, gas will be acoustically pumped into the bubble. The effect is very small per cycle, but is cumulative and becomes significant over many cycles, leading to bubble growth up to the Blake threshold. Since rectified diffusion will increase the size of cavities below the Blake threshold, it represents the "stable cavitation" threshold. Free standing bubbles below this size will not grow during ultrasonic irradiation and will therefore rapidly redissolve. These results can be graphically represented in Fig. 3, where the various domains of bubble dynamics are represented in terms of bubble radius and acoustic pressure.

The dynamic process of bubble collapse has been observed by Lauterborn and others by ultrahigh speed photography (10^5 frames/second) of laser generated cavitation (41). As seen in Fig. 4, the comparison between theory and experiment is remarkably good. These results were obtained in silicone oil, whose high viscosity is responsible for the spherical rebound of the collapsed cavities. The agreement between theoretical predictions and the experimental observations of bubble radius as a function of time are particularly striking.

Given this detailed understanding of the dynamics of cavitation, the relevant question for the chemist lies in the actual mechanisms responsible for sonochemical reactions in homogeneous media. Historically, there have been two separate proposals: "hot-spot" pyrolysis (42,43) and electrical discharge (44,45). The implosive collapse of a bubble will obviously produce adiabatic heating of its contents: estimates of the conditions so induced are in the thousands of degrees and thousands of atmospheres, as discussed shortly. The several proposals of electrical discharge during cavitation [including more recent suggestions (46,47)] have not been well-developed on a molecular level and recently have been thoroughly rebutted as inconsistent with observed sonochemical reactivities and sonoluminescent behavior (48,49). Two other, more limited mechanisms for homogeneous sonochemistry have been suggested. The cleavage of very large polymers involves direct mechanical cleavage either by shock waves generated during transient cavitation or by the intense accelerations caused by the sound field itself ($\sim 10^5$ g at 500 KHz) (50). Secondary reactions with high energy species produced from solvent sonolysis also contributes to polymer degradation. Finally, generally small rate enhancements (51–53) [<20%, although there is one report of a 10-fold increase (54)] of solvolysis reactions have been reported and have been interpreted in terms of a disruption of the solvent structure (55) by the ultrasonic irradiation. The details of this proposed mechanism remain undiscussed.

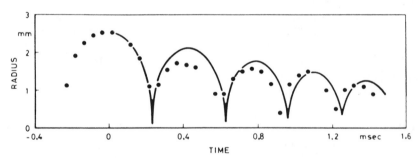

FIG. 4. Dynamics of bubble motion. Laser-induced cavitation in silicone oil: upper portion is the experimental observations at 75,000 frames/second; lower curve compares the experimentally observed radius versus theory. [W. Lauterborn (41).]

The high temperatures and pressures created during transient cavitation are difficult both to calculate and to determine experimentally. The simplest models of collapse, which neglect heat transport and the effects of condensable vapor, predict maximum temperatures and pressures as high as 10,000 K and 10,000 atmospheres. More realistic estimates from increasingly sophisticated hydrodynamic models yield estimates of ~5000 K and ~1000 atmospheres with effective residence times of <100 nseconds, but the models are very sensitive to initial assumptions of the boundary conditions (30–32).

There are only four experimental determinations of cavitational conditions. The first involves spectral analysis of sonoluminescent emission (56), for example, of excited state Na atoms generated upon sonolysis of aqueous NaCl solutions. The results of these studies give estimates of effective temperatures in the range of 3400 K; the assumption is made, however, that the site of luminescence is within the cavitation event. Since sodium ions are involatile, however, it seems that the observed luminescence must be due to species formed outside of the original cavitation zone by secondary reactions, perhaps in a heated liquid shell surrounding the cavity (57). The second probe of cavitation conditions, which also relied on sonoluminescence data, utilized the relative emissivity of NO- and NO_2-saturated water and estimated temperatures of ~1000 K in aqueous solutions at 285 K bulk temperature irradiated at 459 MHz (58). A recent analysis (59) of aqueous sonoluminescence in terms of blackbody radiation gives estimates of ~5500 K; since such sonoluminescence had previously been conclusively demonstrated to derive from chemiluminescence of radical recombinations (60), this approach appears without validity. The last experimental determination of cavitational conditions utilizes the comparative-rate, "chemical thermometry" approach, originally used (61) in shock tube experiments. In this work the relative rates of CO dissociation of metal carbonyls were determined as a function of substrate vapor pressure (62,63) and then analyzed using activation parameters previously determined by gas-phase laser pyrolysis. Both a gas-phase and a liquid-phase reaction zone were observed and the latter interpreted in terms of a heated liquid shell as shown in Fig. 5; the ratio of the volumes of the gas to liquid reaction zones was ~10^5. The effective temperatures were determined to be 5200 K for the gas-phase site and ~1900 K for the liquid shell for alkane solvents sonicated under 1 atm Ar at 20 KHz with an overall vapor pressure of 5.0 torr. Given the differences in irradiation

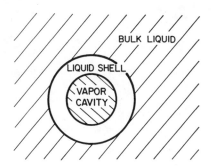

FIG. 5.   The two-site model of sonochemical reaction zones.

conditions used in these various studies, it is not yet possible to determine the extent of real differences among these estimates. Regardless of the details, however, it is clear that cavitational collapse is producing hot-spots with effective temperatures of several thousand degrees.

C. *Cavitation at Surfaces*

When a liquid–solid interface is subjected to ultrasound, transient cavitation still occurs, but with major changes in the nature of the bubble collapse. No longer does spherical implosion of the cavity occur, but instead a markedly asymmetric collapse happens which generates a jet of liquid directed at the surface, as seen in the high speed microphotographs taken by Ellis (*64,65*) and Lauterborn (*66*) and shown in Fig. 6. The tip jet velocities measured by Lauterborn are greater than 100 m/second. The origin of this jet formation is essentially a shaped-charge effect: the rate of collapse is proportional to the local radius of curvature. As collapse of a bubble near a surface begins, it does so with a slight elliptical asymmetry, which is self-reinforcing, and generates the observed jet (*67*) as shown in

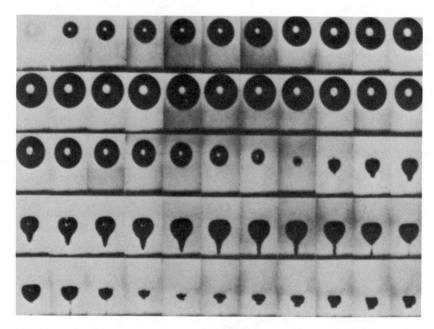

Fig. 6. Cavitation near a surface. Jet formation from laser-induced cavitation in water at 75,000 frames/second. Sequence is from left to right, top to bottom; the solid boundary is at the bottom of each frame. From Ref. 66.

INITIAL SPHERE

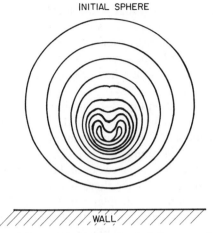

Fɪɢ. 7. Theoretical surface profiles of collapsing bubble near a boundary. Initially, the distance of the bubble's edge from the boundary was $R_0/2$. [After E. A. Neppiras (26).]

Fig. 7. The impingement of this jet can create a localized erosion (and even melting) responsible for surface pitting and ultrasonic cleaning (68–70). A second contribution to erosion created by cavitation involves the impact of shock waves generated by cavitational collapse. The magnitude of such shock waves can be as high as 10^4 atmospheres, which will easily produce plastic deformation of malleable metals (71). The relative magnitudes of these two effects depends heavily on the specific system under consideration.

Acoustic streaming is another nonlinear acoustic phenomenon important to the effect of ultrasound on surfaces (9,72). This time-dependent flow of liquid induced by a high intensity sound field is independent of cavitation. Its origins lie in the conservation of momentum. As a liquid absorbs energy from a propagating acoustic wave, it must also acquire a corresponding momentum, thus creating force gradients and mass transport. Such streaming will occur at moving solid surfaces or at vibrating bubbles. Thus, when a liquid–solid interface is exposed to ultrasound, improved mass transport is expected owing to acoustic streaming. This will occur even when the sound field is a stable standing wave in the absence of cavitation (73).

Enhanced chemical reactivity of solid surfaces are associated with these processes. The cavitational erosion generates unpassivated, highly reactive surfaces; it causes short-lived high temperatures and pressures at the surface; it produces surface defects and deformations; it forms fines and increases the surface area of friable solid supports; and it ejects material in unknown form into solution. Finally, the local turbulent flow associated with acoustic streaming improves mass transport between the liquid phase and the surface, thus increasing observed reaction rates. In general, all of these effects are likely to be occurring simultaneously, and in no case of sonochemical activation of solids have their relative contributions been definitively established.

The effect of ultrasound on liquid–liquid interfaces between immiscible fluids is emulsification. This is one of the major industrial uses of ultrasound (74–76) and a variety of apparatus have been devised which will generate micrometer-sized emulsions (9). The mechanism of ultrasonic emulsification lies in the shearing stresses and deformations created by the sound field of larger droplets. When these stresses become greater than the interfacial surface tension, the droplet will burst (77,78). The chemical effects of emulsification lie principally in the greatly increased surface area of contact between the two immiscible liquids. Results not unlike phase transfer catalysis may be expected.

III

EXPERIMENTAL INFLUENCES ON SONOCHEMISTRY

A. Reactor Design and Configuration

A variety of devices have been used for ultrasonic irradiation of solutions. There are three general designs in use presently: the ultrasonic cleaning bath, the "cup–horn" sonicator, and the direct immersion ultrasonic horn. In all cases the original source of the ultrasound is a piezoelectric material, usually a lead zirconate titanate ceramic (PZT), which is subjected to a high voltage, alternating current with an ultrasonic frequency (roughly 15 KHz to 1 MHz). The piezoelectric source expands and contracts in this electric field and is attached to the wall of a cleaning bath or to an amplifying horn.

The ultrasonic cleaning bath is clearly the most accessible source of laboratory ultrasound and has been used successfully for a variety of liquid–solid heterogeneous sonochemical studies. There are, however, several potential drawbacks to its use. There is no means of control of the acoustic intensity, which will vary from bath to bath and over the lifetime of a single cleaning bath. In addition, their acoustic frequencies are not well controlled and differ from one manufacturer to another, and reproducibility from one bath to another may therefore suffer. Reproducible positioning of the reaction flask in the bath is critical, since standing waves in the bath will create nodal spots where cavitation will not occur (*79*). Similarly, the height of the bath liquid and of the solution within the reaction vessel are extremely important (*79,80*). Temperature control is often neglected with this apparatus. Since the bath temperature can rise >25 K during the course of a long irradiation, this can significantly influence both the intensity of the cavitational collapse and the rate of background thermal reactivity. Thermostating is best done using coolant passed through copper coils suspending in the bath (*not* in contact with the walls). The temperature inside the reaction vessel must be measured directly since it is often warmer than that in the bath itself. Finally, and most critically, the acoustic intensities present in most cleaning baths are only marginal for the generation of cavitation in homogeneous liquids. When solids are present, the weakened tensile strength of the liquid at the interface will allow cavitation at thresholds well below those of simple solutions. Even in the case of heterogeneous sonochemistry, however, the ultrasonic cleaning bath must be viewed as an apparatus of limited capability.

The cup–horn configuration, shown in Fig. 8, was originally designed for cell disruption but has been adopted for sonochemical studies as well (*81*). It has greater acoustic intensities, better frequency control, and potentially better thermostating than the cleaning bath. Again, however, it is very sensitive to the liquid levels and to shape of the reaction vessel. In addition, the reaction vessel faces a size restriction of ~5 cm diameter.

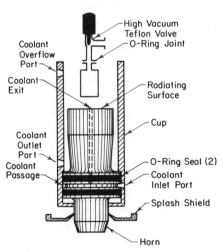

Fig. 8. Cup–horn sonicator. Modification of a design from Heat Systems–Ultrasonics, Inc. (*81,82*).

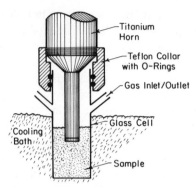

FIG. 9. Direct immersion ultrasonic horn equipped for inert atmosphere work. [Design of K. S. Suslick (*183*).]

Since the ultrasonic radiating surface is not in direct contact with the reaction solution, the acoustic intensities are much lower than those of the direct immersion horn, and so homogeneous sonochemistry is often quite sluggish. On the other hand, there is no possibility of contamination from erosion of the titanium horn.

The most intense source of ultrasound generally used in the chemical laboratory is the direct immersion ultrasonic horn, which we have adapted for inert atmosphere work, as shown in Fig. 9, or for moderate pressures (<10 atmospheres). These devices are available from several manufacturers (*82*) at modest cost and are used primarily for cell disruption. A variety of sizes of power supplies and titanium horns are available, thus allowing flexibility in sample size. The acoustic intensities are easily and reproducibly variable; the acoustic frequency is well controlled, albeit fixed (typically at 20 KHz). Since power levels are quite high, counter-cooling of the reaction solution is essential to provide temperature control; cooling of the piezoelectric ceramic may also be necessary, depending on the configuration. One potential disadvantage in corrosive media is the erosion of the titanium tip; this is generally a very slow process without chemical consequences, given the high tensile strength and low reactivity of Ti metal. This configuration may be used for both homogeneous and heterogeneous sonochemistry.

A rough, but useful, comparison between typical sonochemical and photochemical efficiencies is shown in Table I. As shown, homogeneous sonochemistry is typically *more* efficient than photochemistry, and heterogeneous sonochemistry is several orders of magnitude better. Unlike photochemistry, whose energy inefficiency is inherent in the production of photons, ultrasound can be produced with nearly perfect efficiency from electric power. Still, a primary limitation of sonochemistry remains its

TABLE I

COMPARISONS BETWEEN SONOCHEMICAL AND PHOTOCHEMICAL APPARATUS

	Photochemistry	Homogeneous sonochemistry	Heterogeneous sonochemistry
Source	250 W Quartz–Halogen	200 W Cell disrupter (at 60% power)	150 W Cleaning Bath
Approximate cost (1985)	$1800	$1900	$700
Typical rates	7 μmol/minute	10 μmol/minute	500 μmol/minute
Electrical efficiency	2 mmol/kWh	5 mmol/kWh	200 mmol/kWh

energy inefficiency due to the small fraction of the acoustic power involved in the cavitation events. This might be significantly improved, however, if a more efficient means of utilizing the sound field to generate cavitation can be found.

Large-scale ultrasonic irradiation is extant technology. Liquid processing rates of 200 liters/minute are routinely accessible from a variety of modular, in-line designs with acoustic power of several kW per unit (83). The industrial uses of these units include (1) degassing of liquids, (2) dispersion of solids into liquids, (3) emulsification of immiscible liquids, and (4) large-scale cell disruption (74). While these units are of limited use for most laboratory research, they are of potential importance in eventual industrial application of sonochemical reactions.

B. Extrinsic Variables

Sonochemistry is strongly affected by a variety of external parameters, including acoustic frequency, acoustic intensity, bulk temperature, static pressure, choice of ambient gas, and choice of solvent. These are important considerations in the effective use of ultrasound to influence chemical reactivity, and are also easily understandable in terms of the cavitational hot-spot mechanism. A summary of these effects is given in Table II.

The frequency of the sound field is surprisingly irrelevant to most sonochemistry. Unlike photochemistry, there is no direct coupling of the irradiating field with the molecular species in sonochemistry. The effect of changing sonic frequency is simply one of altering the resonant size of the cavitation event. The overall chemistry is therefore little influenced over the range where cavitation can occur [from tens of hertz to a few megahertz (26)]; observed sonochemical rates may change, but controlled comparisons of efficiency are lacking at this time and will prove difficult to obtain.

TABLE II

THE EFFECTS OF EXTRINSIC VARIABLES OF SONOCHEMISTRY

Extrinisic variable	Physical property	Effect
Acoustic frequency	Period of collapse	Resonant bubble size
Acoustic intensity	Reaction zone size	Cavitation events per volume
Bulk temperature	Liquid vapor pressure	Bubble content, intensity of collapse
	Thermal activation	Enhanced secondary reaction rates
Static pressure	Total applied pressure	Intensity of collapse
	Gas solubility	Bubble content
Ambient gas	Polytropic ratio	Intensity of collapse
	Thermal conductivity	Intensity of collapse
	Chemical reactivity	Primary or secondary sonochemistry
	Gas solubility	Bubble content
Choice of liquid	Vapor pressure	Intensity of collapse
	Surface tension	Transient cavitation threshold
	Viscosity	Transient cavitation threshold
	Chemical reactivity	Primary or secondary sonochemistry

For example, the observed sonochemistry of aqueous solutions is unchanged over this entire range (84). At very high frequencies (above a few megahertz), cavitation ceases, and sonochemistry is generally not observed (85,86). The observed thresholds for cavitation in homogeneous liquids are strongly frequency dependent (8); since homogeneous sonochemistry is generally studied at acoustic intensities well above the threshold, however, this is not a major concern.

Acoustic intensity has a dramatic influence on the observed rates of sonochemical reactions. Below a threshold value, the amplitude of the sound field is too small to induce nucleation or bubble growth. Above the cavitation threshold, increased intensity of irradiation (from an immersion horn, for example) will increase the effective size of the zone of liquid undergoing cavitation, and thus increase the observed sonochemical rate. Furthermore, as the acoustic pressures increase, the range of bubble sizes which will undergo transient cavitation increases (as shown in Fig. 3); this too will increase the observed sonochemical rate. It is often observed experimentally, however, that as one continues to increase acoustic amplitude, rates eventually begin to diminish again (87). Possible explanations for this behavior include bubble shrouding of the sonic horn and overgrowth of bubbles. At high intensities, the cavitation of the liquid near the radiating surface becomes so intense as to produce a shroud of bubbles which will diminish the penetration of the sound into the liquid. Also at high intensities, bubble growth may become so rapid that the bubble grows beyond the size range of transient cavitation before implosive collapse may occur (88).

The effect of the bulk solution temperature lies primarily in its influence on the bubble content before collapse. With increasing temperature, in general, sonochemical reaction rates are *slower*. This reflects the dramatic influence which solvent vapor pressure has on the cavitation event: the greater the solvent vapor pressure found within a bubble prior to collapse, the less effective the collapse. In fact, one can quantitate this relationship rather well (*89*). From simple hydrodynamic models of the cavitation process, Neppiras, for example, derives (*26*) the peak temperature generated during collapse of a gas-filled cavity as

$$T_{max} = T_0 P_a (\gamma - 1)/Q \tag{3}$$

where T_0 is the ambient temperature, P_a the acoustic pressure just prior to collapse, γ the polytropic ratio (the ratio of specific heats, C_p/C_v), and Q the gas pressure in the bubble prior to collapse. In the case of vapor-filled cavities, we may take Q to be roughly approximated by P_v, the vapor pressure of the system, and the maximum temperature of the collapse will be inversely proportional to the vapor pressure of the system. Assuming Arrhenius behavior ($\ln k = \ln A - E_a/RT_{max}$), one expects that the sonochemical rate coefficient should follow

$$\ln k = \ln A - \{E_a/[RT_0 P_a (\gamma - 1)]\}P_v \tag{4}$$

This is only a rough approximation since it neglects the effects of both thermal conductivity and vapor condensation during collapse. Nonetheless, the linear correlation of $\ln k_{obs}$ and P_v is the experimentally observed behavior in a wide range of sonochemical systems in a variety of solvents (*89,90*). It has also been suggested that the effect of ambient temperature lies in the change in ambient gas solubility (*6*), although a plausible mechanism for such an influence was unstated. Recent results in which temperature, solvent vapor pressure, and gas solubility have been varied independently (*89,90*) rule out gas solubility as an important variable. When secondary reactions are being observed (as in secondary corrosion or other thermal chemical reactions occurring after initial acoustic erosion of a passivated surface), then temperature can play its usual role in thermally activated chemical reactions. This explains the occasional observation of increasing rates of corrosion associated with cavitation with increasing temperature (*91*).

Sonochemical yields as a function of increasing static pressure have been reported by different researchers to increase (*6*), to decrease (*92*), and to increase to some point and then decrease (*93*). One would expect that cavitational collapse would increase in intensity with increasing external pressure, since the total imposed pressure at the initiation of collapse would be increased. Given a fixed acoustic intensity, however, nucleation of cavities will no longer occur at some point of increasing ambient pressure, since the acoustic field must overcome the combined tensile strength of the liquid and the applied pressure. In contrast, as one reduces the ambient pressure, eventually one will deactivate the gas-filled crevices which serve as nucleation sites (discussed earlier) and therefore also diminish observed sonochemistry. Further experimental difficulties occur when one attempts to maintain a pressure vessel at constant temperature while under ultrasonic irradiation. It is perhaps not surprising then that the experimental results are conflicting on this question. In reactions which involve the ambient gas directly, enhanced solubility would also play a role in the overall observed rates.

The choice of ambient gas will also have a major impact on sonochemical reactivity. As shown in Eq. (3), the maximum temperature reached during cavitation is strongly dependent on the polytropic ratio ($\gamma = C_p/C_v$) of the ambient gas, which defines the amount of heat released during the adiabatic compression of that gas. This can have a dramatic impact: all other factors being equal the difference between cavitation in the presence of xenon ($\gamma \approx 1.67$) and a freon ($\gamma \approx 1.1$), for example, would yield a ratio of maximum temperatures of sevenfold! Sonochemical rates are also significantly influenced by the ambient gas's thermal conductivity, as shown in Fig. 10, so even the noble gases affect cavitation differently (94,95). The role of thermal transport during cavitational collapse has been long recognized as evidence in favor of the hot-spot mechanism of sonochemistry (94), and recent calculations underscore its effect on conditions generated during cavitational collapse (30,31). In addition, sonochemical reactions will often involve the gases present in the cavitation event (96). For example, H_2, N_2, O_2, and CO_2 are not inert during cavitation and will undergo a variety of redox and radical reactions, as discussed later. Another relevant parameter, gas solubility, has been observed to affect the concentration of cavitation nuclei (97) and, in this way, it may play a role in determining the observed cavitation threshold.

The choice of the solvent has a profound influence over the observed sonochemistry as well. The effect of vapor pressure has already been mentioned. Other liquid properties, such as surface tension and viscosity, will alter the threshold of cavitation (8), but this is generally a minor concern. The chemical reactivity of the solvent is often much more important. As discussed below, aqueous sonochemistry is dominated by secondary reactions of OH· and H· formed from the sonolysis of water vapor in the cavitation zone. No solvent is inert under the high temperature conditions of cavitation: even linear alkanes will undergo pyrolytic-like cracking during high intensity sonication (89). One may minimize this

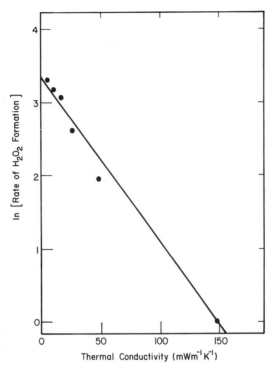

FIG. 10.   Sonochemical rates as a function of ambient gas thermal conductivity. [Replotted data from R. O. Prudhomme (*95*).]

problem, however, by using robust solvents (avoiding halocarbons, in particular) which have low vapor pressures in order to reduce their concentration in the vapor phase of the cavitation event. Furthermore, under the conditions used for heterogeneous sonochemistry, cavitation is primarily at the surface and dominates the observed reactivity. Similarly, one must anticipate secondary solvent reactivity in the trapping of high energy species produced during cavitation.

Thus, the parameters of acoustic intensity, temperature, ambient gas, and solvent choice have strong influences on sonochemical reactions. It is clear that one can fine tune the energetics of cavitation by the use of these variables and hence exercise control on the rates and reaction pathways followed by the associated chemistry. Specific examples will be discussed shortly. Clearly, the thermal conductivity of the ambient gas (e.g., a variable He/Ar atmosphere) and the overall solvent vapor pressure provides easy mechanisms for experimental control of the peak temperatures generated during the cavitational collapse.

IV

OVERVIEW OF THE CHEMICAL EFFECTS OF ULTRASOUND

A. *Aqueous Sonochemistry*

The early studies of the chemical effects of ultrasound have been thoroughly reviewed (5–7). Only the most important and most recent research is mentioned here as needed to provide a perspective on sonochemical reactivity patterns. The sonolysis of water is the earliest and most *exhaustively* studied (3,93,96,98–105). The first observations on the experimental parameters which influence sonochemistry come from these reports. The primary products are H_2O_2 and H_2, and various data supported their formation from the intermediacy of hydroxyl radicals and hydrogen radicals:

$$H_2O\text{-)-)-)} \twoheadrightarrow OH\cdot + H\cdot \ \rightarrow \ H_2O_2 + H_2 \tag{5}$$

Spin trapping experiments (106) have recently provided the final definitive evidence for the intermediacy of OH· and H·, but data have also suggested the formation of a variety of other high energy species, depending on conditions (ambient gas, pH, etc.), including $e^-_{(aq)}$ (107) and $HO_2\cdot$ (108). In the presence of O_2, isotope labeling studies show that much of the peroxide derives from the O_2, without O—O bond cleavage, rather than directly from the water. This clearly must involve a redox process of O_2, for example, with H·. Under typical laboratory sonolysis with an immersion horn, the rate of formation of peroxide will be ~ 30 μM/minute.

Given the facile homolytic cleavage of water during ultrasonic irradiation, a wide range of secondary sonochemistry in aqueous solutions would be expected and indeed has been repeatedly observed. Cited in Table III are those aqueous sonochemical reactions in which products have been well characterized; in some cases, trace amounts of other products were reported in addition to the major products listed. An extensive list of oxidations and reductions have been reported of various inorganic species. Various organics have been sonicated either as aqueous solutions or suspensions, generally with a wide range of highly degraded products being formed. Since extremely reactive intermediates are formed at respectable rates from the sonolysis of water itself, it is not surprising to see this general lack of control of sonochemistry in aqueous media. If we consider the nature of the cavitation event, *the high vapor pressure of water, relative to inorganic species or to dilute organic compounds, condemns aqueous sonochemistry to be dominated by secondary chemical reactions* unrelated to the direct processes which such dissolved substrates might undergo had they been the major species found in the collapsing bubble.

TABLE III
AQUEOUS SONOCHEMISTRY

Substrate present	Principal products	Ref.
Gases		
O_2	H_2O_2, O_3	*102,103*
N_2	HNO_2, HNO_3, NH_2OH, NH_3	*96,109*
$N_2 + H_2$	NH_3	*110*
$CO + H_2$	$HCHO$	*109*
$N_2 + (CO, CH_4$, or $HCHO)$	Amino acids	*109,111*
Inorganics		
Br^-, Cl^-	Br_2, Cl_2	*112*
Ce^{4+}	$Ce^{3+} \cdot$	*113*
$Co(NH_3)N_3^{2+}$	$Co^{2+} + N_3 \cdot$	*114*
Fe^{2+}	Fe^{3+}	*115,116*
$Fe(III)(C_2O_4)_3^{3-}$	Fe^{2+}	*117*
H_2S	$H_2 + S_8$	*118*
I^-	I_3^-	*81,85,100*
MnO_4^-	MnO_2	*119*
NO_3^-	NO_2^-	*120*
OsO_4	OsO_2	*121*
PO_3^{2-}	PO_4^{2-}	*122*
Tl^+	Tl^{3+}	*123*
Organics		
CCl_4	Cl_2, CO_2, HCl, C_2Cl_6, $HOCl$	*42,100,124–126*
CH_3I	CH_4, I_2, CH_3OH, HI, C_2H_6	*127*
R_2CHCl	R_2CHOH, HCl	*52–54*
$Cl_3CCH(OH)_2$	HCl	*128,129*
C_6H_5Br	Br^-, C_2H_2	*130,131*
Maleic acid $+ Br_2$	Fumaric acid	*132,133*
CS_2	S_8, H_2S	*134*
$(C_4H_9)_2S$	$(C_4H_9)_2SO$, polymer	*135*
$RCHO$	CO, CH_4, C_2H_4, $C_2H_4O_2$, RCO_2H	*136*
HCO_2^-	CO_2	*137*
C_5H_5N	HCN, C_2H_2, C_4H_2	*131,138*
C_6H_5OH	$C_6H_4(OH)_2$	*139*
$C_6H_5CO_2H$	$C_6H_4(OH)(CO_2H)$	*140*
$C_6H_{11}OH$	C_2H_2	*141*
RCO_2H	CO, CH_4	*136*
RCO_2R'	RCO_2H, $R'OH$	*51,142–146*
RCH_2NH_3	H_2, CH_4, NH_3, $RCHO$, RCH_2OH	*147*
$(CH_2NH_2)_2$	NH_3	*123*
Thymine	Hydroxylated products	*7,148,149*
Uracil	Hydroxylated products	*7,150,151*
Various amino acids	H_2, CO, NH_3, RNH_2, $HCHO$	*152*
Cysteine	Cystine	*152*
Macromolecules		
$C_6H_5CHCH_2$	Polymerization	*153*
$H_2CC(CH_3)(CO_2H)$	Polymerization	*154*
$H_2CCH(CONH_2)$	Polymerization	*155*
Many polymers	Depolymerization	*50*

B. *Nonaqueous Sonochemistry*

Until the past few years, very few examples of homogeneous non-aqueous sonications had been reported. These included the very slow degradation of a few common solvents (*156*) (CH_3CN and CCl_4), the initiation of explosions of tetranitromethane and nitroglycerine (*157*), the sevenfold acceleration (*158*) of the Curtius rearrangement of $C_6H_5CON_3$ to C_6H_5NCO and N_2, and the depolymerization of high molecular weight polymers (*50*). In general, sonochemistry had not been observed in most common, volatile organic solvents (or aqueous solutions with volatile organics). This led to the commonly stated assumption that intense cavitational collapse could only be supported in high tensile strength liquids such as water (*159,160*). As noted earlier, however, the problem is simply that many organic liquids have high vapor pressures, which greatly diminish the intensity of cavitational collapse.

It is now clearly demonstrated through the use of free radical traps that all organic liquids will undergo cavitation and generate bond homolysis, if the ambient temperature is sufficiently low (i.e., in order to reduce the solvent system's vapor pressure) (*89,90,161,162*). The sonolysis of alkanes is quite similar to very high temperature pyrolysis, yielding the products expected (H_2, CH_4, l-alkenes, and acetylene) from the well-understood Rice radical chain mechanism (*89*). Other recent reports compare the sonolysis and pyrolysis of biacetyl (which gives primarily acetone) (*163*) and the sonolysis and radiolysis of menthone (*164*). Nonaqueous chemistry can be complex, however, as in the tarry polymerization of several substituted benzenes (*165*).

By the proper choice of solvent and experimental conditions (i.e., low volatility, highly stable liquids at low temperature: e.g., decane, $-10°$ C), the rates of degradation of nonaqueous liquids can be made quite slow, well below those of water. This is of considerable advantage, since one may then observe the primary sonochemistry of dissolved substrates rather than secondary reactions with solvent fragments. In general, the examination of sonochemical reactions in aqueous solutions has produced results difficult to interpret due to the complexity of the secondary reactions which so readily occur. One may hope to see the increased use of low-volatility organic liquids in future sonochemical studies.

In addition, there are a few examples of heterogeneous nonaqueous sonochemistry, in both liquid–liquid and liquid–solid systems. Two recent reports have utilized ultrasonic agitation in place of or along with phase transfer catalysis: for the preparation of dichlorocarbene from aqueous $NaOH/CHCl_3$ (*166*), and for N-alkylation of amines with alkyl halides (*167*). Along the same lines, several papers have appeared in which ultrasonic irradiation of liquid–solid reactions enhances rates and yields: (1) the deprotonation of dimethylsulfoxide by NaH slurries (*168*); (2) the preparation of thioamides from amides treated with solid P_4S_{10} (*169*); (3) the reduction of aryl halides to arenes with solid lithium aluminum hydride (*170*); (4) the oxidation of secondary alcohols to ketones with solid $KMnO_4$ (*171*); and (5) the synthesis of aromatic acyl cyanides from acid chlorides and solid KCN (*172*). The last of these had led on to an unusual, and unexplained, observation of reaction pathway switching during ultrasonic irradiation (*173*). During ultrasonic irradiation in aromatic solvents, benzyl bromide, KCN, and alumina yields benzyl cyanide; whereas with mechanical agitation one obtains diarylmethane products from Friedel–Crafts attack on the solvent. Apparently, the sonication is deactivating the Lewis acid sites normally present on the alumina which are responsible for the Friedel–Crafts reactivity.

V

ORGANOMETALLIC SONOCHEMISTRY

The effects of high-intensity ultrasound on organometallic systems is an area of only recent investigation; consequently, a limited range of complexes and reactions have been examined. Still, a variety of novel reactivity patterns are beginning to emerge which are distinct from either normal thermal or photochemical activation. Most of the reactions which have been reported are stoichiometric in terms of the metal or metal complex, but a few examples of true sonocatalysis have also appeared. Although there is some overlap, we will divide our discussion into homogeneous and heterogeneous systems, in part because of the distinct nature of the cavitation event in each.

A. *Homogeneous Systems*

1. *Stoichiometric Reactions*

In 1981, the first report on the sonochemistry of discrete organometallic complexes demonstrated the effect of ultrasound on iron carbonyls in alkane solutions (*174*). The transition metal carbonyls were chosen for these initial studies because their thermal and photochemical reactivities have been well characterized. The comparison among the thermal, photochemical, and sonochemical reactions of $Fe(CO)_5$ provides an excellent example of the unique chemistry which homogeneous cavitation can induce. Because of the mechanistic insights which this system has provided, for our present discussion we will focus upon it as an archetype. Thermolysis of $Fe(CO)_5$, for example, gives pyrophoric, finely divided iron powder (175); ultraviolet photolysis (176) yields $Fe(CO)_9$, via the intermediate $Fe(CO)_4$; multiphoton infrared photolysis in the gas phase (177,178) yields isolated Fe atoms. Multiple ligand dissociation, generating $Fe(CO)_3$, $Fe(CO)_2$, etc., is not available from ordinary thermal or photochemical processes but does occur in matrix-isolated (179,180) and gas-phase laser (181,182) photolyses. These observations reflect the dual difficulties inherent in creating controlled multiple ligand dissociation: first, to deliver sufficient energy in a utilizable form and, second, to quench the highly energetic intermediates before complete ligand loss occurs.

During sonolysis in alkane solvents in the absence of alternate ligands, the unusual clusterfication of $Fe(CO)_5$ to $Fe_3(CO)_{12}$ is observed, together with the formation of finely divided iron (174,183). The rate of decomposition is cleanly first order, and the log of the observed first order rate coefficient is linear with the solvent vapor pressure. This is consistent with a simple dissociation process activated by the intense local heating generated by acoustic cavitation. As discussed earlier, the intensity of the cavitational collapse and the maximum temperature reached during such collapse decreases with increasing solvent vapor pressure. Given this method for controlling the conditions generated during cavitation, we would also expect to see the ratio of products vary as a function of solvent vapor pressure. As shown in Fig. 11, this proves to be the case: the ratio of products can be varied over a 100-fold range, with the production of $Fe_3(CO)_{12}$ strongly favored by increasing solvent volatility, as expected, since the sonochemical production of metallic iron requires greater activation energy than the production of $Fe_3(CO)_{12}$.

In order to probe the nature of the reactive site generated during the cavitation event, one may examine the sonochemical rate as a function of the volatility of the *substrate* (63). If one fixes the total solution vapor pressure by using appropriate solvent mixtures and keeps the $Fe(CO)_5$ concentration constant, but changes the $Fe(CO)_5$ vapor pressure by changing the ambient temperature, the observed first order rate coefficient increases linearly with increasing substrate vapor pressure and has a non-zero intercept, as shown in Fig. 12. This is consistent with a two-site model of sonochemical reactivity: the linear dependence on substrate vapor pressure represents the sonochemistry occurring in the gas phase of the cavitation event, and the non-zero intercept demonstrates a liquid-phase sonochemical site, presumably a thin liquid shell surrounding the cavity, as shown in Fig. 5. The importance of substrate volatility is clear, since the predominant site of sonochemistry is gas phase.

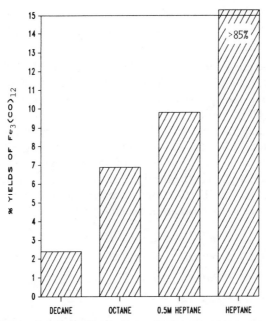

FIG. 11. Relative yields of $Fe_3(CO)_{12}$ versus Fe metal with increasing solvent vapor pressure. [Plotted from data in Ref. *174.*]

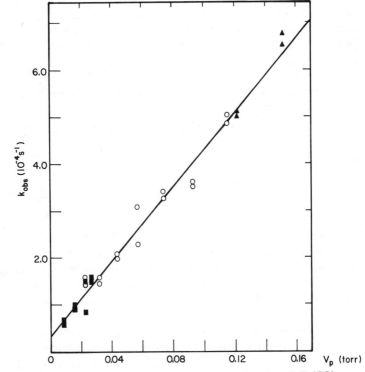

FIG. 12. First order sonochemical rate coefficients as a function of $Fe(CO)_5$ vapor pressure. Total vapor pressure was 5.0 torr *(63)*.

The proposed chemical mechanism by which $Fe_3(CO)_{12}$ is formed during the sonolysis of $Fe(CO)_5$ is shown in Eqs. (6)–(9). $Fe_2(CO)_9$ is not generated during the synthesis of $Fe_3(CO)_{12}$, and sonolysis of $Fe_2(CO)_9$ yields only $Fe(CO)_5$ and finely divided iron. The production of $Fe_3(CO)_{12}$ arises from initial multiple dissociative loss of CO from $Fe(CO)_5$ during cavitation, followed by secondary reactions with excess $Fe(CO)_5$. Ligand trapping studies confirm the formation of $Fe(CO)_3$, but cannot rule out the dimerization of $Fe(CO)_4$ in the localized cavitation site. The reaction of the putative $Fe_2(CO)_8$ with $Fe(CO)_5$ may proceed through initial dissociation in analogy to the matrix isolation reactivity (184) of $Fe(C_4H_4)_2(CO)_4$.

$$Fe(CO)_5 \text{ -)-)-)} \rightarrow Fe(CO)_{5-n} \quad n \text{ CO} \qquad (n = 1\text{--}5) \qquad (6)$$

$$Fe(CO)_3 + Fe(CO)_5 \longrightarrow Fe_2(CO)_8 \qquad (7)$$

$$2\, Fe(CO)_4 \longrightarrow Fe_2(CO)_8 \qquad (8)$$

$$Fe_2(CO)_8 + Fe(CO)_5 \longrightarrow Fe_3(CO)_{12} + CO \qquad (9)$$

In addition to clusterification, ligand substitution also occurs for $Fe(CO)_5$, and in fact for most metal carbonyls. This has proved useful as a mechanistic probe of the reactive species formed during cavitation. Sonication of $Fe(CO)_5$ in the presence of phosphines or phosphites produces $Fe(CO)_{5-n}L_n$ ($n = 1, 2,$ and 3). The ratio of these products is independent of length of sonication; the multiply substituted products increase with increasing initial [L]; $Fe(CO)_4L$ is *not* sonochemically converted to $Fe(CO)_3L_2$ on the timescale of its production from $Fe(CO)_5$. These observations are consistent with the same primary sonochemical event responsible for clusterification

$$Fe(CO)_5 \text{ -)-)-)} \rightarrow Fe(CO)_{5-n} + n \text{ CO} \qquad (n = 1\text{--}5) \qquad (10)$$

$$Fe(CO)_4 + L \longrightarrow Fe(CO)_4L \qquad (11)$$

$$Fe(CO)_3 + L \longrightarrow Fe(CO)_3L \qquad (12)$$

$$Fe(CO)_3 + CO \longrightarrow Fe(CO)_4 \qquad (13)$$

$$Fe(CO)_3L + L \longrightarrow Fe(CO)_3L_2 \qquad (14)$$

Sonochemical ligand substitution readily occurs with a variety of other metal carbonyls, as shown in Table IV. In all cases, multiple ligand substitution originates directly from the parent carbonyl. The rates of sonochemical ligand substitution of the various metal carbonyls follow their relative volatilities, as predicted from the nature of the cavitational collapse.

Another recent example of sonochemical substitution is in the preparation of π-allyllactone(tricarbonyl)iron complexes, which are useful synthetic intermediates in the synthesis of lactones and lactams (185). Upon

TABLE IV

HOMOGENEOUS ORGANOMETALLIC SONOCHEMISTRY

Reactants	Products	Ref.
Clusterification		
$Fe(CO)_5$	$Fe_3(CO)_{12}$, Fe	*174,183*
$Fe_2(CO)_9$	Fe, $Fe(CO)_5$	*174,183*
Ligand substitution[a]		
$Cr(CO)_6 + L$	$Cr(CO)_5L$, $Cr(CO)_4L_2$, $Cr(CO)_3L_3$	*183*
$Mo(CO)_6 + L$	$Mo(CO)_5L$, $Mo(CO)_4L_2$	*183*
$W(CO)_6 + L$	$W(CO)_5L$, $W(CO)_4L_2$	*183*
$Fe(CO)_5 + L$	$Fe(CO)_4L$, $Fe(CO)_3L_2$, $Fe(CO)_2L_3$	*174,183*
$FeCp(CO)_2I + L$	$FeCp(CO)(L)I$	*189*
$Fe_2(CO)_9$ + alkenylepoxide	$Fe(CO)_3(\pi$-allyllactone)	*185*
$Fe_3(CO)_{12} + L$	$Fe(CO)_4L$, $Fe(CO)_3L_2$	*174,183*
$Mn_2(CO)_{10} + L$	$Mn_2(CO)_8L_2$	*183*
$Co(Cp)_2 + CO$	$CoCp(CO)_2$	*191*
Sn_2R_6	$SnR_3\cdot$, $R\cdot$	*193*
Secondary reactions		
$M_2(CO)_{10} + R_3CX$	$M(CO)_5X$ (M = Mn, Re; X = Cl, Br)	*183*
$Co_2(CO)_8$ + alkane	$Co_2(CO)_6(C_2H_2)$, $Co_4(CO)_{10}(C_2H_2)$	*189*
Sonocatalytic reactions		
$Fe_x(CO)_y$ + 1-alkene	*cis-, trans*-2-Alkene	*174,183*
$Ru_x(CO)_y$ + 1-alkene	*cis-, trans*-2-Alkene	*183*
$Mo(CO)_6$ + 1-alkene	*cis-, trans*-2-Alkene	*183*
$Co_2(CO)_8$ + 1-alkene	*cis-, trans*-2-Alkene	*183*

[a] L = various phosphines and phosphites.

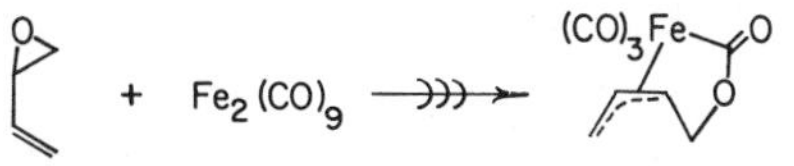

$$\text{epoxide} \quad + \quad Fe_2(CO)_9 \quad \xrightarrow{\;\;\ggg\;\;} \quad (CO)_3Fe \text{ structure} \tag{15}$$

sonication in a cleaning bath, $Fe_2(CO)_9$ slurries in hydrocarbon solutions of alkenyl epoxides rearrange as shown in Eq. (15). The same reaction occurs thermally with $Fe(CO)_4$(tetrahydrofuran), indicating the probable intermediacy of a coordinatively unsaturated (or loosely coordinated) species upon sonication of $Fe_2(CO)_9$. The authors expressed surprise (*185*) that under the conditions $Fe(CO)_5$ and $Fe_3(CO)_{12}$ did not undergo the same reaction, in light of the similarity in their sonocatalytic behavior and their sonochemical substitution with phosphines (*174*). This is clearly due, however, to the low intensities of ultrasound present in these authors' ultrasonic cleaning bath, which are sufficient to induce cavitation in the heterogeneous slurries of $Fe_2(CO)_9$ but which are *not* sufficient in homogeneous solutions of $Fe(CO)_5$ or $Fe_3(CO)_{12}$. Under more intense ultrasonic irradiation, $Fe(CO)_5$, for example, will undergo substitution with alkenes (183).

The sonolysis of $Mn_2(CO)_{10}$ makes for an interesting comparison (186), since either metal–metal (as in photolysis) (187) or metal–carbon (as in moderate temperature thermolysis) (188) bond breakage could occur. Ligand substitution will occur from either route producing the axially disubstituted $Mn_2(CO)_8L_2$. Using benzyl chloride as a trap for the possible intermediacy of $Mn(CO)_5$, the sonochemical substitution of $Mn_2(CO)_{10}$ has been shown to follow the thermal, rather than the photochemical, pathway of dissociative CO loss.

Upon sonication in halocarbon solvents, metal carbonyls undergo facile halogenations (186). The rates of halogenation are solvent dependent, but independent of choice of metal carbonyl or its concentration, and represent the products of secondary reactions occurring from the sonolytic decomposition of the halocarbon solvent, as shown in Eqs. (16)–(20). Alkanes and other halogen atom traps suppress the halogenation of the metal carbonyls.

$$R_3CX \text{ -)-)-)} \rightarrow R_3C\cdot + X\cdot \tag{16}$$

$$2\,R_3C\cdot \longrightarrow R_3CCR_3 \tag{17}$$

$$2\,X\cdot \longrightarrow X_2 \tag{18}$$

$$M_2(CO)_{10} + 2\,X\cdot \longrightarrow 2\,M(CO)_5X \tag{19}$$

$$M_2(CO)_{10} + X_2 \longrightarrow 2\,M(CO)_5X \tag{20}$$

Another example of a secondary sonochemical reaction is the very slow production of acetylene complexes of cobalt carbonyls upon lengthy sonolysis of $Co_2(CO)_8$ in n-alkanes (C_5H_{12} through $C_{10}H_{22}$) (189). The principal products are $Co_2(CO)_6(C_2H_2)$ and $Co_4(CO)_{10}(C_2H_2)$, with small amounts of $Co_4(CO)_{12}$. $Co_4(CO)_{12}$ is an expected product, since it is easily formed upon pyrolysis of $Co_2(CO)_8$. The acetylene of the former complexes originates from the solvent, as confirmed by isotope labeling. Their formation is initially quite surprising, until one notes that their rates of formation are comparable to the slow rate of C_2H_2 formation from sonolysis of the alkane (89) and that cobalt carbonyls undergo facile thermal reactions with alkynes (190). Thus, the origin of this sonochemical alkane activation is *not* in some high energy organometallic fragment, but in the secondary trapping of acetylene sonochemically produced from the alkane.

The sonochemistry of non-carbonyl organometallics has not yet been well developed. Complexes which contain both CO and Cp undergo CO substitution upon sonolysis (189). In preliminary studies of the metallocenes (*191*), $Co(Cp)_2$ has been found to undergo facile ligand substitution during sonication under CO to yield $Co(Cp)(CO)_2$. This reaction under simple thermal conditions (*192*) requires 200 atmospheres CO at 90–150° C and gives low yields; with ultrasonic irradiation, excellent yields are obtained at 3 atmospheres and 20° C. These results are in keeping with the high temperature and pressure conditions generated during acoustic cavitation, and suggest an analogy between sonochemistry and bomb reactions. A recent report describes the sonochemical decomposition of organotin compounds (*193*). Trapping of intermediate radicals by nitroso-durene and analysis by ESR demonstrated alkyl–tin bond cleavage during sonication in benzene solutions. The following reaction scheme [Eqs. (21)–(24)] was suggested to explain the observed ESR spectra. The yields, rates, source of oxidant, or final products in the absence of spin traps were not determined.

$$R_4Sn \text{ -)-)-)} \rightarrow R\cdot + R_3S_n\cdot \qquad (21)$$

$$R\cdot + ONC_6H(CH_3)_4 \longrightarrow ON(R)[C_6H(CH_3)_4] \qquad (22)$$

$$R_3Sn\cdot + C_6H_6 \longrightarrow R_3SnC_6H_6\cdot \qquad (23)$$

$$R_3SnC_6H_6\cdot + ONC_6H(CH_3)_4 + [ox] \longrightarrow \cdot ON(C_6H_4SnR_3)[C_6H(CH_3)_4] + [red] \qquad (24)$$

2. Initiation of Homogeneous Catalysis

Having demonstrated that ultrasound can induce ligand dissociation, the initiation of homogeneous catalysis by ultrasound becomes practical. The potential advantages of such sonocatalysis include (1) the use of low ambient temperatures to preserve thermally sensitive substrates and to enhance selectivity, (2) the ability to generate high energy species unobtainable from photolysis or simple pyrolysis, (3) the mimicry, on a microscopic scale, of bomb reaction conditions, and (4) possible ease of scale-up. The transient, coordinatively unsaturated species produced from the sonolysis of metal carbonyls are likely candidates, since similar species produced photochemically are among the most active catalysts known (*176,194*). The thermal (*195–197*) and photochemical (*198–202*) isomerization of terminal olefins by metal carbonyls have been extensively studied, and provide a useful test case for applications of sonocatalysis.

As shown in Fig. 13, a variety of metal carbonyls upon sonication will catalyze the isomerization of 1-pentene to *cis*- and *trans*-2-pentene (*186*). Initial turnover rates are about 1–100 mol 1-pentene isomerized/mol of precatalyst/hour, and represent rate enhancements of ~10^5 over thermal controls (*174*). The relative sonocatalytic and photocatalytic activities of these carbonyls are in general accord. An exception is $Ru_3(CO)_{12}$, which is

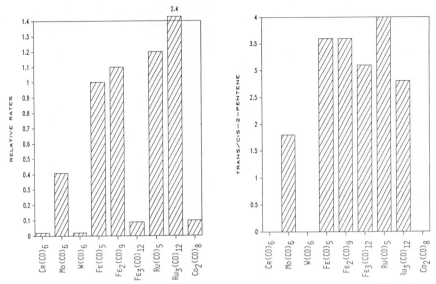

FIG. 13.   Relative rates and trans/cis ratios of the sonocatalysis of 1-pentene isomerization by metal carbonyls. [Plotted from data in Ref. 183.]

relatively more active as a sonocatalyst and shows very different trans/cis ratios under sonolysis; it appears that the catalytic agent produced sonochemically in this case is not that produced photochemically (186). A variety of terminal alkenes will serve as substrates for sonocatalytic isomerization, although increasing steric hindrance, as in 2-ethyl-pent-1-ene and allylbenzene, significantly diminishes the observed rates. Alkenes without β-hydrogens will not serve as substrates, in contrast to the radical chain isomerization of maleic acid which occurs upon ultrasonic irradiation of aqueous Br_2 solutions (132,133).

The exact nature of the catalytic species generated during sonolysis remains unknown. Results are consistent with the generally accepted mechanism for alkene isomerization in analogous thermal (197) and photochemical systems (195). This involves the formation of a hydrido-π-allyl intermediate and alkene rearrangement via hydride migration to form the thermodynamically more stable 2-alkene complex, as shown in a general sense in Eqs. (25)–(29). In keeping with this scheme, sonication of $Fe(CO)_5$ in the presence of 1-pentene and CO does produce $Fe(CO)_4$(pentene), as determined by FTIR spectral stripping (183).

$$M(CO)_n \rightarrow M(CO)_m + (n - m)\ CO \qquad (25)$$

$$M(CO)_m + \text{1-alkene} \rightarrow M(CO)_x(\text{1-alkene}) + (m - x)\ CO \qquad (26)$$

$$M(CO)_x(1\text{-alkene}) \rightarrow M(CO)_x(H)(\pi\text{-allyl}) \tag{27}$$

$$M(CO)_x(H)(\pi\text{-allyl}) \rightarrow M(CO)_x(2\text{-alkene}) \tag{28}$$

$$M(CO)_x(2\text{-alkene}) + 1\text{-alkene} \rightarrow M(CO)_x(1\text{-alkene}) + 2\text{-alkene} \tag{29}$$

B. *Heterogeneous Systems*

1. *Stoichiometric Reactions*

The use of ultrasound to accelerate chemical reactions in heterogeneous systems is becoming increasingly widespread. The physical phenomena which are responsible include the creation of emulsions at liquid–liquid interfaces, the generation of cavitational erosion and cleaning at liquid–solid interfaces, the production of shock wave damage and deformation of solid surfaces, the enhancement in surface area from fragmentation of friable solids, and the improvement in mass transport through turbulent mixing and acoustic streaming. A summary of the heterogeneous systems in which ultrasound has been used is presented in Table V.

In organometallic chemistry, the use of ultrasound in liquid–liquid heterogeneous systems has been limited to Hg. The emulsification of Hg with various liquids dates to the very first reports on sonochemistry (*3,203,204*). The use of such emulsions for chemical purposes, however, was delineated by the extensive investigations of Fry and co-workers (*205–212*), who have reported the sonochemical reaction of various nucleophiles with α,α'-dibromoketones and mercury. The versatility of this reagent is summarized in Eqs. (30)–(36).

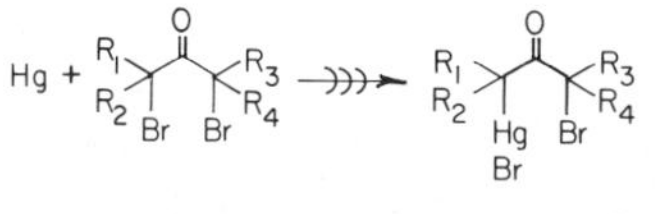

$$\tag{30}$$

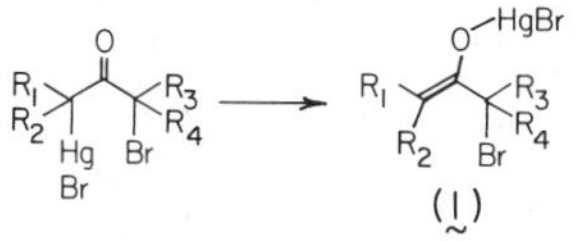

$$\tag{31}$$

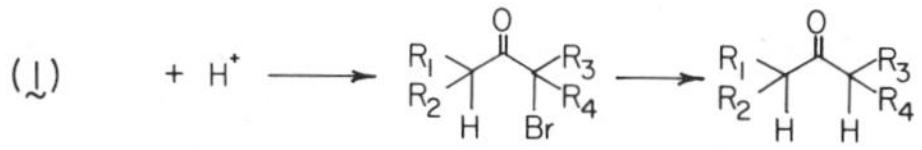

$$\tag{32}$$

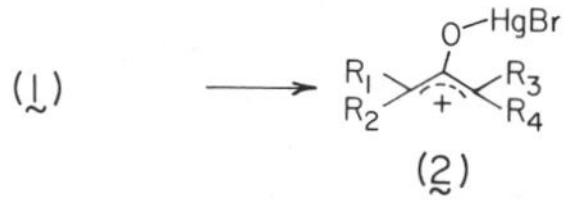

$$\tag{33}$$

$$(\underset{\sim}{2}) \quad + \text{ RCO}_2\text{H} \longrightarrow \quad \begin{array}{c} R_1 \\ R_2 \end{array}\!\!\!\!>\!\!\!\!< \begin{array}{c} R_3 \\ R_4 \end{array} \tag{34}$$

$$(\underset{\sim}{2}) \quad + \text{ ROH} \longrightarrow \quad \begin{array}{c} R_1 \\ R_2 \end{array}\!\!\!\!>\!\!\!\!< \begin{array}{c} R_3 \\ R_4 \end{array} \tag{35}$$

$$(\underset{\sim}{2}) \quad + \text{ } \longrightarrow \quad \begin{array}{c} R_1 \\ R_2 \end{array}\!\!\!\!>\!\!\!\!< \tag{36}$$

There are synthetic advantages to the use of ultrasound in these systems. For example, such Hg dispersions allow reactions with more sterically, hindered ketones yet introduces only one nucleophilic group even in sterically undemanding systems (209). In addition, the reaction given in Eq. (36) represents a convenient one-step synthesis of 1,3-dioxolans (206). The proposed mechanism of these reactions involves nucleophilic attack on the mercurial oxyallyl cation 2. Fry believes that the effect of the ultrasound in this system is a kinetic rate enhancement (205), presumably due to the large surface area of Hg generated in the emulsion. Another reduction carried out in excellent yield by ultrasonically dispersed Hg is shown in Eq. (37); the mechanism by which this occurs is unclear, particularly since no thiobenzaldehyde or 1,2-diphenylthiirane is formed (212). One might speculate that an internal Wurtz coupling is the initial step, forming 1,2-diphenylthiirane, which is rapidly desulfurized to *trans*-stilbene.

$$C_6H_5(H)(Br)CSC(Br)(H)C_6H_5 + Hg \rightarrow \text{ } t\text{-}C_6H_5(H)C\!\!=\!\!C(H)C_6H_5 \tag{37}$$

The effects of ultrasound on liquid–solid heterogeneous organometallic reactions has been a matter of intense recent investigation, particularly in the laboratories of Luche, Boudjouk, and Ishikawa. The first use of ultrasound to prepare organometallic complexes of the main group metals (e.g., lithium, magnesium, and aluminum) from organic halides, however, originates in the seldom cited work of Renaud (213). Grignard reagents and organolithium compounds were formed rapidly, even in wet solvents, from organobromides (but not chlorides), and Al powder reacted with RMgX without the need for Al–Mg alloys. Renaud also found that such enhancements were not observed for Ca, Hg, Zn (but see below), or Be.

The report by Luche and Damiano in 1980 of the use of an ultrasonic cleaner to accelerate lithiation reactions (214) initiated the recent interest.

TABLE V   HETEROGENEOUS ORGANOMETALLIC SONOCHEMISTRY

Organic reactant	Product	Ref.
Hg		
$(R_2BrC)_2CO + R'CO_2H$	$(HR_2C)CO[C(O_2CR')R_2]$	205,207–210
$(R_2BrC)_2CO + R'OH$	$(HR_2C)CO[C(OR')R_2]$	211
$(R_2BrC)_2CO + (H_3C)_2CO$		206
$C_6H_5(H)(Br)CSC(Br)(H)C_6H_5$	$t\text{-}C_6H_5(H)C{=}C(H)C_6H_5$	212
Mg		
$R{-}Br$	$R{-}MgBr$	213,242
$R{-}Br + Al$	AlR_3	213
Li		
$R{-}Br \quad R = (n\text{-}Pr, n\text{-}Bu, Ph)$	$R{-}Li$	214
$R{-}Br + R'R''CO$	$RR'R''COH$	214,215
$R{-}Br + (H_3C)_2NCHO$	$RCHO$	216
$R{-}Br +$ $=O + Cu(I)$		217
$R_3M{-}Cl \quad (M = C, Si, Sn; R = alkyl, aryl)$	R_3MMR_3	218,219
$R_2SiCl_2 \quad (R = arenes)$	$cyclo\text{-}(R_2Si)_3$	219,224
Na		
Arenes	$Na(arene^{-}\cdot)$	225–227
K		
$XH_2C{-}(CH_2)_n{-}CH_2X$	Cycloalkanes	228
LiAlH$_4$		
$R_3M{-}Cl \quad (M = Si, Ge, Sn;$ $X = Cl, NR_2, OR)$	$R_3M{-}H$	220
Zn		
$CF_3I + RR'C{=}O$	$RR'C(OH)CF_3$	229
$CF_3I + R{-}Br + Pd(0)$	$R{-}CF_3$	232
$CF_3I + RC{\equiv}CR + Cu(0)$	$HRC{=}CR(CF_3)$	233
$RR'C{=}O + BrCH_2CO_2R''$	$RR'C(OH)CH_2CO_2R''$	230
$R{-}Br + Li + ZnBr_2 +$ $=O$		235,236
$H_2CI_2 + R_2C{=}CR_2$		237
$1,2\text{-}(BrH_2C)_2C_6H_4 + dienophiles$		238,239
Transition metals		
$RuCl_3 + Zn + 1,5\text{-cyclooctadiene}$	$(\eta^6\text{-}1,3,5\text{-cyclooctadiene})\text{-}$ $(\eta^4\text{-}1,5\text{-cyclooctadiene})Ru(0)$	240
$MCl_5 + Na + CO \quad (M = V, Nb, Ta)$	$M(CO)_6^{-}$	243
$MCl_6 + Na + CO \quad (M = Cr, Mo, W)$	$M_2(CO)_{10}^{2-}$	243
$MnCl_3 + Na + CO$	$Mn(CO)_5^{-}$	243
$FeCl_3 + Na + CO$	$Fe(CO)_4^{2-} + Fe_2(CO)_8^{2-}$	243
$NiCl_2 + Na + CO$	$Ni_6(CO)_{12}^{2-}$	243
$Co(acac)_3 + C_5H_6 + COD + Mg/C_{14}H_{10}$	$Co(Cp)(COD)$	247

Excellent yields of organolithium compounds were found for n-propyl-, n-butyl-, and phenyllithium (61–95%), even at room temperature in wet solvents [Eq. (38)], which has potential utility for large-scale industrial

$$R—Br + Li \rightarrow R—Li \qquad R = Pr, n\text{-}Bu, Ph \tag{38}$$

applications. Lithiation of isopropyl and t-butyl bromides remained sluggish, however.

More impressive is the striking improvement which sonication afforded to the Barbier reaction [Eq. (39)] (214). This one-step coupling of organic

$$R—Li + R'R''CO \rightarrow RR'R''COH \tag{39}$$

halides with carbonyl compounds via magnesium or lithium intermediates is significantly hastened (10–40 minutes) with excellent yields (76–100%) for a wide range of organobromides (including t-butyl and benzyl) and a variety of ketones or aldehydes. This has proved to be the method of choice in the synthesis of complex cyclopentanones via an intramolecular Barbier reaction (215).

Extensions of the use of ultrasound in lithiation reactions have been profitable for a variety of reactions. The Bouveault reaction for the synthesis of aldehydes [Eq. (40)] suffers from side reactions and low yields.

$$R—Li + (H_3C)_2NCHO \rightarrow RCH(OLi)(NCH_3)_2 \rightarrow RCHO \tag{40}$$

Upon sonication in a cleaning bath, mixtures of organic halides, dimethylformamide, and lithium sand in THF give very good yields (67–88%) of aldehydes, although no direct comparison to the simple thermal reaction was made (216). Similar improvements in yields are observed in organocopper conjugate alkylations of enones [Eq. (41)] (217). The formation of the organocopper reagent was accomplished by ultrasonic irradiation of

$$\text{—}\!\!=\!\!O + R—Br + Li + Cu(I) \rightarrow \underset{R}{\text{~~~~~~~}}\!\!=\!\!O \tag{41}$$

alkyl or aryl bromide solutions in the presence of lithium sand and a solution of a Cu(I) salt; temperature control during the sonication is quite important in these reactions to avoid Barbier-type α-alkylations.

Wurtz-type couplings have also been observed upon sonication of lithium in the presence of both organic halides (yields 36–73%) (218) and chlorosilanes or chlorostannanes (yields 42–94%) [Eq. (42)] (219).

$$R_3M—Cl + Li \text{ -)-)-)} \rightarrow R_3MMR_3 \qquad (M = C, Si, Sn; R = alkyl \text{ or } aryl) \tag{42}$$

Lithium wire is acceptable (but higher yields result with lithium sand dispersed in mineral oil) and requires in some cases a small amount of anthracene as electron transfer catalyst. Direct comparisons to reactions run without ultrasound, but under high speed stirring or heating, were not made. In the same vein, the use of low intensity ultrasound for the preparation of main group hydrides from the reaction of the corresponding chlorides with lithium aluminum hydride [Eq. (43)] has been recently reported (220).

$$R_3M\text{---}Cl + LiAlH_4 \text{ -)-)-)} \rightarrow R_3M\text{---}H \qquad (M = Si, Ge, Sn; X = Cl, NR_2, OR) \quad (43)$$

In the case of dichlorosilanes, oligomerization to form cyclopolysilanes occurs in high yields, with the product's ring size dependent upon the steric bulk of the starting silane (219). Boudjouk initially reported (221) the synthesis of West's novel disilene (222) upon sonication of lithium with the highly hindered bis(mesityl)dichlorosilane [Eq. (44)]. It is difficult, however, to obtain consistent results with this sonochemical synthesis of the

$$[2,4,6\text{-}(H_3C)_3C_6H_2]_2SiCl_2 + Li \text{ -)-)-)} \rightarrow R_2Si\text{=}SiR_2 \qquad (44)$$

disilene (223), and the generally observed product is the hexamesityl-cyclotrisilane (224).

The sonochemistry of the other alkali metals is less explored. The use of ultrasound to produce colloidal Na has early origins and was found to greatly facilitate the production of the radical anion salt of 5,6-benzo-quinoline (225) and to give higher yields with greater control in the synthesis of phenylsodium (226). In addition, the use of an ultrasonic cleaning bath to promote the formation of other aromatic radical anions from chunk Na in undried solvents has been reported (227). Luche has recently studied the ultrasonic dispersion of potassium in toluene or xylene and its use for the cyclization of α,ω-difunctionalized alkanes and for other reactions (228).

The effects of ultrasound on zinc reagents has been explored in some detail as well. Ishikawa first examined the use of Zn for trifluoromethyla-tion of carbonyl compounds [Eq. (45)] (229). In this case the choice of

$$RR'C\text{=}O + Zn + CF_3I \text{ -)-)-)} \rightarrow RR'(F_3C)COZnI \xrightarrow{H^+} RR'C(OH)CF_3 \qquad (45)$$

metal was dictated by the relative stability of the perfluoroalkylzinc compounds compared to the corresponding magnesium or lithium species. Good yields (45–86%) were reported for the formation of alcohols from the sonication in a cleaning bath of a mixture of Zn powder with CF_3I in dimethylformamide solutions of several ketones and aldehydes. The closely related Reformatsky reaction [Eq. (46)] has also proven to be

$$RR'C\text{=}O + Zn + BrCH_2CO_2R'' \text{ -)-)-)} \rightarrow RR'C(OH)CH_2CO_2R'' \qquad (46)$$

assisted by low intensity ultrasound (230). Extensive comparison to alternative reaction conditions was made in this thorough study. The use of I_2 or I^- promoters and dioxane as solvent is important for optimal yields. Sonication provided reaction rates and yields (typically 30 minutes and >90% yield) as good or better than the use of activated zinc powders (231) prepared from the reduction of anhydrous $ZnCl_2$. The use of ultrasonically generated organozinc complexes for perfluoroalkylation of allyl, vinyl, and aryl halides with Pd(0) (232) catalyst and of alkynes with Cu(0) catalyst (233) has also been reported.

Organozinc reagents prepared from ultrasonic irradiation of organic halides with Li in the presence of $ZnBr_2$ have recently been used for conjugate addition to α-enones [Eqs. (47) and (48)] (234,235). In the

$$R—Br + Li + ZnBr_2 \text{ -)-)-)} \rightarrow [R_2Zn] \tag{47}$$

$$[R_2Zn] + \overset{}{\diagup\diagdown}=O \longrightarrow R\diagdown\diagup\diagdown=O \tag{48}$$

initial report, reactions were run in an ultrasonic cleaning bath cooled with ice, in dry ether or tetrahydrofuran with $Ni(acac)_2$ as catalyst; it was stated that cavitational effects were probably not involved since such solvents supposedly preclude the occurrence of cavitation (234). In the improved synthesis, however, much more rapid reactions occurred with excellent reproducibility when an immersion horn configuration was used at 0°C with small amounts of tetrahydrofuran dissolved in toluene as solvent (235). Since the rates are improved by the use of less volatile solvents, this sonochemical reaction *probably is due to cavitation*. The efficacy of the 1,4-addition is not hampered by β,β-disubstitution of the enone, in contrast to the use of organocopper reagents. This has led to an elegant synthesis of β-cuparenone in three steps and 50% yield (236).

Low intensity ultrasound has also been applied to the Simmons–Smith cyclopropanation of olefins with zinc–diiodomethane (237). This reaction normally will not occur without activation of mossy Zn with I_2 or Li, and was difficult to scale-up due to delayed initiation. Yields upon sonication are nearly quantitative, activation of the Zn is unnecessary, and no delayed exotherms are observed. In reactions with another class of organic dihalides, ultrasonic irradiation of Zn with α,α'-dibromo-o-xylene has proved a facile way to generate an o-xylylene-like species [Eq. (49)],

$$\text{Zn} + \text{(structure)} \xrightarrow{\text{-)))-}} [\text{(structure)}] \longrightarrow \text{(structure)} \tag{49}$$

which has been trapped by a variety of dienophiles (238). This has found synthetic application in the synthesis of functionalized hexa-hydro-anthracenes and -napthacenes (239).

Finally, an improved synthesis of (η^6-1,3,5-cyclooctatriene)-(η^4-1,5-cyclooctadiene)ruthenium(0) has been reported which utilizes a cleaning bath to hasten the Zn reduction of $RuCl_3$ in the presence of 1,5-cyclooctadiene (240). The use of ultrasound in simple reductions using Zn are a likely area for further development.

In all of the heterogeneous organometallic sonochemistry discussed thus far, the metals used have been *extremely reactive* and easily malleable. The specific origin of the rate and yield improvements has not yet been established in these systems. Faster reaction rates come about in part as a consequence of greater surface area dispersions. The improved mass transport between bulk solution and the reagent surface due to cavitational shock waves and microstreaming are also important contributions. These factors permit the use of lower temperatures, with the subsequent advantages of lessened side reactions and improved reaction control. The importance of lattice defects in initiation of the Grignard reaction (241), for example, may be relevant, since surface damage from cavitation is a probable occurrence. Ultrasonic cleaning of the reactive metal surface to remove passivating impurities (e.g., water, hydroxide, metal halide, or organolithium) must also be important (242).

The activation of less reactive metals remains an important goal which continues to attract major efforts in heterogeneous catalysis, metal vapor chemistry, and organometallic synthesis. Given the extreme conditions generated by acoustic cavitation at surfaces, analogies to autoclave conditions or to metal vapor reactors are not inappropriate. In order to probe the potential generality of ultrasonic activation of heterogeneous reactions, Suslick and Johnson examined (243) the sonochemical reactivity of the normally very unreactive early transition metals with carbon monoxide. Even with the use of "activated," highly dispersed transition metal slurries, as investigated by Rieke (244,245) the formation of the early transition metal carbonyls still require "bomb" conditions (100–300 atm of CO, 100–300°C) (246). The use of ultrasonic irradiation facilitates the reduction of a variety of transition metal salts to an active form that will react at low temperatures with low pressures of CO. Reduction of transition metal halides soluble in tetrahydrofuran or diglyme with Na, using a direct immersion ultrasonic horn under 1–4 atm CO at 10°C, gave fair to good yields of the carbonyl anions for V, Nb, Ta, Cr, Mo, W, Mn, Fe, and Ni [Eqs. (50) and (51)]. Solubility of the metal halide is necessary for effective

$$MCl_5 + Na + CO \text{ -)-)-)} \rightarrow M(CO)_6^- \quad (M = V, Nb, Ta) \tag{50}$$

$$MCl_6 + Na + CO \text{ -)-)-)} \rightarrow M_2(CO)_{10}^{2-} \quad (M = Cr, Mo, W) \tag{51}$$

reaction. An ultrasonic cleaning bath was found to be of only marginal use when compared to the higher intensity immersion horn. Since these reactions are run at low pressures, they may prove uniquely useful in the production of ^{13}CO labeled carbonyl complexes.

The possible mechanisms which one might invoke for the activation of these transition metal slurries include (1) creation of extremely reactive dispersions, (2) improved mass transport between solution and surface, (3) generation of surface hot-spots due to cavitational microjets, and (4) direct trapping with CO of reactive metallic species formed during the reduction of the metal halide. The first three mechanisms can be eliminated, since complete reduction of transition metal halides by Na with ultrasonic irradiation *under Ar*, followed by exposure to CO in the absence or presence of ultrasound, yielded no metal carbonyl (*243*). In the case of the reduction of WCl_6, sonication under CO showed the initial formation of tungsten carbonyl halides, followed by conversion of $W(CO)_6$, and finally its further reduction to $W_2(CO)_{10}^{2-}$, Thus, the reduction process appears to be sequential: reactive species formed upon partial reduction are trapped by CO.

Another recent application to the activation of transition metals was reported (*247*) by Bönnemann, Bogdavovic, and co-workers, in which an extremely reactive Mg species was used to reduce metal salts in the presence of cyclopentadiene, 1,5-cyclo-octadiene, and other ligands to form their metal complexes. The reactive Mg species, characterized as $Mg(THF)_3$ (anthracene), was produced from Mg powder in THF solutions containing a catalytic amount of anthracene by use of an ultrasonic cleaning bath. A plausible scheme for this reaction has been suggested:

$$Mg + C_{14}H_{10} \overset{THF}{-)-)-)\rightarrow} Mg(THF)_3(\eta^2\text{-}C_{14}H_{10}) \tag{52}$$

$$2\ Co(acac)_3 + 3\ Mg(THF)_3(\eta^2\text{-}C_{14}H_{10}) \longrightarrow 2\ Co^* + 3\ Mg^{2+} \tag{53}$$

$$2\ Co^* + 2\ C_5H_6 + 3\ 1,5\text{-}C_8H_{12} \longrightarrow 2\ Co(Cp)(COD) + C_8H_{14} \tag{54}$$

2. *Applications to Heterogeneous Catalysis*

Ultrasonic irradiation can alter the reactivity observed during the heterogeneous catalysis of a variety of reactions. Sonication has shown such behavior (1) by altering the formation of heterogeneous catalysts, (2) by perturbing the properties of previously formed catalysts, or (3) by affecting the reactivity during catalysis. There is an extensive (but little recognized) literature in this area (*248*), most of which is beyond the scope of this review.

In general, however, ultrasonic rate enhancements of heterogeneous catalysis are usually relatively modest (less than 10-fold). The effect of irradiating operating catalysts owes much simply to improved mass transport (249). In addition, increased dispersion during the formation of catalysts under ultrasound (250) [e.g., Ziegler–Natta polymerizations (251)] will enhance reactivity, as will the fracture of friable solids [e.g., noble metals on carbon or silica (191) or malleable metals (252)]. In the case of bulk metal catalysts, the removal of passivating coatings through surface cavitational damage is well established (253–255).

The range of reactions which have been examined is wide (248) and includes hydrogenations (256), ammonia synthesis (257), polymerizations (251), and oxidations (258). Little activity has occurred in this area during the past few years. Recent reports of the effects of sonication on heterogeneous catalysis include the liquefaction of coal by hydrogenation with Cu/Zn (259), the hydrogenation of olefins by formic acid with Pd on carbon (260), and the hydrosilation of 1-alkenes by Pt on carbon (261).

VI

CONCLUDING REMARKS

The use of ultrasound in both homogeneous and heterogeneous reactions will see increasing study. The potential to do high energy chemistry in condensed phases at room temperature remains an attractive feature of sonochemistry. Unique examples of sonochemical reactivity quite different from thermal or photochemical processes have been noted. There are analogies to shock wave and gas-phase pyrolyses, to "bomb" reactions, and to metal vapor chemistry, which will continue to be explored. The use of ultrasound in the synthesis of organometallic species has had particular impact in heterogeneous systems and no doubt will find application in nearly any case where interphase mixing is a problem.

A primary limitation of sonochemistry remains its energy inefficiency. This may be dramatically improved, however, if a more efficient means of coupling the sound field with preformed cavities can be found. The question of selectivity in and control of sonochemical reactions, as with any thermal process, remains a legitimate concern. There are, however, clearly defined means of controlling the conditions generated during cavitational collapse, which permit the variation of product distributions in a rational fashion.

Early in the study of cavitation phenomenon, Minnaert observed that scientists

> have hardly ever investigated the sounds of running water. As a matter of fact we know very little about the murmur of the brook, the roar of the cataract, or the humming of the sea [Ref. *34*].

Be this as it may, we *are* gaining a significant understanding of the chemical consequences of such!

ACKNOWLEDGMENTS

The author sincerely appreciates the efforts of his graduate students and postdoctoral research associates, whose efforts are irreplaceable. In particular, this review has been greatly assisted by D. J. Casadonte, E. B. Flint, D. A. Hammerton, and L. J. Hogenson. The detailed suggestions of Professor J. L. Luche require special thanks. The generosity of funding from the National Science Foundation, the Research Corporation, and the donors of the Petroleum Research Fund, administered by the American Chemical Society, is gratefully acknowledged. The author is a Sloan Foundation Research Fellow and the recipient of a Research Career Development Award of the National Institutes of Health.

REFERENCES

1. J. Thornycroft and S. W. Barnaby, *Inst. C. E.* **122**, 51–102 (1895).
2. Lord Raleigh, *Philos. Mag., Ser. 6* **34**, 94–98 (1917).
3. W. T. Richards and A. L. Loomis, *J. Am. Chem. Soc.* **49**, 3086–3100 (1927).
4. R. W. Wood and A. L. Loomis, *Philos. Mag., Ser. 7* **4**, 417–436 (1927).
5. M. A. Margulis, *Zh. Fiz. Khim.* **50**, 1–18 (1976).
6. P. K. Chendke and H. S. Fogler, *Chem. Eng. J.* **8**, 165–178 (1974).
7. I. E. El'piner, "Ultrasound: Physical, Chemical, and Biological Effects," (trans. F. A. Sinclair). Consultants Bureau, New York, 1964.
8. R. E. Apfel, *in* "Methods of Experimental Physics: Ultrasonics" (P. D.Edmonds, ed.), Vol. 19, pp. 356–411. Academic Press, New York, 1981.
9. J. A. Rooney, *in* "Methods of Experimental Physics: Ultrasonics" (P. D. Edmonds, ed.), Vol. 19, pp. 299–353. Academic Press, New York, 1981.
10. W. L.Nyborg and D. L. Miller, *Appl. Sci. Res.* **38**, 17–24 (1982).
11. P. A. Lewin and L. Bjørnø, *Appl. Sci. Res.* **38**, 25–35 (1982).
12. F. S. Fry (ed.), "Ultrasound: Its Applications in Medicine and Biology." Elsevier, New York, 1978.
13. O. V.Rudenko and S. I. Soluyan, "Theoretical Foundations of Nonlinear Acoustics" (Trans. R. T. Beyer). Consultants Bureau, New York, 1977.
14. R. T. Beyer (ed.), "Non-Linear Acoustics in Fluids." Van Nostrand-Reinhold, Princeton, New Jersey, 1984.
15. L. A. Crum, *I.E.E.E. Ultrasonics Symp.*, 1–11 (1982).
16. M. Greenspan and C. E. Tschiegg, *J. Res. Natl. Bur. Stand., Sect. C* **71**, 299–312 (1967).
17. W. J. Galloway, *J. Acoust. Soc. Am.* **26**, 849–857 (1954).
18. E. N. Harvey, D. K. Barnes, W. D. McElroy, A. H. Whiteley, D. C. Pease, and K. W. Cooper, *J. Cell. Comp. Physiol.* **24**, 1–22 (1944).

19. P. S. Epstein and M. S. Plesset, *J. Chem. Phys.* **18**, 1505–1509 (1950).
20. L. A. Crum, *Nature (London)* **278**, 148–149 (1979).
21. R. H. S. Winterton, *J. Phys. D.: Appl. Phys.* **10**, 2041–2056 (1977).
22. R. E. Apfel, *J. Acoust. Soc. Am.* **48**, 1179–1186 (1970).
23. L. A. Crum and D. A. Nordling, *J. Acoust. Soc. Am* **52**, 294–301 (1972).
24. H. G. Flynn, *in* "Physical Acoustics" (W. P. Mason, ed.), Vol IB, pp. 57–172. Academic Press, New York, 1964.
25. L. A. Crum, *Appl. Sci. Res.* **38**, 101–115 (1982).
26. E. A. Neppiras, *Phys. Rep.* **61**, 159–251 (1980).
27. W. T. Coakley and W. L. Nyborg, *in* "Ultrasound: Its Applications in Medicine and Biology" (F. J. Fry, ed.), Part I, pp. 77 ff. Elsevier, New York, 1978.
28. M. S. Plesset and D. -Y. Hsieh, *Phys. Fluids* **3**, 882–895 (1960).
29. G. J. Lastman and R. A. Wentzell, *J. Acoust. Soc. Am.* **70**, 596–602 (1981), and **69**, 638–642 (1981).
30. M .A. Margulis and A. F. Dmitrieva, *Zh. Fiz. Khim.* **55**, 159–163 (1981).
31. M. A. Margulis and A. F. Dmitrieva, *Zh. Fiz. Khim.* **56**, 323–327 (1982).
32. S. Fujikawa and T. Akamatsu, *J. Fluid Mech.* **97**, 481–512 (1980).
33. W. Lauterborn, *J. Acoust. Soc. Am.* **59**, 283–293 (1976).
34. M. Minnaert, *Philos. Mag., Ser. 7* **16**, 235–248 (1933).
35. F. G. Blake, Jr., Tech Mem. No. 12, Acoustics Research Laboratory, Harvard University, Cambridge, Mass., 1949.
36. B. E. Noltingk and E. A. Neppiras, *Proc. Phys. Soc. London* **B63**, 674–685 (1950).
37. E. A. Neppiras and B. E. Noltingk, *Proc. Phys. Soc. London* **B64**, 1032–1038 (1951).
38. G. W. Willard, *J. Acoust. Soc. Am.* **25**, 669–686 (1953).
39. L. A. Crum and G. M. Hansen, *J. Acoust. Soc. Am.* **72**, 1586–1592 (1982).
40. M. H. Safar, *J. Acoust. Soc. Am.* **43**, 1188–1189 (1968).
41. W. Lauterborn, *Appl. Sci. Res.* **38**, 165–178 (1982) and *Finite-Amplitude Wave Effects in Fluids, Proc. 1973 Symp.* (L. Bjørnø, ed.), pp. 195–202, IPC Science and Technology Press, Guilford, England, 1974.
42. V. Griffing, *J. Chem. Phys.* **18**, 997–998 (1950).
43. V. Griffing, *J. Chem. Phys.* **20**, 939–942 (1952).
44. E. N. Harvey, *J. Am. Chem. Soc.* **61**, 2392–2398 (1939).
45. Y. I. Frenkel, *Zh. Fiz. Khim.* **12**, 305–308 (1940).
46. M. Degrois and P. Baldo, *Acustica* **21**, 222–228 (1969).
47. M. Degrois and P. Baldo, *Ultrasonics* **12**, 25–28 (1974).
48. C. M. Sehgal and R. E. Verrall, *Ultrasonics* **20**, 37 (1982).
49. M. A. Margulis, *Zh. Fiz. Khim.* **55**, 154–158 (1981).
50. A. M. Basedow and K. H. Ebert, *Adv. Polym. Sci.* **22**, 83–148 (1977).
51. E. C. Couppis and G. E. Klinzing, *AIChE J.* **20**, 485–491 (1974).
52. J. P. Lorimer and T. J. Mason, *J. Chem. Soc. Chem. Commun.* 1135–1136 (1980).
53. T. J. Mason, J. P. Lorimer, and B. P. Mistry, *Tetrahedron Lett.* **23**, 5363–5364 (1982).
54. T. J. Mason, J. P. Lorimer, and B. P. Mistry, *Tetrahedron Lett.* **24**, 4371–4372 (1983).
55. T. J. Mason, *Lab. Pract.* **33**, 13–20 (1984).
56. C. Sehgal, R. P. Steer, R. G. Sutherland, and R. E. Verrall, *J. Chem. Phys.* **70**, 2242–2248 (1979).
57. M. Margulis and A. F. Dmitrieva, *Zh. Fiz. Khim.* **56**, 875–877 (1982).
58. C. Sehgal, R. G. Sutherland, and R. E. Verrall, *J. Phys. Chem.* **84**, 396–401 (1980).
59. P. K. Chendke and H. S. Fogler, *J. Phys. Chem.* **87**, 1644–1648 (1983).
60. C. Sehgal, R. G. Sutherland, and R. E. Verrall, *J. Phys. Chem.* **84**, 388–395 (1980).
61. W. Tsang *in* "Shock Waves of Chemistry" (A. Lifshitz, ed.), pp. 59–130. Dekker, New York, 1981.

62. K. S. Suslick and D. A. Hammerton, *IEEE Son. Ultrason. Trans.*, in press (1986).
63. K. S. Suslick, R. E. Cline, Jr., and D. A. Hammerton, *IEEE Ultrason. Symp. Proc.* **4**, in press (1986).
64. T. B. Benjamin and A. T. Ellis, *Philos. Trans. R. Soc. London Ser A* **260**, 221–240 (1966).
65. M. P. Felix and A. T. Ellis, *Appl. Phys. Lett.* **19**, 484–486 (1971).
66. W. Lauterborn and H. Bolle, *J. Fluid Mech.* **72**, 391–399 (1975).
67. M. S. Plesset and R. B. Chapman, *J. Fluid Mech.* **47**, 283–290 (1971).
68. T. J. Bulat, *Ultrasonics* **12**, 59–68 (1974).
69. R. Pohlman, B. Werden, and R. Marziniak, *Ultrasonics* **10**, 156–161 (1972).
70. B. A. Agranat, V. I. Bashkirov, and Y. I. Kitaigorodskii, *in* "Physical Principles of Ultrasonic Technology" (L. Rozenberg, ed.), Vol. 1, pp. 247–330. Plenum, New York, 1973.
71. I. Hansson, K. A. Morch, and C. M. Preece, *Ultrason. Intl.* 267–274 (1977).
72. W. L. Nyborg, *in* "Physical Acoustics" (W. P. Mason, ed.), Vol. 2, Part II, pp. 265–331. Academic Press, New York, 1965.
73. H. V. Fairbanks, T. K. Hu, and J. W. Leonard, *Ultrasonics* **8**, 165 (1970).
74. A. Shoh, *in* "Kirk–Othmer Encyclopedia of Chemical Technology," 3rd ed., Vol. 23, pp. 462–479. Wiley, New York, 1983.
75. M. F. Cracknell, *Contemp. Phys.* **17**, 13–44 (1976).
76. E. J. Murry, *Chem. Technol.* 108–111, 232–234, 376–383 (1975).
77. G. I. Taylor, *Proc. R. Soc. London Ser. A.* **138**, 41–48 (1932).
78. M. K. Li and H. S. Fogler, *J. Fluid Mech.* **88**, 499–511 (1978).
79. A. Weissler and E. J. Hine, *J. Acoust. Soc. Am.* **34**, 130–131 (1962).
80. F. G. P. Aerstin, K. D. Timmerhaus, and H. S. Fogler, *AIChE J.* **13**, 453–456 (1967).
81. K. S. Suslick, P. F. Schubert, and J. W. Goodale, *IEEE Ultrason. Symp. Proc.*, 612–616 (1981).
82. Among others: Heat Systems–Ultrasonics, 1938 New Highway, Farmingdale, NY 11735; Branson Sonic Power, Eagle Rd., Danbury, CT 06810; Sonics & Materials, Kenosia Av., Danbury, CT 06810.
83. Among others: Heat Systems–Ultrasonics, 1983 New Highway, Farmingdale, NY 11735; Branson Cleaning Equipment, Parrott Dr., P.O. Box 768, Shelton, CT 06484; Lewis Corp., 324 Christian St., Oxford, CT 06483.
84. M. Margulis and L. M. Grundel, *Zh. Fiz. Khim.* **56**, 1445–1449, 1941–1945, 2592–2594, 2987–2990 (1982).
85. P. Renaud, *J. Chim. Phys.* **50**, 135 (1953); G. Saracco and F. Arzano, *Chim. Ind.* **50**, 314–318 (1968).
86. R. Busnel, D. Picard, and H. Bouzigues, *J. Chim. Phys* **50**, 97–101 (1953); R. Prudhomme, D. Picard, and R. Busnel, *J. Chim. Phys.* **50**, 107–108 (1953).
87. N. Sata and K. Nakasima, *Bull. Chem. Soc. Jpn.* **18**, 220–226 (1943).
88. V. A. Akulichev, M. G. Sirotyuk and L. D. Rozenberg, *in* "High Intensity Ultrasonic Fields" (L. D. Rozenberg, ed.), pp. 203–420. Plenum, New York, 1971.
89. K. S. Suslick, J. J. Gawienowski, P. F. Schubert, and H. Wang, *J. Phys. Chem* **87**, 2299–2301 (1983).
90. K. S. Suslick, J. J. Gawienowski, P. F. Schubert, and H. H. Wang, *Ultrasonics* **22**, 33–36 (1984).
91. M. Ibishi and B. Brown, *J. Acoust. Soc. Am.* **41**, 568–572 (1967).
92. A. Weissler, *J. Acoust, Soc. Am.* **25**, 651–657 (1953).
93. L. M. Bronskaya, V. S. Vigderman, A. V. Sokol'skaya, and I. E. El'Piner, *Sov. Phys. Acoust.* **13**, 374–375 (1968).

94. M. E. Fitzgerald, V. Griffing, and J. Sullivan, *J. Chem. Phys.* **25**, 926–933 (1956).
95. R. O. Prudhomme, *Bull. Soc. Chim. Biol.* **39**, 425–430 (1957).
96. E. L. Mead, R. G. Sutherland, and R. E. Verrall, *Can. J. Chem.* **54**, 1114–1120 (1975).
97. M. Ceschia and G. Iernetti, *Acustica* **29**, 127–137 (1973).
98. S. E. Bresler, *Zh. Fiz. Khim.* **14**, 309–311 (1940).
99. R. O. Prudhomme and P. Grabar, *J. Chim. Phys.* **46**, 323–331 (1949).
100. O. Lindstrom, *J. Acoust. Soc. Am.* **27**, 654–671 (1955).
101. A. Weissler, *J. Am. Chem. Soc.* **81** 1077–1081 (1959).
102. A. Henglein, *Naturwissenschaften* **43**, 277 (1956); **44**, 179 (1957).
103. M. Del Duca, E. Yeager, M. O. Davies, and F. Hovorka, *J. Acoust. Soc. Am.* **30**, 301–307 (1958).
104. H. Gueguen, *C. R. Hebd. Seanc. Acad. Sci., Paris* **253**, 260–2611, 647–649 (1961).
105. M. A. Margulis and A. N. Mal'tsev, *Zh. Fiz. Khim.* **42**, 1441–1451 (1968).
106. K. Makino, M. Mossoba, and P. Riesz, *J. Am. Chem. Soc.* **104**, 3537–3539 (1982); *J. Phys. Chem.* **87**, 1369–1377 (1983).
107. M. A. Margulis and A. N. Mal'tsev, *Zh. Fiz. Khim.* **42**, 2660–2663 (1968).
108. B. Lippitt, J. M. McCord, and I. Fridovich, *J. Biol. Chem.* **247**, 4688–4690 (1972).
109. A. V. Sokol'skaya, *J. Gen. Chem. USSR* **48**, 1289–1292 (1978).
110. A. V. Sokol'skaya and I. E. El'piner, *Akust. Zhur.* **3**, 293 (1957).
111. A. V. Sokol'skaya, *J. Evol. Biochem. Physiol.* **11**, 446 (1975).
112. H. Gueguen, *Ann. Chim.* **8**, 667–713 (1963).
113. M. A. Margulis, *Zh. Fiz. Khim.* **50**, 2271–2274 (1976).
114. D. Rehorek and E. G. Janzen, *Z. Chem.* **24**, 228–229 (1984).
115. M. A. Margulis, *Zh. Fiz. Khim.* **50**, 2267–2270 (1976).
116 C. Sehgal, R. G. Sutherland, and R. E. Verrall, *J. Phys. Chem.* **84**, 2920–2922 (1980).
117. A. N. Mal'tsev and M. A. Margulis, *Akust. Zh.* **14**, 295–297 (1968).
118. G. Cauwet, C. M. Coste, H. Knoche, and J. P. Longuemard, *Bull. Soc. Chim. Fr.* 45–48 (1976).
119. H. Beuthe, *Z. Phys. Chem.* **A163**, 161–171 (1933).
120. M. A. Margulis, *Zh. Fiz. Khim.* **48**, 2812–2818 (1974).
121. W. Wawrzyczek and D. Tylzanowska, *Nature (London)* **194**, 571–572 (1962).
122. M. Haissinsky and A. Mangeot, *Nuov. Cim.* **4**, 1086–1095 (1956).
123. M. Anbar and I. Pecht, *J. Phys. Chem.* **68**, 352–355 (1964); **71**, 1246–1249 (1967).
124. A. Kling and R. Kling *C. R. Hebd. Seanc. Acad. Sci., Paris* **223**, 1131–1133 (1946).
125. B. H. Jennings and S. N. Townsend, *J. Phys. Chem.* **65**, 1574–1579 (1961).
126. P. K. Chendke and H. S. Fogler, *J. Phys. Chem.* **87**, 1362–1369 (1983).
127. T. M. Tuszynski and W. F. Graydon, *I&EC Fundam.* **7**, 396–400 (1968).
128. Y. Sakai, Y. Sadaoka, and Y. Takamaru, *J. Phys. Chem.* **81**, 509–511 (1977).
129. B. Prasad and P. K. Sharma, *Chim. Anal.* **51**, 619–622 (1969).
130. L. Zechmeister and L. Wallcave, *J. Am. Chem. Soc.* **77**, 2853–2855 (1955).
131. L. Zechmeister and E. F. Magoon, *J. Am. Chem. Soc.* **78**, 2149–2150 (1956).
132. I. E. Elpiner, A. V. Sokolskaya, and M. A. Margulis, *Nature (London)* **208**, 945–946 (1965).
133. V. L. Starchesvskii, T. V. Vasilina, L. M. Grindel, M. A. Margulis, and E. N. Mokryi, *Zh. Fiz. Khim.* **58**, 1940–1944 (1984).
134. R. O. Prudhomme and P. Grabar, *Bull. Soc. Chim. Biol.* **29**, 122–130 (1958).
135. L. A. Spurlock and S. B. Reifsneider, *J. Am. Chem. Soc.* **92**, 6112–6117 (1970).
136. S. B. Reifsneider and L. A. Spurlock, *J. Am. Chem. Soc.* **95**, 299–305 (1973).
137. M. A. Margulis, *Zh. Fiz. Khim.* **50**, 2531–2535 (1976).

138. D. L. Currell and L. Zechmeister, *J. Am. Chem. Soc.* **80**, 205–208 (1958).
139. J. W. Chen, J. A. Chang, and G. V. Smith, *Chem. Eng. Progr. Symp. Ser.* **67**, 18–29 (1966).
140. A. Weissler, *J. Am. Chem. Soc.* **71**, 419–421 (1949).
141. D. L. Currell and S. S. Nagy, *J. Acoust. Soc. Am.* **44**, 1201–1203 (1968).
142. I. Miyagawa, *J. Soc. Org. Synth. Chem.* (*Jpn.*) **7**, 167–172 (1949).
143. J. W. Chen, W. M. Kalback, *I. & E. C. Fundam.* **6**, 175–178 (1967).
144. D. S. Kristol H. Klotz, and R. C. Parker, *Tetrahedron Lett.* **22**, 907–908 (1981).
145. T. E. Needham and R. J. Gerraughty; *J. Pharm. Sci.* **58**, 62–64 (1969).
146. S. Moon, L. Duchin and J. V. Cooney, *Tetrahedron Lett.* **41**, 3917–3920 (1979).
147. R. G. Fayter, Jr. and L. A. Spurlock, *J. Acoust. Soc. Am.* **56**, 1461–1468 (1974).
148. E. L. Mead, R. G. Sutherland, and R. E. Verrall, *Can. J. Chem.* **53**, 2394–2399 (1975); *J. Chem. Soc. Chem. Commun.* 414–415 (1973).
149. C. M. Sehgal and S. Y. Wang, *J. Am. Chem. Soc.* **103**, 6606–6611 (1981).
150. T. J. Yu, R. G. Sutherland, and R. E. Verrall, *Can. J. Chem.* **58**, 1909–1919 (1980).
151. J. R. McKee, C. L. Christman, W. D. O'Brien, Jr., and S. Y. Wang, *Biochemistry* 4651–4654 (1977).
152. W. H. Staas and L. A. Spurlock, *J. Chem. Soc. Perkin Trans. I* **1**, 1675–1679 (1975).
153. A. A. Berlin and B. S. El'tsefon, *Khim. Nauka. Prom.* **2**, 667–668 (1957).
154. P. Alexander and M. Fox, *J. Polymer Sci.* **12**, 533–541 (1954).
155. A. Henglein and R. Schulz, *Z. Naturforsch.* **7B**, 484–485 (1952).
156. A. Weissler, I. Pecht, and M. Anbar; *Science* **150**, 1288–1289 (1965).
157. V. E. Gordeev, A. I. Serbinov, and Y. K. Troshin, *Dokl. Akad. Nauk SSSR* **172**, 383–385 (1967).
158. C. W. Porter and L. Young, *J. Am. Chem. Soc.* **60**, 1497–1500 (1938)
159. N. Berkowitz and S. C. Srivastava, *Can. J. Chem.* **41**, 1787–1793 (1963).
160. S. Prakash and J. D. Pandey, *Tetrahedron* **21**, 903–908 (1965).
161. I. Rosenthal, M. M. Mossoba, and P. Riesz, *J. Magn. Res.* **45**, 359–361 (1981).
162. C. Sehgal, T. J. Yu, R. G. Sutherland, and R. E. Verrall, *J. Phys. Chem.* **86**, 2982–2986 (1982).
163. F. Soehnlen, D. Perrin, D. Masson, and M. L. Gaulard, *Ultrason. Int.* 603–605 (1979).
164. T. G. Kachakhidze, *Nauchn. Tr. Gruz. Politeth. Inst. Lenina* 45–48 (1980).
165. a. G. K. Diedrich, P. Kruus, and L. M. Rachlis, *Can. J. Chem.* **50**, 1743–1750 (1972); D. J. Donaldson, M. D. Farrington, and P. Kruus, *J. Phys. Chem.* **83**, 3130–3135 (1979).
166. S. L. Regen and A. Singh, *J. Org. Chem.* **47**, 1587–1588 (1982).
167. R. S. Davidson, A. M. Patel, A. Safdar, and D. Thornthwaite, *Tetrahedron Lett.* **24**, 5907–5910 (1983).
168. K. Sjoberg, *Tetrahedron Lett.* 6383–6384 (1966).
169. S. Raucher and P. Klein, *J. Org. Chem.* **46**, 3558–3559 (1981).
170. B. H. Han and P. Boudjouk, *Tetrahedron Lett.* **23**, 1643–1646 (1982).
171. J. Yamawaki, S. Sumi, T. Ando, and T. Hanafusa, *Chem. Lett.* 379–380 (1983).
172. T. Ando, T. Kawate, J. Yamawaki, and T. Hanafusa, *Synthesis* 637–638 (1983).
173. T. Ando, S. Sumi, T. Kawate, J. Ichihara, and T. Hanafusa, *J. C. S. Chem. Commun.* 439–440 (1984).
174. K. S. Suslick, P. F. Schubert, and J. W. Goodale, *J. Am. Chem. Soc.* **103**, 7342–7344 (1981).
175. H. E. Carlton and J. H. Oxley, *AIChE J.* **11**, 79 (1965).
176. G. L. Geoffroy and M. S. Wrighton, "Organometallic Photochemistry." Academic Press, New York, 1979.

177. Y. Langsam and A. M. Ronn, *Chem. Phys.* **54**, 277–290 (1981).

178. K. E. Lewis, D. M. Golden, and G. P. Smith, *J. Am. Chem. Soc.* **106**, 3905–3912 (1984).

179. M. Poliakoff and J. J. Turner, *J. Chem. Soc. Faraday Trans.* 2 **70**, 93–99 (1974).

180. M. Poliakoff, *J. Chem. Soc. Dalton Trans.* 210–212 (1974).

181. G. Nathanson, B. Gitlin, A. M. Rosan, and J. T. Yardley, *J. Chem. Phys.* **74**, 361–369, 370–378 (1981).

182. Z. Karny, R. Naaman, and R. N. Zare, *Chem. Phys. Lett.* **59**, 33–37 (1978).

183. K. S. Suslick, J. W. Goodale, P. F. Schubert, and H. H. Wang, *J. Am. Chem. Soc.* **105**, 5781–5785 (1983).

184. I. Fischler, K. Hildenbrand, and E. Koerner von Gustorf, *Angew. Chem. Int. Ed. Engl.* **14**, 54 (1975).

185. A. M. Horton, D. M. Hollishead, and S. V. Ley, *Tetrahedron* **40**, 1737–1742 (1984).

186. K. S. Suslick and P. F. Schubert, *J. Am. Chem. Soc.* **105**, 6042–6044 (1983).

187. H. B. Abrahamson and M. S. Wrighton, *J. Am. Chem. Soc.* **99**, 5510–5512 (1977).

188. N. Coville, A. M. Stolzenberg, and E. L. Muetterties, *J. Am. Chem. Soc.* **105**, 2499–2500 (1983).

189. K. S. Suslick and H. H. Wang, unpublished results.

190. R. S. Dickson and P. J. Fraser, *Adv. Organomet. Chem.* **12**, 323–377 (1974).

191. K. S. Suslick and R. E. Johnson, unpublished results.

192. E. O. Fischer and R. Jira, *Z. Naturforsch.* **10b**, 355 (1955).

193. D. Rehorek and E. G. Janzen, *J. Organomet. Chem.* **268**, L35–39 (1984).

194. J. C. Mitchener and M. S. Wrighton, *J. Am. Chem. Soc.* **103**, 975–977 (1981).

195. G. W. Parshall, "Homogeneous Catalysis." Wiley (Interscience), New York, 1980.

196. C. Masters, "Homogeneous Transition-Metal Catalysis." Chapman & Hall, New York, 1981.

197. C. P. Casey and C. R. Cyr, *J. Am. Chem. Soc.* **95**, 2248–2253 (1973).

198. J. L. Graff, R. D. Sanner, and M. S. Wrighton, *Organometallics* **1**, 837–842 (1982).

199. D. B. Chase and F. J. Weigert, *J. Am. Chem Soc.* **103**, 977–978 (1981).

200. G. L. Swartz and R. J. Clark, *Inorg. Chem.* **19**, 3191–3195 (1980).

201. M. Wrighton, G. S. Hammond, and H. B. Gray, *J. Organomet. Chem.* **70**, 283–301 (1974).

202. F. Asinger, B. Fell, and K. Schrage, *Chem. Ber.* **98**, 372–386 (1965).

203. C. Bondy and K. Söllner, *Trans. Faraday Soc.* **31**, 835–846 (1935).

204. E. C. Marboe and W. A. Weyl, *J. Appl. Phys.* **21**, 937–938 (1950).

205. A. J. Fry and D. Herr, *Tetrahedron Lett.* **40**, 1721–1724 (1978).

206. A. J. Fry, G. S. Ginsburg, and R. A. Parente, *J. Chem. Soc. Chem. Commun.* 1040–1041 (1978).

207. A. J. Fry and J. P. Bujanauskas, *J. Org. Chem.* **43**, 3157–3163 (1978).

208. A. J. Fry and G. S. Ginsburg, *J. Am. Chem. Soc.* **101**, 3927–3932 (1979).

209. A. J. Fry and A. T. Lefor, *J. Org. Chem.* **44**, 1270–1273 (1979).

210. A. J. Fry, W. A. Donaldson, and G. S. Ginsburg, *J. Org. Chem.* **44**, 349–352 (1979).

211. A. J. Fry and S. S. Hong, *J. Org. Chem.* **46**, 1962–1964 (1981).

212. A. J. Fry, K. Ankner, and V. Hana, *Tetrahedron Lett.* **22**, 1791–1794 (1981).

213. P. Renaud, *Bull. Soc. Chim. Fr., Ser.* 5 **17**, 1044–1045 (1950).

214. J. L. Luche and J. C. Damiano, *J. Am. Chem. Soc.* **102**, 7926–7927 (1980).

215. B. M. Trost and B. P. Coppola, *J. Am. Chem. Soc.* **104**, 6879–6881 (1982).

216. C. Petrier, A. L. Gemal, and J. L. Luche, *Tetrahedron Lett.* **23**, 3361–3364 (1982).

217. J. L. Luche, C. Petrier, A. L. Gemal, and N. Zikra, *J. Org. Chem.* **47**, 3805–3806 (1982).

218. P. Boudjouk and B. H. Han, *Tetrahedron Lett.* **22**, 3813–3814 (1981).
219. B. H. Han and P. Boudjouk, *Tetrahedron Lett.* **22**, 2757–2758 (1981).
220. E. Lukevics, V. N. Gevorgyan, and Y. S. Goldberg, *Tetrahedron Lett.* **25**, 1415–1416 (1984).
221. P. Boudjouk, B. H. Han, and K. R. Anderson, *J. Am. Chem. Soc.* **104**, 4992–4993 (1982).
222. R. West, M. J. Fink, and J. Michl, *Science* **214**, 1343–1344 (1981).
223. P. Boudjouk, *J. Chem. Educ.,* in press (1985).
224. S. Masamune, S. Murakami, and H. Lobita, *Organometrics* **2**, 1464–1466 (1983).
225. W. Slough and A. R. Ubbelohde, *J. Chem. Soc.* 918–919 (1957).
226. M. W. T. Pratt and R. Helsby, *Nature (London)* **184**, 1694–1695 (1959).
227. T. Azuma, S. Yanagida, H. Sakurai, S. Sasa, and K. Yoshino, *Synth. Commun.* **12**, 137–140 (1982).
228. J. L. Luche, C. Petrier, and C. Dupuy, *Tetrahedron Lett.* **25**, 753–756 (1984).
229. T. Kitazume and N. Ishikawa, *Chem. Lett.* 1679–1680 (1981).
230. B. H. Han and P. Boudjouk, *J. Org. Chem.* **47**, 5030–5032 (1982).
231. R. D. Rieke and S. J. Uhm, *Synthesis* 452–453 (1975).
232. T. Kitazume and N. Ishikawa, *Chem. Lett.* 137–140 (1982).
233. T. Kitazume and N. Ishikawa, *Chem. Lett.* 1453–1454 (1984).
234. J. L. Luche, C. Petrier, J. P. Lansard, and E. A. Greene, *J. Org. Chem.* **48**, 3837–3839 (1983).
235. C. Petrier, J. L. Luche, and C. Dupuy, *Tetrahedron Lett.* **25**, 3463–3466 (1984).
236. A. E. Greene, J. P. Lansard, J. L. Luche, and C. Petrier, *J. Org. Chem.* **49**, 931–932 (1984).
237. O. Repic and S. Vogt, *Tetrahedron Lett.* **23**, 2729–2732 (1982).
238. B. H. Han and P. Boudjouk, *J. Org. Chem.* **47**, 751–752 (1982).
239. S. Chew and R. J. Ferrier, *J. Chem. Soc. Chem. Commun.* 911–912 (1984).
240. K. Itoh, H. Nagashima, T. Ohshima, N. Oshima, and H. Nishiyama, *J. Organomet. Chem.* **272**, 179–188 (1984).
241. C. L. Hill, J. B. Van der Sande, and G. M. Whitesides, *J. Org. Chem.* **45**, 1020–1028 (1980).
242. J. D. Sprich and G. S. Lewandos, *Inorg. Chim. Acta* **76**, L241–242 (1983).
243. K. S. Suslick and R. E. Johnson, *J. Am. Chem. Soc.* **106**, 6856–6858 (1984).
244. R. D. Rieke, K. Ofele, and E. O. Fischer, *J. Organomet. Chem.* **76**, C19–21 (1974); R. C. Rieke, *Acc. Chem. Res.* **10**, 301–306 (1977).
245. G. L. Rochfort and R. D. Rieke, *Inorg. Chem.* **23**, 787–789 (1984).
246. G. Wilkinson, F. G. A. Stone, and E. W. Abel (eds.), "Comprehensive Organometallic Chemistry: The Synthesis, Reactions, and Structures of Organometallic Compounds," Vol. 3–6, p. 8. Pergamon, Oxford, 1982.
247. H. Bönnemann, B. Bogdanović, R. Brinkman, D. W. He, and B. Spliethoff, *Angew. Chem. Intl. Ed. Engl.* **22**, 728 (1983).
248. A. N. Mal'tsev, *Z. Fiz. Khim.* **50**, 1641–1652 (1976).
249. W. Lintner and D. Hanesian, *Ultrasonics* **15**, 21–26 (1977).
250. O. V. Abramov and I. I. Teumin, *in* "Physical Principles of Ultrasonic Technology" (L. D. Rosenberg, ed.), Vol. 2, pp. 145–273. Plenum, New York, 1973.
251. F. Radenkov, Kh. Khristov, R. Kircheva, and L. Petrov, *Khim. Ind.* **49**, 11–13 (1977).
252. A. S. Kuzharov, L. A. Vlasenko, and V. V. Suchkov, *Zh. Fiz. Khim.* **58**, 894–896 (1984).
253. R. T. Knapp, J. W. Dailey, and F. G. Hammitt, "Cavitation." McGraw-Hill, New York, 1970.

254. R. Walker and C. T. Walker, *Nature (London) Phys. Sci.* **244.** 141–142 (1973); *Nature (London)* **250,** 410–411 (1974).
255. R. C. Alkire and S. Perusich, *Corros. Sci.* **23,** 1121–1132 (1983).
256. L. Wen-Chou, A. N. Mal'tsev, and N. I. Kobozev, *Z. Fiz. Khim.* **38,** 80 (1964).
257. A. N. Mal'tsev and I. V. Solov'eva, *Zh. Fiz. Khim.* **44,** 1092–1095 (1970).
258. J. W. Chen. J. A. Chang, and G. V. Smith, *Chem. Eng. Prog. Symp. Ser.* **67,** 18–26 (1971).
259. M. Nakanishi, *Jpn. Patent* 81, 127, 684, Oct. 6, 1981.
260. P. Boudjouk and B. H. Han, *J. Catal.* **79,** 489–492 (1983).
261. B. H. Han and P. Boudjouk, *Organometrics* **2,** 769–771 (1983).

ADVANCES IN ORGANOMETALLIC CHEMISTRY, VOL. 25

Carbene and Carbyne Complexes of Ruthenium, Osmium, and Iridium

MARK A. GALLOP and WARREN R. ROPER

Department of Chemistry
University of Auckland
Private Bag, Auckland, New Zealand

I

INTRODUCTION

The importance of transition metal carbene complexes (compounds with formal $M{=}C$ bonds) and of transition metal carbyne complexes (compounds with formal $M{\equiv}C$ bonds) is now well appreciated. Carbene complexes are involved in olefin metathesis (*1*) and have many applications in organic synthesis (*2*), while carbyne complexes have similar relevance to

Copyright © 1986 by Academic Press, Inc.
All rights of reproduction in any form reserved.

acetylene metathesis (*3*) and are important building-blocks in metal cluster synthesis (*4*). Numerous reviews have dealt with various aspects of this chemistry (*5*), and the justification for this article dealing with carbene and carbyne complexes of just three Group 8 elements, Ru, Os, and Ir, is that these elements have provided the first substantial body of examples of dihalocarbene complexes. These ligands are especially interesting for the synthetic possibilities provided by the presence of two good leaving groups on the carbene carbon atom. This can lead, for example, to carbyne complexes through reaction with lithium reagents. Moreover, some of the difluorocarbene complexes have been isolated in two different oxidation states of the metal thus offering an interesting comparison of carbene reactivity as a function of oxidation state. The same comparison is also possible with several of the carbyne complexes. In addition, complexes of the rare terminal methylene ligand ($=CH_2$) have been isolated for each of the elements, Ru, Os, and Ir.

Finally, the possibility of building the M=C bond into an unsaturated metallacycle where there is the possibility for electron delocalization has been realized for the first time with the characterization of "osmabenzene" derivatives. For these reasons then, it seemed worthwhile to review the carbene and carbyne chemistry of these Group 8 elements, and for completeness we have included discussion of other heteroatom-substituted carbene complexes as well. We begin by general consideration of the bonding in molecules with multiple metal–carbon bonds.

II

BONDING MODELS AND REACTIVITY PATTERNS FOR TRANSITION METAL CARBENE AND CARBYNE COMPLEXES

The wealth of empirical information collected for transition metal carbene and carbyne complexes may be best interpreted within the framework of sound theoretical models for these compounds. Perhaps the most significant contribution made by the theoretical studies of carbene and carbyne complexes concerns an understanding of the reactivity patterns they display. In this section the relationship between bonding and reactivity is examined, with particular emphasis being given to the ways in which studies of Ru, Os, and Ir compounds have helped unify the bonding models applied to seemingly diverse types of carbene and carbyne complexes.

A. Carbene Complexes

1. Bonding Models

Theoretical studies of metal carbene complexes have been undertaken by several groups. Rather than attempting to analyze the complexes as a whole, it is more convenient to consider the carbene and metal fragments separately and then to analyze the effects of combining them.

a. The Carbene Fragment. The frontier orbitals of the singlet state of the simplest carbene, methylene, are

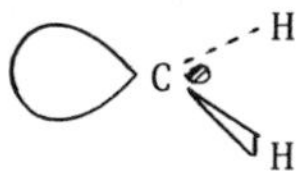

The highest occupied molecular orbital (HOMO) is a doubly occupied sp^2 hybrid, while the lowest unoccupied molecular orbital (LUMO) is a carbon p_π orbital.

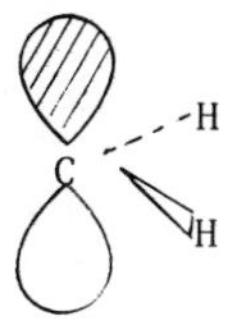

The effects of varying the carbene substituent should be considered. Introduction of an electronegative group like OR or NR_2 will (i) inductively remove electron density via the σ-bonding orbitals, and (ii), because of the filled p orbitals on the heteroatom, π-donation to the LUMO of the carbene is possible, the effect will be to raise the LUMO in energy and increase the electron density on carbon. These effects will operate synergistically, and the balance will depend on the nature of the substituents. Mulliken population analyses for several free carbenes have indicated that the carbene carbon is typically negatively charged (6).

b. Coordination to a Metal. Orbitals of the metal fragment having appropriate symmetry combine with orbitals of the carbene moiety forming σ and π components of the metal–carbon multiple bond. In complexes where the metal–carbon π-interaction is not particularly strong, the metal–carbon π^* orbital is low-lying and is largely localized on the carbon atom (C_α) (7–11). This orbital is the complex LUMO. The HOMO is an

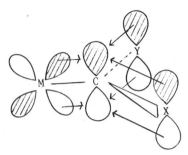

FIG. 1.   Competitive π-donation to vacant C $2p_\pi$ orbital.

orbital, largely metal-based, that is not associated with the metal–carbon bond.

When the π-interaction between metal and carbene fragments is increased, the metal–carbon $2p_\pi$ orbital is pushed up higher in energy and is no longer significantly localized on C_α. In complexes of this type, the LUMO is an orbital not associated with the metal–carbon bond (10).

Substituent Effects. Substitution of the groups attached to C_α with π-donors such as nitrogen decreases the π-backbonding from the metal. The filled nonbonding p orbital on the nitrogen donates electron density to the vacant p orbital on C_α, raising the energy of the LUMO of the carbene fragment and thus decreasing the metal–carbon interaction. An alternative view is that the π-donor substituent and the metal compete to donate electron density into the unfilled carbon p orbital, and, in cases with good π-donors like nitrogen, the metal is less effective (see Fig. 1).

Charge Distributions. Mulliken population analyses have been used to calculate approximate charge distributions in a number of carbene complexes (see, for example, Refs. 10 and 12) and several conclusions can be drawn:

1. The charge on C_α is usually negative, and will be most negative in complexes where the metal–carbon π-interaction is greatest.
2. The metal–carbon bond is typically polarized $M^{\delta+}$—$C^{\delta-}$.
3. C_α is more negatively charged than any carbonyl carbons present in the complex.
4. Changing the complex from neutral to cationic makes C_α slightly positive.

Carbene Complex Geometries. Molecular orbital studies of the various conformations of several transition metal–carbene complexes have been undertaken by the groups of Fenske and Hoffmann (8,13). Of the two

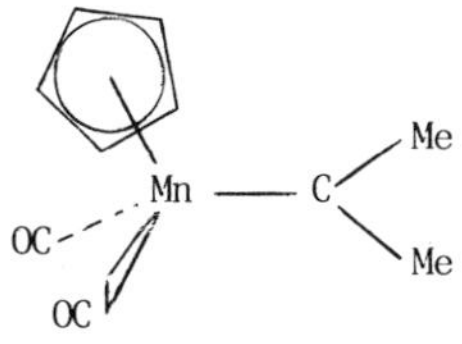

(1a) 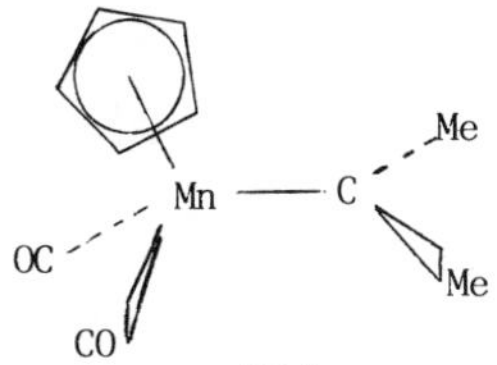   (1b)

limiting positions of the carbene ligand in $(\eta^5\text{-}C_5H_5)(CO)_2Mn{=}CMe_2$ **(1)**, the vertical orientation **(a)** is calculated to be more stable than the horizontal one **(b)** by 9 kcal mol^{-1} (8). While Hoffman has explained this conformational preference in terms of maximum metal–ligand overlap, Fenske has applied the criterion of minimum orbital energy to rationalize the observed structure.

2. Reactivity Patterns

Development of the chemistry of metal carbenes commenced with the characterization of complexes such as $(CO)_5W{=}C(OMe)Ph$ by Fischer and his co-workers (e.g., Ref. 14 and references therein). The series of hydrocarbon-substituted carbene compounds discovered later by the Schrock group (e.g., Ref. 15 and references therein) appeared to be so different from the other known carbene complexes that they were placed in a different class altogether. Overemphasis of this distinction has persisted in the literature of carbene complexes, even to the present day.

The division arose from the apparent paradoxical reactivity of these compounds. The carbene ligands in the complexes of the Group 6a and 7a metals reacted with nucleophilic reagents while those in the complexes of the early transition metals, tantalum and niobium, reacted with electrophilic reagents. The discovery of carbene complexes of divalent ruthenium and osmium in which the carbene ligand is electrophilic, and of zerovalent complexes in which the same carbene ligand bound to the same metal has much reduced electrophilicity (or even nucleophilicity), has served to link the two disparate categories given above. It is possible to explain the chemical differences in all carbene complexes in terms of the one bonding model as follows.

The importance of both frontier orbital-controlled and electronic charge-controlled factors in determining chemical reactivity has been recognized (16). These concepts are the key to interpreting two types of reactivity expected for carbene complexes, i.e., reactions with nucleophilic and electrophilic reagents:

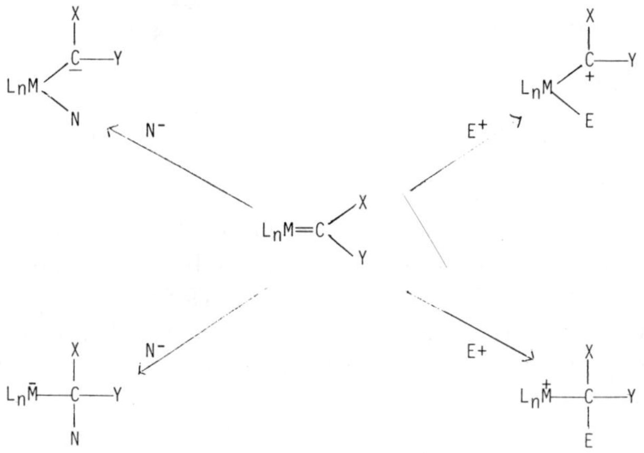

a. Reactions with Nucleophiles.   Nucleophilic addition to both the metal and C_α could be anticipated. Although the metal–carbon bond polarization is $M^{\delta+}$—$C^{\delta-}$, direct addition to the metal in an 18-electron complex is never observed (ligand substitution reactions via coordinatively unsaturated intermediates, however, are well known).

Self-consistent field molecular orbital calculations by Fenske and co-workers have confirmed that nucleophilic additions to Fischer and related complexes [e.g., $(CO)_5Cr=CXY$, $(\eta^5\text{-}C_5H_5)(CO)_2Mn=CXY$], are frontier orbital-controlled rather than charge-controlled reactions (7–9). Interaction of the HOMO of the nucleophile with the carbene complex LUMO (localized on C_α) destroys the metal–carbon π-interaction and converts the bond to a single one.

Nucleophilic attack at the carbene carbon should be most likely when the metal–carbon π^* orbital (LUMO) is low-lying and most localized on C_α, and when the negative charge on this atom is minimized. These conditions are satisfied when (i) the metal center is electron-deficient, i.e., cationic or containing electron-withdrawing ligands (e.g., CO), or (ii) the carbene substituents are electron-withdrawing without being good π-donors (e.g., halogens). Thus nucleophilic attack at C_α is a characteristic reaction of the many Group 6a and 7a carbene complexes which contain electron-withdrawing carbonyl ligands or are cationic, e.g.,

$$(CO)_5Cr=C(OMe)Ph \xrightarrow{\text{RNH}_2} (CO)_5Cr=C(NHR)Ph \qquad \text{(Ref. 17)}$$

$$[(\eta^5\text{-}C_5H_5)(NO)(PPh_3)Re=CH_2]^+ \xrightarrow{\text{MeLi}} (\eta^5\text{-}C_5H_5)(NO)(PPh_3)Re\text{—}CH_2Me$$
$$\text{(Ref. 18)}$$

The susceptibility of d^6 Ru, Os, and Ir halocarbene complexes to nucleophilic substitution reactions (see Section V) can also be understood in terms of this model, e.g.,

$$Cl_2(CO)(PPh_3)_2Ru=CCl_2 \xrightarrow{\text{HSCH}_2\text{CH}_2\text{SH}} Cl_2(CO)(PPh_3)_2Ru=\overline{CSCH_2CH_2S}$$

(Ref. 19)

b. Reactions with Electrophiles. As the metal–carbon π-bonding interaction in a carbene complex increases, the π^* orbital is both destabilized and delocalized. Electrophilic attack, rather than frontier orbital-controlled nucleophilic attack, then, is expected in compounds where the metal–carbon π-interaction is particularly significant.

Experimental evidence shows that both the metal and C_α can be the sites of electrophilic attack. Electrophiles would be expected to add to C_α when this atom is most negatively charged and when the π-bonding orbital is most heavily concentrated there. These criteria are met in complexes where the metal center is electron-rich and where the carbene substituents are not good π-donors, e.g.,

$$(\eta^5\text{-}C_5H_5)_2MeTa=CH_2 \xrightarrow{\text{AlMe}_3} (\eta^5\text{-}C_5H_5)_2MeTa=CH_2 \rightarrow AlMe_3 \quad \text{(Ref. 15)}$$

(2)

$$Cl(NO)(PPh_3)_2Os=CH_2 \xrightarrow{\text{HCl}} Cl_2(NO)(PPh_3)_2Os—CH_3 \qquad \text{(Ref. 20)}$$

(3)

$$(CO)_2(PPh_3)_2Ru=CF_2 \xrightarrow{\text{HCl}} Cl(CO)_2(PPh_3)_2Ru—CF_2H \qquad \text{(Ref. 21)}$$

(4)

The tantalum center in the Schrock methylene complex **2** is electron rich by virtue of the electron-releasing ligands coordinated to it, while the osmium and ruthenium center in the methylene **3** and difluorocarbene **4** species are electron-rich because they have d^8 electron configurations in neutral complexes.

c. Effect of Metal Basicity. A very important conclusion arising from this discussion is that the mode of reactivity of the metal–carbon bond in carbene complexes toward nucleophilic and electrophilic reagents is a sensitive function of the metal electron density. The effect on carbene reactivity of changes in metal oxidation state is not readily observed because few complexes are known with the same carbene ligand bound to a given metal in two different oxidation states. However, two examples which clearly illustrate the effect of metal basicity on reactivity are provided by comparison of the d^6 complexes $RuCl_2(=CF_2)(CO)(PPh_3)_2$ and $[OsI(=CH_2)(CO)_2(PPh_3)_2)]^+$ with their d^8 counterparts $Ru(=CF_2)(CO)_2(PPh_3)_2$ and $Os(=CH_2)Cl(NO)(PPh_3)_2$.

The difluorocarbene complex **5** reacts rapidly with nucleophiles but is completely unreactive toward electrophilic reagents *(22)*, e.g.,

$$Cl_2(CO)(PPh_3)_2Ru=CF_2 \xrightarrow{\text{MeOH}} Cl_2(CO)(PPh_3)_2Ru=CF(OMe)$$

$$(5)$$

The more electron-rich complex $Ru(=CF_2)(CO)_2(PPh_3)_2$ **4** reacts only slowly with some nucleophiles but rapidly with electrophiles *(21)* (e.g., see reaction with HCl above).

The intermediacy of cationic d^6 osmium methylene complex **6** is implicated in the reactions of $OsI(CH_2I)(CO)_2(PPh_3)_2$ with nucleophiles *(23)*, e.g.,

$$I(CO)_2(PPh_3)_2Os—CH_2I \rightarrow [I(CO)_2(PPh_3)_2Os=CH_2]^+I^{-*} \xrightarrow{\text{PPh}_3}$$

$$(6)$$

$$[I(CO)_2(PPh_3)_2Os—CH_2—PPh_3]I$$

$$* = \text{postulated intermediate}$$

This reaction closely resembles the formation of ylide complex **8** from the stable rhenium methylene **7** *(24)*:

$$[(\eta^5\text{-}C_5H_5)(NO)(PPh_3)Re=CH_2]PF_6 \xrightarrow{\text{PPh}_3} [(\eta^5\text{-}C_5H_5)(NO)(PPh_3)Re—CH_2—PPh_3]PF_6$$

$$(7) \qquad\qquad\qquad\qquad (8)$$

The nucleophilic reactivity of the d^8 methylene complex $Os(=CH_2)Cl(NO)(PPh_3)_2$ **(3)** with HCl was noted above.

The conclusion that there should not be a sharp distinction between the electrophilic and nucleophilic character of the metal–carbon double bond, but rather a spectrum of reactivity, also follows from the bonding model. There is some experimental evidence for this:

1. While the tantalum methylene complex **2** reacts with MeI *(15)*, the osmium complex **3** is quite unreactive, suggesting greater nucleophilicity of the former compound.
2. The rhenium complex **7** has been shown to react with both electrophilic and nucleophilic reagents *(24,25)*:

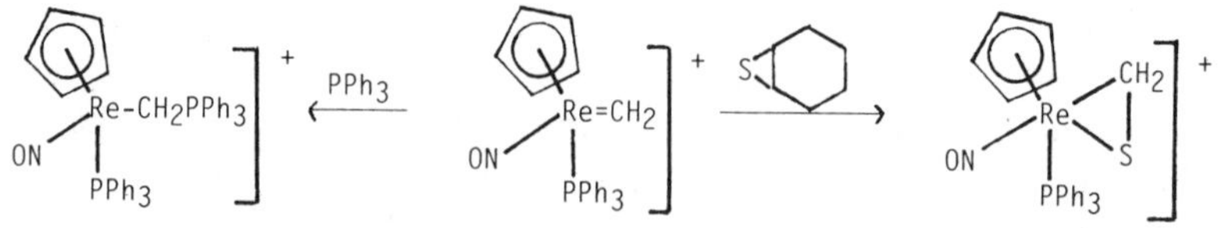

$$(7)$$

3. The cationic methylene complexes $[(\eta^5\text{-}C_5H_5)(CO)_3M{=}CH_2]^+$ (M = Mo, W) react rapidly with nucleophiles (e.g., olefins) but are unreactive toward electrophiles (26).

Thus the reactivity of transition metal–carbene complexes, that is, whether they behave as electrophiles or nucleophiles, is well explained on the basis of the frontier orbital theory. Studies of carbene complexes of ruthenium and osmium, by providing examples with the metal in either of two oxidation states [Ru(II), Os(II); Ru(0), Os(0)], help clarify this picture, and further illustrations of this will be found in the following sections.

It should be noted that an alternative viewpoint which has been suggested to account for the existence of electrophilic and nucleophilic metal carbenes is that each arises from different bonding schemes (27). Electrophilic, 18-electron metal carbenes can be considered as bonding between singlet metal and singlet carbene fragments whereas nucleophilic, often electron-deficient metal carbenes have bonding between triplet metal and triplet carbene fragments. In this view singlet-bonding metal carbenes are characterized by strong π-accepting ligands on the metal and by heteroatomic or phenyl substituents on the carbene. On the other hand, triplet-bonding metal carbenes are characterized by the lack of strong π-accepting ligands on the metal and by hydrogen or alkyl skubstituents on the carbene.

B. Carbyne Complexes

The chemistry of transition metal–carbyne complexes is rather less developed than the chemistry of carbene complexes. This is almost certainly because reactions which form new carbyne complexes are relatively rare when compared with those forming metal carbenes. The few theoretical studies of carbyne complexes which are available indicate that close parallels exist between the bonding in carbene and carbyne compounds. These parallels also extend to chemical reactivity, and studies of Group 8 complexes again prove instructive.

1. Bonding Models

a. The Carbyne Fragment. The carbyne ligand is a regarded as a 3-electron donor, and in this sense it is comparable to the linear nitrosyl ligand. The value of this analogy will become apparent in Section VI. Orbitals of the carbyne fragment have usually been determined by treating the fragment as CR^+. On this basis the orbitals available for bonding to the metal are

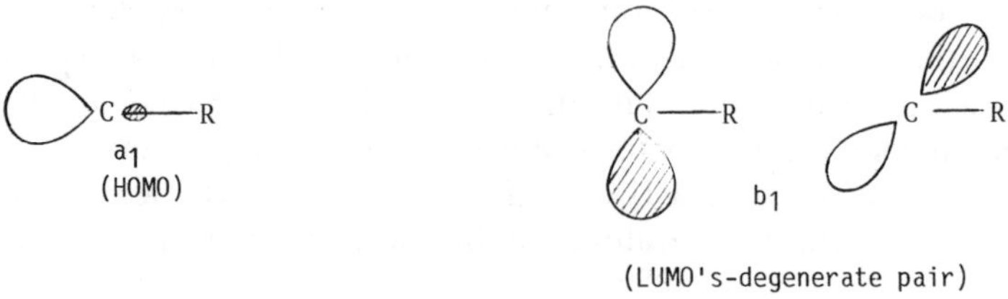

a_1
(HOMO)

b_1

(LUMO's-degenerate pair)

The HOMO is a doubly occupied sp hybrid of σ-symmetry while a degenerate pair of (orthogonal) orbitals of π-symmetry are the LUMO's of the carbyne fragment.

Heteroatomic Substituents. The effect of attaching a heteroatomic substituent to the carbyne carbon, for example, NMe_2, is to remove the degeneracy of the π-type LUMO's (28). There is extensive mixing (π-donation) from the lone pair on nitrogen into one of the p_π orbitals on carbon. The π^*-orbital for the $N \rightarrow C$ π-interaction is moved to higher energy, and the other p_π orbital becomes the sole LUMO.

The effect of an aryl ring as substituent is similar to that of an amino group, since it too has the potential to π-donate, but the net effect is much smaller.

b. Complex Formation. Combination of the frontier orbitals of the carbyne fragment with appropriate metal fragment orbitals leads to the formation of one σ and two π bonds in the carbyne complex. Calculations by Fenske (28,29) on Fischer-type carbynes indicate that the energy-level patterns of the orbitals are very similar to those of the related carbene complexes. The HOMO is metal fragment based and is not associated with the triple bond, while the LUMO is one of the π^* orbitals of the metal–carbon bond.

In the cationic complex $[(\eta^5\text{-}C_5H_5)(CO)_2Mn\equiv CMe]^+$ the lack of symmetry in the metal fragment means that the two π-bonds are not degenerate. The percentage electron density of the two π-orbitals on the carbyne carbon have been calculated as 33 and 35%—relatively large figures which clearly support the triple bond formulation (28).

Substituent Effects. When the carbyne substituent is changed to a good π-donor such as NMe_2, the nonequivalence of the π-orbital becomes more marked. This is indicated by a greater difference between the overlap populations for the two π-orbitals (29). In terms of resonance formalism, the effect can be considered to be an increased contribution of hybrid B to the description of the metal–carbon interaction:

$$M\equiv C-NMe_2 \leftrightarrow \bar{M}=C=\overset{+}{N}Me_2$$

A   B

The implication is that the stronger the π-donating ability of the substituent, the weaker the π-accepting ability of the carbyne ligand.

The change to a silicon-based substituent group, e.g., $SiMe_3$, has the opposite effect. The introduction of two more orbitals of π-symmetry appropriate for bonding stabilizes the metal–carbon interaction and increases the percentage electron density of the π-orbital on the carbyne ligand (28).

Varying the Metal Fragment. The close relationship between octahedral carbene and cationic carbyne complexes of Group 6a and 7a metals is evident. However, when neutral carbyne complexes are considered, the analogy is not as close because the LUMO is not always one of the metal–carbon π^*-antibonding orbitals. Thus, for the $Cr(CO)_4Br$ fragment the CPh carbyne leads to a π^* LUMO, but for the CMe and $CNMe_2$ carbynes the LUMO's are concentrated on the carbonyl fragments.

Charge. The small amount of charge distribution data for carbyne complexes (based on Mulliken population analyses) indicates that the metal–carbon bond is generally polarized $M^{\delta+}-C^{\delta-}$ and that the carbyne carbon is always more negative than adjacent carbonyl carbons (28,30). These conclusions are directly analogous to those derived for carbene complexes.

2. Reactivity Patterns

In view of the similarities between the bonding models for carbene and carbyne complexes it is not surprising that similar patterns of reactivity should be observed for these compounds. Thus nucleophilic and electrophilic additions to the metal–carbon triple bond are anticipated under appropriate circumstances, and both orbital and electrostatic considerations will be expected to play a role.

a. Nucleophilic Addition. Nucleophilic reagents can attack the carbyne complex at at least three positions (1), (2), and (3):

$M \equiv C - R$

↑ ↑ ↑

(1) (2) (3)

The model predicts that frontier orbital-controlled addition at position (1) should be favorable for octahedral or pseudooctahedral complexes of

Group 6a and 7a metals, particularly when cationic (*28,29*). However, the presence of an equilibrium such as:

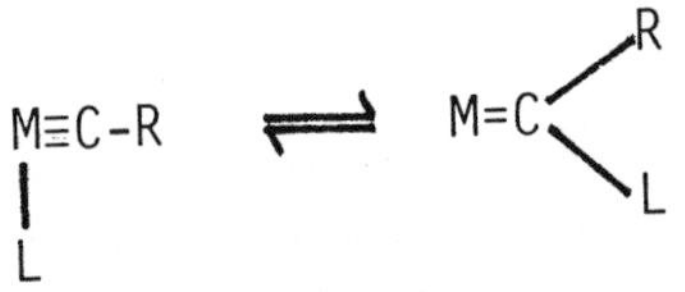

can make experimental determination of the initial site of nucleophilic attack ambiguous.

In some cases the site of attack is in no doubt, e.g., **9** adds PMe_3 at the carbyne carbon in an orbital-controlled reaction (*29,31*):

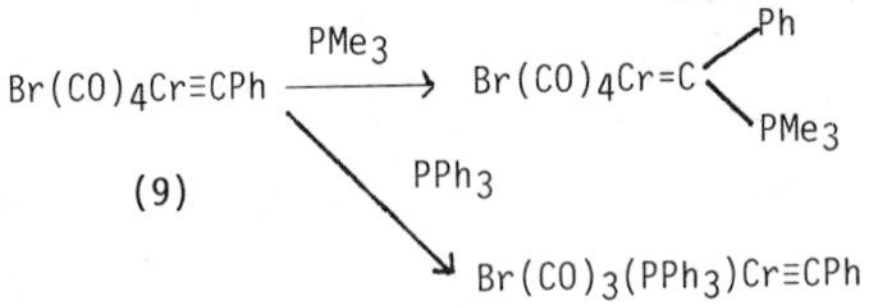

The site can be very susceptible to the nature of the attacking nucleophile, with PPh_3 displacing CO from the metal in **9** rather than adding to the carbyne carbon (note than PPh_3 attack at the carbyne carbon, followed by CO loss and PPh_3 migration, is unreasonable since the labilizing effect of the carbyne ligand on the carbonyls would be reduced on coordination of a base).

Many anionic nucleophiles add to the carbyne carbon in cationic Fischer complexes, and these reactions can be used to synthesize carbene ligands unavailable by other routes. The stepwise reduction of a metal–carbon triple bond has been demonstrated (*32*):

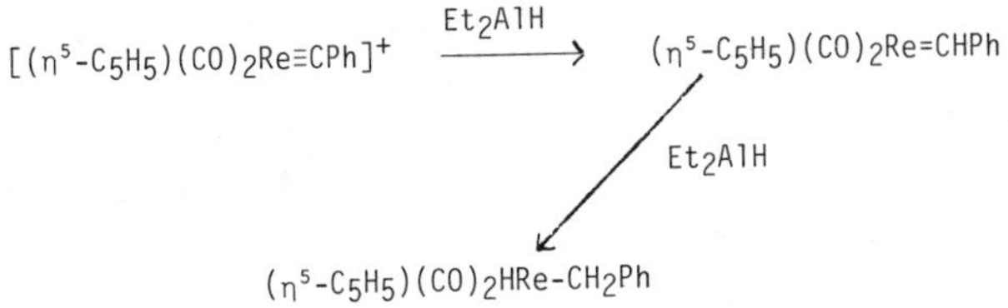

Hydride addition to the cationic Os(O) carbyne complex **10** occurs at the para position of the aryl ring rather than at the carbyne carbon, affording the vinylidene complex **11** (*33*):

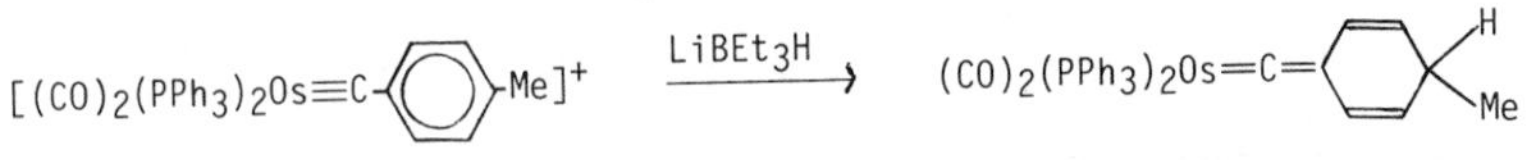

(10) (11)

Calculations for the model compound $[(CO)_2(PPh_3)_2Fe{\equiv}CPh]^+$ indicate that the LUMO is a low-lying metal carbyne π^*-orbital localized on the aryl ring and most concentrated at the para position. If these conclusions are applicable to the osmium compound **10** the hydride addition would appear to be a frontier orbital-controlled reaction.

b. Electrophilic Addition. The addition of electrophiles to carbyne complexes can also occur at three distinct sites:

The Carbyne Carbon. Protic and Lewis acids can add to the carbyne carbon of complexes containing electron-rich metal centers:

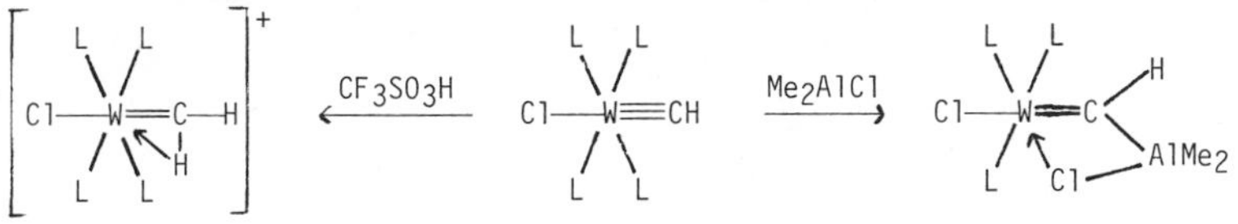

(12)   (Ref.34,35)

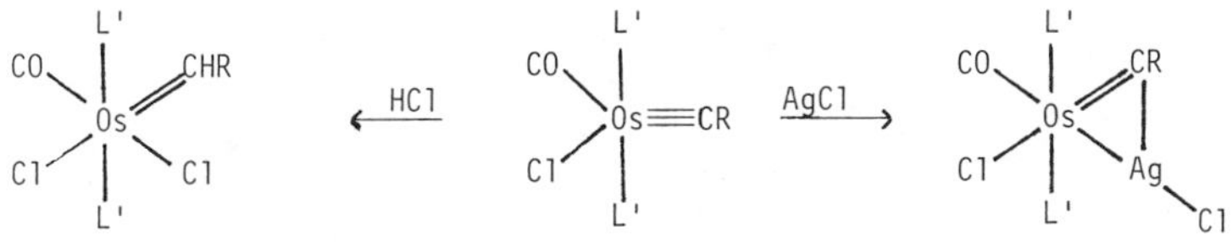

(13)   (Ref.36,40)

L = PMe₃, L' = PPh₃, R = p-tolyl

Protonation of **12** yields a compound best described as a face-protonated methylidyne complex, the tungsten–carbon bond length lying in the range observed for a triple bond (*28*). Protonation of the osmium compound **13** yields a true carbene complex, which for R = Ph has been characterized by X-ray crystallography (see Sections IV and VI).

The Metal. Proton addition to the metal rather than to the carbyne carbon has been observed, e.g.,

$$(\eta^5\text{-}C_5H_5)[P(OMe)_3]_2Mo\equiv CCH_2CMe_3 \xrightarrow{HBF_4} [(\eta^5\text{-}C_5H_5)[P(OMe)_3]_2HMo\equiv CCH_2CMe_3]BF_4$$

(Ref.37)

The Carbyne Substituent. One example of protonation at a carbyne substituent group has been recorded (*38*):

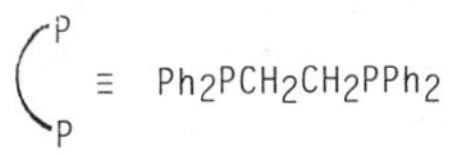

c. Effect of Metal Basicity. That the reactivity of a metal–carbon multiple bond can be reversed by changing the electron density at the metal center is again well illustrated by Group 8a complexes. Osmium dihapto thioacyl complexes have been prepared by electrophilic addition to the electron-rich Os(O) carbyne **13** (*36*) and by nucleophilic addition to the electron-deficient Os(II) carbyne **14** (*39*):

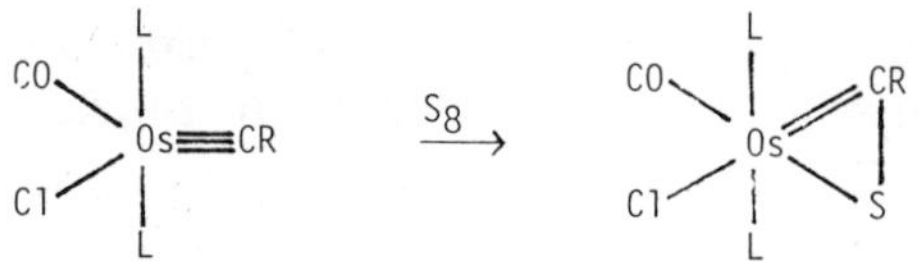

(13)

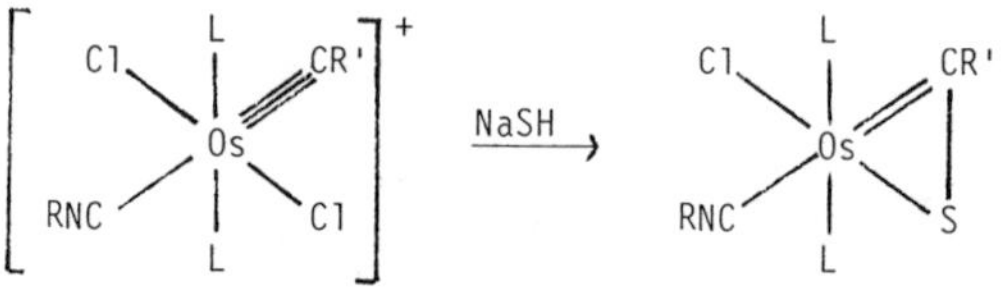

(14)

L = PPh₃, R = p̲-tolyl, R' = C₆H₄NMe₂-4

The similarity, then, between carbene and carbyne complex chemistry of Group 8a transition metals, as well as of Group 6a and 7a metals, is apparent.

III

N-, O-, S-, AND Se-SUBSTITUTED CARBENE COMPLEXES

The development of the chemistry of carbene complexes of the Group 8a metals, Ru, Os, and Ir, parallels chemistry realized initially with transition metals from Groups 6 and 7. The pioneering studies of E. O. Fischer and co-workers have led to the characterization of many hundreds of carbene complexes in which the heteroatoms N, O, and S are bonded to the carbene carbon atoms. The first carbene ligands coordinated to Ru, Os, and Ir centers also contained substituents based on these heteroatoms, and in this section the preparation and properties of N-, O-, S-, and Se-substituted carbene complexes of these metals are detailed.

A. Synthesis

1. Electrophilic Addition to Coordinated Imidoyl, Acyl, Thioacyl, Dithioester, and Diselenoester Ligands

The ruthenium formimidoyl complex **15** is converted to neutral and cationic compounds containing the secondary carbene ligands CHNHR and CHN(Me)R by protonation and methylation, respectively (*41,42*):

$(\eta^2\text{-}O_2CMe)(CO)(PPh_3)_2Ru\text{-}CH\text{=}NR$ $\xrightarrow{\text{HClO}_4}$ $[(\eta^2\text{-}O_2CMe)(CO)(PPh_3)_2Ru\text{=}CHNHR]^+$

(15) $\xrightarrow{\text{Me}_2SO_4}$

$[(\eta^2\text{-}O_2CMe)(CO)(PPh_3)_2Ru\text{=}CHNMeR]^+$

R = p-tolyl

The formimidoyl group in $Ru(CH{=}NR)(\eta^1\text{-}O_2CMe)(CO)(CNR)(PPh_3)_2$ is readily methylated by MeI, with iodide replacing both acetate and triphenylphosphine (*43*):

$(\eta^1\text{-}O_2CMe)(CO)(CNR)(PPh_3)_2Ru\text{-}CH\text{=}NR$ $\xrightarrow{\text{MeI}}$ $I_2(CO)(CNR)(PPh_3)Ru\text{=}CH\text{-}NMeR$

R = p-tolyl

The osmium formimidoyl **16** is reversibly protonated to a secondary carbene complex (*44*):

$Cl(CO)_2(PPh_3)_2Os\text{-}CH\text{=}NMe$ $\underset{\text{OH}^- \text{ or MeNH}_2}{\overset{\text{H}^+}{\rightleftarrows}}$ $[Cl(CO)_2(PPh_3)_2Os\text{=}CHNHMe]^+$

(16)

Alkylation of an anionic acylmetallate is the second step in the classic Fischer synthesis of carbene complexes. The same type of reaction can be performed with stable neutral acyl complexes, producing cationic carbene compounds. Compound **17** has been prepared from a ruthenium acetyl complex by direct methylation and by protonation with subsequent methylenation (*45*):

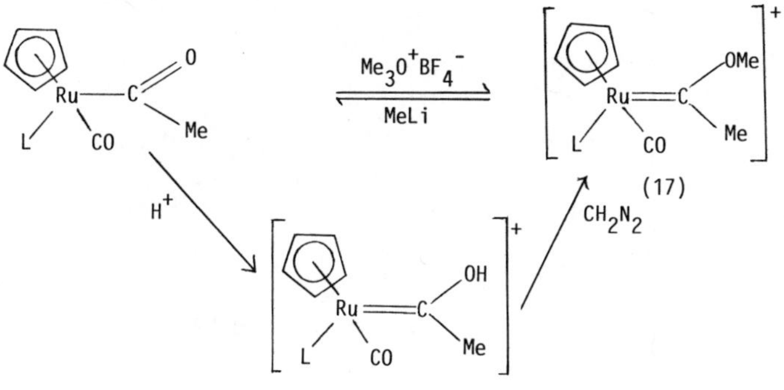

L = PPh$_3$, CO, P(c-hex)$_3$

The osmium hydroxycarbene complex **18** is formed in an acid-assisted migratory-insertion reaction of $OsClEt(CO)_2(PPh_3)_2$. This alkyl compound results from reaction of the ethylene adduct **19** with one equivalent of acid (*46*).

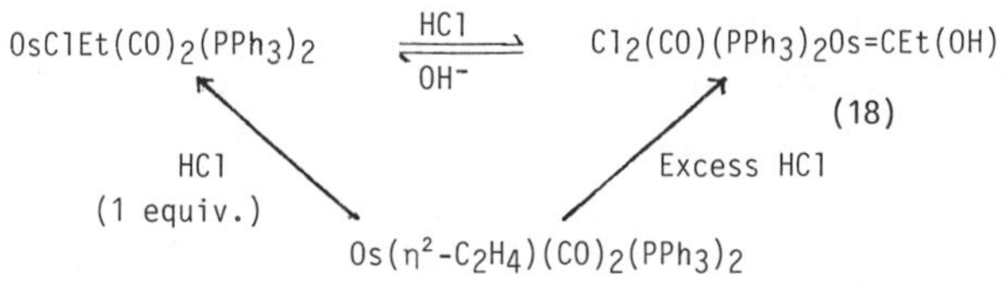

$$OsClEt(CO)_2(PPh_3)_2 \quad \underset{OH^-}{\overset{HCl}{\rightleftharpoons}} \quad Cl_2(CO)(PPh_3)_2Os=CEt(OH)$$

(18)

HCl
(1 equiv.)

Excess HCl

$$Os(\eta^2\text{-}C_2H_4)(CO)_2(PPh_3)_2$$

(19)

Intramolecular acyl alkylation affords cyclic carbene ligands, e.g.,

$$(\eta^5\text{-}C_5H_5)Ru(CO)_2^- + ClCH_2CH_2CH_2\underset{O}{\overset{\|}{C}}Cl \rightarrow (\eta^5\text{-}C_5H_5)(CO)_2Ru\underset{O}{\overset{\|}{C}}CH_2CH_2CH_2Cl$$

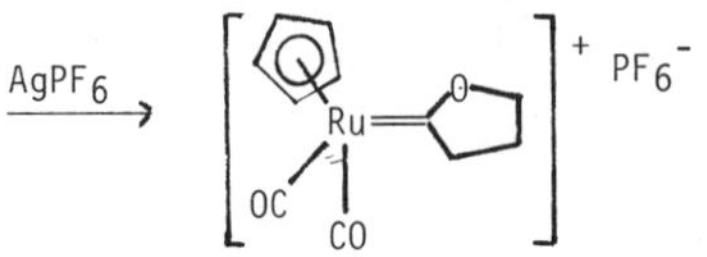

(Ref.47)

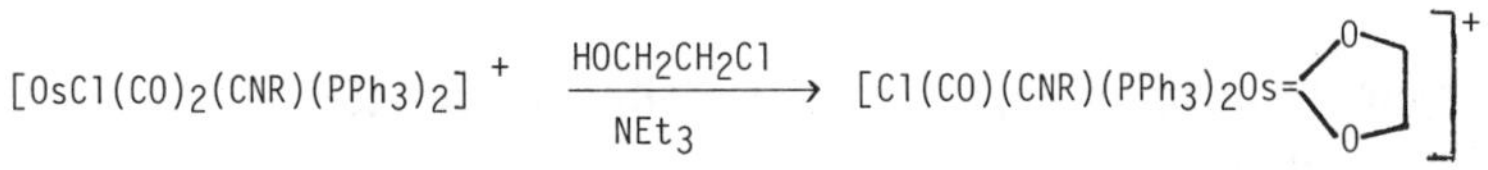

(Ref.55)

R = p-tolyl

Alkylation of thioacyl ligands provides a route to thiocarbene complexes. Faraone and co-workers have prepared iridium thiocarbene complexes **20** and **21** by methyl iodide addition to Ir(I) thiocarbonyl

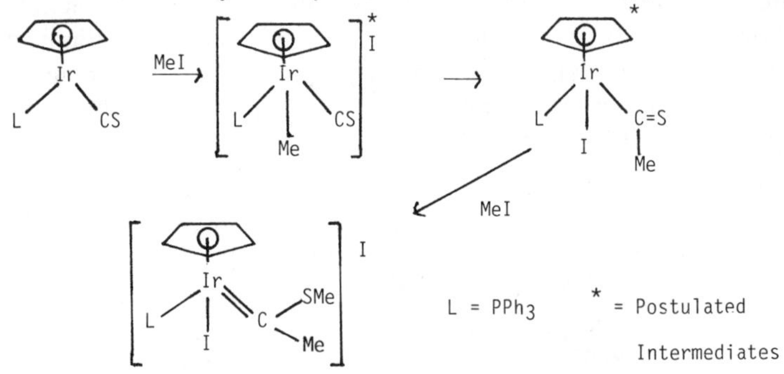

L = PPh₃   * = Postulated

Intermediates

(20)

$$Ir(C_6F_5)(CS)(PPh_3)_2 \quad \xrightarrow{MeI} \quad I_2(C_6F_5)(PPh_3)_2Ir=CMe(SMe)$$

(21)

compounds, and the intermediacy of thioacyl species in these reactions seem reasonable (48,49). The related secondary carbene ligand CHSMe has been generated in an osmium complex by methylation of a thio-formyl species (44):

$$Cl(CO)_2(PPh_3)_2Os-CH=S \xrightarrow{CF_3SO_3Me} [Cl(CO)_2(PPh_3)_2Os=CHSMe]CF_3SO_3$$

Methylation of the dihaptothioacyl complex **22** affords compound **23** containing a bidentate carbene ligand, which on reaction with chloride ion leads to the neutral monodentate carbene complex **24** (50,51). The chelate carbene complex **26** is generated in a novel interligand reaction from the thiocarboxamidothiocarbonyl cation **25**. The thiocarbonyl carbon acts as the electrophilic component in this reaction, and **26** is further alkylated to a bidentate dicarbene species (52).

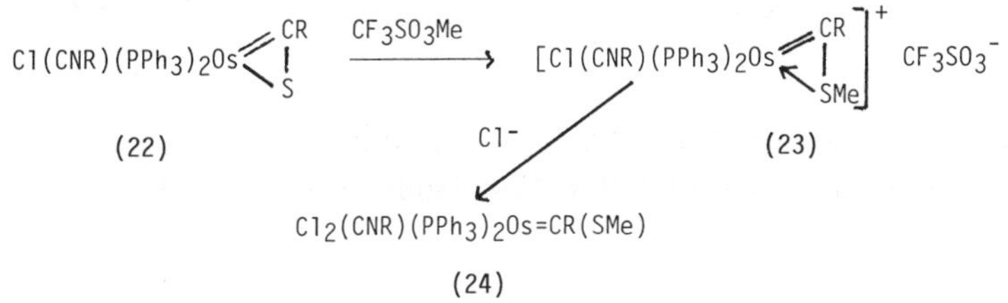

(22) → (23) → (24)

R = p-tolyl

(25) → (26)

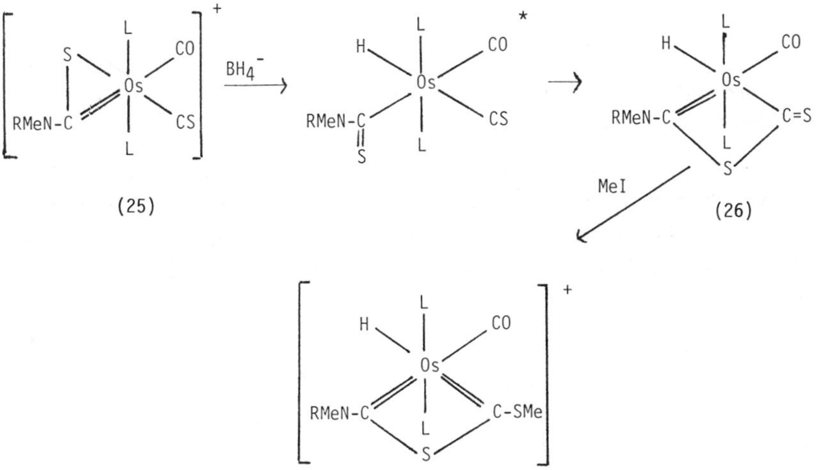

* = Postulated Intermediate

R = p-tolyl

dihapto-CS$_2$ adducts of ruthenium and osmium have been alkylated to dithiocarbene complexes, and the intermediacy of dithioester species seems very likely (*53,54*). Cyclic carbene ligands are formed when dihaloalkanes are used as the electrophilic species in these reactions:

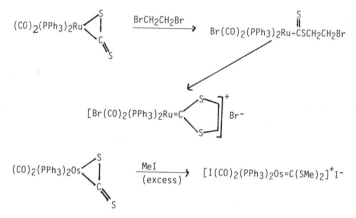

1,3-Diselenolan-2-ylidene complexes of ruthenium have been prepared in analogous reactions of ruthenium CSe$_2$ adducts with 1,2-dibromoethane (*53*), e.g.,

2. Nucleophilic Addition to Coordinated Isocyanide and Vinylidene Ligands

Transition metal isocyanide complexes can undergo reactions with nucleophiles to generate carbene complexes. Pt(II) and Pd(II) complexes have been most extensively investigated, and the range of nucleophilic reagents employed in these reactions has included alcohols, amines, and thiols (*56*):

$$L_nM(CNR) \xrightarrow{R_mXH} L_nM=C\begin{array}{l} {}^{NHR} \\ {}_{XR_m} \end{array}$$

$$X = N, O, S$$

The reactions of [M(CNMe)$_6$]$^{2+}$ (M = Ru, Os) with amines have been studied. Compound **27** reacts with ethylamine to afford the carbene

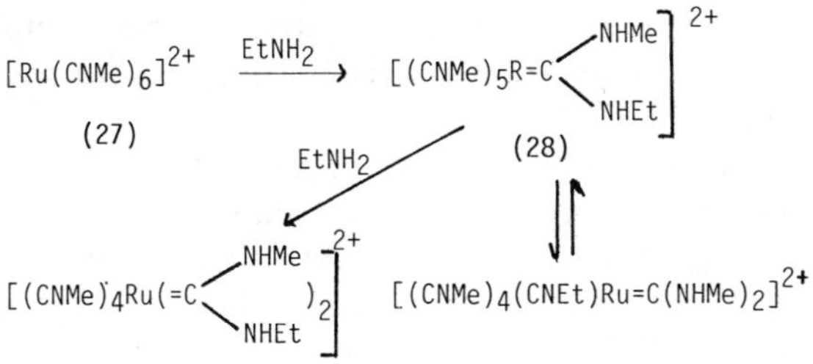

(27) (28) (29) (30)

complex **28** (*57*). On prolonged treatment with excess ethylamine, the *cis*-biscarbene **29** can be isolated (*58*). Compound **28** rearranges on warming, probably via an intramolecular cyclization, to form **30** (*57*). The cyclic carbene chelate complex **31** is generated in the reaction of **27** with hydrazine (*58*):

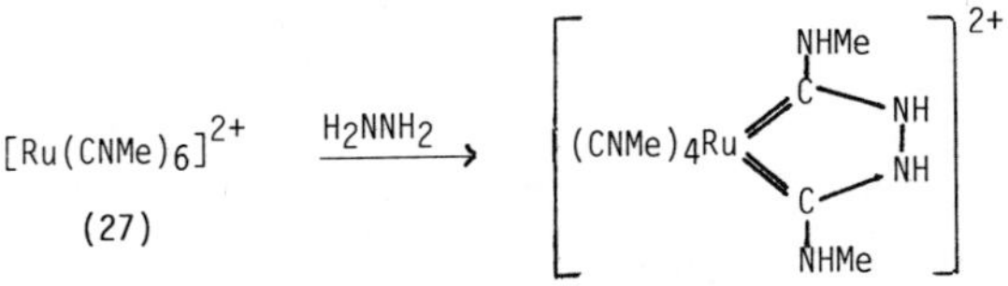

(27) (31)

The osmium analog of **27** reacts with methylamine to give mixtures of *cis*-, *bis*-, and *mer*-triscarbene complexes (*59*).

Os(II) complexes of tosylmethylisocyanide react with aldehydes and ketones in the presence of sodium methoxide, producing cyclic carbene complexes (*66*). Compound **32** undergoes a rapid reaction with benzaldehyde and NaOMe, yielding the oxazol-2-ylidene complex **33** and eliminating *p*-toluene sulfinic acid (*66*):

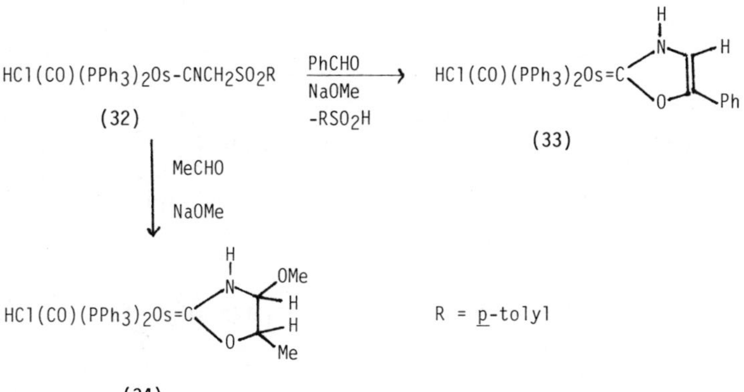

(32) (33) (34)

The same reaction with acetaldehyde produces carbene complex **34** with a saturated ring, i.e., an oxazolidin-2-ylidene ligand with the tosyl group replaced by methoxide.

Clark and co-workers have reported reactions of Ir(III) cations with terminal alkynes in methanol in which alkoxycarbene complexes are formed (*60*). By analogy with a more extensively studied Pt(II) system (*61*), it has been concluded that cationic vinylidene complexes, e.g., **35**, are reaction intermediates, e.g.,

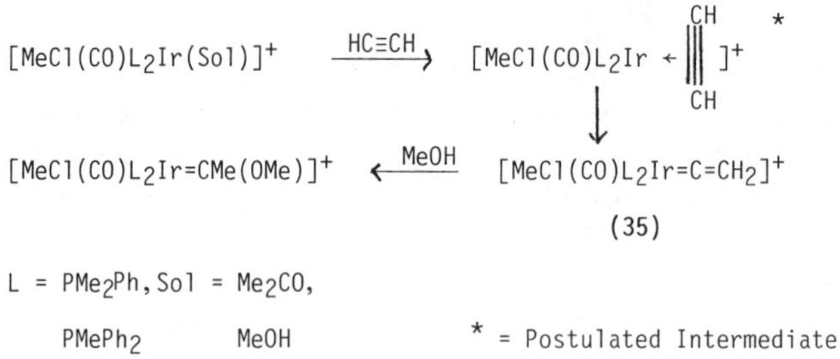

$$[MeCl(CO)L_2Ir(Sol)]^+ \xrightarrow{HC \equiv CH} [MeCl(CO)L_2Ir \leftarrow \overset{CH}{\underset{CH}{\|}}]^{+ \, *}$$

$$[MeCl(CO)L_2Ir=CMe(OMe)]^+ \xleftarrow{MeOH} [MeCl(CO)L_2Ir=C=CH_2]^+$$

$$(35)$$

$$L = PMe_2Ph, \; Sol = Me_2CO,$$

$$PMePh_2 \qquad MeOH \qquad\qquad * = \text{Postulated Intermediate}$$

3. *From Electron-Rich Olefins*

Electron-rich olefins such as **36** have been used by Lappert in the synthesis of a great number of mono-, bis-, tris-, and tetrakiscarbene complexes from various transition metal species (*62*). Ru, Os, and Ir carbene complexes have been prepared from reactions with these olefins, e.g.,

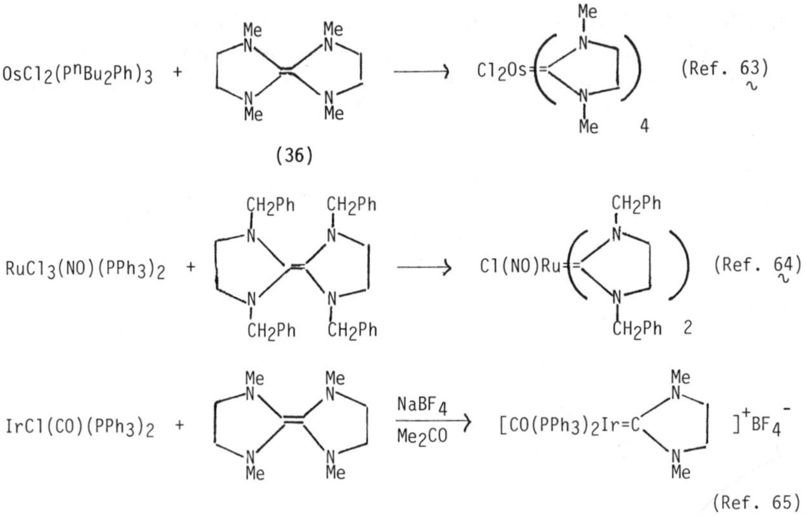

$$OsCl_2(P^nBu_2Ph)_3 \; + \qquad\qquad \longrightarrow \qquad\qquad (Ref. \; 63)$$

$$(36)$$

$$RuCl_3(NO)(PPh_3)_2 \; + \qquad\qquad \longrightarrow \qquad\qquad (Ref. \; 64)$$

$$IrCl(CO)(PPh_3)_2 \; + \qquad\qquad \xrightarrow[Me_2CO]{NaBF_4} \; [CO(PPh_3)_2Ir=C \cdots]^+ BF_4^-$$

$$(Ref. \; 65)$$

Ruthenium complexes containing ortho-metallated *N*-arylcarbene ligands have been generated from reactions employing tetraaryl-substituted electron-rich olefins. For example, olefin **37** reacts with $RuCl_2(PPh_3)_3$ with elimination of PPh_3 and HCl, forming the 5-coordinate carbene complex **38** (*67*):

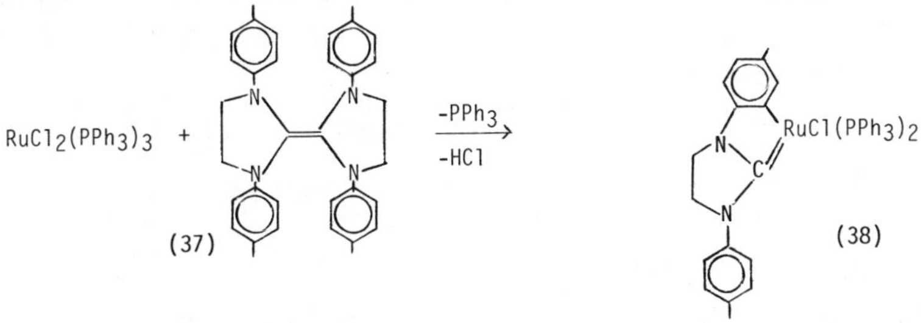

Formation of a trinuclear ruthenium carbene complex via the olefin scission reaction has also been noted (*68*):

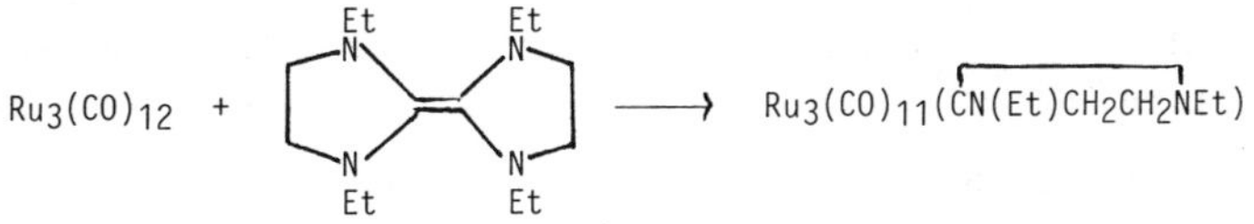

4. *From Organic Salts or Salt-Like Reagents*

Ru, Os, and Ir carbene complexes have been prepared from reactions of anionic or low-valent metal complexes with some organic salts or neutral compounds with highly ionic bonds. Oxidative addition of halothiazole and -oxazole species to $IrCl(CO)(PMe_2Ph)_2$ affords Ir(III) complexes which on protonation yield cationic carbenes (*69*), e.g.,

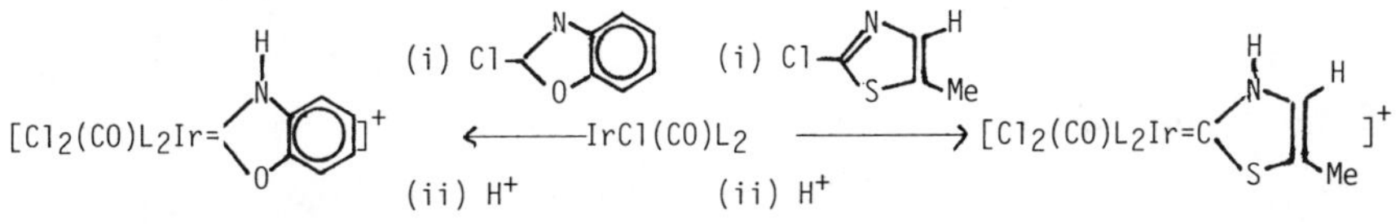

L = PMe_2Ph

d^{10} Ir and Os carbonyl anions react with thiazolium and pyridinium tetrafluoroborate salts giving cationic and neutral carbene complexes respectively, e.g.,

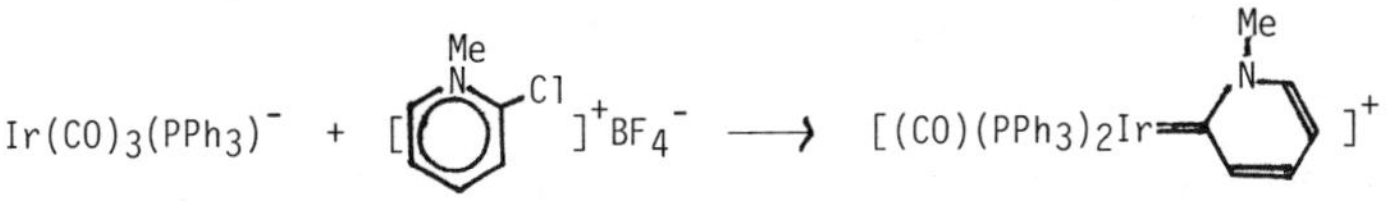

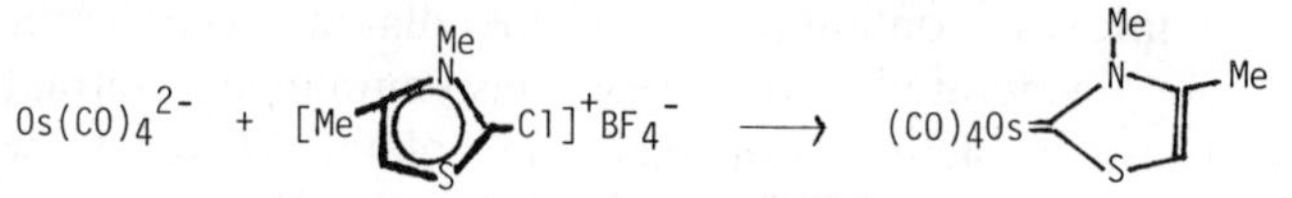

$$Os(CO)_4^{2-} + [Me-\text{(ring)}-Cl]^+BF_4^- \longrightarrow (CO)_4Os\text{(ring)} \quad (Ref. 71)$$

Lappert has described the three-fragment oxidative addition of (halomethylene) dimethylammonium halides to Ru and Ir d^8 complexes (62,72). The products are compounds containing the secondary carbene ligand $CHNMe_2$, e.g.,

$$Ru(CO)_3(PPh_3)_2 \xrightarrow{[NMe_2(CHCl)]Cl} Cl_2(CO)(PPh_3)_2Ru = CHNMe_2$$

$$IrCl(CO)(PPh_3)_2 \xrightarrow{[NMe_2(CHCl)]Cl} [Cl_2(CO)(PPh_3)_2Ir = CHNMe_2]^+Cl^-$$

5. From Halogenocarbene Complexes

Halide displacement from the carbene ligands of Ru, Os, and Ir halocarbene complexes by N-, O-, and S-based nucleophiles frequently leads to the formation of new heteroatom-substituted carbene complexes. This important class of reactivity will be discussed in more detail in Section V,D, but it is appropriate here to illustrate the scope of this method with several examples:

N-Based Nucleophiles

$$Cl_2(CO)(PPh_3)_2Ru = CF_2 \xrightarrow{Me_2NH} Cl_2(CO)(PPh_3)_2Ru = CFNMe_2 \quad (Ref. 22)$$

$$Cl_2(CO)(PPh_3)_2Os = CCl_2 \xrightarrow{Me_2NH} Cl_2(CO)(PPh_3)_2Os = CClNMe_2 \quad (Ref. 39)$$

$$Cl_3(PPh_3)_2Ir = CCl_2 \xrightarrow{H_2NCH_2CH_2NH_2} Cl_3(PPh_3)_2Ir = \overline{CNHCH_2CH_2NH} \quad (Ref. 73)$$

O-Based Nucleophiles

$$Cl_2(CO)(PPh_3)_2Ru = CF_2 \xrightarrow{MeOH} Cl_2(CO)(PPh_3)_2Ru = CF(OMe) \quad (Ref. 22)$$

$$Cl_2(CO)(PPh_3)_2Ru = CF_2 \xrightarrow{HOCH_2CH_2OH} Cl_2(CO)(PPh_3)_2Ru = \overline{COCH_2CH_2O}$$

$$(Ref. 22)$$

S-Based Nucleophiles

$$Cl_2(CO)(PPh_3)_2Ru = CCl_2 \xrightarrow{MeSH} Cl_2(CO)(PPh_3)_2Ru = C(SMe)_2 \quad (Ref. 39)$$

$$Cl_2(CO)(PPh_3)_2Os = CFCl \xrightarrow{NaSEt} Cl_2(CO)(PPh_3)_2Os = CFSEt \quad (Ref. 74)$$

$$Cl_3(PPh_3)_2Ir = CCl_2 \xrightarrow{HSCH_2CH_2SH} Cl_3(PPh_3)_2Ir = \overline{CSCH_2CH_2S} \quad (Ref. 73)$$

6. N-, O-, S-, and Se-Substituted Carbene Complexes by Substituents

Table I summarizes the range of N-, O-, S-, and Se-substituted carbene complexes of Ru, Os, and Ir that have been prepared to date. While not

TABLE I
N-, O-, S-, AND Se-SUBSTITUTED CARBENE COMPLEXES BY SUBSTITUENTS

Substituent atoms	Carbene ligand[a,b]	Examples	Method[c] of preparation	Ref.
N, H	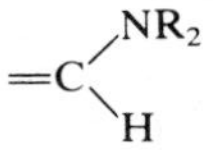 $C{=}$ with NR_2, H	$RuCl(\eta^1\text{-}O_2CMe)(CHNH\text{-}p\text{-tolyl})(CO)(PPh_3)_2$	I	41
		$IrCl_3(CHNMe_2)(PPh_3)_2$	IV	72
		$OsCl_2(CHNMe_2)(CO)(PPh_3)_2$	V	39
N, C	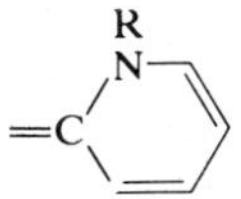 $C{=}$ with R, N (pyridine ring)	$[Ir(CO)(CN(Me)CHCHCHCH)(PPh_3)_2][BF_4]$	IV	70
	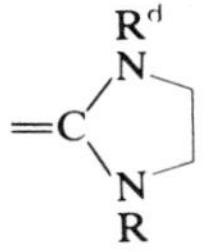	$RuCl_2(CN(Et)CH_2CH_2NEt)_4$	III	76
N, N	$C{=}$ with R^d, N, N, R (ring)	$[Ir(CO)(CN(Me)CH_2CH_2NMe)(PPh_3)_2][BF_4]$	III	65
		$RuCl(CN(C_6H_4\text{-}4\text{-}Me)CH_2CH_2NC_6H_3\text{-}4\text{-}Me)(PPh_3)_2$	III	67
		$RuCl_2(CO)(CNHCH_2CH_2NH)(PPh_3)_2$	V	19
	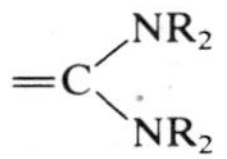 $C{=}$ with NR_2, NR_2	$[Ru(C(NHMe)(NHEt))(CNMe)_5][PF_6]_2$	II	57
		$[Os(C(NHMe)_2)(CNMe)_4][PF_6]_2$	II	59
	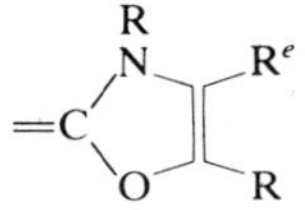	$OsHCl(CNHCH{=}C(Ph)O)(CO)(PPh_3)_2$	II	66
N, O	$C{=}$ with R, R^e, N, O, R (ring)	$[IrCl_2(CO)(CNC_6H_4O)(PMe_2Ph)_2][BF_4]$	IV	69
		$RuCl_2(CNHCH_2CH_2O)(CO)(PPh_3)_2$	V	19

(continued)

TABLE I (continued)

Substituent atoms	Carbene ligand[a,b]	Examples	Method[c] of preparation	Ref.
N, S	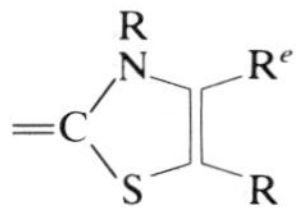 ring: R–N, R^e, R, S (=C)	Os(CN(Me)C(Me)=CHS)(CO)$_4$	IV	71
		[IrCl$_2$(CO)(CNHCH=C(Me)S)(PMe$_2$Ph)$_2$][BF$_4$]	IV	69
O, H	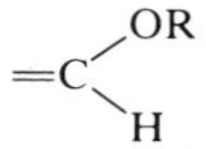 =C with OR, H	[OsCl(CHOMe)(CO)(CN-p-tolyl)(PPh$_3$)$_2$]CF$_3$SO$_3$	I	44
O, C	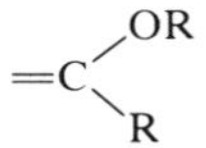 =C with OR, R	[(η-C$_5$H$_5$)Ru(CO)(PPh$_3$)(C(Me)OMe)][BF$_4$]	I	45
		[IrCl(Me)(CO)(C(Me)OMe)(PMe$_2$Ph)$_2$](PF$_6$)	II	60
	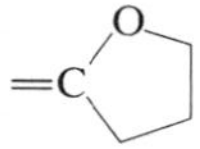 furanyl ring (O) =C	[(η-C$_5$H$_5$)Ru(CO)$_2$(COCH$_2$CH$_2$CH$_2$)][PF$_6$]	I	47
		[IrCl(Me)(CO)(COCH$_2$CH$_2$CH$_2$)(PMePh$_2$)$_2$][PF$_6$]	II	60
O, O	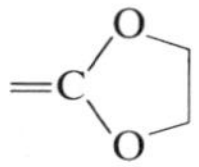 dioxolane ring (O, O) =C	[OsCl(CO)(CN-p-tolyl)(COCH$_2$CH$_2$O)(PPh$_3$)$_2$][ClO$_4$]	I	55
		RuCl$_2$(CO)(COCH$_2$CH$_2$O)(PPh$_3$)$_2$	V	22
S, H	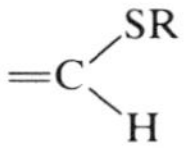 =C with SR, H	[OsCl(CO)$_2$(CHSMe)(PPh$_3$)$_2$][CF$_3$SO$_3$]	I	44

	Structure	Compound		Ref.
S, C	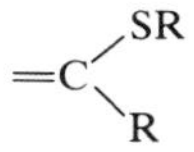 $=C\langle^{SR}_{R}$	$[(\eta\text{-}C_5H_5Ir(PPh_3)(C(Me)SMe)]I$	I	48
S, S	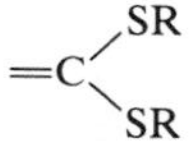 $=C\langle^{SR}_{SR}$	$[OsI(CO)_2(C(SMe)_2)(PPh_3)_2]I$	I	54
		$RuCl_2(CO)(C(S\text{-}p\text{-tolyl})_2)(PPh_3)_2$	V	19
	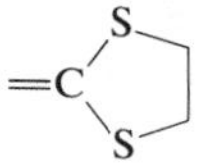	$[OsBr(CO)_2(CSCH_2CH_2S)(PPh_3)_2]Br$	I	53
		$RuCl_2(CO)(CSCH_2CH_2S)(PPh_3)_2$	V	19
Se, Se	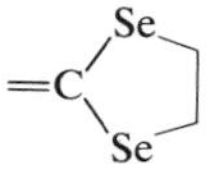	$RuBr_2(CO)(CSeCH_2CH_2Se)(PPh_3)_2$	I	53

[a] R = H, alkyl, or aryl.
[b] Bond localization depicted.
[c] I   = Electrophilic addition to coordinated imidoyl, acyl, thioacyl, dithioester, and diselenoester ligands.
   II = Nucleophilic addition to coordinated isocyanide and vinylidene ligands.
   III = From electrophilic olefins.
   IV = From organic salts or salt-like reagents.
   V = From halogenocarbene complexes.
[d] Unsaturated ligands of this form also known.
[e] Saturated ligands of this form also known.

comprehensive, it serves to illustrate the structural and preparative diversity of compounds in this class.

B. Structure and Spectroscopic Properties

1. Structure

Table II summarizes the structural studies of Ru, Os, and Ir carbene complexes with the heteroatom substituents N, O, S, or Se. These X-ray data clearly illustrate two features of the bonding in these compounds:

1. The bond order of the metal–carbene carbon linkage is not significantly greater than unity, with the ruthenium–carbon distances lying in the range observed for a variety of Ru complexes with σ-bonded carbon ligands (77). The short ruthenium–carbene carbon distances observed in the ortho-metallated N-arylcarbene complexes have been attributed to the constraints imposed by a chelating system with a small "bite" angle, rather than to any multiple character in these bonds (67).

2. There is considerable multiple character in the carbene carbon–X(Y) bonds. An internal comparison of carbon–X(Y) bond lengths in each of the structures tabulated reveals a substantial shortening of carbene carbon–X(Y) relative to other carbon–X(Y) bonds in the molecule. Carbene carbon–nitrogen distances are found to be shorter in the mono-N-substituted carbenes than in the bis-N-substituted compounds.

These observations are compatible with the model for the carbene complex presented in Section II,A. Both metal and π-donor substituents compete to donate electron density to unfilled carbene p_z orbitals, and with good π-donors such as nitrogen, the metal is less effective. In terms of resonance formalism, the resonance hybrid **39** makes a more significant contribution than **40** to the structure of the carbene ligands in these compounds. Similar conclusions are reached when the structures of Group 6, 7, and other Group 8 heteroatom-substituted carbene complexes are considered.

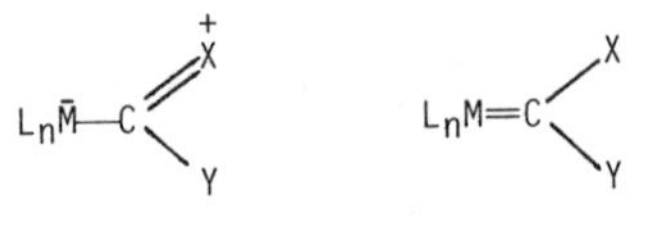

$$L_n\overset{-}{M}-C\overset{\overset{+}{X}}{\underset{Y}{\big\langle}} \qquad L_nM=C\overset{X}{\underset{Y}{\big\langle}}$$

(39)   (40)

TABLE II

HETEROATOM-SUBSTITUTED CARBENE COMPLEXES: SOME STRUCTURAL PARAMETERS[a]

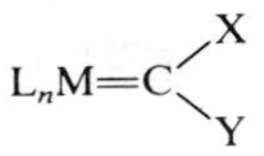

$$L_nM=C{\overset{X}{\underset{Y}{\diagup}}}$$

L_nM	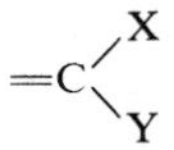 $=C{\overset{X}{\underset{Y}{\diagup}}}$	M—C	C—X	C—Y	Ref.
$RuI_2(CN\text{-}p\text{-tolyl})(CO)(PPh_3)_2$	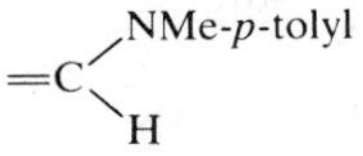 $=C{\overset{NMe\text{-}p\text{-tolyl}}{\underset{H}{\diagup}}}$	2.045	1.28	—	43
$[Ru(NH_3)_4(CO)][PF_6]_2$	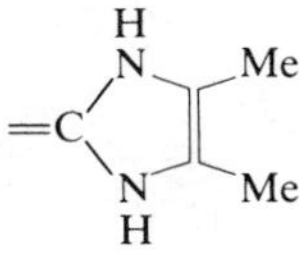 imidazoline (Me, Me; H—N, N—H)	2.128	1.347	1.356	75
$RuCl_2(\overline{CNEtCH_2CH_2NEt})_3$	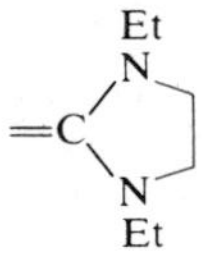 ring (Et—N, N—Et)	2.105(5)[b]	1.349(11)[b]		76
$RuRCl(PEt_3)_2{}^c$	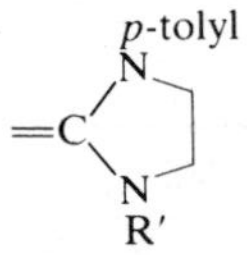 ring (p-tolyl—N, N—R')	1.911(9)	1.39	1.34	67

(continued)

TABLE II (*continued*)

L_nM	$=C\!\begin{smallmatrix}X\\Y\end{smallmatrix}$	M—C	C—X	C—Y	Ref.
RuRCl(CO)(PEt$_3$)$_2$[d]	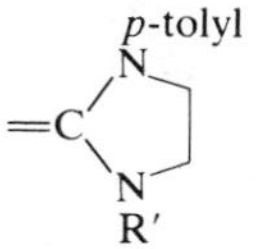 $=C\!<$ (p-tolyl-N · · · N-R' ring)	1.989	1.34	1.35	67
OsRH(CO)(PPh$_3$)$_2$[e]	$=C\!\begin{smallmatrix}\text{NMe-}p\text{-tolyl}\\ \text{SC(S)R'}\end{smallmatrix}$	2.18	1.24	1.67	52

[a] Distances are in angstroms.
[b] Average distance.
[c] RuCl[CN(p-tolyl)CH$_2$CH$_2$NC$_6$H$_3$Me-4](PEt$_3$)$_2$.
[d] Ru(CO)Cl[CN(p-tolyl)CH$_2$CH$_2$NC$_6$H$_3$Me-4](PEt$_3$)$_2$.
[e] Os(CS$_2$CNMe-p-tolyl)H(CO)(PPh$_3$)$_2$.

2. Spectroscopic Properties

Some spectroscopy parameters of selected N-, O-, S-, or Se-substituted carbene complexes are listed in Table III.

^{13}C-NMR chemical shifts are a useful diagnostic tool for carbene complexes, with C_{carb} being substantially deshielded. The C_{carb} resonance is seen to shift to higher fields as the electron deficiency of the metal center is increased (i.e., with electron-withdrawing ligands and in cationic complexes).

^{1}H-NMR is useful in the characterization of Ru, Os, and Ir secondary carbene complexes, the α-proton also resonating at low field. The two signals observed for the CHOMe resonance in $[OsCl(CHOMe)(CO)(CNR)(PPh_3)_2]CF_3SO_3$ have been ascribed to the presence of geometrical isomers for the carbene ligand (44), i.e.,

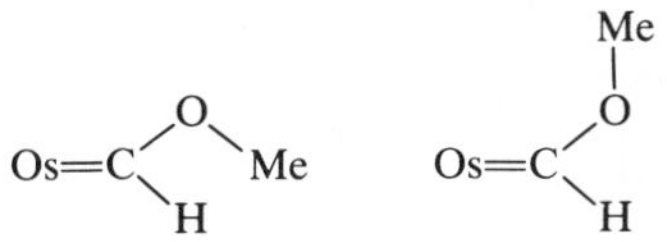

^{1}H-NMR studies of oligocarbene Ru(II) complexes indicate a substantial barrier to rotation about the metal–carbene carbon and nitrogen–R bonds. This restricted rotation is thought to arise as a consequence of intramolecular non-bonding cis interactions of the carbene nitrogen–R substituents, and not because of any significant double bond character in ruthenium–carbene carbon (76).

N-Substituted carbene complexes show $\nu(CN)$ absorption in the 1470–1620 cm^{-1} range of the IR spectrum. These data are consistent with the crystallographic evidence for substantial carbon–nitrogen multiple bonding in these compounds.

C. Reactivity

1. Nucleophilic Substitution Reactions at C_α

The theoretical model developed for the carbene complex becomes of value when the reactivity of these compounds is considered. As noted in Section II,A,2, nucleophilic attack at the carbene carbon should be a significant reaction and would be most likely in complexes where the negative charge on this atom is minimized, that is, (i) when the metal center is electron-deficient (i.e., cationic or containing electron-withdrawing spectator ligands) or (ii) when the carbene substituents are electron-withdrawing, without being particularly good π-donors.

TABLE III

HETEROATOM-SUBSTITUTED CARBENE COMPLEXES: SOME SPECTROSCOPIC PARAMETERS

$$L_nM=C\begin{smallmatrix}X\\Y\end{smallmatrix}$$

L_nM	$=C\begin{smallmatrix}X\\Y\end{smallmatrix}$	$\delta(C)^a$	$\delta(H)^a$	$\nu(CN)^b$	Ref.
RuCl$_2$(CO)(PPh$_3$)$_2$	$=C\begin{smallmatrix}NMe_2\\H\end{smallmatrix}$		10.90	1571	20
RuCl$_2$(CO)(PEt$_3$)$_2$	$=C\begin{smallmatrix}NMe_2\\H\end{smallmatrix}$	248.5	10.53	1553	62
[OsCl(CO)$_2$(PPh$_3$)$_2$][ClO$_4$]	$=C\begin{smallmatrix}NHMe\\H\end{smallmatrix}$		10.13	1610	44
RuCl$_2$(CO)(PPh$_3$)$_2$	$=C\begin{smallmatrix}NH\text{-}p\text{-tolyl}\\H\end{smallmatrix}$			1540	42
[IrCl$_2$(CO)(PPh$_3$)$_2$]Cl	$=C\begin{smallmatrix}NMe_2\\H\end{smallmatrix}$	192.7	9.47	1612	72
RuI$_2$(CO)(CN-p-tolyl)(PPh$_3$)	$=C\begin{smallmatrix}NMe\text{-}p\text{-tolyl}\\H\end{smallmatrix}$			1520	43

Compound	Structure	NMR	IR	Ref.
[Ru(CNMe)$_5$][PF$_6$]$_2$	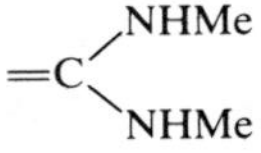 C=(NHMe)(NHMe)		1568	57
RuCl$_2$(CN(ME)CH$_2$CH$_2$NMe)$_3$	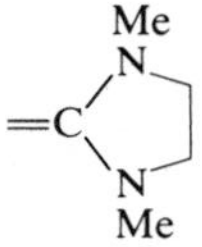 (Me)N—C—N(Me) ring		1485	63
RuCl$_2$(CN(Et)CH$_2$CH$_2$NEt)$_3$	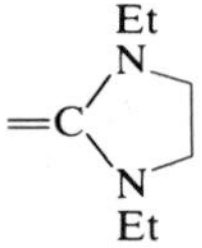 (Et)N—C—N(Et) ring	227.8	1470	63
RuCl$_2$(CO)(CN(Et)CH$_2$CH$_2$NEt)$_2$	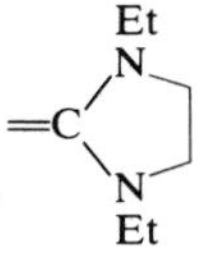 (Et)N—C—N(Et) ring	218.5, 212.7	1490	63
[OsCl(NO)(CN(Me)CH$_2$CH$_2$NMe)$_3$][BF$_4$]$_2$	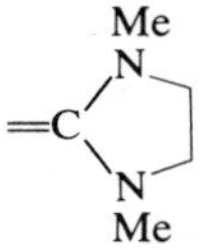 (Me)N—C—N(Me) ring	177.6	1535, 1515	64
RuRCl(PEt$_3$)$_2$[c]	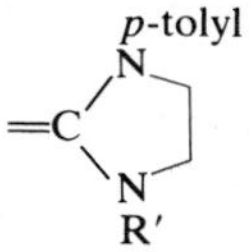 (p-tolyl)N—C—N(R') ring	223.3	1519	67

(continued)

TABLE III (continued)

$$L_nM = C\begin{smallmatrix} X \\ \\ Y \end{smallmatrix}$$

L_nM	$=C\begin{smallmatrix} X \\ Y \end{smallmatrix}$	$\delta(C)^a$	$\delta(H)^a$	$\nu(CN)^b$	Ref.
RuRCl(CO)(PEt₃)₂^d	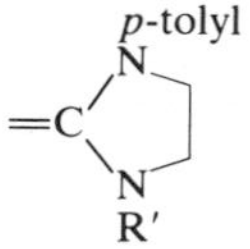 =C with N(p-tolyl)—N(R′) imidazolidine	219.3		1518	67
[OsCl(CO)(CNR)(PPh₃)₂]CF₃SO₃	=C(OMe)(H)		12.03,^{e} 13.05		44
[OsCl(CO)₂(PPh₃)₂]CF₃SO₃	=C(SMe)(H)		14.70		44

^{a} Chemical shifts in ppm.
^{b} In cm^{-1}. All bands medium-strong intensity.
^{c} RuCl[CN(p-tolyl)CH₂CH₂NC₆H₃Me-4](PEt₃)₂.
^{d} RuCl(CO)[CN(p-tolyl)CH₂CH₂NC₆H₃Me-4](PEt₃)₂.
^{e} Two isomers.

Indeed, nucleophilic attack at C_α is the most widely observed single reaction of Fischer carbenes. Substituent substitution is favored when one of the substituents on the carbene carbon is a good leaving group, e.g., halide, alkoxide, etc. Aminolysis of complex **41** typifies this mode of reaction (*17*),

$$(CO)_5Cr=CPh(OMe) \xrightarrow{\text{RNH}_2} (CO)_5Cr=CPh(NHR)$$

(41)

The driving force for this transformation is the fact that the less electronegative nitrogen atom is a better π-donor than oxygen and can form a stronger bond with the carbene carbon atom. Hence displacement of alkoxide by amines and thiols is commonly observed, but the reverse reactions are seldom seen.

The reactivity displayed by the heteroatom-substituted Ru, Os, and Ir carbene complexes discussed in this section toward nucleophilic reagents contrasts sharply with that described for the Fischer compounds. The reactions of these Group 8 complexes are almost exclusively restricted to the metal–ligand framework, with only two related substituent substitution reactions being reported (*44*):

$$[Cl(CO)_2(PPh_3)_2Os=CH(SMe)]^+ \xrightarrow[\text{-MeSH}]{\text{MeNH}_2} [Cl(CO)_2(PPh_3)_2Os=CH(NHMe)]^+$$

$$\downarrow \begin{matrix} H_2O \\ -MeSH \end{matrix} \qquad\qquad \downarrow \begin{matrix} MeNH_2 \\ -H^+ \end{matrix}$$

$$Cl(CO)_2(PPh_3)_2Os-CH=O \qquad Cl(CO)_2(PPh_3)_2Os-CH=NMe$$

There are probably several factors responsible for this reactivity difference:

1. Fischer carbenes characteristically contain a number of electron-withdrawing carbonyl ligands while the typical Ru, Os, or Ir carbene complexes described above frequently contain several σ-donor ligands. The metal centers in these former compounds, then, are rather electron-deficient, with nucleophilic attack at C_α being a favorable reaction.

2. Many of the N-, O-, S-, or Se-substituted carbene complexes of Ru, Os and Ir contain cyclic carbene ligands. Nucleophilic ring opening of such a ligand would generate a nucleophilic center within easy bonding distances of C_α and the reverse, ring-closing reaction may readily occur.

3. N-Substituted carbene ligands predominate amongst these Ru, Os, and Ir complexes, and thus substituent substitution reactions may be unfavorable on thermodynamic grounds.

2. *Ligand Substitution Reactions*

A great number of ligand substitution reactions of N-, O-, S-, or Se-substituted Ru, Os, and Ir carbene complexes have been reported (e.g., Refs. *43,63,64,67,* and *71*). These reactions include displacement of both carbene and noncarbene ligands, with carbon monoxide, isocyanides, phosphines, phosphites, *N*-donor ligands (e.g., pyridine), and halides being successfully used as substituting ligands. Scheme 1 illustrates these modes of reactivity for the oligocarbene complex, RuCl₂(C̅N(Me)CH₂CH₂NMe)₄ (*63*):

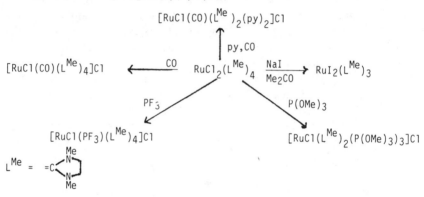

SCHEME 1.   Ligand substitution reactivity of RuCl₂(C̅N(Me)CH₂CH₂NMe)₄.

3. *Oxidative Addition Reactions*

A number of oxidative additions to Ru, Os, and Ir carbene complexes in which the carbene ligand is preserved is known. Some selected examples below illustrate the scope of this type of transformation:

A one-electron oxidation of the Ru(II) imidazolylidene complex **42**, generating the paramagnetic carbene complex **43**, has been noted (*75*):

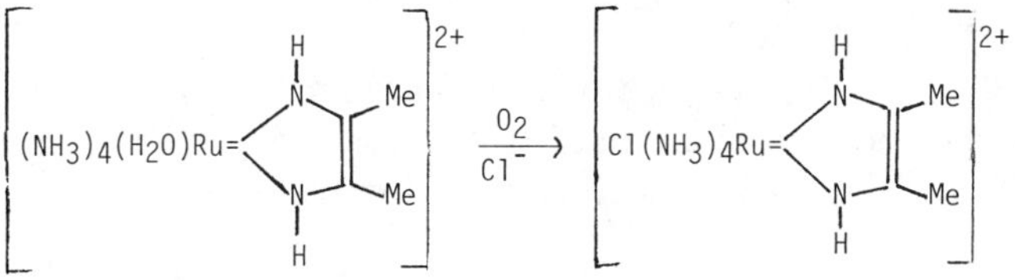

(42)   (43)

4. Carbene Ligand Rearrangement

Rearrangement of the ruthenium (diaminocarbene) isocyanide complex **28** has been noted above. Migration of the carbene substituent group is thought to occur via an intramolecular cyclization reaction (*57,58*):

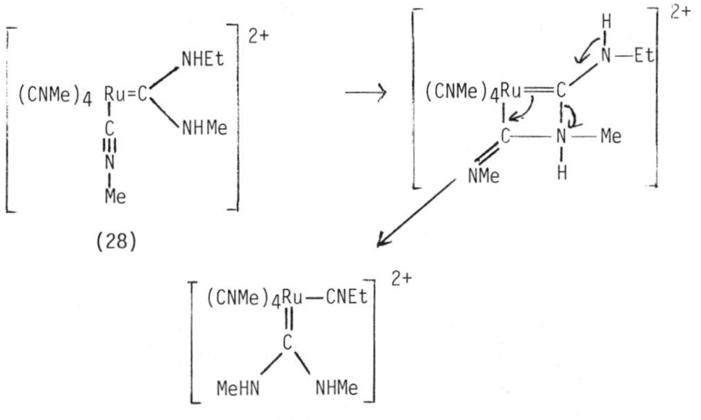

(28)

(30)

IV

NONHETEROATOM-SUBSTITUTED CARBENE (ALKYLIDENE) COMPLEXES

Although transition metal alkylidene complexes, i.e., carbene complexes containing only hydrogen or carbon-based substituents, were first recognized over 15 years ago, it is only relatively recently that Ru, Os, and Ir alkylidene complexes have been characterized. Neutral and cationic complexes of these Group 8 metals are known for both metal electron configurations d^8 and d^6. The synthesis, structural properties, and reactivity of these compounds are discussed in this section.

A. Synthesis

1. d^8 Alkylidene Complexes

The reaction of low-valent Ru, Os, and Ir substrates with diazoalkanes provides the most general route to d^8 alkylidene complexes of these metals. Some features of this reaction are outlined below.

a. The Diazoalkane Route. The ability of numerous transition metal complexes to catalyze the decomposition of nearly all aliphatic diazo compounds, R_2CN_2, has been known for many years. The intermediacy of reactive metal carbene species has been proposed in several of these reactions (see, e.g., Refs. *78–81*). More recently, the formation of stable, isolable carbene complexes from diazoalkanes has been achieved in a limited number of systems, e.g.,

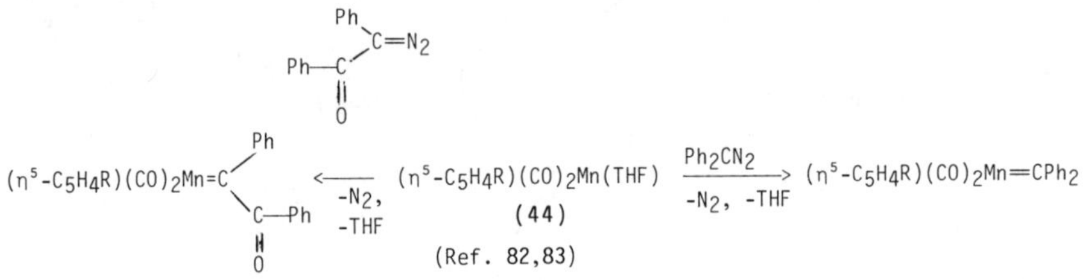

R = H, Me.

Terminal alkylidene complexes generated in this manner can react further with the organometallic substrate when steric interactions are not unfavorable, forming dinuclear species. Herrmann has used this methodology extensively in the preparation of bridging alkylidene complexes (*84*).

The success of such a synthesis relies on two principal factors:

1. The presence of a two-electron donor ligand in the organometallic substrate that can be readily displaced by the CRR' fragment of the diazo precursor is required. Alternatively, a coordinatively unsaturated species may be a suitable reactant.
2. The reaction conditions should be as mild as possible to prevent uncontrolled decomposition of the diazoalkane and carbene complex product.

Reactions with OsCl(NO)(PPh₃)₃. The zero-valent osmiumnitrosyl complex $OsCl(NO)(PPh_3)_3$ (**45**) readily dissociates one PPh_3 ligand, yielding a coordinatively unsaturated species that undergoes many reactions characteristic of the isoelectronic Vaska compound, $IrCl(CO)(PPh_3)_2$ (**46**) (*39*). Compound **45** has proven to be a versatile substrate in the synthesis of d^8

alkylidene complexes via the diazo reaction. It reacts rapidly at room temperature with diazomethane (20) and diazoethane (85) forming stable terminal methylene and ethylidene complexes 47 and 48, respectively. Room temperature reactions of 45 with p-tolyldiazomethane and ethyl diazoacetate also afford zero-valent alkylidene complexes 49 and 50.

$$\text{OsCl(NO)(PPh}_3)_3 \xrightarrow[\text{-PPh}_3, \ -N_2]{\overset{R}{\underset{H}{>}}C=N_2} \text{Cl(NO)(PPh}_3)_2\text{Os}=C\overset{R}{\underset{H}{<}}$$

(45)

(47) R=H   (49) R=p-tolyl

(48) R=Me   (50) R= -C-OEt
$\qquad\qquad\qquad\qquad\qquad\quad \underset{O}{\overset{\|}{}}$

Terminal methylene complexes are relatively rare—less than 10 such compounds have been isolated and about as many again have been characterized by spectroscopic techniques only. The methylene complexes previously reported fall into two groups, (i) neutral complexes of the early transition metals (e.g., Ti, Ta) and (ii) cationic complexes of the later transition metals (e.g., Re, Fe). The osmium complex 47 is important, then, as it is a new example extending the neutral group to the later transition metals. Compound 47 is the prototype for the series $Os(=CHR)Cl(NO)(PPh_3)_2$ and is one of only three terminal methylene complexes to be structurally characterized by X-ray crystallography (see Section IV,B).

There have been few previous reports of complexes containing the secondary alkylidene ligands =CHMe and =CH-p-tolyl, and 50 is the first characterized complex to contain the carbethoxymethylene ligand, =CH(CO₂Et).

Reactions with Other Substrate Complexes. The ruthenium analog of 45, coordinatively unsaturated $RuCl(NO)(PPh_3)_2$, also yields methylene and ethylidene complexes on treatment with diazomethane and diazoethane (39,85).

$$RuCl(NO)(PPh_3)_2 \xrightarrow[-N_2]{RCHN_2} Ru(=CHR)Cl(NO)(PPh_3)_2$$

$$R = CH_2CH_3$$

The reaction of Vaska's compound (46) with diazomethane was reported in 1966 (81). A diethyl ether suspension of 46 afforded chloromethyl complex 52 on treatment with CH_2N_2 at −30°C, and a d^8 iridium methylene complex 51 was proposed as the likely intermediate:

$$IrCl(CO)(PPh_3)_2 \xrightarrow[-N_2, \ Et_2O, \ -30°C]{CH_2N_2} [Ir(=CH_2)Cl(CO)(PPh_3)_2]^* \rightarrow Ir(CH_2Cl)(CO)(PPh_3)_2$$

(46)   (51)   (52)

* = postulated intermediate

A recent reinvestigation of this reaction has substantiated the intermediacy of the 5-coordinate methylene complex **51**. Thermally unstable **51** is formed when a THF solution of **46** is treated with CH_2N_2 at $-60°C$ (*86*). The iodo analog $Ir(=CH_2)I(CO)(PPh_3)_2$ has been similarly prepared and is somewhat more stable than **51**, with iodide migrating less readily to the carbene carbon than chloride (*86*).

The reactions of other reactive zero-valent Group 8 substrates $M(CO)_2(PPh_3)_3$ (M = Ru, Os), $RuH(NO)(PPh_3)_3$, and $Ir(NO)(PPh_3)_3$ with $RCHN_2$ (R = H, Me, *p*-tolyl, CO_2Et) failed to yield carbene complex products, indicating that very specific properties of the substrate complex are required for successful isolation of alkylidene complexes.

A Mechanism for Alkylidene Formation. There is no unambiguous example of free-carbene capture by a metal substrate, and the mild reaction conditions used in the generation of these carbene complexes from diazoalkanes suggests that such a mechanism is highly unlikely here. Transition metal diazoalkane complexes, then, are almost certainly implicated as intermediates in these reactions.

Formal 1,3-dipolar addition of a diazoalkane to an unsaturated metal fragment ML_n would produce **53**. Such a species would be expected to be unstable with respect to nitrogen extrusion, and would seem a plausible intermediate in the diazo reaction:

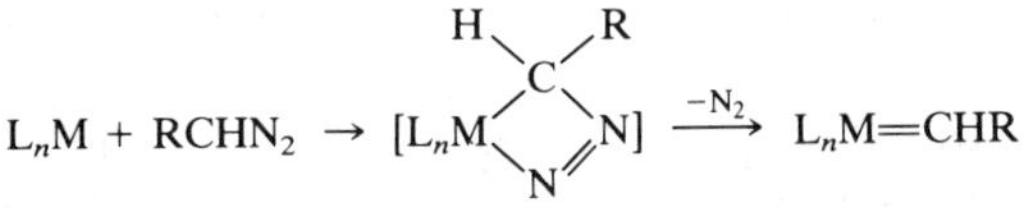

$$L_nM + RCHN_2 \rightarrow [L_nM\diagup\diagdown N] \xrightarrow{-N_2} L_nM{=}CHR$$

(53)

Similar metallacyclic species have been proposed as intermediates in other reactions of metal compounds with diazoalkanes (*87,88*). The presence of unfavorable steric interactions in an intermediate such as **53** could well explain the failure to observe any reaction of diaryldiazoalkane with **45** (*85*).

b. Other Routes to d^8 Alkylidene Complexes. Shaw and co-workers have described the reaction of $IrCl_3$ with the bidentate phosphine $^tBu_2P(CH_2)_5P^tBu_2$ to form $IrHCl[^tBu_2P(CH_2)_2CH(CH_2)_2P^tBu_2]$, (**54**) (*89*). Compound **54** reversibly eliminates hydrogen on thermolysis at reduced pressure, yielding the Ir(I) carbene complex **55**:

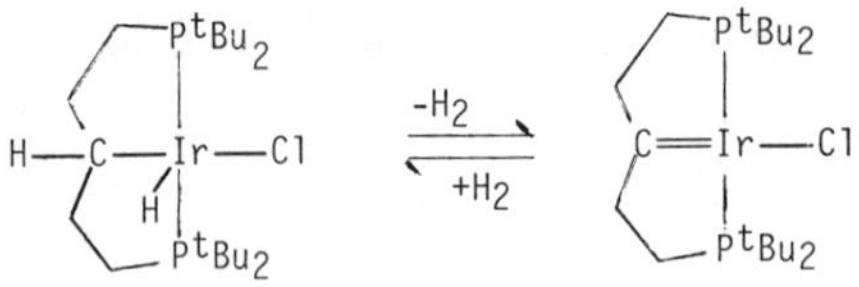

(54)   (55)

The (methoxymethyl) ethyleneiridium(I) complex **56** reacts with electrophilic reagents (e.g., Me_3SiOTf) to generate the hydridoallyl complex **58**. Suspected intermediates in this reaction include a cationic d^8 methyleneethylene species **57**, which rapidly converts to a metallacyclobutane species and subsequently undergoes β-elimination to form **58** (*90*). When **56**, with C_2H_4 replaced by p-tol-C≡C-p-tol, is treated with Me_3SiBr, a stable metallacyclobutene complex results (*91*). An iridium(I) methylene complex is again suggested as an intermediate.

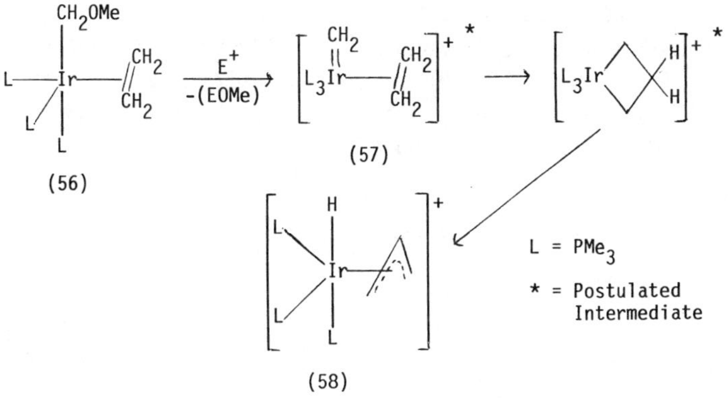

2. d^6 Alkylidene Complexes

Ru, Os, and Ir alkylidene complexes with the metal electron configuration d^6 have been prepared by two distinct routes. The addition of mineral and organic acids to electron-rich Ru and Os carbyne complexes affords stable carbene complexes (see also Section VI,C,1,a) while hydroxy-, alkoxy-, and halomethyl complexes of Os and Ir behave as precursors to highly reactive d^6 methylene species. These latter compounds have not been isolated, but their intermediacy is inferred from the subsequent nucleophilic additions to carbene carbon.

From Carbyne Complexes. Addition of HCl across the metal–carbon triple bonds of Ru and Os d^8 arylcarbyne complexes yield stable, neutral secondary alkylidene complexes:

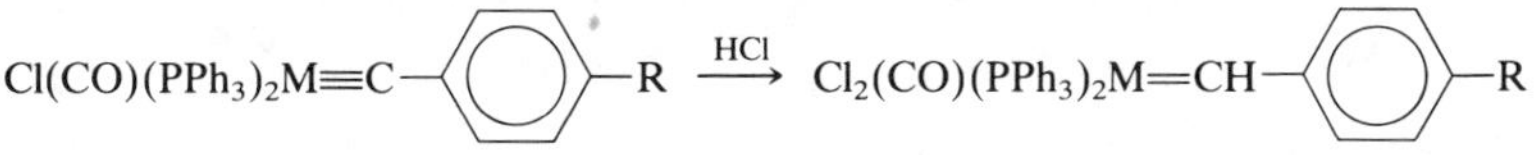

59	M = Os, R = H	(Ref. *39*)
	R = Me	(Ref. *36*)
	M = Ru, R = OMe	(Ref. *39*)

Compound **59** has been characterized by X-ray crystallography (*39*), and a description of the molecular structure is given in Section IV,B.

If organic acids such as MeCOOH are used in place of HCl, cationic carbene complexes are formed (39), e.g.,

$$Cl(CO)(PPh_3)_2Os\equiv CPh \xrightarrow[\text{NaClO}_4]{\text{MeCO}_2\text{H}} [(\eta^2\text{-O}_2CMe)(CO)(PPh_3)_2Os=CHPh]ClO_4$$

Related cationic complexes are readily formed by abstraction of chloride from the neutral carbene complexes.

From Hydroxy-, Alkoxy-, and Halomethyl Complexes. Hydroxy-, alkoxy-, and halomethyl ligands have proved to be incipient methylene functions in several d^6 Os and Ir complexes.

The dihaptoformaldehyde complex $Os(\eta^2\text{-CH}_2O)(CO)_2(PPh_3)_2$ reacts with hydrogen halides, affording hydroxymethyl species. Further reaction leads to the formation of halomethyl complexes, probably via the intermediacy of methylene complexes (60) (73):

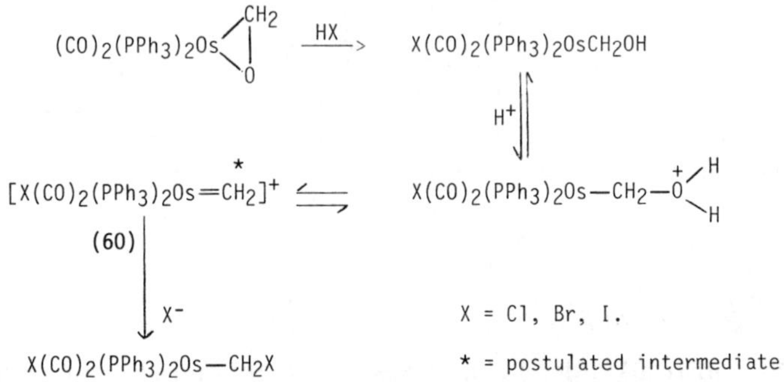

The spectroscopically characterized iron methylene complex **63** is formed in a closely related reaction of an alkoxymethyl compound (92):

$$(dppe)(\eta^5\text{-C}_5\text{H}_5)Fe\text{—CH}_2\text{—OEt} \xrightarrow{\text{CF}_3\text{CO}_2\text{H}} [(dppe)(\eta^5\text{-C}_5\text{H}_5)Fe=CH_2]^+$$

(63)

$$dppe = Ph_2PCH_2CH_2PPh_2$$

The reactivity of the iodomethyl complex **62** suggests the importance of the equilibrium:

$$I(CO)_2(PPh_3)_2Os\text{-CH}_2I \rightleftharpoons [I(CO)_2(PPh_3)_2Os=CH_2]I$$

(62)   (61)

For example (23),

$$[I(CO)_2(PPh_3)_2OsCH_2PPh_3]I \xleftarrow{\text{PPh}_3} I(CO)_2(PPh_3)_2Os\text{—CH}_2I \xrightarrow{\text{ROH}}$$
$$I(CO)_2(PPh_3)_2Os\text{—CH}_2OR$$

(62)   R = Me, Et

The chloromethyl complex $OsCl_2(CH_2Cl)(NO)(PPh_3)_2$ (64) undergoes similar reactions to its dicarbonyl counterpart (20).

The case for a highly electrophilic methylene intermediate in these reactions is strengthened by comparisons with the stable cationic rhenium methylene 65 (24) (see also Section II,A above):

$$[(\eta^5\text{-}C_5H_5)(NO)(PPh_3)Re\text{-}CH_2PPh_3]^+ \xleftarrow{\text{PPh}_3} [(\eta^5\text{-}C_5H_5)(NO)(PPh_3)Re\text{=}CH_2]^+$$

$$|\,(65)$$

$$\downarrow \text{NaOMe}$$

$$(\eta^5\text{-}C_5H_5)(NO)(PPh_3)Re\text{—}CH_2OMe$$

Methyl migration to the electrophilic methylene carbon of a cationic Ir(III) complex (66) to form an iridium ethyl complex has been reported (93):

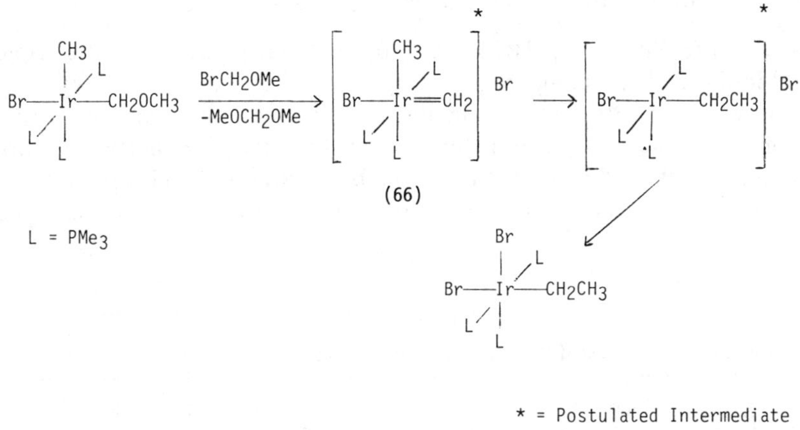

(66)

L = PMe₃

* = Postulated Intermediate

In contrast, nucleophilic tantalum alkylalkylidene complexes show no tendency to undergo this migration reaction (15).

B. Structure and Spectroscopic Properties

1. Structure

X-Ray crystallographic studies of the methylene complex $Os(\text{=}CH_2)Cl(NO)(PPh_3)_2$ (47) and the phenylcarbene complex $OsCl_2(CO)(\text{=}CHPh)(PPh_3)_2$ (59) have been undertaken to fully characterize these compounds. The only other terminal methylene complexes to be crystallographically characterized are $(\eta^5\text{-}C_5H_5)_2MeTa\text{=}CH_2$ (94) and

$[(\eta^5\text{-}C_5Me_5)(NO)(P(OPh)_3)Re=CH_2]PF_6$ (95), while crystal structures of three other phenylcarbene complexes have been reported (96–98).

Coordination about the osmium in **47** is best regarded as distorted trigonal bipyramidal with axial triphenylphosphine ligands. The distortion is toward a square pyramidal geometry with an apical nitrosyl ligand. Coordination about Os in the six-coordinate phenylcarbene complex is octahedral, as expected.

a. The Metal–Carbon Bond. The metal–carbon bond lengths of carbene complexes **47** and **59** and some related compounds are listed in Table IV. The metal–carbene carbon distances are similar in the methylene and phenylcarbene complexes and are comparable to the bond length found in the vinylidene complex **67**. These lengths lie in the range observed for osmium–carbon in carbonyl complexes, which also may be regarded as having double-bond character. The osmium–carbon distances are significantly shorter than the characterized Os—C single bond in $\overline{OsCl(CO)_2(o\text{-}PPh_2C_6H_4CHC_6H_4PPh_2\text{-}o)}$ (**69**) but longer than the triple bond in the carbyne complex (**68**).

It is interesting to note that the decrease in metal electron density that accompanies the change from five- to six-coordinate geometry does not have a detectable effect on the metal–carbene carbon bond length in these complexes. The metal–carbyne carbon bond in several osmium carbyne

TABLE IV

METAL–CARBON BOND LENGTHS IN OSMIUM CARBENE AND RELATED COMPOUNDS

Compound[a]	Ligand	M—C (Å)	Ref.
$Cl(NO)L_2Os=CH_2$ (**47**)	CH_2	1.92(1)	20
$Cl_2(CO)L_2Os=CHPh$ (**59**)	CHPh	1.94(1)	39
	CO	1.83(2)	
 $(CO)_2L_2Os=C$ (structure) Me H (**67**)	Vinylidene	1.90(1)	33
	CO	1.93(1)	
$Cl(CO)L_2Os\equiv C$  —Me (**68**)	CPh	1.77(2)	36
$\overline{OsCl(CO)_2(o\text{-}PPh_2C_6H_4CHC_6H_4PPh_2\text{-}o)}$ (**69**)	Alkyl	2.215(8)	99

[a] $L = PPh_3$.

complexes shows a similar lack of sensitivity to changes in metal electron density (see Section, VI,B).

b. Carbene Ligand Conformation. The conformations of the methylene ligand in 47 and the phenylcarbene ligand in 59 are also noteworthy.

A simple symmetry-based analysis of the bonding orbitals for the carbene and metal fragments of a trigonal bipyramidal d^8 complex suggests that the vertical orientation (i) should be preferred (100):

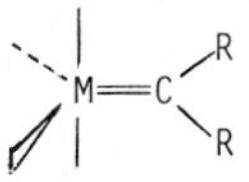

(i) 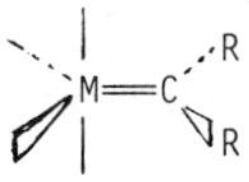   (ii)

While steric factors would lead to orientation (ii) being favored, the observation that the methylene ligand in 47 lies in the vertical plane indicates the importance of the former analysis.

The carbene ligand in 59 is coplanar with the equatorial plane of the complex, the phenyl ring also lying approximately in the same plane. This is the conformation that would be predicted on both steric and electronic grounds.

The Os—C_{carb}Ph angle of 139(1)° in 59 is substantially increased from the angle of 120° expected for an sp^2 hybridized carbon atom. This distortion is probably a consequence of both steric interactions between the phenyl ring and other equatorial ligands and the minimal steric demands of the proton (the other carbene substituent).

c. Trans Influence. The trans influence of the phenylcarbene ligand in 59 is marked, the trans metal–chlorine bond length being about 0.07 Å longer than the cis bond length. This result is in agreement with the observed lability of the trans chloride ligand.

2. Spectroscopic Properties

The value of ^{1}H-NMR and ^{13}C-NMR spectroscopy in characterizing transition metal carbene complexes was noted in Section III,B,2. The carbene carbon resonance is invariably found at low field (200–400 ppm) in the ^{13}C-NMR spectrum, while protons attached to C_α in 18-electron primary and secondary carbene complexes also resonate at low fields. NMR data for some Ru, Os, and Ir alkylidene complexes and related compounds are given in Table V.

TABLE V

^{1}H AND ^{13}C-NMR DATA FOR ALKYLIDENE COMPLEXES

Compound[a]	=CHR $(\delta)^b$	=CHR $(\delta)^b$	Ref.
Cl(NO)L$_2$Os=CH$_2$	13.81 (t)		20
Cl(NO)L$_2$Ru=CH$_2$	13.30 (t)		39
I(CO)L$_2$Ir=CH$_2$	12.88 (t)		86
(η^5-C$_5$H$_5$)$_2$MeTa=CH$_2$	10.11 (s)	224	101
Cl(NO)L$_2$Os=CHMe	14.90 (m)	227	85
[(η^5-C$_5$H$_5$)(CO)LFe=CHMe]$^+$	17.94 (q)	380	102
Cl(NO)L$_2$Os=CH-p-tolyl	15.92 (t)		85
Cl$_2$(CO)L$_2$Os=CH-p-tolyl	18.05 (t)		36
[Cl(CO)(MeCN)L$_2$Os=CHPh]$^+$	17.82 (t)		39
(η^5-C$_5$H$_5$)(CH$_2$Ph)Ta=CHPh	6.73 (s)	221	96
Cl(NO)L$_2$Os=CHCO$_2$Et	13.85 (t)		85

[a] L = PPh$_3$
[b] Chemical shifts in ppm.

C. Reactivity

The effect of metal basicity on the mode of reactivity of the metal–carbon bond in carbene complexes toward electrophilic and nucleophilic reagents was emphasized in Section II above. Reactivity studies of alkylidene ligands in d^8 and d^6 Ru, Os, and Ir complexes reinforce the notion that electrophilic additions to electron-rich compounds and nucleophilic additions to electron-deficient compounds are the expected patterns. Notable exceptions include addition of CO and CNR to the osmium methylene complex 47. These latter reactions can be interpreted in terms of non-innocent participation of the nitrosyl ligand.

1. Electrophilic Addition Reactions

The characteristic reactivity of neutral d^8 alkylidene complexes of Ru, Os, and Ir is with electrophilic reagents. The osmium methylene 47 reacts with the widest range of electrophiles, the most significant reactions being summarized in Scheme 2.

a. Protic Acids. The osmium methyl complex 70 is formed immediately in the reaction of 47 with HCl (20). Ruthenium and iridium methyl complexes are formed analogously from the methylene complexes Ru(=CH$_2$)Cl(NO)(PPh$_3$)$_2$ and Ir(=CH$_2$)I(CO)(PPh$_3$)$_2$ (39,86), while HCl addition to the osmium alkylidenes 48, 49, and 50 affords stable alkyl

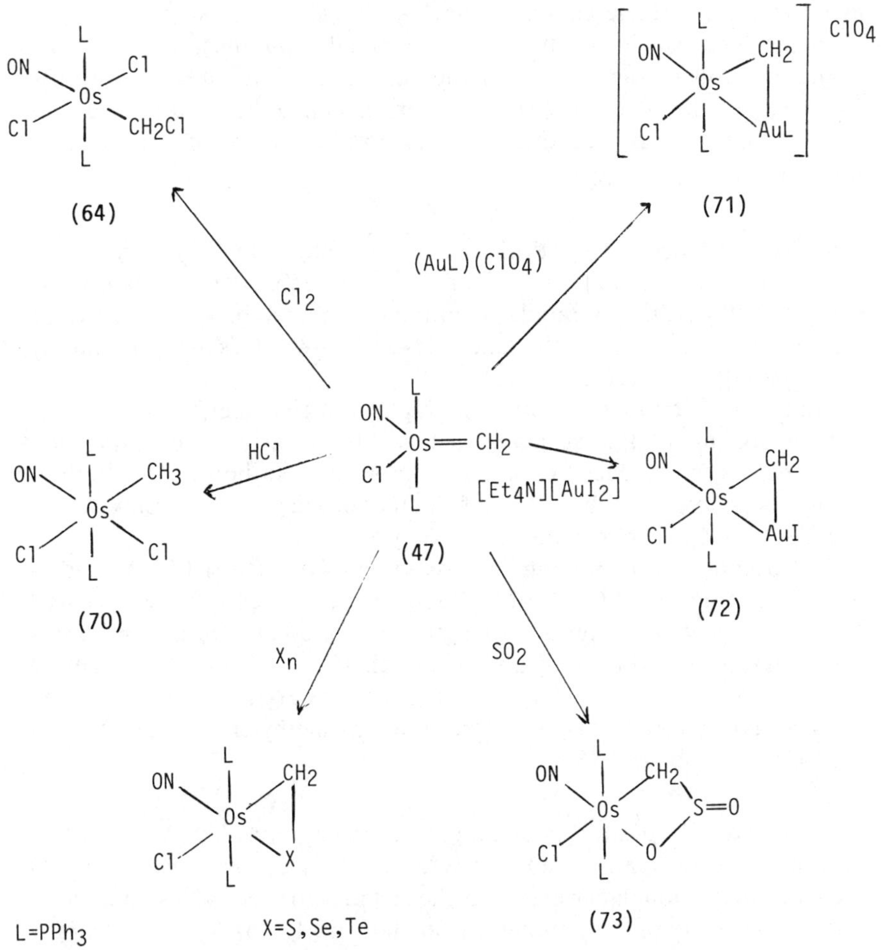

Scheme 2. Electrophilic addition to $Os(=CH_2)Cl(NO)(PPh_3)_2$.

complexes (85):

$$Cl(NO)(PPh_3)_2Os=CHR \xrightarrow{HCl} Cl_2(NO)(PPh_3)_2Os-CH_2R$$

(48) R = Me
(49) R = p-tolyl
(50) R = CO_2Et

The initial site of protonation in these reactions has not been unambiguously determined. Alkyl ligand formation by protonation at the metal followed by a rapid 1,2-proton shift to the alkylidene ligand is equally as plausible as direct protonation at C_α. As the metal electron density increases in a carbene complex, the HOMO of the metal–carbon bond becomes increasingly localized on the metal, and thus, in a frontier orbital-controlled reaction, protonation at the metal would be most likely. In a charge-controlled reaction, however, proton addition to the carbene carbon would be expected, this being the most negatively charged center in the molecule.

b. Gold Adducts. $[AuPPh_3]^+$ has been used as an H^+ synthon in a number of reactions in cluster chemistry (e.g., *103*). The reaction of **47** with $[AuPPh_3][ClO_4]$ is rapid, generating a cationic bridging methylene complex **71**. The neutral methylene-bridged complex **72** is formed similarly from $[Et_4N][AuI_2]$ (*20*).

Compound **72** has been crystallographically characterized. The most striking feature of this structure is how little structural reorganization accompanies this adduct formation. The osmium–carbon bond length at 1.90(2) Å is not detectably altered from that of the parent complex **47** and the C—Os—C angle is increased by only 8°.

The osmium carbyne complex $Os(\equiv C\text{-}p\text{-tolyl})Cl(CO)(PPh_3)_2$ under-goes an analogous adduct-forming reaction with AgCl (*40*) (see Section VI,C). The metal–carbon bond lengthening on adduct formation here is significant, in accord with the observation that this carbyne adduct is more stable than the methylene analog. $Ir(\equiv CH_2)I(CO)(PPh_3)_2$ and $Ru(\equiv CH_2)Cl(NO)(PPh_3)_2$ also form bridging methylene complexes with gold electrophiles (*39,86*).

c. Sulfene Complexes. Compound **47** reacts immediately with SO_2, affording the osmium sulfene complex **73** (*104*). The structure of **73** indicates that a nonplanar metallacyclic ligand is formed in this interaction (*104*). The length of the osmium–carbon bond [2.115(6) Å] is substantially greater than the bond length in **47** and approaches characterized Os—C single bond distances (see Table IV). This SO_2 adduct contrasts with the gold adduct: while the gold atom is only weakly bound in **72**, the SO_2 interacts strongly with the metal–carbon bond in **73**.

The ruthenium and iridium methylene complexes and the osmium alkylidenes **48–50** also react with SO_2 to afford sulfene complexes (*104*).

d. Chalcoformaldehyde Complexes. Compound **47** undergoes slow reactions with elemental sulfur, selenium, and tellurium to yield thio-, seleno-, and telluroformaldehyde complexes (*20*). The other member of this series, $Os(\eta^2\text{-}CH_2O)Cl(NO)(PPh_3)_2$, has been prepared directly from CH_2O and **45** (*39*).

e. Reactions with Halogens.   Chlorine addition across the Os=C bond of **47** affords the chloromethyl complex **64** (*20*). Reactions of **64** with nucleophiles can be thought of as proceeding via an intermediate cationic methylene complex $[OsCl_2(NO)(=CH_2)(PPh_3)_2]^+$ (**74**) (see below).

2. Nucleophilic Addition Reactions

a. Halomethyl Ligands as Methylene Precursors.   It has been noted above that the reactions of halomethyl complexes **62** and **64** with nucleophiles may be thought of as proceeding via cationic d^6 methylene complexes $[OsI(CO)_2(=CH_2)(PPh_3)_2]^+$ (**61**) and $[OsCl_2(NO)(=CH_2)$ $(PPh_3)_2]^+$ (**74**). Two further examples which clearly illustrate the change in reactivity of the metal–carbon double bond that can be brought about by varying the basicity of the metal center are

$$OsI(CH_2I)(CO)_2(PPh_3)_2 \xrightarrow{NaBH_4} OsI(CH_3)(CO)_2(PPh_3)_2 \qquad (Ref.\ 105)$$

$$(62)$$

$$\xrightarrow{XH^-}$$

$$(X = S,\ Se,\ Te) \rightarrow Os(\eta^2\text{-}CH_2X)(CO)_2(PPh_3)_2 \quad (Ref.\ 106)$$

Thus methyl and chalcoformaldehyde complexes of osmium are accessible by both electrophilic addition to a neutral d^8 methylene complex and nucleophilic addition to a cationic d^6 methylene complex.

The insertion reactivity of the electrophilic iridium methylene species **57** was noted above. A cationic iridium methylene complex is also the likely intermediate in the thermal rearrangement of $Ir(=CH_2)I(CO)(PPh_3)_2$ to the ortho-metallated ylide complex **75** (*86*).

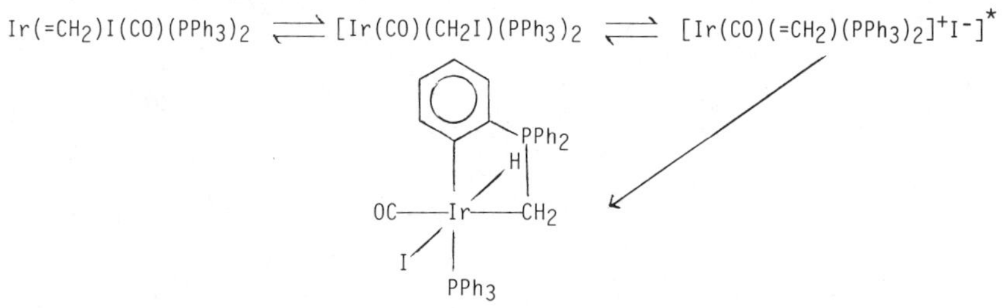

$$Ir(=CH_2)I(CO)(PPh_3)_2 \rightleftharpoons [Ir(CO)(CH_2I)(PPh_3)_2 \rightleftharpoons [Ir(CO)(=CH_2)(PPh_3)_2]^+I^-]^*$$

(75)   * Postulated Intermediates

b. Nucleophilic Addition to Os(=CH$_2$)Cl(NO)(PPh$_3$)$_2$.   The nucleophilicity of the metal–carbon bond in **47** has been stressed in the previous sections. At first sight it seems a little surprising, then, that nucleophilic

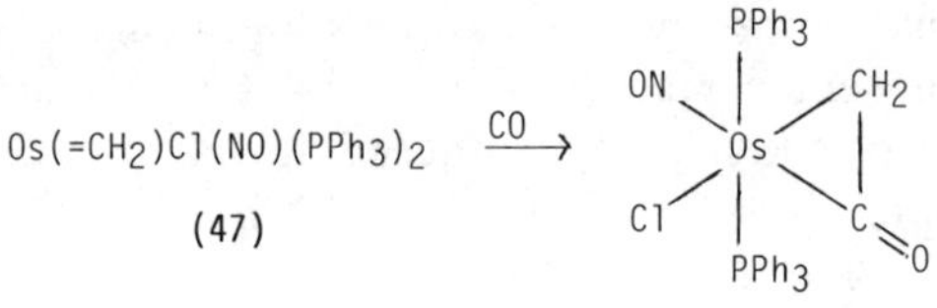

$$Os(=CH_2)Cl(NO)(PPh_3)_2 \xrightarrow{CO}$$

(47)

(76)

reagents such as CO and CNR should also react with this compound. Compound **47** rapidly adds CO, affording the ketene complex **76** (*39*).

Carbonylation of the manganese carbene complex $(\eta^5\text{-}C_5H_5)$ $(CO)_2Mn{=}CPh_2$ to an analogous ketene complex has been reported (*107*). Compound **76** is also formed directly in the reaction of ketene with **45** (*39*).

It is reasonable to postulate the coordination of CO to osmium prior to the formation of the ketene ligand in **76**, i.e.,

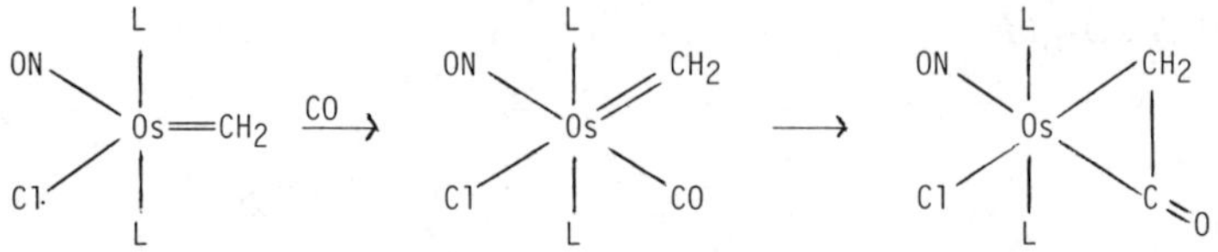

$$\xrightarrow{CO}$$

(47)   (76)

Such a mechanism would have to involve the nitrosyl ligand acting in a non-innocent manner, changing from a three-electron donor to a one-electron donor in the intermediate complex. Such participation of the nitrosyl ligand has precedent in related systems (*108*).

A similar mechanism may be invoked to explain the reaction of **47** with isocyanides in which ketenimine complexes are formed (*39*):

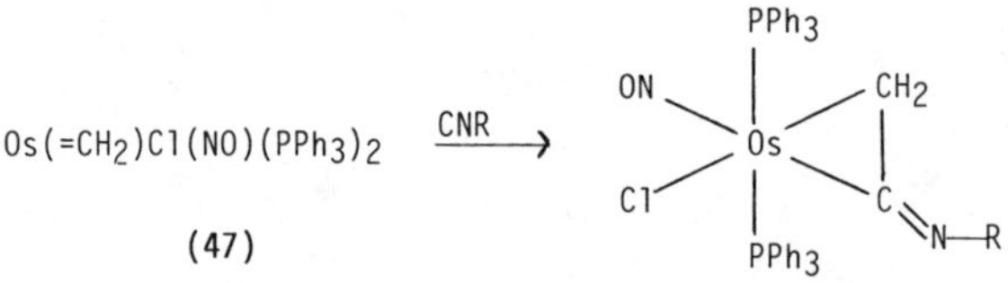

$$Os(=CH_2)Cl(NO)(PPh_3)_2 \xrightarrow{CNR}$$

(47)

R = p-tolyl, p-chlorophenyl

Other workers have also observed the interaction of an isocyanide with a carbene ligand to yield a coordinated ketenimine (*109*).

V

DIHALO- AND MONOHALOCARBENE COMPLEXES

The first transition metal dichlorocarbene complex was reported in 1977. This was a derivative of *meso*-tetraphenylporphyrinatoiron(II), $Fe(TPP)(=CCl_2)(H_2O)$, which resulted from the reaction of Fe(TPP) with CCl_4 in the presence of an excess of reducing agent (*110*). A difluorocarbene complex of molybdenum, $[(\eta^5-C_5H_5)(CO)_3Mo=CF_2]^+$, followed in 1978 but this complex could not be isolated, characterization being by ^{19}F- and ^{13}C-NMR spectroscopy (*111*). In 1980 a stable dichloro-carbene complex of osmium(II) was described (*112*) and since then a large number of dihalocarbene complexes of ruthenium, osmium, and iridium has been prepared. The striking feature of these compounds is the readiness with which they undergo substitution reactions of the carbene ligand. This reactivity results from the combination of an electrophilic carbene center and the presence of good leaving groups as carbene substituents. Various monohalocarbene complexes are accessible through these substitution reactions. In this section the syntheses, structural features and reactions of dihalo- and monohalocarbene complexes are discussed.

A. *Synthesis of Dihalocarbene Complexes*

1. *Modification of Trihalomethyl Derivatives, $L_nM—CX_3$*

This is the principal method of synthesis. In the special case when X = F a number of reagents is effective for fluoride abstraction to produce initially a cationic CF_2 intermediate,

$$L_nM—CF_3 \xrightarrow{-F^-} [L_nM=CF_2]^+$$

but for X = Cl, very few CCl_3 derivatives have been isolated, chloride loss being spontaneous or perhaps a rearrangement process occurring

$$L_nM—CCl_3 \xrightarrow{-L} L_{n-1}ClM=CCl_2$$

a. CF_2 Complexes from CF_3 Precursors. CF_3 derivatives of the transition metals constitute a well-studied group of compounds, and many good synthetic methods are available (*113,114*). For Group 8 metals a suitable reagent is $Hg(CF_3)_2$ which via an oxidative addition process leads cleanly to CF_3 derivatives (22), e.g.,

$$Ru(CO)_2(PPh_3)_3 + Hg(CF_3)_2 \rightarrow Ru(CF_3)(HgCF_3)(CO)_2(PPh_3)_2 + PPh_3$$

A derived Ru(II) CF_3 complex, $Ru(CF_3)Cl(CO)(MeCN)(PPh_3)_2$, is a good substrate for generation of a CF_2 complex using either anhydrous HCl gas (22) or Me_3SiCl as fluoride-abstracting reagents,

$$Ru(CF_3)Cl(CO)(MeCN)(PPh_3)_2 \xrightarrow{HCl} RuCl_2(=CF_2)(CO)(PPh_3)_2 + HF$$

or

$$Ru(CF_3)Cl(CO)(MeCN)(PPh_3)_2 \xrightarrow{Me_3SiCl} RuCl_2(=CF_2)(CO)(PPh_3)_2 + Me_3SiF$$

When $Cd(CF_3)_2$ is used instead of $Hg(CF_3)_2$ to introduce a CF_3 group the reaction proceeds a step further producing a zero-valent CF_2 complex (21).

$$Ru(CO)_2(PPh_3)_3 + Cd(CF_3)_2 \rightarrow Ru(=CF_2)(CO)_2(PPh_3)_2 + PPh_3 + CF_3CdF$$

The simple oxidative addition product $Ru(CF_3)(CdCF_3)(CO)_2(PPh_3)_2$ is a likely intermediate in this reaction. Several stable, zero-valent CF_2 complexes have been isolated using $Cd(CF_3)_2$, and they are included in Table VII (see Section V,C,2,b).

b. *CCl_2 Complexes from Oxidative Addition of CCl_4 to a Zero-Valent Ru Complex.* Reaction between $Ru(CO)(PPh_3)_3$ (115) and CCl_4 gives a number of different products, but one of them, in low yield, is $RuCl_2(=CCl_2)(CO)(PPh_3)_2$ (116). The product can be seen as arising from "three-fragment oxidative addition" or the rearrangement of the simple oxidative addition product $Ru(CCl_3)Cl(CO)(PPh_3)_3$. CCl_4 is also the source of the CCl_2 ligand in the iron tetraphenylporphyrin series of carbene complexes (110).

c. *CCl_2 Complexes from Reactions with $Hg(CCl_3)_2$.* Many organo-mercury compounds are effective reagents for the transfer of organic groups to transition metals. $Hg(CCl_3)_2$ is no exception and with suitable transition metal substrates leads directly to CCl_2 complexes in high yield (112), e.g.,

$$OsHCl(CO)(PPh_3)_3 + Hg(CCl_3)_2 \rightarrow OsCl_2(=CCl_2)(CO)(PPh_3)_2 + PPh_3 + Hg + CHCl_3$$

Other CCl_2-complexes which have been prepared using $Hg(CCl_3)_2$ include $RuCl_2(=CCl_2)(CO)(PPh_3)_2$ (19), $OsCl_2(=CCl_2)(CS)(PPh_3)_2$ (39) and $IrCl_3(=CCl_2)(PPh_3)_2$ (73). The trichloromethyl derivatives which are presumed to be intermediates in these reactions have not been isolated, but an iridium trichloromethyl complex, $Ir(CCl_3)Cl_2(CO)(PMe_2Ph)_2$, is known (117).

d. CCl_2 Complexes from Reaction of $L_nM—CF_3$ with BCl_3. $BX_3(X = Cl, Br, I)$ converts L_nMCF_3 to L_nMCX_3 (118), and when advantage is taken of this reaction to convert a ruthenium(II) CF_3 complex to a ruthenium(II) CCl_3 complex the isolated product is a CCl_2 complex (119):

$$Ru(CF_3)Cl(CO)_2(PPh_3)_2 + BCl_3 \rightarrow Ru(CCl_3)Cl(CO)_2(PPh_3)_2 \rightarrow$$

$$RuCl_2(=CCl_2)(CO)(PPh_3)_2$$

The suggested intermediate $Ru(CCl_3)Cl(CO)_2(PPh_3)_2$ could not be detected.

2. Modification of a Complex Already Incorporating a Dihalocarbene Ligand

The high reactivity of dihalocarbene ligands makes it unlikely that many examples of this method will be recognized; however, the following two process should be noted.

a. Cation Generation from CCl_2 Complexes using Silver Salts. The neutral CCl_2 complexes $RuCl_2(=CCl_2)(CO)(PPh_3)_2$ and $OsCl_2(=CCl_2)(CO)(PPh_3)_2$ both react with $AgClO_4$ in the presence of MeCN to form the stable, but very sensitive to hydrolysis, salts $[RuCl(=CCl_2)(CO)(MeCN)(PPh_3)_2]ClO_4$ (39) and $[OsCl(=CCl_2)(CO)(MeCN)(PPh_3)_2]ClO_4$ (116).

b. Conversion of CCl_2 Complex to CFCl Complex. The fluorinating ability of $Cd(CF_3)_2$ can be used in forming the mixed dihalocarbene complex $OsCl_2(=CFCl)(CO)(PPh_3)_2$ directly from $OsCl_2(=CCl_2)(CO)(PPh_3)_2$ (74).

B. Synthesis of Monohalocarbene Complexes

1. Modification of α-Dihaloalkyl Derivatives, $L_nM—CX_2R$

Just as trihalomethyl derivatives lead to dihalocarbene complexes through halide loss it is to be expected that α-dihaloalkyl derivatives should be good precursors of monohalocarbene complexes. In practice, comparatively few α-dihaloalkyl derivatives are known, and the few attempts which have been made with Group 8 metals to utilize this approach have all failed to give isolable monohalocarbene complexes. However, the derived products (see Section V,D,2) indicate the generation of monohalocarbene complexes as intermediates.

2. Modification of Carbyne Complexes

Halide addition to a cationic carbyne complex, $[L_nM\equiv CR]^+$, or halogen oxidation of a low oxidation state carbyne complex are both potential routes to monohalocarbene species. Examples of the first process are well known for carbyne complexes from Groups 6 and 7 of the periodic table (120), e.g.,

$$[(\eta^5\text{-}C_5H_5)(CO)_2Mn\equiv CR]^+ + F^- \rightarrow (\eta^5\text{-}C_5H_5)(CO)_2Mn=CFR$$

and in Group 8 the reaction

$$Cl(CO)(PPh_3)_2Os\equiv CR + Cl_2 \rightarrow Cl_2(CO)(PPh_3)_2Os=CClR$$

is an example of the second process (36). It is possible that this last reaction proceeds through a transient cationic carbyne intermediate, $[Os(\equiv CR)Cl_2(CO)(PPh_3)_2]^+$.

3. Oxidative Addition of [Me$_2$NCCl$_2$]Cl

The reaction of $[Me_2NCCl_2]Cl$ with various transition metal substrates has proved to be a useful method of synthesizing chloroaminocarbene complexes (121). A Group 8 example is

$$RhCl(PPh_3)_3 + [Me_2NCCl_2]Cl \rightarrow RhCl_3(=CClNMe_2)(PPh_3)_2 + PPh_3$$

4. Modification of a Complex-Already Incorporating a Monohalocarbene Ligand

Monohalocarbene ligands often remain unchanged while the substitution of other ligands at the metal center occurs. In the case of $RuCl_2(=CClNMe_2)(CO)(PPh_3)_2$ the carbene ligand sufficiently labilizes the trans Cl ligand that reaction with CN-p-tolyl leads to direct formation of the cationic complex $[RuCl(=CClNMe_2)(CO)(CN\text{-}p\text{-tolyl})(PPh_3)_2]^+$ (19). Likewise, the reaction of $OsCl_2(=CFNMe_2)(CO)(PPh_3)_2$ with AgSbF$_6$ in MeCN leads to chloride removal from the metal and formation of $[OsCl(=CFNMe_2)(CO)(MeCN)(PPh_3)_2]^+$ (74).

c. Structure and Spectroscopic Properties

1. Structure

X-ray crystallographic studies have been carried out on four dihalo-carbene complexes and one monohalocarbene complex of Group 8 metals. Some important structural parameters are presented in Table VI. The geometries of the six-coordinate complexes are essentially octahedral and of the five-coordinate complexes essentially trigonal bipyramidal. In every structure the two triphenylphosphine ligands are trans. None of the carbon–halogen distances determined so far suggest any substantial π-interaction between halogen and carbene carbon, except perhaps in $Os(=CF_2)Cl(NO)(PPh_3)_2$. One structure, $Ru(=CF_2)(CO)_2(PPh_3)_2$, has exceptionally long C—F bond distances.

a. The Metal–Carbon Bond. All the metal–carbon distances listed in Table VI are significantly shorter than M—C single bonds and are in the range expected for M=C double bonds (see Section IV,B,1). In the case of $Os(=CF_2)Cl(NO)(PPh_3)_2$ an interesting comparison can be made with $Os(=CH_2)Cl(NO)(PPh_3)_2$. Replacement of H-substituents on the cabene by F-substituents increases the osmium–carbon bond distance from 1.92(1) to 1.967(4) Å. In terms of the bonding model discussed in Section II,A this suggests that, in this molecule at least, fluorine is functioning as a weak π-donor. The measured C—F bond distances also support this idea.

b. The Carbon–Halogen Bond. The C—Cl distance of 1.721(5) Å found in $IrCl_3(=CCl_2)(PPh_3)_2$ is very close to the C—Cl distances in vinyl chloride [1.728(7) Å] and in 1,1'-dichloroethylene [1.710(10) Å]. There is therefore no reason to believe that chlorine functions as a π-donor, and the very short iridium–carbon distance of 1.872(7) Å is compatible with this view. The observed C—F distances (Table VI) spread over a rather wide range. In Section V,C,2 it will be noted that $\nu(C—F)$ values also show a wide spread and that they correlate with the observed C—F distances. In the case of $Ru(=CF_2)(CO)_2(PPh_3)_2$ the C—F bond distances [1.36(1) and 1.37(1) Å] seem long when compared with tetrafluoroethylene [1.313(10) Å] and carbonylfluoride [1.3156(5) Å] but are less than the C—F bond length of 1.392(6) Å found in $(\eta^5\text{-}C_5H_5)(CO)_2Mn=CFPh$ (*120*). It is interesting that of the three zero-valent CF_2 structures, the angle F—C—F is smallest [88.7(9)°] for $Ru(=CF_2)(CO)_2(PPh_3)_2$ where the metal–carbon distance is shortest and C—F longest and largest [105.0(7)°] for $Os(=CF_2)Cl(NO)(PPh_3)_2$ where M=C is longest and C—F shortest.

TABLE VI

Dihalo- and Monohalocarbene Complexes: Some Structural Parameters

Compound[a]	M—C (Å)	C—F (Å)	C—Cl (Å)	angle F—C—F or Cl—C—Cl or F—C—O	Ref.
IrCl₃(=CCl₂)L₂	1.872(7)		1.721(5)	107.5(4)	73
Ru(=CF₂)(CO)₂L₂	1.83(1)	1.36(1), 1.37(1)		88.7(9)	21
Os(=CF₂)(CO)₂L₂	1.884(13)	1.331(18), 1.407(17)		101.4(12)	119
Os(=CF₂)Cl(NO)L₂	1.967(4)	1.278(11), 1.285(11)		105.0(7)	119
RuCl₂(=CFOCH₂CMe₃)(CO)L₂	1.914(5)	1.307(6)		107.4(5)	74

[a] L = PPh₃.

c. *Carbene Ligand Orientation.* For each of the five-coordinate CF_2 complexes the carbene plane is perpendicular to the equatorial plane of the trigonal bipyramid. As discussed in Section IV,B,1, this is the preferred orientation for maximum π-overlap. For six-coordinate, octahedral complexes, conformational preferences are expected to be less marked, and in $RuCl_2(=CFOCH_2CMe_3)(CO)(PPh_3)_2$ the carbene ligand lies in the equatorial plane while in $IrCl_3(=CCl_2)(PPh_3)_2$ the carbene is twisted 24.4° from the equatorial plane.

d. *trans-Influence.* The dihalo- and monohalocarbene ligands all exhibit a pronounced trans influence on metal–halogen bonds. In $IrCl_3(=CCl_2)(PPh_3)_2$ the Ir—Cl distance trans to CCl_2 is 2.407(2) Å compared with 2.359(1) Å for the mutually trans Cl's; in $RuCl_2(=CFOCH_2CMe_3)(CO)(PPh_3)_2$, Ru—Cl trans to the carbene is 2.48(1) Å compared with 2.435(1) Å trans to CO. This bond-weakening effect translates into chemical lability, and examples of halide replacement from the metal have already been mentioned in Sections V,A,2 and V,B,4.

2. Spectroscopic Properties

a. *^{13}C-NMR Spectroscopy.* The ^{13}C chemical shift for the carbene carbon has been measured in only two halocarbene complexes, $OsCl_2$-$(=CCl_2)(CO)(PPh_3)_2$ (δ 222.5 ppm) (*112*) and $[RuCl(=CClNMe_2)\cdot(CO)(CN-p\text{-tolyl})(PPh_3)_2]^+$ (δ 222.28 ppm) (*39*). The values are close to those found for other carbene complexes (see Section IV,B,2).

b. *Infrared Spectrscopy.* For most of the dihalo- and monohalocarbene complexes which have been prepared it is possible to identify IR absorptions corresponding to $\nu(C—F)$ and $\nu(C—Cl)$. These data are collected in Table VII. The dihalocarbene complexes are associated with two IR absorptions; for the difluorocarbene complexes, where a greater variety of compound types is available, e.g., in different oxidation states for the same metal, the $\nu(C—F)$ absorptions occur over the wide range 1210–980 cm^{-1}. Where structural data are available to complement the IR data in every case shorter C—F bond distances are associated with higher values of $\nu(C—F)$ and vice versa. In terms of the reactivity patterns to be discussed in the next section, high $\nu(C—F)$ values are associated with the most electrophilic carbene centers and low $\nu(C—F)$ values with the most nucleophilic carbene centers. In this respect, $\nu(C—F)$ values for difluorocarbene complexes are like the $\nu(CO)$ values for metal carbonyls in being a probe of the electronic interaction between metal and ligand.

TABLE VII

Dihalo- and Monohalocarbene Complexes: IR Data

Compound[a]	ν(C—F) (cm^{-1})	ν(C—Cl) (cm^{-1})	Ref.
$RuCl_2(=CCl_2)(CO)L_2$		860, 790	19
$[RuCl(=CCl_2)(CO)(MeCN)L_2]^+$		865, 812	39
$OsCl_2(=CCl_2)(CO)L_2$		880, 780	112
$[OsCl(=CCl_2)(CO)(MeCN)L_2]^+$		890, 790	116
$OsCl_2(=CCl_2)(CS)L_2$		864, 770	39
$IrCl_3(=CCl_2)L_2$		880, 810	73
$OsCl_2(=CFCl)(CO)L_2$	1124	885	74
$[OsCl(=CFCl)(CO)(MeCN)L_2]^+$	1125	909	74
$RuCl_2(=CF_2)(CO)L_2$	1210, 1155		22
$Ru(=CF_2)Cl(NO)L_2$	1092, 980		21
$Os(=CF_2)(CO)_2L_2$	1142, 1022		119
$Os(=CF_2)(CO)(CS)L_2$	1122, 990		119
$Os(=CF_2)Cl(NO)L_2$	1100, 990		119
$RuCl_2(=CFNMe_2)(CO)L_2$	1154, 1020		119
$Ru(=CF_2)CO)_2L_2$	1026		22
$[RuCl(=CFNMe_2)(CO)(CN\text{-}p\text{-tolyl})L_2]^+$	1038		22
$RuCl_2(=CFOMe)(CO)L_2$	1060		22
$RuCl_2(=CClNMe_2)(CO)L_2$		795	19
$[RuCl(=CClNMe_2)(CO)(CN\text{-}p\text{-tolyl})L_2]^+$		814	19
$OsCl_2(=CClNMe_2)(CO)L_2$		790	39
$IrCl_3(=CClNMe_2)L_2$		820	73

[a] L = PPh_3.

D. Reactivity

The facility with which electrophilic halocarbene complexes undergo substitution reactions makes them extremely versatile synthetic intermediates, and this section summarizes these synthetic possibilities. Scheme 3 illustrates the usefulness of $RuCl_2(=CCl_2)(CO)(PPh_3)_2$. When the ligands are bound to electron-rich metal centers the electrophilicity is much reduced and interaction of the M=C function with some electrophiles can be observed.

1. Reactions with Nucleophiles

a. O-, S-, Se-, and Te-Containing Nucleophiles. All dihalocarbene ligands react with H_2O to produce a carbonyl ligand. A similar reaction with a monohalocarbene ligand will produce an acyl ligand. H_2S and H_2Se react with dichlorocarbene ligands to give thiocarbonyl and selenocarbonyl ligands, respectively. Although reaction with H_2O is very fast, no reaction

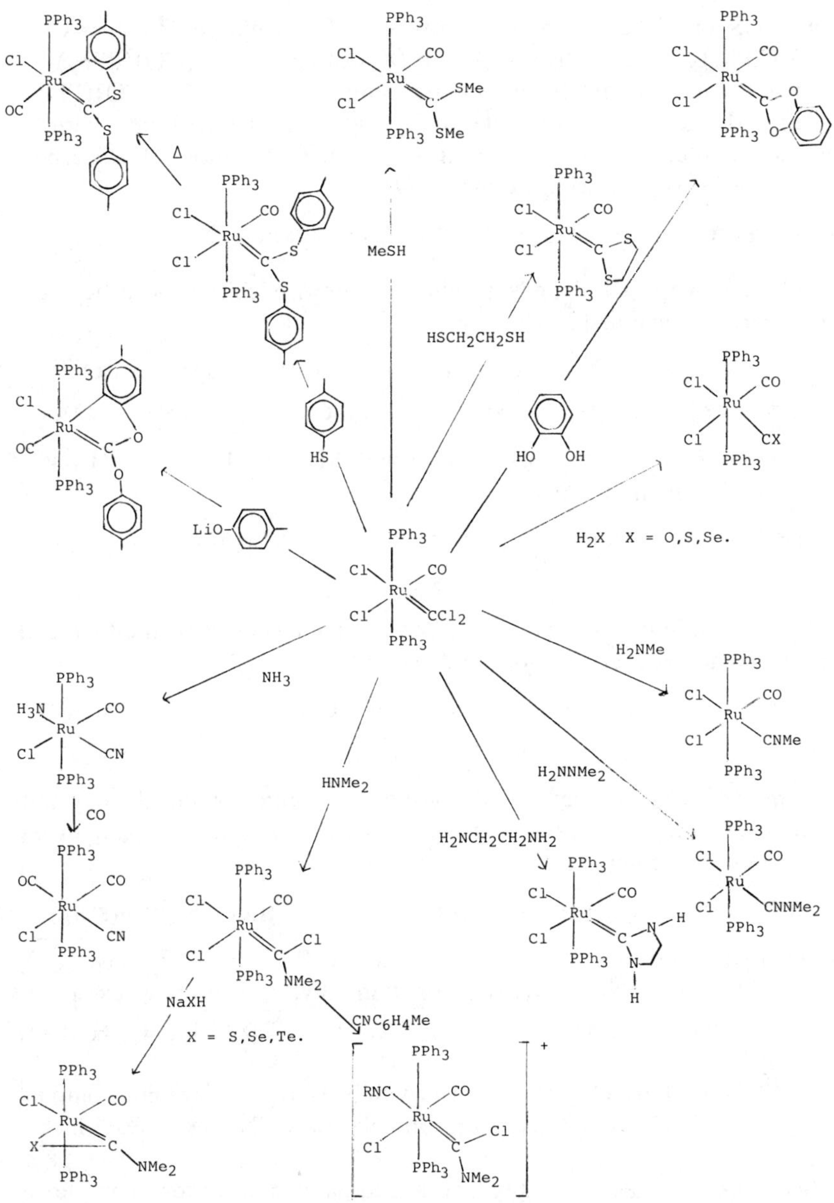

Scheme 3. Reactions of $RuCl_2(=CCl_2)(CO)(PPh_3)_2$.

with H_2S or H_2Se is observed for the CF_2 ligand in $RuCl_2(=CF_2)$-$(CO)(PPh_3)_2$. TeH^- also reacts with $OsCl_2(=CCl_2)(CO)(PPh_3)_2$ to give a tellurocarbonyl ligand in the complex $OsCl_2(CTe)(CO)(PPh_3)_2$ (112). Alcohols react with the CCl_2 ligand giving carbonyl with no detectable intermediates, but CF_2 ligands yield stable fluoroalkoxycarbene ligands as in the following example (74).

$$RuCl_2(=CF_2)(CO)(PPh_3)_2 + HOCH_2CMe_3 \rightarrow RuCl_2(=CFOCH_2CMe_3)(CO)(PPh_3)_2 + HF$$

Thiols react with CCl_2 ligands, producing, stepwise, chlorothiocarbene and dithiocarbene ligands (39,19), e.g.,

$$IrCl_3(=CCl_2)(PPh_3)_2 + MeSH \rightarrow IrCl_3[=CCl(SMe)](PPh_3)_2 + HCl$$

$$RuCl_2(=CCl_2)(CO)(PPh_3)_2 + 2 MeSH \rightarrow RuCl_2[C(SMe)_2](CO)(PPh_3)_2 + 2 HCl$$

Monochlorocarbene ligands also react with SH^-, SeH^-, and TeH^- as in the following example (19),

$$RuCl_2(=CClNME_2)(CO)(PPh_3)_2 + XH^- \rightarrow \overline{RuCl(CXNMe_2)}(CO)(PPh_3)_2 + Cl^-$$

$$X = S, Se, Te$$

b. N-Containing Nucleophiles. Primary amines react with either CCl_2 or CF_2 complexes, giving isocyanide complexes (112), e.g.,

$$OsCl_2(=CCl_2)(CO)(PPh_3)_2 + 3 RNH_2 \rightarrow OsCl_2(CNR)(CO)(PPh_3)_2 + 2 RNH_3Cl$$

$$R = Me, \textit{n}\text{-Bu}, \textit{p}\text{-tolyl}$$

The method can be used for the synthesis of functionalized isocyanide complexes, e.g., reaction with ethanolamine produces a 2-hydroxy-ethylisocyanide complex,

$$RuCl_2(=CCl_2)(CO)(PPh_3)_2 + NH_2CH_2CH_2OH \rightarrow RuCl_2(CNCH_2CH_2OH)(CO)(PPh_3)_2 + 2 HCl$$

which rearranges to a cyclic carbene complex, $RuCl_2(=\overline{CNHCH_2CH_2O})$-$(CO)(PPh_3)_2$ (19). Similarly, reaction with ethylenediamine gives directly a cyclic diaminocarbene ligand in $RuCl_2(=\overline{CNHCH_2CH_2NH})(CO)$-$(PPh_3)_2$ (19).

$RuCl_2(=CCl_2)(CO)(PPh_3)_2$ reacts with NH_3 to give a cyanide complex, $RuCl(CN)(CO)(NH_3)(PPh_3)_2$, presumably *via* a CNH complex which is deprotonated (19).

Secondary amines invariably give mixed haloaminocarbene complexes. The following examples are typical of this process (19,74).

$$RuCl_2(=CCl_2)(CO)(PPh_3)_2 + Me_2NH \rightarrow RuCl_2(=CClNMe_2)(CO)(PPh_3)_2 + HCl$$

$$OsCl_2(=CFCl)(CO)(PPh_3)_2 + Me_2NH \rightarrow OsCl_2(=CFNMe_2)(CO)(PPh_3)_2 + HCl$$

The presence of the amino-substituent must so reduce the electrophilicity of the carbene that no further reaction with the amine occurs.

c. C-Containing Nucleophiles.   In the case of aromatic lithium reagents this reaction leads to carbyne complexes and will be discussed in Section VI,A,1.

d. BH_4^-.   In appropriate circumstances it should be possible to convert dihalocarbene ligands to dihalomethyl ligands by hydride addition, but so far only the complete reduction of the CCl_2 ligand to the CH_3 ligand has been observed (*116*).

$$[OsCl(=CCl_2)(CO)(MeCN)(PPh_3)_2]^+ + \text{excess } BH_4^- \rightarrow OsH(CH_3)(CO)(H_2O)(PPh_3)_2$$

For a chloroaminocarbene complex it has been possible to replace Cl by H using $NaBH_4$ (*39*):

$$OsCl_2(=CClNMe_2)(CO)(PPh_3)_2 + BH_4^- \rightarrow OsCl_2(=CHNMe_2)(CO)(PPh_3)_2$$

2. Metallacycle Formation Involving Electrophilic Carbene Addition to a Benzene Ring of the PPh₃ Ligand

For very electrophilic carbene ligands bound to a metal center which also has coordinated an aromatic phosphine ligand,there is the possibility of the following intramolecular substitution reaction leading to a metallacycle:

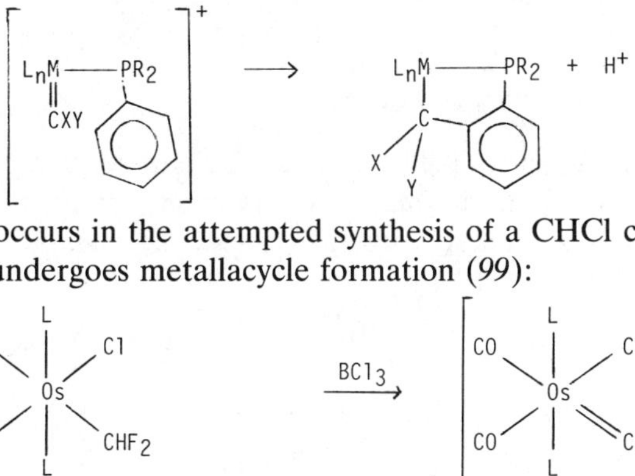

An example occurs in the attempted synthesis of a CHCl complex which immediately undergoes metallacycle formation (*99*):

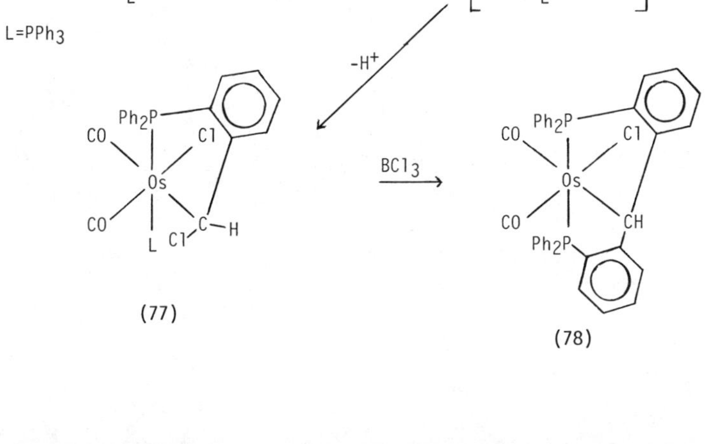

L=PPh₃

(77)

(78)

Excess BCl_3 leads to the metallabicycle **78** by further carbene generation (Cl^- abstraction from **77** and substitution at a benzene ring on a second PPh_3 ligand. Compounds **77** and **78** have been characterized by X-ray crystal structure determination (*99*). This is likely to be a quite general reaction. The iridium complex, $[IrCl_2(=CCl_2)(CO)(PPh_3)_2]^+$, the cationic analog of $OsCl_2(=CCl_2)(CO)(PPh_3)_2$ (with enhanced carbene electrophilicity because of the positive charge) immediately forms metallacycles of the same type (*122*).

3. Reactions with Electrophiles

The CF_2 ligand in the zero-valent complex $Ru(=CF_2)(CO)_2(PPh_3)_2$ (*21*) reacts only very slowly with H_2O to form $Ru(CO)_3(PPh_3)_2$. This contrasts with $RuCl_2(=CF_2)(CO)(PPh_3)_2$ which hydrolyzes instantly to $RuCl_2(CO)_2(PPh_3)_2$. The Ru=C bond in this zero-valent complex is sufficiently electron-rich to interact with electrophiles, as is demonstrated in the complexes formed with Ag(I) and also with AuI. The structure of the latter complex, $\overline{Ru(CF_2AuI)}(CO)_2(PPh_3)_2$ is presumably similar to that which has been confirmed by X-ray study for $\overline{Os(CH_2AuI)}Cl(NO)(PPh_3)_2$ (*20*). A further electrophile which reacts with $Ru(=CF_2)(CO)_2(PPh_3)_2$ is HCl, which gives $Ru(CF_2H)Cl(CO)_2(PPh_3)_2$ (*21*).

4. Migratory Insertion Reactions

Dihalocarbene ligands, like other neutral 2-*e* donor carbon ligands, are expected to participate in migratory-insertion reactions when bound adjacent to a σ-bound alkyl or hydride ligand. An example is provided by the following reaction (*119*):

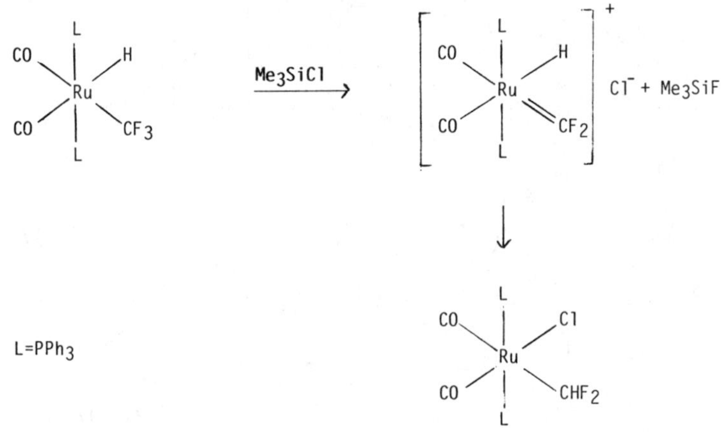

$L=PPh_3$

VI

CARBYNE COMPLEXES OF RUTHENIUM AND OSMIUM

Transition metal carbyne complexes are still relatively uncommon as only a few synthetic approaches to these compounds has proved generally applicable. In addition to making the initial characterization (*123*), the Fischer group has made the largest contribution to carbyne complex chemistry, with some 200 mononuclear complexes of Group 6 and 7 metals having been prepared.

More recently, Schrock has reported the formation of coordinatively unsaturated Ta and W carbyne complexes (*124*). Like unsaturated carbene complexes, these carbyne compounds are now established as being active intermediates in a number of catalytic reactions. The discovery of acetylene metathesis reactions catalyzed by carbyne complexes (*3*), for example, has generated considerable interest in this class of compound.

Carbyne complex chemistry of osmium and ruthenium is discussed in this section. These studies demonstrate clearly the parallels that exist between the metal–carbon bonds in carbene and carbyne complexes and again emphasize the importance of metal basicity in determining complex reactivity.

A. Synthesis

There are, broadly speaking, three general routes to transition metal carbyne complexes:

1. The introduction of a new carbyne ligand to the metal in a single step.
2. The modification of an existing ligand present in the complex.
3. The modification of existing carbyne complexes.

The direct formation route makes use of a "metathesis-type" reaction using acetylenes as the organic substrates (*3,125*). Although the scope of this method has not yet been fully explored, it seems unlikely that it will find widespread application.

The modification of existing groups into carbyne ligands has been the chief method of forming new carbyne complexes (*126*). Carbene complexes have proved to be useful substrates in these reactions, undergoing various rearrangements and substituent abstractions to afford carbyne compounds. The formation of osmium and ruthenium arylcarbynes from dihalogenocarbene complexes exemplifies this method.

Other ligands that have been successfully converted to coordinated carbynes include thiocarbonyls, isocyanides, and vinyl groups. The potential for α-metalladiazoalkane species, i.e., compounds of the form R—C—ML$_n$, to behave as carbyne complex precursors has not been fully

$$R-\underset{\underset{N_2}{\|}}{C}-ML_n$$

appreciated previously, and the reactivity of an α-osmiumdiazo compound is discussed below.

Reactions of carbyne complexes that maintain the integrity of the metal–carbon triple bond form the third route to new carbynes. Substituent modification, ligand exchange, oxidation, and reduction reactions have all been reported (see, e.g., Ref. 126).

Specific routes to Ru and Os carbyne complexes will now be detailed.

1. Modification of Existing Ligands

a. Carbyne Complexes from Dihalogenocarbenes

The Aryl Lithium Reaction. Halogenocarbene complexes would appear to be well suited as carbyne complex precursors, with the carbene substituents being very good leaving groups. Indeed, the most intriguing single reaction of Os and Ru dichlorocarbene complexes is with aryl lithium reagents to form zero-valent carbyne complexes:

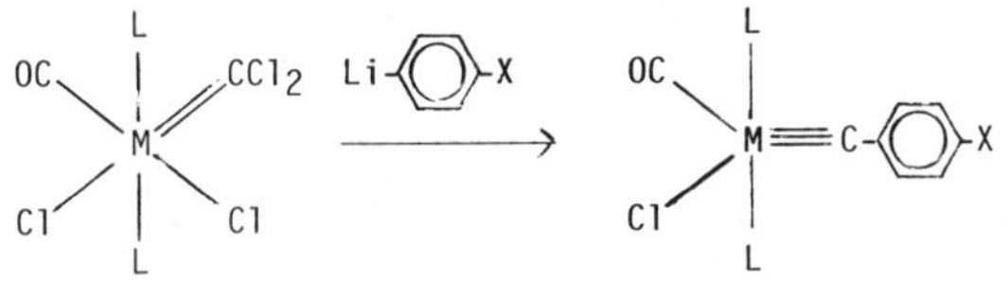

M=Os, X=Me **(79)** (Ref. 36)   M=Ru, X=H **(82)** (Ref. 39)

X=H **(80)** (Ref. 131)   X=OMe **(83)** (Ref. 39)

X=NMe$_2$ **(81)** (Ref. 131)   X=NMe$_2$ **(84)** (Ref. 39)

Compound **79** has been characterized by X-ray crystallography, and the molecular structure is discussed in Section VI,B.

When the reaction was attempted with Cl$_2$C=IrCl$_3$(PPh$_3$)$_2$ a crystalline solid, which was a green color typical of osmium carbyne complexes, could be observed at $-78°C$, but attempts to characterize this solid at room temperature were unsuccessful (39). No well-defined carbyne complexes of iridium have been reported to date.

When other lithium reagents such as LiMe were used in these reactions, product complexes, observable at low temperatures, could not be isolated.

Mechanisms for Carbyne Formation. Two possible mechanisms for the aryl lithium reaction can be proposed (see Scheme 4). Since two equiva-

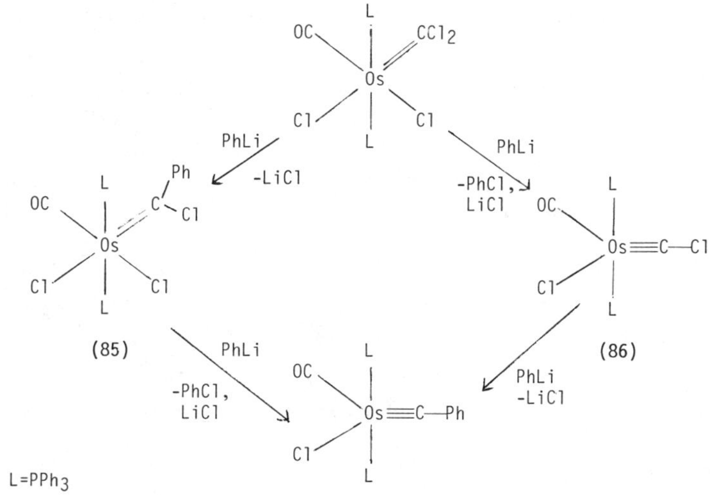

Scheme 4. Synthesis of osmium carbyne.

lents of ArLi are required for the reaction, at least two distinct steps are likely to be involved: (i) a reduction step in which formally the metal center is reduced by removal of Cl_2 from the complex, and (ii) a substitution step in which a chloride substituent is replaced by an aryl group.

It was noted in Section V,B that the chlorophenyl carbene complex 85 can be prepared by chlorine addition to carbyne complex 80. Treatment of 85 with one equivalent of PhLi does not afford 80, suggesting that the reaction sequence is reduction/substitution rather than substitution/reduction. The recent report (127) of a nucleophilic displacement reaction of the molybdenum chlorocarbyne complex 87 with PhLi to generate phenylcarbyne complex 88 suggests that the intermediacy of the chlorocarbyne complex 86 in the above mechanism is not unreasonable.

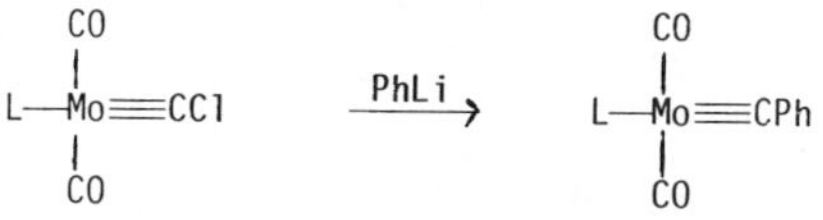

$L = HB(3,5\text{-}Me_2\text{-}C_3HN_2)_3$

b. *α-Metalladiazoalkanes as Carbyne Complex Precursors.* Dinitrogen loss from a diazoalkyl group coordinated to an unsaturated transition metal center would appear to offer an attractive route to carbyne complexes, i.e.,

$$L_nM-CR=N_2 \xrightarrow{-N_2} L_nM-\ddot{C}-R \rightarrow L_nM\equiv CR$$

Herrmann has reported the reaction of the mercury diazo compound $Hg(CN_2CO_2Et)_2$ with $Mn(CO)_5Br$ to afford the biscarbyne-bridged dimer **90** (*128*). The intermediacy of a terminal mononuclear carbyne complex **89** is strongly implicated here:

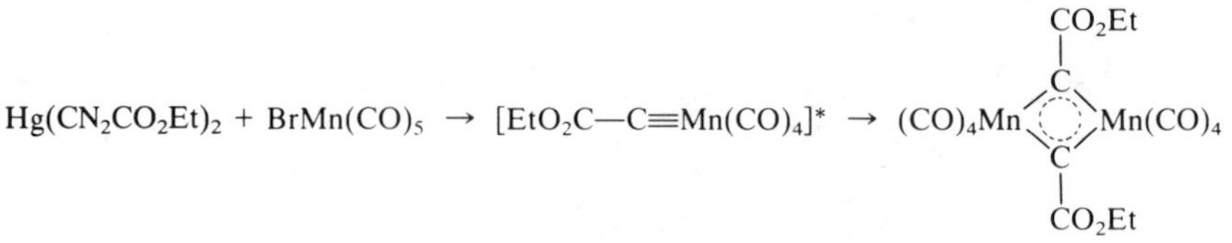

$$Hg(CN_2CO_2Et)_2 + BrMn(CO)_5 \rightarrow [EtO_2C-C\equiv Mn(CO)_4]^* \rightarrow (CO)_4Mn\underset{\underset{CO_2Et}{|}}{\overset{\overset{CO_2Et}{|}}{\overset{C}{\underset{C}{\bigcirc}}}}Mn(CO)_4$$

(89)   (90)

* = postulated intermediate

The synthesis of several palladium(II) complexes containing α-diazoalkyl ligands has been noted (*129*) but no details of the reactivity of these compounds were recorded.

The α-osmiumdiazo compound **91** decomposes in a thermal reaction to yield the metallacyclic complex **93** (*130*). This resembles the electrophilic carbene insertion reaction forming $OsCl(CO)_2(PPh_2C_6H_4CHCl)(PPh_3)$ **(77)** (see Section V,D,2), and we suggest that a similar insertion reaction of an electrophilic, cationic osmium carbyne **92** is the key step in this transformation. An X-ray structure determination has confirmed the formulation of **93**.

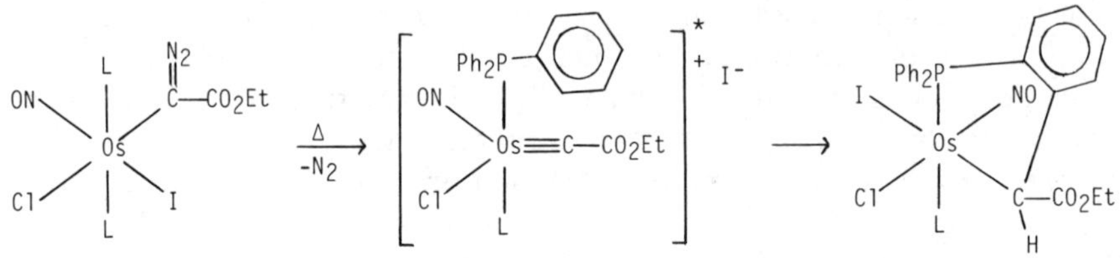

(91)   (92)   (93)

L=PPh₃   * = Postulated Intermediate

2. Modification of Existing Carbyne Complexes

a. Cationic d^8 Carbyne Complexes. One of the general properties of the d^8 Os and Ru carbyne complexes is the lability of the chloride ligand,

and cationic complexes may be readily generated, e.g.,

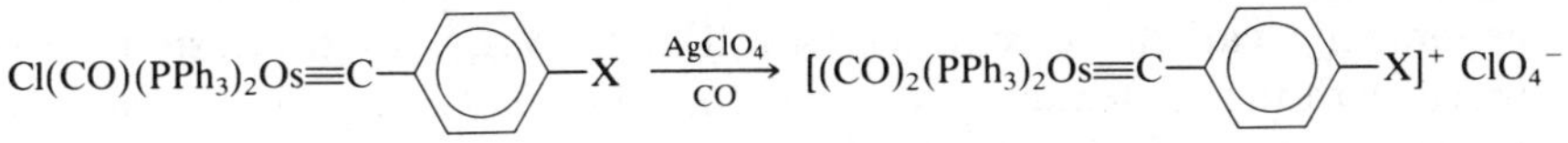

$$Cl(CO)(PPh_3)_2Os\equiv C\text{—}\bigcirc\text{—}X \xrightarrow[CO]{AgClO_4} [(CO)_2(PPh_3)_2Os\equiv C\text{—}\bigcirc\text{—}X]^+ ClO_4^-$$

(94)   X = Me   (Ref. 33)
        X = NMe$_2$ (Ref. 39)

b. Oxidation of the Metal by Oxygen. The analogy between the carbene and linear nitrosyl ligands as three-electron donors CR^+ and NO^+ was noted in Section II. Thus the Os(O) nitrosyl complex $OsCl(CO)(NO)(PPh_3)_2$ (95) is isoelectronic with the neutral Os d^8 arylcarbyne complexes 79–81. The usefulness of the analogy is evident in the reaction of these carbyne complexes with oxygen:

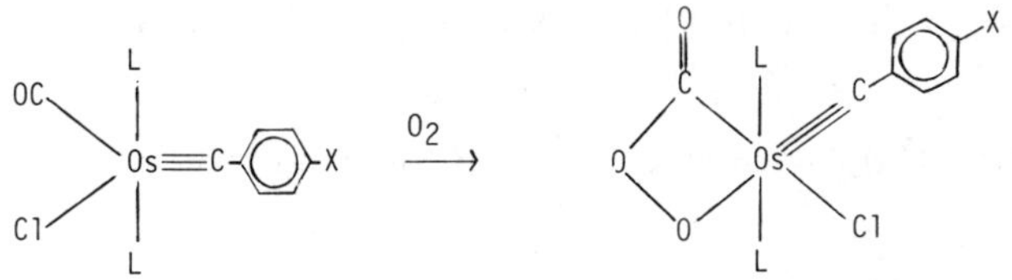

L=PPh$_3$ X=H (96)   (Refs. 131,39)

                                   X=NMe$_2$ (97)   (Ref. 131)

The unusual peroxycarbonyl ligand in these complexes was first characterized in $Os\overline{(C(O)OO)}Cl(NO)(PPh_3)_2$, the product of oxygenation of 95 (39).

The peroxycarbonyl ligand is cleaved from 96 by reaction with HCl, and the octahedral, d^6 carbyne complex 98 can be isolated. Similar treatment of 97 affords cationic complex 99 (131).

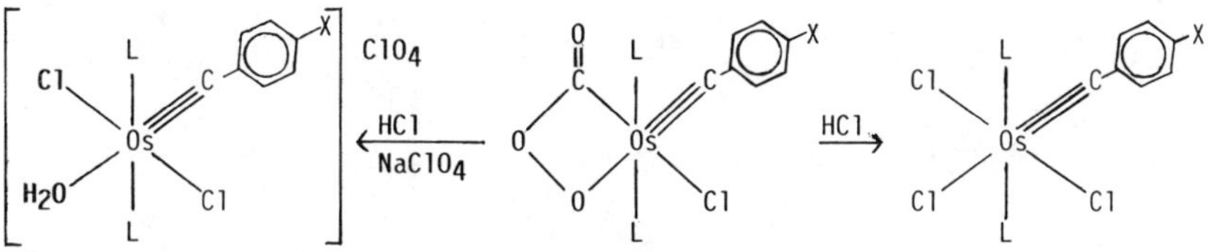

(99)   X=H (96)   X=H (98)

X=NMe$_2$ (99)   X=NMe$_2$ (97)

Two derivatives of 99, one neutral, $OsCl_2(SCN)(\equiv C\text{—}C_6H_4NMe_2)\text{-}$ $(PPh_3)_2$ (100), and one cationic, $[OsCl_2(CN\text{-}p\text{-tolyl})(\equiv C\text{—}C_6H_4NMe_2)\text{-}$ $(PPh_3)_2]^+$ (101), have been characterized by X-ray crystallography (131).

c. Oxidation of the Metal by Iodine.   Addition of chlorine and iodine across the osmium–carbon triple bond of **80** affords reactive halogenophenyl carbene complexes (see Section V,B). In contrast, iodine reacts with the ruthenium phenylcarbyne complex **82** to yield the cationic d^6 compound **102**. CO is lost from **102** on thermolysis, and a neutral carbyne complex can be isolated (*39*).

$$Cl(CO)(PPh_3)_2Ru \equiv CPh \xrightarrow{I_2} [ClI(CO)(PPh_3)_2Ru \equiv CPh]I \xrightarrow{\Delta, -CO} ClI_2(PPh_3)_2Ru \equiv CPh$$

   (82)   **(102)**

Oxidation of a cationic d^8 ruthenium carbyne complex by iodine has also been observed (*39*):

$$[(CO)_2(PPh_3)_2Ru \equiv CPh]ClO_4 \xrightarrow[-CO]{I_2} [I_2(CO)(PPh_3)_2Ru \equiv CPh]ClO_4$$

                                                        (103)

Compound **103** reacts with PhLi to afford the five-coordinate carbyne complex **104** (*39*):

$$[I_2(CO)(PPh_3)_2Ru \equiv CPh]ClO_4 \xrightarrow{PhLi} I(CO)(PPh_3)_2Ru \equiv CPh$$

            (103)   **(104)**

This reaction is significant as it is evidence that PhLi can behave as a reducing agent in the manner proposed in the mechanism for carbyne complex formation above.

B. *Structure and Spectroscopic Properties*

1. *Structure*

The carbyne syntheses described above have provided a set of neutral and cationic complexes in two different oxidation states:

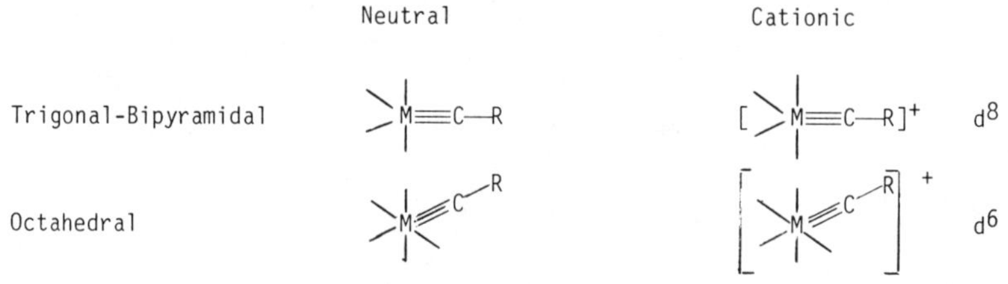

	Neutral	Cationic	
Trigonal-Bipyramidal	$>\!\!\overset{\mid}{\underset{\mid}{M}}\!\!\equiv\!\! C\!-\!R$	$[>\!\!\overset{\mid}{\underset{\mid}{M}}\!\!\equiv\!\! C\!-\!R]^+$	d^8
Octahedral	$>\!\!\overset{\mid}{\underset{\mid}{M}}\!\!\overset{C}{\diagdown}\!\!{}^{-R}$	$\left[>\!\!\overset{\mid}{\underset{\mid}{M}}\!\!\overset{C}{\diagdown}\!\!{}^{-R}\right]^+$	d^6

To fully characterize and obtain structural information for the carbyne complexes, X-ray crystal structure analyses of $OsCl(CO)(\equiv C-C_6H_4Me)(PPh_3)_2$ (**79**), $OsCl_2(SCN)(\equiv C-C_6H_4NMe_2)(PPh_3)_2$ (**100**), and $[OsCl_2(CN\text{-}p\text{-}tolyl)(\equiv C-C_6H_4NMe_2)(PPh_3)_2]ClO_4$ (**101**) have been carried out. Selected structural data are summarized in Table VIII.

TABLE VIII

Structural Data for Osmium Carbyne Complexes

Complex[a]	M≡C (Å)	≡C—C (Å)	M≡C—C (°)	Cl—M≡C (°)	P—M—P	Ref.
OsCl(CO)(≡C—C₆H₄Me)L₂ (79)	1.78(2)	1.44(2)	165(2)	133.0(6)	174.1(1)	36
OsCl₂(SCN)(≡C—C₆H₄NMe₂)L₂ (100)	1.75(1)	1.41(2)	169.5(9)	92.1(3)	174.0(1)	131
				102.2(3)		
[OsCl₂(CN-p-tolyl)(≡C—C₆H₄NMe₂)L₂]⁺ (101)	1.78(1)	1.39(2)	174(1)	95.9(5)	177.7(2)	131
				104.4(5)		

[a] L = PPh₃.

The osmium–carbyne carbon bond lengths for the three complexes do not differ significantly, and reference to Table IV indicates that these distances are distinctly shorter than the characterized metal–carbon double bonds of osmium carbene and carbonyl complexes. In both osmium alkylidene and carbyne complexes, then, the metal–carbon multiple bond lengths are largely insensitive to changes in the metal electron density (cf. Section IV,B).

The geometry about the carbyne carbon is approximately linear in **79**, **100**, and **101**. M≡C—C angles are close to 180° in all other crystallographically characterized carbyne complexes.

The chloride ligands are substantially bent away from the triple bond in each of **79**, **100**, and **101**. The bending of carbonyl ligands away from the carbyne ligands in R—C≡M(CO)$_4$X (X = halogen) has been attributed to direct π-donation by the halogen into the π-system of the carbonyls (*29*). Such an explanation is not tenable for the osmium complexes, and it is more likely here that the bending occurs to relieve the steric strain arising from the very short metal–carbon triple bonds.

2. Spectroscopic Properties

a. ^{13}*C-NMR Spectroscopy*. ^{13}C-NMR spectroscopy is a particularly useful method for investigating and characterizing carbyne complexes. The carbyne carbon resonance is found typically at low field, the majority of complexes giving signals in the range 250–350 ppm. ^{13}C-NMR data for osmium carbyne complexes **94** and **101** are given in Table IX.

b. Vibrational Spectroscopy.   Detailed infrared and Raman spectroscopic studies of the Fischer carbyne complexes Me—C≡W(CO)$_4$X (*132*) and Ph—C≡W(CO)$_4$X (*133*) have been undertaken. The metal–carbon triple bond stretching mode is seen in the IR and Raman spectra, and the frequencies lie in the 1250–1400 cm^{-1} region. The analysis of the phenyl carbyne system has indicated that the metal–carbon vibration is strongly coupled with the skeletal deformation modes of the phenyl ring.

Infrared absorptions in the 1300–1400 cm^{-1} region are observed for all the Ru and Os aryl carbyne complexes. It is likely that these absorptions correspond to combinations of metal–carbon and phenyl ring modes. Additional IR absorptions in the 1550–1600 cm^{-1} region are also observed for these complexes.

Carbonyl absorptions in the neutral, five-coordinate Ru and Os carbynes occur at unusually low wavenumbers, suggesting that it is appropriate to regard these molecules as zerovalent complexes. A significant increase in ν(CO) for the d^8 dicarbonyl cations is noted. IR data for selected carbyne complexes are given in Table IX.

TABLE IX

Spectroscopic Data for Ru and Os Carbyne Complexes

Complex[a]	≡C—R (δ)[b]	ν(carbyne)[c] (i)	ν(carbyne)[c] (ii)	ν(CO)[c]	Ref.
OsCl(CO)(≡C—C$_6$H$_4$Me)L$_2$ (**79**)		1595	1358	1864	36
OsCl(CO)(≡C—Ph)L$_2$ (**80**)		1584	1355	1860	131
RuCl(CO)(≡C—C$_6$H$_4$NMe$_2$)L$_2$ (**84**)		1590	1370	1865	39
[Os(CO)$_2$(≡C—C$_6$H$_4$Me)L$_2$]$^+$ (**94**)	331.0(t)	1592	1375	2010, 1944	33
Os(C(O)OO)Cl(≡C—Ph)L$_2$ (**96**)		1585	1395		39
OsCl$_3$(≡C—Ph)L$_2$ (**98**)		1585	1410		39
[OsCl$_2$(CN-p-tolyl)(≡C—C$_6$H$_4$NMe$_2$)L$_2$]$^+$ (**101**)	303.38(t)	1600	1415		131

[a] L = PPh$_3$.
[b] Chemical shifts in ppm.
[c] In cm^{-1}.

C. Reactivity

The similarity between the bonding models for transition metal carbene and carbyne complexes was noted in Section II. That the reactivity of the metal–carbon double and triple bonds in isoelectronic carbene and carbyne complexes should be comparable, then, is not surprising. In this section, the familiar relationship between metal–carbon bond reactivity and metal electron density is examined for Ru and Os carbyne complexes.

1. Reactions with Electrophiles

a. *Protic Acids.* Proton addition to electron-rich, neutral d^8 aryl carbyne complexes of Ru and Os to afford alkylidene complexes was noted in Section IV,A,2. In contrast to the neutral complex **80**, the cationic Os(O) complex **94** does not react with HCl in a thermal reaction. Compound **94** does add HCl under photolytic conditions, however, yielding the neutral phenylcarbene complex $OsCl_2(CO)(=CHPh)(PPh_3)_2$ (*134*):

$$[(CO)_2(PPh_3)_2Os\equiv CPh]^+ \xrightarrow[HCl]{h\nu} [Cl(CO)_2(PPh_3)_2Os=CHPh]^{+*} \xrightarrow[-CO]{Cl^-}$$

$$\text{(94)} \qquad\qquad \text{(105)}$$

$$Cl_2(CO)(PPh_3)_2Os=CHPh$$

$$= *\text{postulated intermediate}$$

The carbyne/nitrosyl analogy may be useful again in understanding this photochemical reaction. In nitrosyl chemistry, photoinduced metal to ligand charge transfer (CT) in a trigonal bipyramidal complex terminates in an excited state which will undergo a structural change to square pyramidal geometry with a bent NO ligand in the apical position (*135*). Photoinduced electron transfer from Os(O) to the carbyne ligand in **94** with concomitant structural reorganization may be proposed here, and the relaxed CT state can be envisaged as a coordinatively unsaturated octahedral Os(II) complex **106** containing a deprotonated phenylcarbene ligand (i.e., a bent carbyne). HCl addition to **106** would then generate **105**, which could undergo substitution of CO by chloride to yield the neutral phenyl carbene complex.

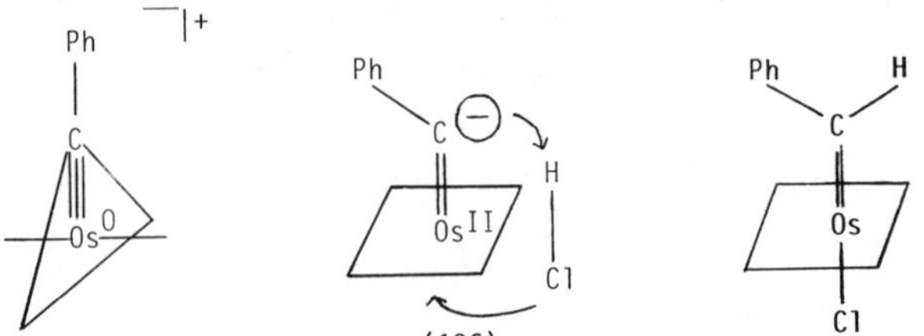

(106)

b. *Group I Metal Halides.*   Mixed dimetallacyclopropene species are formed from the reactions of the neutral five-coordinate carbyne complexes with Group I metal halides, e.g.,

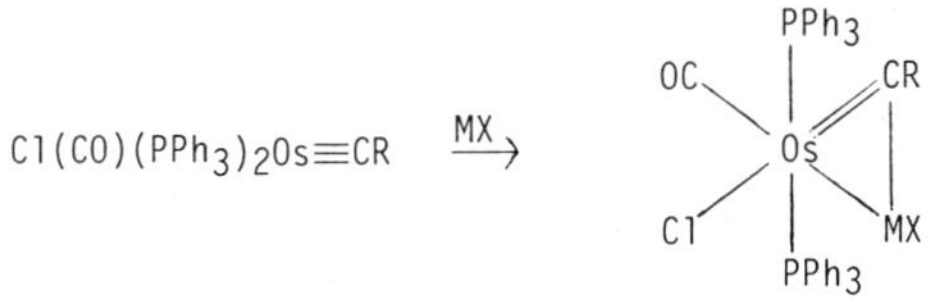

Cl(CO)(PPh₃)₂Os≡CR $\xrightarrow{MX}$

MX=AuCl   (Ref. 40)

AgCl   (107)

CuI

R=p-tolyl

The X-ray structure determination of **107** reveals that the osmium–carbon bond length is increased by 0.07 Å on going from the parent carbyne complex **79** to the silver adduct **107**. This may be contrasted with the weaker interaction between the metal–carbon bond and the AuI fragment in Os(CH₂AuI)Cl(NO)(PPh₃)₂ (see Section IV,C,1).

c. *Chalcogens.*   The reaction of **79** with elemental sulfur, selenium, and tellurium affords the corresponding dihaptochalcoacyl species (*36*). Entirely analogous reactivity of the osmium methylene complex Os(=CH₂)Cl(NO)(PPh₃)₂ has been discussed in Section IV,C,1.

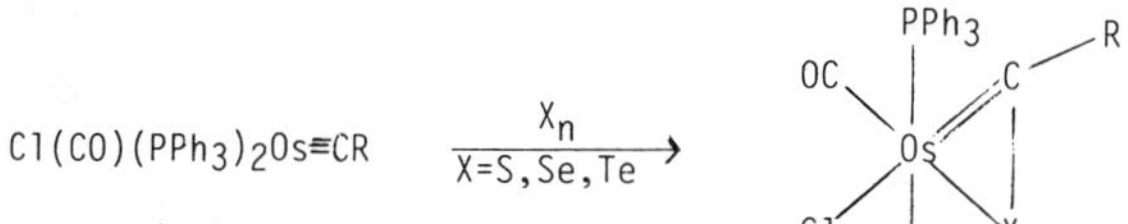

Cl(CO)(PPh₃)₂Os≡CR $\xrightarrow[X=S,Se,Te]{X_n}$

(**79**)

d. *Sulfur Dioxide.*   The phenylcarbyne complex **80** reacts immediately with SO₂ to give a 1:1 adduct (**108**) (*39*):

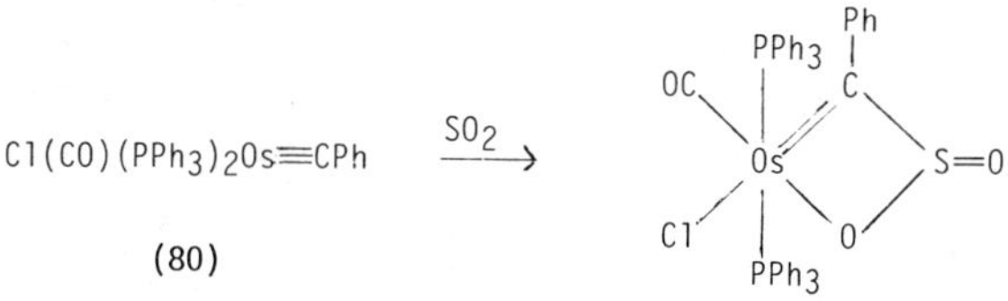

Cl(CO)(PPh₃)₂Os≡CPh $\xrightarrow{SO_2}$

(**80**)

(**108**)

Unlike the analogous SO₂ adducts of the d^8 Ru, Os, and Ir methylene complexes, **108** is rather unstable, decomposing to OsClPh(CO)₂(PPh₃)₂ on exposure to moisture.

e. Halogens.   The formation of chloroarylcarbene complexes by chlorine addition to zero-valent Os carbyne complexes has been discussed in Section V,B,2. The reaction of iodine with Ru(0) carbyne complexes affords oxidized carbyne species (see Section VI,A,2).

2. Reactions with Nucleophiles

The octahedral Ru(II) and Os(II) carbyne complexes are isoelectronic with the Group 6 compounds R—C≡M(CO)$_4$X prepared by Fischer and co-workers. Many of the Fischer compounds, particularly the cationic complexes, are susceptible to nucleophilic attack at the carbyne carbon (see Section II,B,2), and similar reactivity might be anticipated for the Group 8 carbynes.

That the neutral d^6 Ru and Os carbyne complexes have proved to be unreactive towards nucleophilic reagents suggests the relative electron richness of the MX$_3$(PPh$_3$)$_2$ (M = Ru, Os) fragments when compared with the M(CO)$_4$X residues in the Fischer complexes. The cationic Ru(II) and Os(II) compounds, however, are readily attacked by appropriate nucleophiles, e.g.,

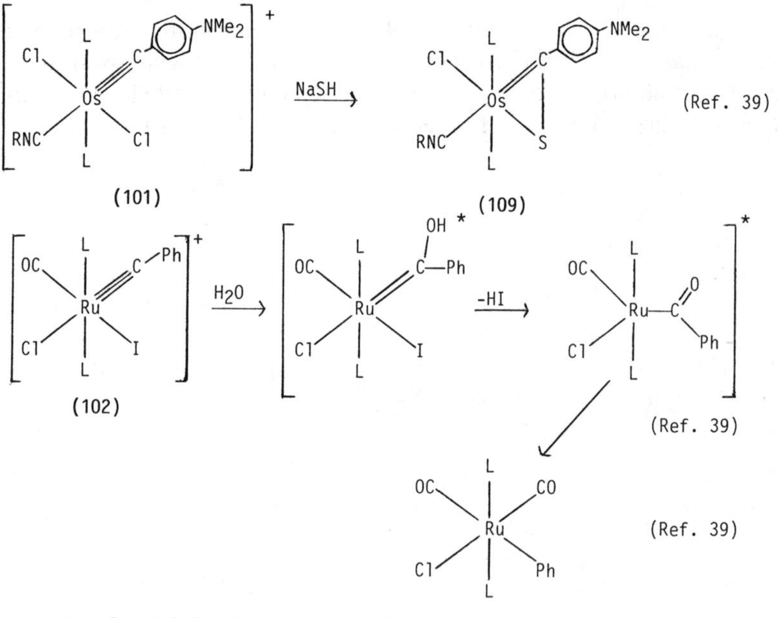

L = PPh$_3$, R=p-tolyl,   * = Postulated Intermediate.

Formation of the thioacyl complex **109** may be contrasted with the preparation of analogous chalcoacyl compounds by electrophilic addition to the zero-valent carbyne complex **79** (see above).

The most electrophilic site in the cationic d^8 Ru and Os arylcarbyne complexes is not the carbyne carbon, but the para position of the aryl ring. As noted in Section II,B,2, hydride reduction of these compounds affords zero-valent vinylidene species:

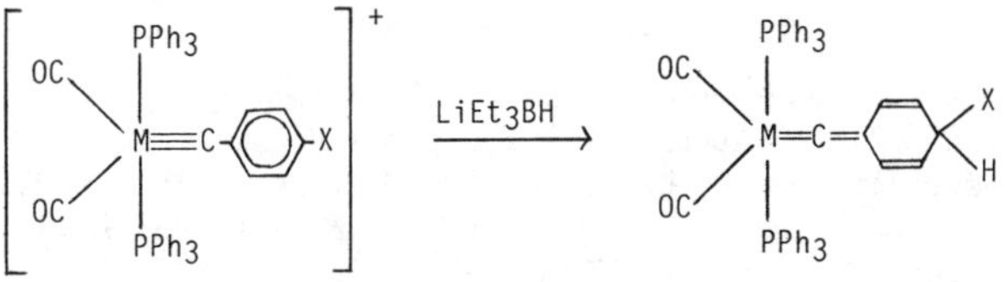

M=Os, X=Me (110)   (Ref. 33)

M=Ru, X=H   (Ref. 39)

Compound **110** has been structurally characterized by X-ray crystallography.

VII

CONCLUSION

The foregoing sections reveal that M=C and M≡C bonds are now well established features of the chemistry of Ru, Os, and Ir. Many exciting possibilities exist for using these functions in further reactions. For example, cyclization reactions involving the M≡C function together with one or two alkyne molecules could lead to unsaturated four- and six-membered metallacycles, and, for the six-membered rings particularly, the opportunity of electron delocalization would exist. Molecules of this type ("metalla-benzenes") have now been recognized (although they result from cyclization reactions involving carbyne precursors rather than pre-formed carbyne complexes), and their number is likely to increase (*136*). Examples are also known for the four-membered class ("metallacyclobutadienes") (*137*).

There is some advantage in considering these molecules with M=C and M≡C bonds alongside the longer known molecules with M=O, M=NR, and M≡N bonds. The widespread occurrence of molecules in which Group 8 metals participate in multiple bonding to first-row main group elements suggests that it may be worthwhile to seek molecules with M=S, M=PR, and even $M=SiR_2$ bonds as well. By making use of the steric protection afforded to these multiple linkages by two mutually trans triphenylphosphine ligands the chances of success are increased. Some tentative experiments aimed at isolating stable molecules of the type $L_nM{=}PR$ have been described (*138*), and there will no doubt be further developments in this direction.

REFERENCES

1. R. H. Grubbs, *in* "Comprehensive Organometallic Chemistry" (G. W. Wilkinson, F. G. A. Stone, and E. W. Abel, eds.), Vol. 8, p. 499. Pergamon, Oxford, 1982.
2. S. Brandt and P. Helquist, *J. Am. Chem. Soc.* **101**, 6473 (1979); M. Brookhart, D. Timmers, J. R. Tucker, G. D. Williams, G. R. Husk, H. Brunner, and B. Hammer *J. Am. Chem. Soc.* **105**, 6721 (1983); W. D. Wulff and P. C. Tang, *J. Am. Chem. Soc.* **106**, 434 (1984); M. A. McGuire and L. S. Hegedus, *J. Am. Chem. Soc.* **104**, 5538 (1982); S. L. Buchwald and R. H. Grubbs, *J. Am. Chem. Soc.* **105**, 5490 (1983); K. H. Dotz, R. Dietz, A. von Imhof, H. Lorentz, and G. Huttner, *Chem. Ber.* **109**, 2003 (1976).
3. J. H. Wengrovius, J. Sancho, and R. R. Schrock, *J. Am. Chem. Soc.* **103**, 3932 (1981).
4. F. G. A. Stone, *Angew. Chem., Int. Ed. Engl.* **23**, 89 (1984).
5. E. O. Fischer, *Pure Appl. Chem.* **24**, 407 (1970); **30**, 353 (1972); F. A. Cotton and C. M. Lukehart, *Prog. Inorg. Chem.* **16**, 487 (1972); F. J. Brown *Prog. Inorg. Chem.* **27**, 1 (1980); D. J. Cardin, B. Cetinkaya, and M. F. Lappert, *Chem. Rev.* **72**, 545 (1972); E. O. Fischer, *Adv. Organometal. Chem.* **14**, 1 (1976); R. R. Schrock, *Acc. Chem. Res.* **12**, 98 (1979); R. R. Schrock, *Science* **219**, 13 (1983).
6. T. F. Block and R. F. Fenske, *J. Organometal. Chem.* **139**, 235 (1977). .
7. T. F. Block, R. F. Fenske, and C. P. Casey, *J. Am. Chem. Soc.* **98**, 441 (1976).
8. N. M. Kostić and R. F. Fenske, *J. Am. Chem. Soc.* **104**, 3879 (1982).
9. N. M. Kostic and R. F. Fenske, *Organometallics* **1**, 974 (1982).
10. R. J. Goddard, R. Hoffman, and E. D. Jemmis, *J. Am. Chem. Soc.* **102**, 7667 (1980).
11. H. Nakatsuji, J. Ushio, S. Han, and T. Yonezawa, *J. Am. Chem. Soc.* **105**, 426 (1983).
12. T. F. Block and R. F. Fenske, *J. Am. Chem. Soc.* **99**, 4321 (1977).
13. B. E. R. Schilling, R. Hoffmann, and D. L. Lichtenberger, *J. Am. Chem. Soc.* **101**, 585 (1979).
14. E. O. Fischer, *Adv. Organometal. Chem.* **14**, 1 (1976).
15. R. R. Schrock, *Acc. Chem. Res.* **12**, 98 (1979).
16. K. Fukui, *Angew. Chem. Int. Ed. Engl.* **21**, 801 (1982).
17. H. Werner, E. O. Fischer, B. Heckl, and C. G. Kreiter, *J. Organometal. Chem.* **28**, 367 (1971).
18. J. H. Merrifield, C. E. Strouse, and J. A. Gladysz, *Organometallics* **1**, 1204 (1982).
19. W. R. Roper and A. H. Wright, *J. Organometal. Chem.* **233**, C59 (1982).
20. A. F. Hill, W. R. Roper, J. M. Waters, and A. H. Wright, *J. Am. Chem. Soc.* **105**, 5939 (1983).
21. G. R. Clark, S. V. Hoskins, T. C. Jones, and W. R. Roper, *J. Chem. Soc. Chem. Commun.* 719 (1983).
22. G. R. Clark, S. V. Hoskins, and W. R. Roper, *J. Organometal. Chem.* **234**, C9 (1982).
23. G. R. Clark, C. E. L. Headford, K. Marsden, and W. R. Roper, *J. Organometal. Chem.* **231**, 335 (1982).
24. W. Tam, G. Lin, W. Wong, W. A. Kiel, V. K. Wong, and J. A. Gladysz, *J. Am. Chem. Soc.* **104**, 141 (1982).
25. W. E. Buhro, A. T. Patton, C. E. Strouse, and J. A. Gladysz, *J. Am. Chem. Soc.* **105**, 1056 (1983).

26. S. E. Kegley, M. Brookhart, and G. R. Husk, *Organometallics* **1,** 760 (1982).
27. T. E. Taylor and M. B. Hall, *J. Am. Chem. Soc.* **106,** 1576 (1984).
28. N. M. Kostić and R. F. Fenske, *J. Am. Chem. Soc.* **103,** 4677 (1981).
29. N. M. Kostić and R. F. Fenske, *Organometallics* **1,** 489 (1982).
30. J. Ushio, H. Nakatsuji, and T. Yonezawa, *J. Am. Chem. Soc.* **106,** 5892 (1984).
31. W. R. Kreissl, W. Uedelhoven, and G. Kreis, *Chem. Ber.* **111,** 3283 (1978).
32. E. O. Fischer and A. Frank, *Chem. Ber.* **111,** 3740 (1978).
33. W. R. Roper, J. M. Waters, L. J. Wright, and F. Van Meurs, *J. Organometal Chem.* **201,** C27 (1980).
34. S. J. Holmes, D. N. Clark, H. W. Turner, and R. R. Schrock, *J. Am. Chem. Soc.* **104,** 6322 (1982).
35. S. J. Holmes, R. R. Schrock, M. R. Churchill, and H. J. Wasserman, *Organometallics* **3,** 476 (1984).
36. G. R. Clark, K. Marsden, W. R. Roper, and L. J. Wright, *J. Am. Chem. Soc.* **102,** 6570 (1980).
37. M. Bottrill and M. Green, *J. Am. Chem. Soc.* **99,** 5795 (1977).
38. A. J. L. Pombeiro and R. L. Richards, *Transition Met. Chem.* **5,** 55 (1980).
39. W. R. Roper and A. H. Wright, unpublished observations.
40. G. R. Clark, C. M. Cochrane, W. R. Roper, and L. J. Wright, *J. Organometal. Chem.* **199,** C35 (1980).
41. D. F. Christian, G. R. Clark, W. R. Roper, J. M. Waters, and K. R. Whittle, *J. Chem. Soc., Chem. Commun.* 458 (1972).
42. D. F. Christian and W. R. Roper, *J. Organometal. Chem.* **80,** C35 (1974).
43. D. F. Christian, G. R. Clark, and W. R. Roper, *J. Organometal. Chem.* **81,** C7 (1974).
44. T. J. Collins and W. R. Roper, *J. Organometal. Chem.* **159,** 73 (1978).
45. M. L. H. Green, L. C. Mitchard, and M. G. Swanwick, *J. Chem. Soc. A* 794 (1971).
46. K. R. Grundy and W. R. Roper, *J. Organometal. Chem.* **216,** 255 (1981).
47. C. H. Game, M. Green, J. R. Moss, and F. G. A. Stone, *J. Chem. Soc. Dalton Trans.* 351 (1974).
48. F. Faraone, G. Tresoldi, and G. A. Loprete, *J. Chem. Soc. Dalton Trans.* 933 (1979).
49. G. Tresoldi, F. Faraone, and P. Piraino, *J. Chem. Soc. Dalton Trans.* 1053 (1979).
50. G. R. Clark, T. J. Collins, K. Marsden, and W. R. Roper, *J. Organometal. Chem.* **259,** 215 (1983).
51. G. R. Clark, T. J. Collins, K. Marsden, and W. R. Roper, *J. Organometal. Chem.* **157,** C23 (1978).
52. G. R. Clark, T. J. Collins, D. Hall, S. M. James, and W. R. Roper, *J. Organometal. Chem.* **141,** C5 (1977).
53. T. J. Collins, K. R. Grundy, W. R. Roper, and S. F. Wong, *J. Organometal. Chem.* **107,** C37 (1976).
54. K. R. Grundy, R. O. Harris, and W. R. Roper, *J. Organometal. Chem.* **90,** C34 (1975).
55. K. R. Grundy and W. R. Roper, *J. Organometal. Chem.* **113,** C45 (1976).
56. H. Fischer, *in* "The Chemistry of the Metal–Carbon Bond" (F. R. Hartley and S. Patai, eds.), Vol. 1, p. 185. Wiley (Interscience), New York, 1982.
57. D. J. Noonan and A. L. Balch, *J. Am. Chem. Soc.* **95,** 4769 (1973).
58. D. J. Noonan and A. L. Balch, *Inorg. Chem.* **13,** 921 (1974).
59. J. Chatt, R. L. Richards, and G. H. D. Royston, *J. Chem. Soc. Dalton Trans.* 1433 (1973).
60. H. C. Clark and L. E. Manzer, *J. Organometal. Chem.* **47,** C17 (1973).
61. M. H. Chisholm and H. C. Clark, *J. Am. Chem. Soc.* **94,** 1532 (1972).
62. M. F. Lappert, *J. Organometal. Chem.* **100,** 139 (1975).

63. P. B. Hitchcock, M. F. Lappert, and P. L. Pye, *J. Chem. Soc. Dalton Trans.* 826 (1978).
64. M. F. Lappert and P. L. Pye, *J. Chem. Soc. Dalton Trans.* 837 (1978).
65. B. Cetinkaya, P. Dixneuf, and M. F. Lappert, *J. Chem. Soc. Dalton Trans.* 1827 (1974).
66. K. R. Grundy and W. R. Roper, *J. Organometal. Chem.* **91,** C61 (1975).
67. P. B. Hitchcock, M. F. Lappert, P. L. Pye, and S. Thomas, *J. Chem. Soc. Dalton Trans.* 1929 (1979).
68. M. F. Lappert and P. L. Pye, *J. Chem. Soc. Dalton Trans.* 2172 (1977).
69. P. J. Fraser, W. R. Roper, and F. G. A. Stone, *J. Chem. Soc. Dalton Trans.* 102 (1974).
70. P. J. Fraser, W. R. Roper, and F. G. A. Stone, *J. Chem. Soc. Dalton Trans.* 760 (1974).
71. M. Green, F. G. A. Stone, and M. Underhill, *J. Chem. Soc. Dalton Trans.* 939 (1975).
72. A. J. Hartshorn, M. F. Lappert, and K. Turner, *J. Chem. Soc. Dalton Trans.* 348 (1978).
73. G. R. Clark, W. R. Roper, and A. H. Wright, *J. Organometal. Chem.* **236,** C7 (1982).
74. S. V. Hoskins, R. A. Pauptit, W. R. Roper, and J. M. Waters, *J. Organometal. Chem.* **269,** C55 (1984).
75. R. J. Sundberg, R. F. Bryan, I. F. Taylor, Jr., and H. Taube, *J. Am. Chem. Soc.* **96,** 381 (1974).
76. P. B. Hitchcock, M. F. Lappert, and P. L. Pye, *J. Chem. Soc. Chem. Commun.* 644 (1976).
77. J. A. Moreland and R. J. Doedens, *Inorg. Chem.* **15,** 2486 (1976) and references therein.
78. P. Yates, *J. Am. Chem. Soc.* **74,** 5376 (1952).
79. C. Rüchardt and G. N. Schrauzer, *Chem. Ber.* **93,** 1840 (1960).
80. H. Werner and J. H. Richards, *J. Am. Chem. Soc.* **90,** 4976 (1968).
81. F. D. Mango and I. Dvoretzky, *J. Am. Chem. Soc.* **88,** 1654 (1966).
82. W. A. Herrmann, *Angew Chem. Int. Ed. Engl.* **13,** 599 (1974).
83. W. A. Herrmann, *Chem. Ber.* **108,** 486 (1975).
84. W. A. Herrmann, *Adv. Organometal. Chem.* **20,** 159 (1982).
85. M. A. Gallop and W. R. Roper, unpublished observations; M. A. Gallop, M. Sc. thesis University of Auckland, 1984.
86. G. R. Clark, W. R. Roper, and A. H. Wright, *J. Organometal. Chem.* **273,** C17 (1984).
87. J. Clemens, R. E. Davis, M. Green, J. D. Oliver, and F. G. A. Stone, *J. Chem. Soc. Chem. Commun.* 1095 (1971).
88. P. A. Chaloner, G. D. Glick, and R. A. Moss, *J. Chem. Soc. Chem. Commun.* 880 (1983).
89. H. D. Empsall, E. M. Hyde, R. Markham, W. S. McDonald, M. C. Norton, B. L. Shaw, and B. Weeks, *J. Chem. Soc. Chem. Commun.* 589 (1977).
90. D. L. Thorn, *Organometallic* **1,** 879 (1982).
91. J. C. Calabrese, D. C. Roe, D. L. Thorn, and T. H. Tulip, *Organometallics* **3,** 1223 (1984).
92. M. Brookhart, J. R. Tucker, T. C. Flood, and J. Jensen, *J. Am. Chem. Soc.* **102,** 1203 (1980).
93. D. L. Thorn and T. H. Tulip, *J. Am. Chem. Soc.* **103,** 5984 (1981).
94. L. J. Guggenberger and R. R. Schrock, *J. Am. Chem. Soc.* **97,** 6578 (1975).
95. A. T. Patton, C. E. Strouse, C. B. Knobler, and J. A. Gladysz, *J. Am. Chem. Soc.* **105,** 5804 (1983).

96. L. W. Messerle, P. Jennische, R. R. Schrock, and G. Stucky, *J. Am. Chem. Soc.* **102**, 6744 (1980).
97. J. A. Marsella, K. Folting, J. C. Huffman, and K. G. Caulton, *J. Am. Chem. Soc.* **103**, 5596 (1981).
98. W. A. Kiel, G. V. Lin, A. G. Constable, F. B. McCormick, C. E. Strouse, O. Eisenstein, and J. A. Gladyz, *J. Am. Chem. Soc.* **104**, 4865 (1982).
99. S. V. Hoskins, C. E. F. Rickard, and W. R. Roper, *J. Chem. Soc. Chem. Commun.* 1000 (1984).
100. T. A. Albright, *Tetrahedron* **38**, 1339 (1982).
101. R. R. Schrock and P. R. Sharp, *J. Am. Chem. Soc.* **100**, 2389 (1978).
102. M. Brookhart, J. R. Tucker, and G. R. Husk, *J. Am. Chem. Soc.* **105**, 258 (1983).
103. B. F. G. Johnson, D. A. Kaner, J. Lewis, P. R. Raithby, and M. J. Taylor, *J. Chem. Soc. Chem. Commun.* 314 (1982).
104. W. R. Roper, J. M. Waters, and A. H. Wright, *J. Organometal. Chem.* **275**, C13 (1984).
105. C. E. L. Headford, and W. R. Roper, *J. Organometal. Chem.* **198**, C7 (1980).
106. C. E. L. Headford and W. R. Roper, *J. Organometal Chem.* **244**, C53 (1983).
107. W. A. Herrmann and J. Plank, *Angew. Chem. Int. Ed. Engl.* **17**, 525 (1978).
108. F. Basolo, *Chemtech.* 54 (1983).
109. a. R. Aumann and E. O. Fischer, *Chem. Ber.* **101**, 954 (1968).
    b. C. G. Kreiter and R. Aumann, *Chem. Ber.* **111**, 1223 (1978).
110. D. Mansuy, M. Lange, J.-C. Chottard, P. Guerin, P. Morliere, D. Brault, and M. Rougé, *J. Chem. Soc. Chem. Commun.* 648 (1977); D. Mansuy, *Pure Appl. Chem.* **52**, 681 (1980).
111. D. L. Reger and M. D. Dukes, *J. Organometal. Chem.* **153**, 67 (1978).
112. G. R. Clark, K. Marsden, W. R. Roper, and L. J. Wright, *J. Am. Chem. Soc.* **102**, 1206 (1980).
113. P. M. Treichel and F. G. A. Stone, *Adv. Organometal Chem.* **1**, 143 (1964).
114. J. A. Morrison, *Adv. Inorg. Radiochem.* **27**, 293 (1983).
115. W. R. Roper and L. J. Wright, *J. Organometal. Chem.* **234**, C5 (1982).
116. W. R. Roper and L. J. Wright, unpublished observations.
117. A. J. Deeming and B. L. Shaw, *J. Chem. Soc. A.* 1128 (1969).
118. T. G. Richmond and D. F. Shriver, *Organometallics* **3**, 305 (1984).
119. S. V. Hoskins and W. R. Roper, unpublished observations.
120. E. O. Fischer, W. Kleine, W. Schambeck, and U. Schubert, *Z. Naturforsch. B: Anorg. Chem., Org. Chem.* **36B**, 1575 (1981).
121. A. J. Hartshorn, M. F. Lappert, and K. Turner, *J. Chem. Soc. Chem. Commun.* 929 (1975).
122. G. R. Clark, T. R. Greene, and W. R. Roper, *J. Organometal. Chem.* **293**, C25 (1985).
123. E. O. Fischer, G. Kreis, C. G. Kreiter, J. Muller, G. Huttner, and H. Lorenz, *Angew. Chem. Int. Ed. Engl.* **12**, 564 (1973).
124. L. J. Guggenberger and R. R. Schrock, *J. Am. Chem. Soc.* **97**, 2935 (1975).
125. R. R. Schrock, M. L. Listemann, and L. G. Sturgeoff, *J. Am. Chem. Soc.* **104**, 4291 (1982).
126. U. Schubert, *in* "The Chemistry of the Metal–Carbon Bond" (F. R. Hartley and S. Patai, eds.), Vol. 1, p. 233. Wiley (Interscience), New York, 1982.
127. T. Desmond, F. J. Lalor, G. Ferguson, and M. Parvez, *J. Chem. Soc. Chem. Commun.* 75 (1984).
128. W. A. Herrmann, *Angew. Chem. Int. Ed. Eng.* **13**, 812 (1974).
129. S. I. Murahashi, Y. Kitani, T. Hosokawa, K. Miki, and N. Kasai, *J. Chem. Soc. Chem. Commun.* 450 (1979).

130. M. A. Gallop, C. E. F. Rickard, and W. R. Roper, *J. Chem. Soc. Chem. Commun.* 1002 (1984).
131. G. R. Clark, N. R. Edmonds, R. A. Pauptit, W. R. Roper, J. M. Waters, and A. H. Wright, *J. Organomet. Chem.* **244**, C57 (1983).
132. N. Q. Dao, E. O. Fischer, W. R. Wagner, and D. Neugebauer, *Chem. Ber.* **112**, 2552 (1979).
133. N. Q. Dao, E. O. Fisher, and C. Kappenstein, *Nouv. J. Chim.* **4**, 85 (1980). ᷄.
134. A. Vogler, J. Kisslinger, and W. R. Roper, *Z. Naturforsch. B: Anorg. Chem., Org. Chem.* **38B**, 1506 (1983).
135. a. J. H. Enemark and R. D. Feltham, *Coord. Chem. Rev.* **13**, 339 (1974).
     b. W. Evans and J. I. Zink, *J. Am. Chem. Soc.* **103**, 2635 (1981).
136. G. P. Elliott, W. R. Roper, and J. M. Waters, *J. Chem. Soc. Chem. Commun.* 811 (1982).
137. G. P. Elliott and W. R. Roper, *J. Organometal. Chem.* **250**, C5 (1983).
138. D. S. Bohle and W. R. Roper, *J. Organometal. Chem.* **273**, C4 (1984).

ADVANCES IN ORGANOMETALLIC CHEMISTRY, VOL. 25

Borabenzene Metal Complexes

GERHARD E. HERBERICH and HOLGER OHST

Institut für Anorganische Chemie
Rheinisch–Westfälische Technische Hochschule
Aachen, Federal Republic of Germany

I

INTRODUCTION

The discovery of **1** (*1*), in 1970, opened a new and fascinating chapter of organometallic chemistry. This cation was the first compound derived from the hypothetical borabenzene **2** and the first complex of a classical boron–carbon ligand. Since then approximately 100 borabenzene derivatives, mainly complexes of 3d metals, have been characterized. Other unsaturated boron–carbon systems have been shown to act as ligands to metals (*2*). This development has also strongly stimulated the challenging quest for the simple species **2–5**.

199

Copyright © 1986 by Academic Press, Inc.
All rights of reproduction in any form reserved

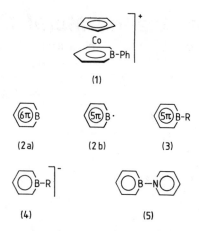

Several earlier reviews have treated borabenzene metal complexes in various contexts and in varying depth (2–5). This article places emphasis on recent results and perspectives while still being comprehensive.

II

NOMENCLATURE

The nomenclature of boron compounds involves some intricacies. IUPAC rules allow the terms borabenzene or borinine for **2**; the older name borin has become obsolete with the recent revision of the extended Hantzsch–Widman system (6). Anions **4** are termed boratabenzene ions; an alternative would be borininate instead of the earlier borinate (7).

The coordinated ligand is called η^6-boratabenzene by Chemical Abstracts; however, as ligands in sandwich-type complexes should be named as neutral entities according to the IUPAC rules, η^6-bora-3,5-cyclohexadien-2-yl would be the correct name. For the sake of simplicity we prefer boratabenzene in the following.

III

HISTORICAL PERSPECTIVE

The early sixties saw a broad and intense interest in the making of the heretofore unknown molecular compounds with a boron–metal bond (8). In 1963, the first compounds with boron–metal σ-bonds,

$Cp_2WH_2 \cdot BF_3$ (9) and $(Me_2N)_2BMn(CO)_5$ (10), were described. The synthesis, in 1965, of $(NMe_4)_2[Fe(C_2B_9H_{11})_2]$ (11) opened the stupendous world of the metallacarbaboranes (12–14). Attempts to formally replace hydrocarbon ligands such as C_6H_6 and C_5H_5 with isoelectronic boron–nitrogen ligands were based on theories as the benzene–borazine analogy (15) and the concept of "borazaro" compounds (16) and led, in 1967, to the discovery of the borazine complex $[Me_3B_3N_3Me_3]Cr(CO)_3$ (17,18) and of a compound $Fe[(MeN)_2(BPh)_2N]_2$ termed "inorganic ferrocene" (19). These two compounds were the first boron-containing π-complexes.

Interestingly, the chemistry of complexes with boron–nitrogen ligands has remained very limited. The higher electronegativity of nitrogen as compared to carbon and boron results in a more localized electron distribution in the π-electron system as well as in the σ-bond skeleton of these ligands. As a consequence, overlap between metal and ligand orbitals is reduced. Indeed, in a comparative thermochemical study of $(Me_3B_3N_3Me_3)$ $Cr(CO)_3$ and $(Me_6C_6)Cr(CO)_3$ the bond enthalpy contributions $[D$ $(Cr—Me_3B_3N_3Me_3) \approx 105$ kJ mol^{-1} and D $(Cr—C_6Me_6) = 206$ kJ mol$^{-1}]$ demonstrated the lower stability of the borazine–chromium bond (20).

In 1970, the synthesis of the orange–red sandwich cation 1 from cobaltocene and $PhBCl_2$ (1) marked a further starting point in the chemistry of boron metal compounds. The presence of a planar benzenoid C_5H_5B ligand moiety in 1 was deduced from ^1H and ^{11}B NMR data (1). This was made ironclad by two X-ray structure determinations which revealed typical centrosymmetric sandwich structures for the 19-e complexes $Co(C_5H_5BOMe)_2$ (6) (21,22) and $Co(C_5H_5BMe)_2$ (7) (22) as shown in Fig. 1.

A second and independent entry into the field was provided by Ashe and Shu with their synthesis of lithium 1-phenylboratabenzene 8 in THF

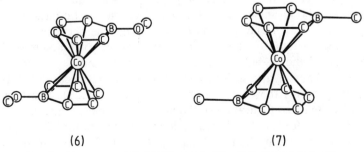

(6)   (7)

FIG. 1.   Structures of $Co(C_5H_5BOMe)_2$ (6) and $Co(C_5H_5BMe)_2$ (7).

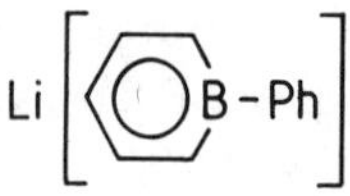

$$\text{Li} \left[\langle\bigcirc\rangle\text{B-Ph} \right]$$

(8)

solution (*23*). This work verified an earlier prediction that boratabenzene ions should be aromatic systems (*16*).

Borabenzene itself (**2**) has remained elusive. A qualitative contention that **2** should be a highly reactive species that can be stabilized by Lewis bases (*7*) has been reinstated by semiempirical and *ab initio* calculations (*24,25*). Moreover, the calculation predicts a singlet ground state (**2a**) and a singlet (**2a**)–triplet (**2b**) separation of approximately 1 eV (*25*).

In mass spectra of borabenzene metal complexes, ions of type $[M(C_5H_5B)]^+$ are commonly observed; $[Co(C_5H_5B)]^+$ decays via loss of an uncharged C_5H_5B fragment as evidenced by the observation of a metastable ion peak (*7*).

Recently the pyridine adduct of borabenzene **5** has been synthesized as a yellow crystalline solid (*26*). Its structural characterization by X-ray methods provides the first structural data of an uncomplexed borabenzene ring (*26*).

IV

STRUCTURE AND BONDING

A. Structural Investigations

All borabenzene–metal complexes investigated structurally so far show very similar patterns for the ligand geometry (Table I) and for the metal–ligand bonding (Table II); only the cobalt complex **6** deserves separate consideration (see below).

The C_5H_5BR ligand is essentially planar, and the intraring bond distances and bond angles show a common characteristic pattern. Comparison with borabenzene–pyridine (**5**) shows that the perturbation of the borabenzene system by the metal entails mainly two manifestations: lengthening of the B—C-2 and C-3—C-4 bonds and reduction of the C-2—B—C-6 angle. In addition, small ring puckerings occur with torsional angles up to 8°; these vary from case to case (*22,28*), and thus are induced by the d^n configuration of the metal, the counter-ligands, and the crystal packing forces. In all cases, however, the boranediyl group is slightly bent away from the metal, e.g., in **11** by 1.4° relative to the plane through the carbon atoms C-2 to C-6 (*29*) and by 6.5° in **9** (*27*).

TABLE I

THE LIGAND GEOMETRY

Compound	Distances (average, pm)			Angles (average, °)				Ref.
	B—C-2	C-2—C-3	C-3—C-4	C-2—B—C-6	B—C-2—C-3	C-2—C-3—C-4	C-3—C-4—C-5	
$(C_5H_5BMe)V(CO)_4$ (9)	150.0(10)	137.8(12)	138.0(12)	111.8(7)	122.3(8)	120.7(7)	121.5(8)	27
$(C_5H_5BPh)Mn(CO)_3$ (10)	152.5(10)	141.0(10)	141.5(10)	113.8(7)	120.9(7)	121.9(8)	120.5(8)	28
$(C_5H_5BPh)Co(CO)_2$ (11)	152.5(2)	138.6(2)	140.2(2)	112.7(1)	122.1(1)	120.1(1)	122.9(1)	29
$[(C_5H_5BMe)Fe(CO)_2]_2$ (12)[a]	150.9(11)	140.4(12)	138.7(9)	113.3(10)	121.4(7)	120.2(8)	123.0(9)	28
$Co(C_5H_5BMe)_2$ (7)	151.4(8)	138.8(8)	141.2(8)	114.5(4)	121.1(5)	120.9(5)	121.3(5)	22
$Co(C_5H_5BOMe)_2$ (6)	152.5(7)	139.4(9)	141.8(9)	114.8(5)	120.4(6)	121.7(6)	121.2(7)	21,22
$C_5H_5B—NC_5H_5$ (5)	147.3(4)	139.5(4)	137.8(4)	119.1(2)	117.8(2)	121.7(2)	122.0(2)	26

[a] **cis** Isomer.

TABLE II
The Metal–Ligand Bond

Compound	Distances (average, pm)						Δ^b	Ref.
	M—B	M—C-2	M—C-3	M—C-4	M—Cp^a	M—CO		
(C$_5$H$_5$BMe)V(CO)$_4$ (9)	249.6(8)	234.2(7)	228.6(7)	226.5(7)	226.7^c	192.9(6)	26	27
(C$_5$H$_5$BPh)Mn(CO)$_3$ (10)	238.5(8)	225.0(7)	217.5(7)	215.5(8)	216.5^d	179.5(8)	34	28
(C$_5$H$_5$BPh)Co(CO)$_2$ (11)	227.4(2)	217.5(2)	212.2(2)	205.1(2)	208.8^e	174.1(3)		29
[(C$_5$H$_5$BMe)Fe(CO)$_2$]$_2$ (12)^f	230.9(12)	219.2(7)	218.0(7)	214.1(8)	210.9^{f,g}		22	28
Co(C$_5$H$_5$BMe)$_2$ (7)	228.3(5)	221.8(5)	216.9(6)	207.5(5)	209.6^h		28	22
Co(C$_5$H$_5$BOMe)$_2$ (6)	234.8(8)	224.0(5)	215.3(6)	205.7(6)	209.6^h		39	21,22

^{a} M—C bond length in analogous Cp complex.
^{b} Slip distortion; $\Delta \equiv d(M'\!-\!C\text{-}4)$, where M' is the projection of the metal atom onto the ring plane.
^{c} CpV(CO)$_2$(dppe); Ref. 30.
^{d} Ref. 31.
^{e} (C$_5$Me$_5$)Co(CO)$_2$; Ref. 32.
^{f} cis Isomer.
^{g} Ref. 33.
^{h} Ref. 34.

In terms of bond lengths M—C and M—B, the interaction of the metal with the C_5H_5BR ligand seems to be comparatively weak, certainly weaker than in the closely related Cp analogs. Although the distance M—C-4 tends to be slightly shorter than the M—C separation in corresponding Cp compounds, all other M—C bonds are considerably longer, typically by 8–12 pm. The boron is within bonding distance of the metal, but the M—B separations are longer than in metallacarboranes; the 13-vertex compound $CpCo(C_2B_{10}H_{12})$ with Co—B bond distances of 209.3(3), 219.9(6), and 220.3(4) pm (35) may serve as an example for comparison with the borabenzene cobalt complex 11.

In bonding to the ligand the metal is always shifted away from the boron and interacts most strongly with C-4. This slip distortion is larger than expected considering the difference between the covalent radii of boron and carbon alone (22), and thus must partially be of electronic origin (cf. refs. 36 and 37).

Especially in the methoxy complex 6 the Co—B distance is significantly longer than in the analogous methyl compound 7 (22). It has also been noted that the B—O bond length in 6 indicates some double bonding between boron and oxygen (22). In other words, the π interaction between boron and the exocyclic methoxy substituent is antibonding with respect to the Co—B bonding and thus enforces an unusually pronounced slip distortion in 6.

B. Bonding Considerations

The bonding situation in borabenzene–metal complexes has been treated by several authors and with varying methods and intentions (38–41). The complexes are partitioned in boratabenzene ions and positive complex fragments or metal ions. The π MO's of the C_5H_5BR ligand (Fig. 2) may

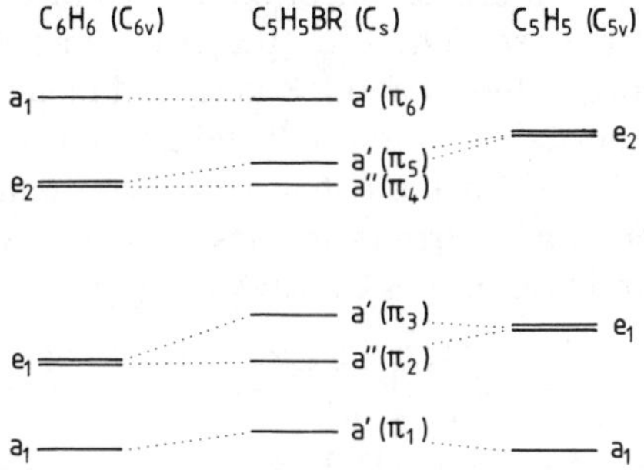

FIG. 2. Qualitative π MO's of the C_5H_5BR ligand in comparison with the C_6H_6 and Cp systems.

qualitatively be derived from benzene by substitution of a B atom for a C atom (39–41). This perturbation lifts the degeneracy of the e_1 and e_2 sets in benzene, raises the energy of all MO's of a' symmetry, and causes orbital polarizations. Alternatively, the π MO's may also be constructed from those of a pentadienyl moiety and a boranediyl group (38).

The interaction of the C_5H_5BR ligand with the metal has been investigated for the bis(ligand) complexes $FeL_2(d^6)$ and $CoL_2(d^7)$ (with $L = C_5H_5BH$, C_5H_5BMe), and for the isoelectronic series $[CrL_2]^-$, MnL_2, and $[FeL_2]^+$ (d^5, with $L = C_5H_5BH$), using INDO SCF MO methods (40,41). The overall bonding situation is similar to that in metallocenes and can be characterized by the following salient features: (1) The one-electron energy sequence of the mainly $3d$ metal levels is $d_{x^2-y^2} \sim d_{xy} < d_{z^2} < d_{xz} \sim d_{yz}$, very similar to that found in ferrocene. (2) For the Co systems the splitting of the d_{xz} and d_{yz} levels which are degenerate in axial symmetry is found to be small (≤ 3000 cm^{-1}). (3) The dominant bonding interaction is due to π donation from a'' (π_2) and a' (π_3) into the d_{xz} and d_{yz} orbitals. (4) The low symmetry causes considerable intra-d-orbital mixing, especially between the $d_{x^2-y^2}$ and d_{xy} orbitals. (5) Because of the more electropositive nature of the boron, the σ orbitals of the ring are higher in energy, and bonding interactions with metal d orbitals are quite noticeable.

The above results justify a simple ligand field model for the bis(ligand) metal species which is based on the assumption of pseudoaxial symmetry (40). This model allowed a consistent reinterpretation of an early ESR study of $Co(C_5H_5BPh)_2$ (13) (42); the reinterpretation was later confirmed by additional and more sophisticated ESR work on 13 (43,44).

A second MO study on FeL_2 and CoL_2 (with $L = C_5H_5BH$) using the Fenske–Hall method arrives at a similar description; the slip distortion and the rotational barrier for ring rotation were also analyzed (39). Metal slippage assures maximum overlap with the ligand orbitals; the overall effect is produced by superposition of several contributions, and a naive approach is not warranted. The C_{2h} (trans) rotamer of $Fe(C_5H_5BH)_2$ is calculated to be more stable by about 8 kcal/mol than the C_{2v} (cis) and C_2 (gauche, 90°) rotamers. Thus the predicted ground state conformation is identical with that found for the cobalt complexes 6 and 7; experimental information on rotational barriers in these systems is not available.

The electronic structures of the C_5H_5BMe complexes 9, 14–16 have been

B–Me	B–Me	B–Me	B–Me
V	Mn	Co	Co
(CO)$_4$	(CO)$_3$	(CO)$_2$	
(9)	(14)	(15)	(16)

(9)   (14)   (15)

$V_4 = 11.4$ kJ/mol $V_3 = 28.9$ kJ/mol $V_2 = 46.6$ kJ/mol

FIG. 3. Calculated ground state conformations and rotational barriers for **9**, **14**, and **15**.

investigated in a recent INDO SCF MO study with the aim to compare measured and calculated He(I) PE spectra (*38*).

Koopman's theorem is found to be valid only in the case of the vanadium (d^4) complex **9**. The amount of orbital reorganization is increasing considerably in the series **9–14–15–16**. It is also found that strong metal–ligand interaction combined with low symmetry leads to extensive delocalization of the outer valence MO's; especially in the cylobutadiene complex **16**, only two orbitals with >80% metal character are found for which the convenient term essential $3d$ metal orbital would be justified.

Predicted ground state conformations and rotational barriers for **9**, **14** and **15** are given in Fig. 3. The same conformations are found in the solid state for the phenyl analogs **10** (*28*) and **11** (*29*). The crystal structure of **9**, however, shows a torsion of the $V(CO)_4$ group of 34.6° relative to the minimum of the internal rotation potential (*27,38*). This may well be due to crystal packing forces as the rotational barrier is rather low in this case (*38*). Experimentally hindered rotation could not be detected by NMR spectroscopy; even in the most favorable case **15**, only one sharp ^{13}C-NMR line for the two CO groups was observed at −95°C (Bruker WH 270, 67.88 MHz, toluene) which gives an upper limit of ~32 kJ/mol for the barrier height (*45*).

C. Physical Properties

1. NMR Spectra

Borabenzene complexes show ^{1}H-NMR spectra of the $ABB'CC'$ type. Some typical data are collected in Table III. The protons H-2/H-6 always resonate at higher field than H-3/H-5 and are easily recognized by their prominent doublet structure. The proton H-4 is extremely variable in its position which ranges from lowest to highest relative to the other borabenzene protons; especially in all true d^8 complexes (such as, e.g., **15** but not **16**) it is the most shielded proton. The coupling constants do not vary significantly. They are $J_{23} = 9.0$ and $J_{34} = 6.0$ Hz, somewhat smaller than

TABLE III

SOME ^{1}H AND ^{11}B CHEMICAL SHIFTS

Compound	$\delta(^1H)$ (ppm)				$\delta(^{11}B)$ (ppm)	Solvent	Ref.
	H-2/H-6	H-3/H-5	H-4				
V(CO)$_4$(C$_5$H$_5$BMe) (9)	5.15	6.10	6.10		28.7	d_6-Acetone	27
Mn(CO)$_3$(C$_5$H$_5$BMe) (14)	4.23	6.12	5.61		26.8	d_6-Acetone	46
Co(CO)$_2$(C$_5$H$_5$BMe) (15)	5.59	6.31	5.33		23.2	d_6-Acetone	47
Range ^{a}	6.11–3.75	6.31–4.91	6.47–3.99		28.7–14.4	—	
V(CO)$_4$(C$_5$H$_5$BPh) (17)	5.72	6.29	6.29		26.0	d_6-Acetone	27
Mn(CO)$_3$(C$_5$H$_5$BPh) (10)	4.97	6.36	5.89		24.6	d_6-Acetone	48
Co(CO)$_2$(C$_5$H$_5$BPh) (11)	6.19	6.65	5.64		20.4	d_6-Acetone	49
Range ^{b}	6.19–4.83	6.9 –5.66	6.9 –4.19		26.0–12.4	—	

^{a} For all C$_5$H$_5$BMe complexes of transition metals.
^{b} For all C$_5$H$_5$BPh complexes.

in alkali metal boratabenzenes [$8:J_{23} = 10.0$, $J_{34} = 6.5$ Hz (23)], and small additional couplings ($J_{24} \approx 1$–2 Hz) have occasionally been discerned.

Boron resonances (Table III) appear in the range expected for sandwich-type complexes with metal–boron bonding (for reference data see Ref. 50). In cases where pairs of B-methyl and B-phenyl derivatives (as, e.g., 9 and 17) are known, the [11]B resonance of the methyl compound is at lower field by 2.0–3.7 ppm (average, 2.6 ppm).

Few ^{13}C-NMR data are available (46,51,52); they provide a simple means to elucidate substituent positions in ring-substituted borabenzene derivatives (46,52).

2. IR Spectra

IR spectra of mixed (boratabenzene)carbonylmetal complexes (Table IV) were used early on to compare the donor–acceptor properties of borabenzene and cyclopentadienyl ligands, respectively (48). Owing to the lower effective symmetry, C_5H_5BR complexes often show a greater number of $\nu(CO)$ bands than the corresponding Cp compounds. In addition, the $\nu(CO)$ bands appear at higher frequencies and couple less strongly in the C_5H_5BR complexes than in the Cp compounds.

Interpretation of these two observations along accepted lines of reasoning (55,56) indicates that bora-3,5-cyclohexadien-2-yl ligands are weaker donors and stronger acceptors than the cyclopentadienyl ring (48). This is largely related to the ligand ring size (40); the orbitals responsible for the ligand back-bonding ability are higher in energy in the five-membered Cp ring than in the six-membered C_5H_5BR ligands [see Fig. 2, e_2 for Cp,

TABLE IV

SELECTED $\nu(CO)$ DATA

Compounds	$\nu(CO)$ (cm^{-1})	$\overline{\nu(CO)}$ (cm^{-1})	Solvent	Ref.
V(CO)$_4$(C$_5$H$_5$BPh) (17)	2037, 1977, 1945	1986	Pentane	27
V(CO)$_4$(C$_5$H$_5$BMe) (9)	2035, 1973, 1945	1984	Pentane	27
V(CO)$_4$(C$_5$H$_5$)	2031, 1931	1964	Hexane	53
Mn(CO)$_3$(C$_5$H$_5$BPh) (10)	2039, 1974, 1960	1991	Hexane	48
Mn(CO)$_3$(C$_5$H$_5$BMe) (14)	2030, 1965, 1960	1985	Pentane	46
Mn(CO)$_3$(C$_5$H$_5$)	2028, 1944	1972	Hexane	48
Co(CO)$_2$(C$_5$H$_5$BPh) (11)	2043, 1990	2017	Hexane	49
Co(CO)$_2$(C$_5$H$_5$BMe) (15)	2039, 1986	2013	Hexane	47
Co(CO)$_2$(C$_5$H$_5$)	2036, 1974	2005	Cyclohexane	54

$a''(\pi_4)$ and $a'(\pi_5)$ for C_5H_5BR]. In addition, it is seen that the more electronegative B-phenyl substituent enhances back-bonding slightly as compared to a B-methyl group.

3. Miscellaneous Data

Ionization potentials have been determined from mass spectroscopic data for $CoCp_2$ and its borabenzene analogs (such as **6** and **7**) (57), and for $FeCp_2$ and its borabenzene analogs by $He(I)$ ionization (58). Complete $He(I)$ PE spectra have been recorded for $V(CO)_4(C_5H_5BMe)$ (**9**), $Mn(CO)_3(C_5H_5BMe)$ (**14**), $Co(CO)_2(C_5H_5BMe)$ (**15**), and $Co(C_4H_4)$ (C_5H_5BMe) (**16**) (38). In all cases, formal substitution of a Cp ring with a C_5H_5BR ring effects an increase of the ionization potential. This increase amounts to 0.4–0.6 eV per ring in the Co series, to 0.4 eV per ring in the Fe series, and to 0.25 eV in the Mn series where, e.g., the vertical ionization potential is 8.05 eV for $Mn(CO)_3Cp$ (59) and 8.3 eV for **14** (38). These observations are again evidence for the more pronounced acceptor character of the borabenzene ligand. The same conclusion has been reached from a Mössbauer spectroscopic comparison between $FeCp_2$ and $Fe(C_5H_5BPh)_2$ (**18**) (58).

V

SYNTHETIC METHODS

The characterized borabenzene metal compounds are listed in Table V. In this section the various entries into borabenzene–metal chemistry are described; Sections VI and VII treat the reactions of these compounds.

A. The Cobaltocene Route

The reaction of cobaltocene with organoboron dihalides RBX_2 (R = Me, Ph and X = Cl, Br mainly) and boron trihalides (BCl_3, BBr_3) leads essentially to three types of (boratabenzene) cobalt complexes, **19, 20,** and **21** (7,57). $CoCp_2$ plays a dual role; in part it acts as a reductant, in part it

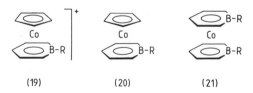

(19)   (20)   (21)

TABLE V
Characterized Borabenzene Metal Compounds

Compound	Method of synthesis[a]	Characterization[b]	Ref.
Lithium			
Li[C₅H₅BPh][c]	C, I	H, B	23,60
Potassium			
K[C₅H₅BR], R = Me[c]	I, J	H	60,61
R = Ph[c]	I, J	H	60,61
Vanadium			
V(CO)₄(C₅H₅BR), R = Me	B	A, mp, MS, X, H, B, IR, PE	27,38
R = Ph	B	A, mp , MS, H, B, IR	27
V(C₅H₅BR)₂, R = Me	B	A, mp, MS, μ, EC	27,62
R = Ph	B	A, mp, MS, μ	27
V(C₅H₅BR)(C₇H₇), R = Me	H	A, mp, MS, X, μ, EC	27
R = Ph	H	A, mp, MS, μ, EC	27
Chromium			
CrH(CO)₃(C₅H₅BMe)	F	MS, H, IR	63
Na[Cr(CO)₃(C₅H₅BMe)]·2[O(C₂H₄)₂O]	B	A, mp, H, B, IR, EC	63
[PPh₄][Cr(CO)₃(C₅H₅BMe)]	B	A, mp, H, B, IR, EC	63
Cr(C₅H₅BR)₂, R = Me	B	A, mp, MS, IR, μ, EC	62,64
R = Ph	B	A, mp, MS, IR, μ, EC	62,64
Hg[Cr(CO)₃(C₅H₅BMe)]₂	F	A, mp, MS, H, B, IR	63

(*Continued*)

TABLE V (continued)

Compound	Method of synthesis[a]	Characterization[b]	Ref.
Manganese			
Mn(CO)$_3$(C$_5$H$_5$BR), R = Me	E	A, bp, MS, H, B, C, IR, PE	38,46
R = Ph	E	A, mp, MS, X, H, B, IR	28,48
Mn(CO)$_3$[2-(MeCO)C$_5$H$_4$BMe]	M	A, mp, MS, H, B, C, IR	46
Mn(CO)$_3$(4-MeC$_5$H$_4$BPh)	C	A, mp, MS, H, B, IR	65
Rhenium			
Re(CO)$_3$(4-MeC$_5$H$_4$BPh)	C	A, mp, MS, H, B, IR	65
Iron			
FeCp(C$_5$H$_5$BR), R = Me	E	A, mp, MS, H, B, EC	62,66
R = Ph	E	A, mp, MS, H, B, EC	62,66
FeCp(2-MeC$_5$H$_4$BPh)	O	A, mp, MS, H, B, C	52
Fe(C$_5$H$_5$BR)$_2$, R = H	K	MS, X, H, B, IR	67
R = Me	B, E, K	A, mp, MS, H, B, UV, IR, IP, Möss, EC	58,62,67,68
R = Bu^t	B, K	mp, MS, H, B, UV, IP	58,67
R = Ph	B, E	A, mp, MS, H, B, UV, IR, IP, Möss, EC	58,62,68
R = OMe	C	MS, H, B	67
K[Fe(C$_5$H$_5$BPh)$_2$]^c	D	ESR	62
Fe(C$_5$H$_5$BMe)[2-(MeCO)C$_5$H$_4$BMe]	M	MS, H, IR	58,66
Fe(4-MeC$_5$H$_4$BPh)$_2$	C	A, mp, MS, H, B	65
Fe(3,4-Me$_2$C$_5$H$_3$BPh)$_2$^{d}	C	A, mp, MS, H	65
[Fe(C$_5$H$_5$BMe)(PhMe)]PF$_6$	P	A, mp, H, B	66
[Fe(CO)$_2$(C$_5$H$_5$BR)]$_2$(Fe—Fe), R = Me^e	E	A, mp, MS, X, H, IR	28,68
R = Ph	C, E	A, mp, MS, H, B, IR	29,68

212

Ruthenium

Ru(C$_5$H$_5$BR)$_2$, R = Me	B	A, mp, MS, H, B	61
R = Ph	B	A, mp, MS, H, B	61
[Ru(C$_5$H$_5$BPh)(C$_6$H$_6$)]PF$_6$	B	A, mp, H, B	69

Osmium

Os(C$_5$H$_5$BPh)$_2$	B	A, mp, MS, H, B	61

Cobalt

Co(CO)$_2$(C$_5$H$_5$BR), R = Me	B, G	A, bp, MS, H, B, IR, PE	38,47
R = Ph	B, C, G	A, mp, MS, X, H, B, IR	29,47,49
Co(C$_2$B$_9$H$_{11}$)(C$_5$H$_5$BPh)	A	A, mp, MS, H, B, IR	70
Co(C$_4$H$_4$)(C$_5$H$_5$BR), R = Me	B	A, bp, MS, H, B, PE	38,47
R = Ph	B	A, mp, MS, H, B	47
Co(C$_4$Me$_4$)(C$_5$H$_5$BR), R = Me	B	A, mp, MS, H, B	71
R = Ph	B	A, mp, MS, H, B	71
Co(C$_4$Me$_4$)[2-(OHC)C$_5$H$_4$BMe]	M	A, mp, MS, H, B, IR	71
Co(C$_4$Me$_4$)[2-(MeCO)C$_5$H$_4$BR], R = Me	M	A, mp, MS, H, B, IR	71
R = Ph	M	A, mp, MS, H, B, IR	71
Co(C$_4$Me$_4$)[2,6-(MeCO)$_2$C$_5$H$_3$BMe]	M	A, mp, MS, H, B, IR	71
Co(C$_4$Ph$_4$)(C$_5$H$_5$BR), R = Me	G, H	A, mp, MS, H, IR	71,72
R = Ph	H	A, mp, MS, H, B	71
Co(C$_5$H$_5$)(C$_5$H$_5$BR), R = Me	A	A, mp, MS, IP, EC	7,57,73
R = Ph	A	A, mp, MS, IP	7,57,69,73
R = Mes	A	A, mp, MS	74
[Co(C$_5$H$_5$)(C$_5$H$_5$BR)]PF$_6$, R = Me	D	A, mp, H	7
R = Ph	A, D	A, mp, H, B, EC	1,7,69
R = CH$_2$Ph	A	A, H	74
R = Mes	D	A, H	74
[Co(C$_5$H$_5$)(4-MeC$_5$H$_4$BMe)]PF$_6$	P	A, mp, H, B	75
Co(C$_5$H$_6$)(C$_5$H$_5$BPh)	N	A, mp, MS, H, B	69

(continued)

TABLE V (continued)

Compound	Method of synthesis[a]	Characterization[b]	Ref.
Co(C$_5$H$_5$BR)$_2$, R = Me	A	A, mp, MS, X, IP, EC	7,22,57,73
R = Ph	A	A, mp, MS, UV, IP, ESR, EC	7,40,42–44,57,73
R = CH$_2$Ph	A	A, mp, MS	74
R = OMe	K	mp, X, IP, μ	21,22,57
[Co(C$_5$H$_5$BPh)$_2$]PF$_6$	D	A, mp, H, B	69,76
Na[Co(C$_5$H$_5$BPh)$_2$][c]	D	μ	60
[PPh$_4$][Co(C$_5$H$_5$BPh)$_2$]	D	A, mp, μ	60
Co(C$_5$H$_5$BMe)(4-PhC$_5$H$_5$BMe)	N	A, mp, MS, H, B	60
Co(C$_5$H$_5$BMe)(6-PhC$_5$H$_5$BMe)	N	A, mp, MS, H, B	60
Co(C$_5$H$_5$BPh)(4-HC$_5$H$_5$BPh)	L	A, mp, MS, H, B, IR	69
Co(C$_5$H$_5$BPh)(6-HC$_5$H$_5$BPh)	L, N	A, mp, MS, H, B, IR	60,69
Co(C$_5$H$_5$BPh)(C$_7$H$_8$)[f]	G	A, mp, MS, H, B	60
Co(C$_5$H$_5$BR)(COD), R = Me	G	A, mp, MS, H, B	71
R = Ph	C, G	A, mp, MS, H, B, Co	60,71,77,101
Co$_2$[μ-(1-4-η: 1-4-η-C$_4$Me$_4$)](C$_5$H$_5$BR)$_2$(Co—Co), R = Me	H	A, mp, MS, H, B	71
R = Ph		A, mp, MS, H, B	71
Rhodium			
[Rh(C$_5$Me$_5$)(C$_5$H$_5$BPh)]PF$_6$	B	A, mp, H, B	69
[Rh(C$_5$Me$_5$)(4-MeC$_5$H$_4$BMe)]PF$_6$	P	H, B	75
Rh(C$_5$Me$_5$H)(C$_5$H$_5$BPh)	L	A, mp, MS, H, B, IR	69
Rh(C$_5$H$_5$BR)(COD), R = Me	B	A, mp, MS, H, B	61
R = Ph	B	A, mp,MS, H, B	61,78

214

Iridium			
[Ir(C$_5$Me$_5$)(C$_5$H$_5$BPh)]PF$_6$	B	A, mp, H, B	69
Nickel			
[Ni(CO)(C$_5$H$_5$BMe)]$_2$(Ni—Ni)	E	A, mp, MS, H, B, C, IR	47
Ni(C$_5$H$_5$BPh)(4,5-H$_2$C$_5$H$_5$BPh)	C	A, mp, MS, X, H, B, C, IR	29
Ni(C$_5$H$_5$BPh)(5,6-H$_2$C$_5$H$_5$BPh)	C	A, mp, MS, H, B, C	29
Ni(C$_5$H$_5$BPh)1(1,4,5-η^3-C$_8$H$_{13}$)	C	A, mp, MS, H, B, C	29
Platinum			
PtMe$_3$(C$_5$H$_5$BPh)	B	A, mp, MS, H, B	61
Mercury			
HgCl(C$_5$H$_5$BR), R = Me[c]	B	H, B	27
R = Ph[c]	B	H, B	27
Thallium			
Tl(C$_5$H$_5$BR), R = Me	B	A, mp, MS, H, B, C, UV	51
R = Ph	B	A, mp, MS, H, B, C, UV	51,69

[a] A, Boranediyl insertion; B, via boratabenzene ions; C, via bora-2,5-cyclohexadienes; D, electron transfer reaction; E, ligand transfer reaction; F, electrophilic addition to metal; G, substitution of a borabenzene ligand; H, substitution of ancillary ligands; I, reductive degradation; J, cyanide degradation; K, nucleophilic substitution at boron; L, nucleophilic addition to ligand; M, electrophilic substitution at borabenzene ligand; N, electrophilic addition to ligand; O, ring contraction; P, ring-member substitution.

[b] A, Elemental analysis; X, X-ray structure determination; H, B, C, Co: ^{1}H, ^{11}B, ^{13}C, and ^{59}Co-NMR spectra, respectively; IP, ionization potential; PE, He(I) photoelectron spectrum; μ, magnetic moment; Möss, Mössbauer data; EC, electrochemical data.

[c] Observed in solution, not isolated.

[d] Stereochemistry unknown.

[e] **cis** Isomer in solid state.

[f] C$_7$H$_8$, norbornadiene.

215

undergoes insertion of a boranediyl group RB yielding primary products of type **20**. Complexes **20** may then undergo a second, much slower boranediyl insertion.

The boranediyl insertion with BCl_3 and BBr_3 gives products with 1-haloboratabenzene ligands which easily undergo nucleophilic substitution at boron; complex **6** has been made in this way (57,79). However, these reactions have never been published in detail.

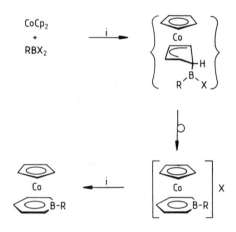

Scheme 1.   i, $+CoCp_2/-[CoCp_2]X$.

This unique formation of a borabenzene ring skeleton is thought to proceed as delineated in Scheme 1. A more detailed explanation together with comments on stoichiometry, reaction conditions, and by-products may be found elsewhere (2,7). However, it should be noted that Scheme 1 is based on the well-known reducing power of cobaltocene and its ability to add radicals efficiently (80), both properties being intimately connected with the uncommon 19-*e* configuration of $CoCp_2$.

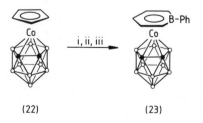

Scheme 2.   i, $NaC_{10}H_8$; ii, $PhBCl_2$; iii, H_2O, O_2.

In principle other highly reducing 19-*e* complexes should exhibit reactivity similar to cobaltocene. This challenging hypothesis has apparently been tested in only one case. The metallacarbaborane **22** can be converted to the boratabenzene analog **23** by reduction and subsequent treatment with PhBCl$_2$ (Scheme 2) (*70*).

B. *Alkali Metal Boratabenzenes*

Alkali metal boratabenzenes have a wide synthetic applicability just like alkali metal cyclopentadienides. Two syntheses have been developed: Ashe's synthesis via organotin intermediates (*23*) and our cyanide degradation of bis (boratabenzene) cobalt complexes (*61*).

1. *The Organotin Route*

The first approach to lithium boratabenzenes (Scheme 3) was an elegant adaption of a more general synthetic strategy (*81*). Hydrostannation of 1,4-pentadiyne (*82*) with Bu$_2$SnH$_2$ gives the stanna-2,5-cyclohexadiene **24** (*83,84*), which on treatment with PhBBr$_2$ produces

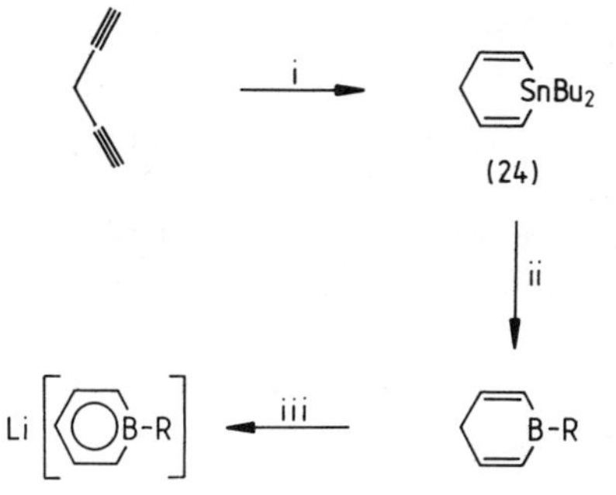

SCHEME 3. i, Bu$_2$SnH$_2$; ii, +RBBr$_2$/−Bu$_2$SnBr$_2$; iii, LiBu^t in pentane/THF.

1-phenylbora-2,5-cyclohexadiene **25** (*23*). Deprotonation of **25** with a sterically hindered base such as LiBu^t then affords a THF solution of **8** (*23*). Later 1-phenylboratabenzene was shown to be a considerably weaker base in THF solution than cyclopentadienide ion (*85*). In later ramifications of this work analogs of **8** with R = Me (*58*), Bu^t (*58*), and OMe (*26*) have been produced in solution but none of these compounds has been characterized.

2. Degradation of (Boratabenzene)metal Complexes

Alkali metal boratabenzenes may be liberated from bis (boratabenzene) cobalt complexes **7** and **13** by reductive degradation with elemental Li, sodium amalgam, or Na/K alloy (*60*), or alternatively by degradation with cyanides (*61*). The latter method has been developed in detail (Scheme 4). It produces spectroscopically pure (¹H-NMR control) solutions of the products **26**; the excess alkali metal cyanide and the undefined cyanocobalt compounds produced are essentially insoluble in acetonitrile.

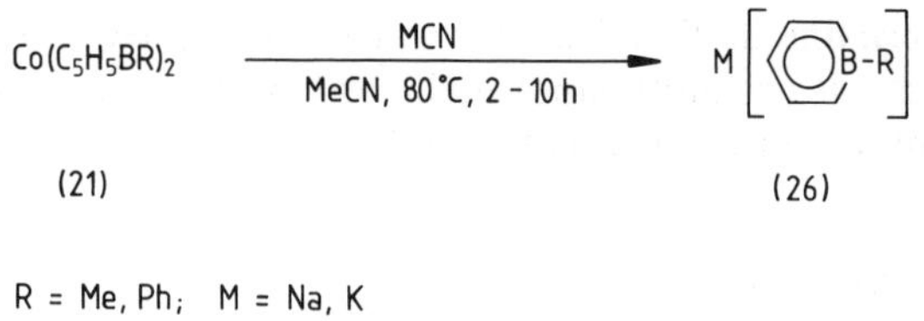

$$Co(C_5H_5BR)_2 \xrightarrow[\text{MeCN, 80 °C, 2 - 10 h}]{\text{MCN}} M\left[\begin{array}{c} \bigcirc B-R \end{array}\right]$$

(21)   (26)

R = Me, Ph;   M = Na, K

SCHEME 4

3. Benzoannulated Systems

The easily obtainable 1,2-dihydro-2-boranaphthalene **27** can be deprotonated by lithium 2,2,6,6-tetramethylpiperidide (LiTMP) to give the 2-boratanaphthalene **28**, characterized via its derivative **29** (Scheme 5) (*26*).

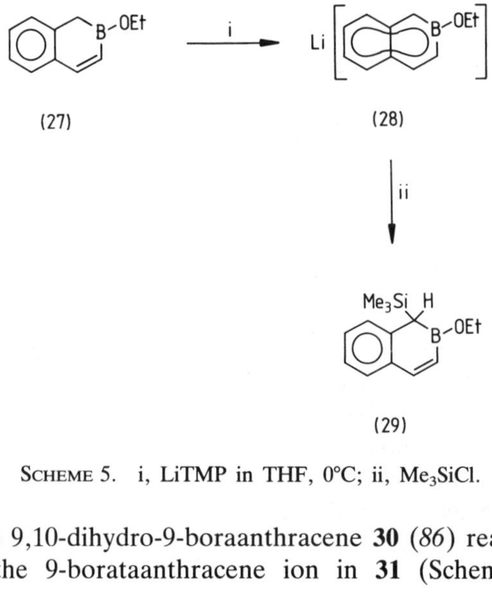

(27)   (28)

(29)

SCHEME 5. i, LiTMP in THF, 0°C; ii, Me₃SiCl.

Likewise the 9,10-dihydro-9-boraanthracene **30** (*86*) readily deprotonates to give the 9-borataanthracene ion in **31** (Scheme 6) (*86,87*).

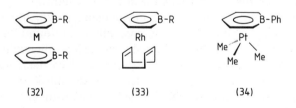

SCHEME 6

The acidity of **30** has been measured with reference to Streitwieser's acidity scale (*87*); a pK_A value of 15.8 for **30** as compared to 22.8 for fluorene and 18.5 for 9-phenylfluorene demonstrates a significant increase in acidity (*87*).

In closely related work the 9-phenyl analog of **31** was described with $\delta(^{11}B) = -10$ ppm (*88*); this value has aroused some scepticism (*89*) since **8** [$\delta(^{11}B) = 27$ ppm] and other borataaromatic systems have $\delta(^{11}B) = 22$–32 ppm (*75,90,91*).

C. Complexes from Boratabenzenes

Simple transition metal halides react cleanly with alkali metal borabenzenes. In this way sandwich-type complexes **32** of V (*27*), Cr (*64*), Fe (*58*), Ru (*61*), and Os (*61*) have been made. The corresponding nickel complexes seem to be nonexistent, quite in contrast to NiCp$_2$; in attempted preparations, mixtures of diamagnetic C—C linked dimers were obtained (*29*). In the manganese case, high sensitivity to air and water has precluded preparative success until now. Some organometallic halides have added further variations to the main theme. The complexes **33** of Rh and **34** of Pt were obtained from [(COD)RhCl]$_2$ and [Me$_3$PtI]$_4$, respectively (*61*).

R = Me, Ph

On treatment with TlCl, alkali metal boratabenzenes afford the corresponding Tl compounds (*51*). The lower reactivity of these can be essential for some syntheses. Some of the rare cationic borabenzene complexes **35–37** could be made using thallium boratabenzenes as reagents. Similarly, (C$_4$Me$_4$)Co(CO)$_2$I yielded the mixed sandwich complexes **38** and **39** in excellent yields (*71*).

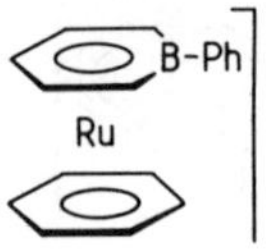

(35)

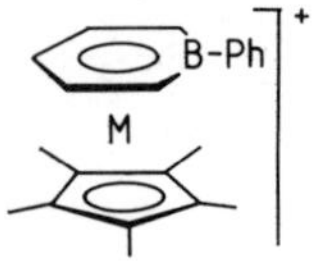

(36): M = Rh
(37): M = Ir

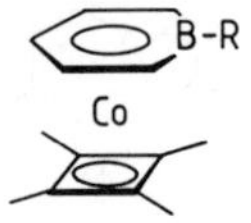

(38): R = Me
(39): R = Ph

In some cases metal carbonyl compounds may be combined with boratabenzenes. Thus, $Na[C_5H_5BMe]$ smoothly adds a $Cr(CO)_3$ group from $Cr(CO)_3(NH_3)_3$, giving the salt **40** (*63*). Treatment of sodium boratabenzenes with $HgCl_2$ and then with $Na[V(CO)_6] \cdot 2[O(C_2H_4OMe)_2]$ produces the complexes $V(CO)_4$ (C_5H_5BR) (**9** and **17**) (*27*). These on heating in cycloheptatriene form the paramagnetic 17-*e* complexes **41** (with an effective magnetic moment $\mu_{eff} = 1.69$ B.M.) and **42** ($\mu_{eff} = 1.72$ B.M.) (*27*).

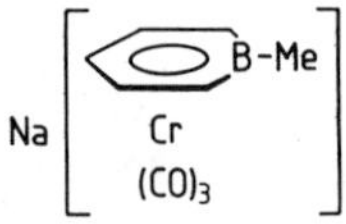

(40)

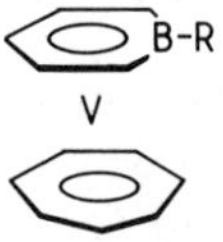

(41): R = Me
(42): R = Ph

D.   *Complexes from Bora-2,5-cyclohexadienes*

It is well known that, when treated with complex substrates, cyclopentadiene can form complexes with cyclopentadiene, cyclopentadienyl, and even cyclopentenyl ligands. The same possibilities are found for bora-2,5-cyclohexadienes, but with the additional complexity of ligand isomerism.

The bora-2,5-cyclohexadienes **25**, **43**, and **44** have been used in experimental studies. Photochemical reaction of **25** and **43** with $Fe(CO)_5$ produces the robust complexes **45** (*29*) and **46** (*92*) with η^5-divinylborane structures (*92*), whereas thermally induced complex formation of **44** with $Fe_2(CO)_9$ is accompanied by ligand isomerization and affords complex **47** (*67*). We note in passing that $\delta(^{11}B) = 38.8$ ppm for **47** is at rather low field. The only strictly comparable boraolefin known is 1-methoxy-6-(trimethylsilyl)bora-2,4-cyclohexadiene [$\delta(^{11}B) = 47.1$ ppm] (*26*). On this basis, the high field shift upon complexation (only 8.3 ppm) indicates weak Fe–B interaction.

Formation of borabenzene ligands requires more forcing conditions. Prolonged irradiation of **45** slowly produces [Fe(CO)$_2$(C$_5$H$_5$BPh)]$_2$ (**48**) (*29*), and pyrolysis affords bis(boratabenzene)iron derivatives in all cases (*65,67,68*).

The reaction of Co$_2$(CO)$_8$ with **25** is complicated at room temperature, but above 60°C Co(CO)$_2$(C$_5$H$_5$BPh) (**11**) is obtained as the only complex product (*29*). Ni(CO)$_4$ reacts with **25** in boiling hexane to give the (cyclopentadienyl)(cyclopentenyl)nickel analog **49** [δ(^{11}B) = 24.4 ppm, one broad signal]; this complex contains a novel boron–carbon ligand with

strong Ni—B interaction (*29*). At higher temperatures, in boiling benzene or heptane, **49** isomerizes to **50** [δ(^{11}B) = 24.0 ppm for the boratabenzene ligand and δ(^{11}B) = 41.5 ppm for the bora-3-cyclohex-2-yl ligand] with one very weak Ni—B interaction (*29*).

Bora-2,5-cyclohexadienes have a much greater synthetic potential than is apparent from the examples given so far. This may be exemplified by two recent reactions. Reductive complex formation in the system Co(acac)$_3$/COD/**25**/Mg/THF affords complex **51** via the organotin route (*77*) while an earlier synthesis used the cobaltocene route (*60*). Ni(COD)$_2$ very cleanly forms the (η^3-1,4,5-cyclooctenyl)nickel complex **52** (*29*).

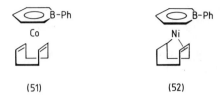

(51)   (52)

E. *Miscellaneous Syntheses*

Ring-member substitution, a very characteristic reaction of some 18-*e* borabenzene complexes (see Section VII,B), can also occur with 1,4-dibora-2,5-cyclohexadiene complexes. The cobalt complex **53** cleanly reacts with MeCOCl/AlCl$_3$ to give the cation **54** (Scheme 7) (*75*). The Rh complex (C$_5$Me$_5$)Rh[MeB(CHCH)$_2$BMe] reacts analogously (*75*).

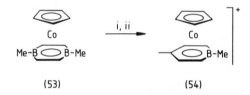

(53)   (54)

SCHEME 7. i, MeCOCl/AlCl$_3$, CH$_2$Cl$_2$, 0°C; ii, H$_2$O.

Under mild conditions 1-phenyl-4,5-dihydro-1*H*-borepine (**55**) undergoes complex formation without rearrangement (*92,93*). However, under forcing conditions, typically in boiling mesitylene, skeletal rearrangements take place and borole as well as borabenzene ligands are formed (*52,94*).

(55)   (56)

In the system [Fe(CO)$_2$Cp]$_2$/**55** the main products are FeCp$_2$, the borabenzene derivative **56**, and two triple-decked complexes CpFe[μ-(2-EtC$_4$H$_3$BPh)]FeCp and CpFe[μ-(2-ViC$_4$H$_3$BPh)]FeCp with a central borole ligand (*52*).

VI

REACTIONS AT THE METAL CENTER

A. Electron Transfer Reactions

Electrochemical methods have been used to study the redox properties of (boratabenzene)metal complexes (62,73). Table VI gives some typical data together with information on the corresponding Cp compounds. These data show that in going from a Cp complex to the analogous C_5H_5BR compound all redox potentials are shifted anodically. This shift has been termed the borinato shift (62), and later, when the more general nature of the phenomenon was recognized, the term borylene shift was proposed (95). The difference between the B-methyl and B-phenyl compounds is one order of magnitude less important. However, the C_5H_5BPh ligand is again seen to be slightly more electron withdrawing than the C_5H_5BMe ligand.

As a consequence of the more electron withdrawing character of the boratabenzene ligand, anionic borabenzene complexes are greatly stabilized as compared to their Cp counterparts. The bis(boratabenzene)metal complexes 32 of V (62), Cr (62), Fe (below −10°C) (62), and Co (173) show fully reversible one-electron reductions at easily accessible cathodic potentials.

Chemically the iron complex 18 is reduced by K/Na alloy in THF to give a green solution of the salt 57. The d^7 anion in 57 has been characterized by its ESR spectrum in frozen solution (62). Similarly, on treatment with sodium amalgam, the cobalt complexes 7 and 13 yield dark brownish-red solutions of 58 and 59, respectively. A surprisingly robust PPh_4^+ salt 60 (mp 158–159°C) could be isolated. Solution and solid state magnetic measurements confirm the presence of two unpaired electrons in these 20-e species as in $NiCp_2$ (60).

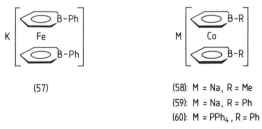

(57)

(58): M = Na, R = Me
(59): M = Na, R = Ph
(60): M = PPh₄, R = Ph

By contrast, cationic borabenzene complexes are destabilized as compared to the corresponding Cp compounds. As a consequence only six

TABLE VI
SELECTED ELECTROCHEMICAL DATA

Compound	Process +/0			Process 0/−			Ref.
	$\bar{E}$ (V)[a]	Solvent	Δ (V)[b]	$\bar{E}$ (V)	Solvent	Δ (V)	
VCp$_2$	−0.7[c]	THF		≈-3[c]	THF		62
V(C$_5$H$_5$BMe)$_2$	$(E_\text{p}^a +0.76)$[d]	CH$_2$Cl$_2$	0.7	−1.34	CH$_2$Cl$_2$	≈0.8	62
CrCp$_2$	−0.61	CH$_2$Cl$_2$		$(E_\text{p}^c -2.34)$[b]	THF		62
Cr(C$_5$H$_5$BMe)$_2$	+0.45	CH$_2$Cl$_2$	0.57	−1.26	CH$_2$Cl$_2$	0.63	
Cr(C$_5$H$_5$BPh)$_2$	+0.53	CH$_2$Cl$_2$		−1.01	CH$_2$Cl$_2$		
FeCp$_2$	+0.48	CH$_2$Cl$_2$		−2.93[e]	DMF		62
Fe(C$_5$H$_5$BMe)$_2$	+1.10	CH$_2$Cl$_2$	0.32			0.58	
Fe(C$_5$H$_5$BPh)$_2$ (18)	+1.13	CH$_2$Cl$_2$		−1.77	THF		
CoCp$_2$	−0.95	AN		−1.88[f]	AN		73
Co(C$_5$H$_5$BMe)$_2$ (7)	−0.02	AN	0.50	−1.25	AN	0.38	
Co(C$_5$H$_5$BPh)$_2$ (13)	+0.05	AN		−1.11	AN		
[Cr(CO)$_3$Cp]$^-$				−0.28	AN		63
[Cr(CO)$_3$(C$_5$H$_5$BMe)]$^-$				+0.01	AN	0.29	

[a] All potentials versus the saturated calomel electrode (SCE).
[b] Borylene shift per ring.
[c] From Ref. 96.
[d] Irreversible process at room temperature.
[e] From Ref. 97.
[f] Polarographic half-wave potential, from Ref. 98.

nontrivially different cationic borabenzene complexes are known (see Table V). Electrochemical generation of cations is often followed by irreversible reactions (62); if oxidation occurs at >0.1 V versus SCE the cations produced are not expected to be isolable species (62).

The 19-e complexes $CoCp(C_5H_5BR)$ (type 20) are easily oxidized to the corresponding cations $[CoCp(C_5H_5BR)]^+$ (type 19); chemically, oxidation with $FeCl_3$ in Et_2O/H_2O is very convenient [R = Me, Ph (7); R = CH_2Ph, Mes (74)]. These cations 19 are stable in acidic or neutral media. Basic conditions induce a boranediyl extrusion as exemplified for $[CoCp(C_5H_5BPh)]^+$ (1) in Scheme 8. In the presence of an additional oxidant, $[CoCp_2]^+$ and a derivative of $PhB(OH)_2$ are formed (69). In the absence of an additional oxidant part of the cation 1 is reduced (69).

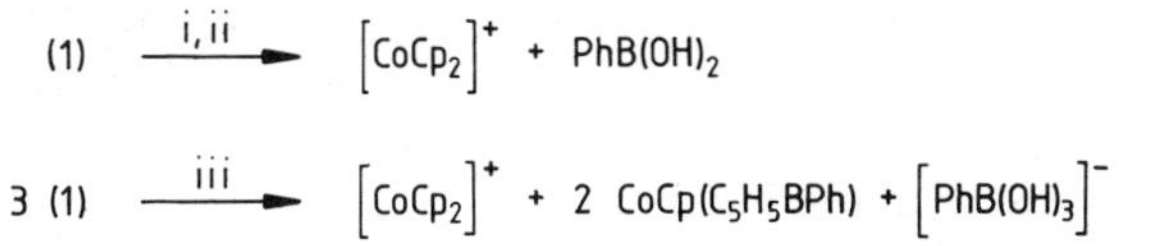

$$(1) \xrightarrow{\text{i, ii}} [CoCp_2]^+ + PhB(OH)_2$$

$$3 \ (1) \xrightarrow{\text{iii}} [CoCp_2]^+ + 2 \ CoCp(C_5H_5BPh) + [PhB(OH)_3]^-$$

SCHEME 8. i, Et_2NH, O_2 in CH_2Cl_2, 0°C; ii, H_2O; iii, KOH(s) in THF, 0°C.

The analogous oxidation of $Co(C_5H_5BPh)_2$ (13) occurs at a less cathodic potential (73). Oxidation with $[FeCp_2]PF_6$ in CH_2Cl_2 gives the cation $[Co(C_5H_5BPh)_2]^+$ (61) as salt 61·(PF_6^-), when traces of water are rigorously excluded (69); in the presence of water oxidation of 13 results in boranediyl extrusion and formation of cation 1 (Scheme 9) (69). Obviously cation 61 is strongly electrophilic for hard nucleophiles such as H_2O but not for $FeCp_2$.

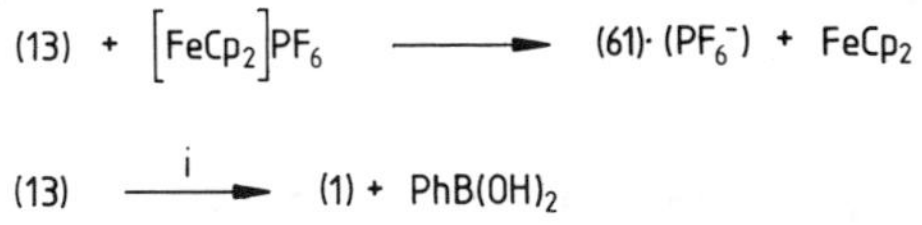

$$(13) + [FeCp_2]PF_6 \longrightarrow (61) \cdot (PF_6^-) + FeCp_2$$

$$(13) \xrightarrow{\text{i}} (1) + PhB(OH)_2$$

SCHEME 9. i, $FeCl_3$(aq) in Et_2O/H_2O.

In most cases oxidation of uncharged borabenzene complexes produces cations which can only be observed electrochemically. The iron compounds 62 and 63 may serve as an example. Oxidation is fully reversible in rigorously dried CH_2Cl_2 but irreversible in more basic solvents such as THF and acetonitrile (62). Preparative oxidation with Ce(IV) salts cleanly produces monosubstituted ferricenium cations 64 (Scheme 10) (66). In contrast to the above mentioned boranediyl extrusions, the substituent at boron is retained here in the newly formed cyclopentadienyl ring.

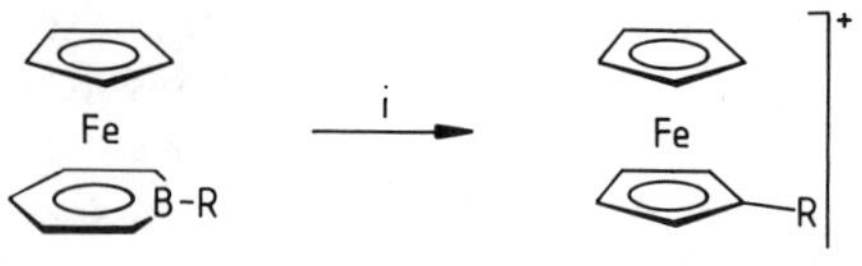

(62): R = Me   (64)
(63): R = Ph

SCHEME 10.   i, $(NH_4)_2Ce(NO_3)_6$, $CH_2Cl_2/MeOH$, $-20°C$.

B. Fission of the Metal–Ligand Bond

Borabenzene metal complexes, like their cyclopentadienyl counterparts, are not readily amenable to breaking of the metal–ligand bond. Such bond breaking will occur more easily in complexes with antibonding electrons. By fortuitous chance the cobaltocene route to borabenzene chemistry (Section V,A) provided 19-e complexes with the useful property of inherently weakened metal–ligand bonds. Thus most of this section centers on the reactivity of $Co(C_5H_5BMe)_2$ (7) and $Co(C_5H_5BPh)_2$ (13).

1. Ligand Transfer Reactions

The known reactions involving transfer of a C_5H_5BR ligand are collected in Table VII. Alternative syntheses are available for the bis(borata-benzene)iron complexes 65 and 18 only (58). In the system 7/$Ni(CO)_4$ the main product is $Co(CO)_2(C_5H_5BMe)$ (15), and 66 is obtained as a thermally unstable by-product (47). In solution, 66 adopts a folded, doubly CO-bridged structure with a dihedral angle of ~121° for the $Ni_2(CO)_2$ moiety (47).

TABLE VII

LIGAND TRANSFER REACTIONS

Reactants	Temp. (°C)	Product	Yield (%)	Ref.
7/$Mn_2(CO)_{10}$	160	$Mn(CO)_3(C_5H_5BMe)$ (14)	67	46
13/$Mn_2(CO)_{10}$	140	$Mn(CO)_3(C_5H_5BPh)$ (10)	884	48
7/$Fe_2(CO)_9$	110	$[Fe(CO)_2(C_5H_5BMe)]_2$ (12)	70	68
13/$Fe_2(CO)_9$	110	$[Fe(CO)_2(C_5H_5BPh)]_2$ (48)	81	68
7/$[Fe(CO)_2Cp]_2$	165	$FeCp(C_5H_5BMe)$ (62)	84	66
13/$[Fe(CO)_2Cp]_2$	165	$FeCp(C_5H_5BPh)$ (63)	95	66
$[Fe(CO)_2(C_5H_5BMe)]_2$ (12)	230	$Fe(C_5H_5BMe)_2$ (65)	68[a]	68
$[Fe(CO)_2(C_5H_5BPh)]_2$ (48)	230	$Fe(C_5H_5BPh)_2$ (18)	87[a]	68
7/$Ni(CO)_4$	80	$[Ni(CO)(C_5H_5BMe)]_2$ (66)	5.5	47

[a] One-pot synthesis, yield based on 7 and 13, respectively.

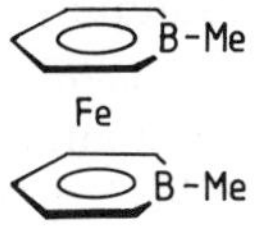

(65) 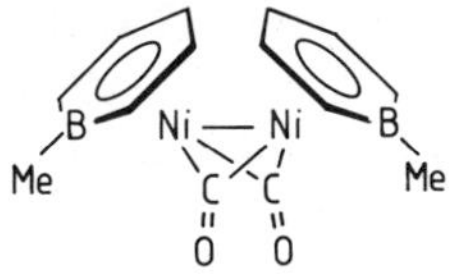   (66)

2. Substitution of a Boratabenzene Ligand

The bis(boratabenzene)cobalt complexes **7** and **13** may also undergo substitution of a C_5H_5BR ligand. With $Ni(CO)_4$ in refluxing toluene, the substitution products $Co(CO)_2(C_5H_5BR)$ **15** and **11** are formed (47,49). The cyanide degradation is another important example (Section V,B,2).

Labilization of the metal–ligand bond is more pronounced in the 20-*e* species $M[Co(C_5H_5BR)_2]$ **58** and **59**. These salts are stable in THF at room temperature for several days, but decompose in acetonitrile within hours. In the presence of alkali metal total reductive degradation is achieved, whereas in the absence of strong reductants a dismutation takes place (Scheme 11) (60). In the presence of suitable back-bonding ligands such as COD and norbornadiene substitution of a boratabenzene ligand with a diene is effected; for instance, $Co(C_5H_5BPh)(COD)$ **(51)** and its methyl analog have been obtained in this way (60,71).

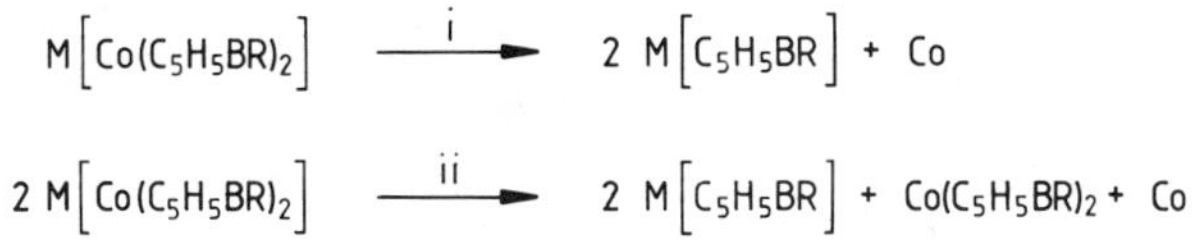

$$M\left[Co(C_5H_5BR)_2\right] \xrightarrow{\ i\ } 2\ M\left[C_5H_5BR\right] + Co$$

$$2\ M\left[Co(C_5H_5BR)_2\right] \xrightarrow{\ ii\ } 2\ M\left[C_5H_5BR\right] + Co(C_5H_5BR)_2 + Co$$

SCHEME 11.   i, $NaHg_x$ in MeCN; ii, in MeCN, 20°C.

VII

REACTIONS AT THE COORDINATED LIGAND

A. Nucleophilic Attack

In contrast to the highly reactive organoboranes, borabenzene metal complexes are surprisingly inert toward nucleophiles. However, cationic complexes may undergo nucleophilic addition reactions, and nucleophilic substitution has been observed with compounds having a hydrogen or an electronegative substituent at boron.

1. Nucleophilic Addition Reactions

Hydride addition to $[CoCp(C_5H_5BPh)]^+$ **(1)** occurs at C-2 and C-4 as well as at the Cp ligand. The products **67** with an (η^5-1-bora-2,4-

pentadiene)-type ligand, **68** with an (η^5-divinylborane)-type ligand, and the cyclopentadiene complex **69** are obtained in yields of 12, 30, and 51%, respectively (69). Similar hydride additions are found for the cations $[Co(C_5H_5BPh)_2]^+$ (**61**) and $[(C_5Me_5)Rh(C_5H_5BPh)]^+$ (**36**) (69).

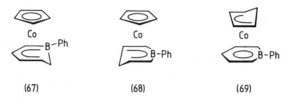

| (67) | (68) | (69) |

Quite in contrast to, e.g., $[CoCp_2]^+$, borabenzene metal cations show a pronounced affinity toward hard nucleophiles such as amines, OH^-, and to some extent even F^- and H_2O. Qualitatively this affinity increases in the order $CoCp_2]^+ \ll$ **36** < **1** < **61** (69). $[CoCp(C_5H_5BPh)]^+$ (**1**) adds tertiary amines at boron. With pyridine, the pyridinioboratacyclohexadienyl complex **70** is formed ($K = 174 \pm 5$ liters mol^{-1}, in MeCN, 20°C), which can be isolated from CH_2Cl_2 as PF_6^- salt (69). The similar rhodium and iridium cations **36** and **37** form the stable cyanide adducts **71** and **72** (69).

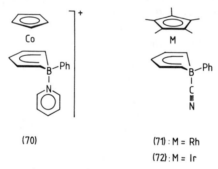

| (70) | (71) : M = Rh |
| | (72) : M = Ir |

More often, nucleophilic addition at boron can only be observed spectroscopically and is followed by more or less fast boranediyl extrusion (see Section VI,A) (69). A speculative mechanism has been proposed which links this boranediyl extrusion with the boranediyl insertion described earlier (Section V,A) (2,69).

2. *Nucleophilic Substitution at Boron*

Substitution of a boron-bonded group in a sandwich-type complex has first been observed in 1971 for (1-haloboratabenzene)cobalt complexes (57,79). The reaction of $CoCp_2$ with a large excess of BBr_3 in toluene produces a dark red solution [presumably of $Co(C_5H_5BBr)_2$, cf. Section V,A] which slowly deposits a yellow material, $[Co(C_5H_5BBr)_2]BBr_4$ (**73**); hydrolysis of this solid gives a highly acidic solution (pH < 1) containing

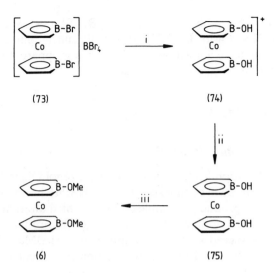

(73)   (74)

(6)   (75)

SCHEME 12. i, H_2O, 0°C; ii, NaOH(aq), pH $\approx$ 3–5; iii, MeOH.

the cation **74** (Scheme 12) (*79*). At pH > 3 the paramagnetic solid **75** precipitates from the solution (cf. Section VI,A and Scheme 8); recrystallization from MeOH then affords the crystalline complex **6** (*79*).

Complex $Fe(C_5H_5BOMe)_2$ (**76**) (obtained from **47** by pyrolysis) reacts with $LiAlH_4$ to afford **77**, the only B-unsubstituted boratabenzene complex known; this in turn can be alkylated by $LiBu^t$ to give the unexpected substitution product **78** (Scheme 13) (*67*).

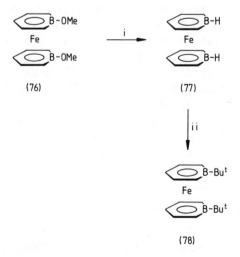

(76)   (77)

(78)

SCHEME 13. i, $LiAlH_4$; ii, $LiBu^t$.

Related work, e.g., on $Fe(CO)_3[Et_2C_2(BI)_2S]$ (99) and $CoCp[XB-(CHCH)_2BX]$ (X = OMe, Cl) (75), shows that nucleophilic substitution reactions certainly have a much wider scope than is apparent from the few examples known in borabenzene chemistry.

B. Electrophilic Attack

1. Electrophilic Addition Reactions

Addition of electrophiles to a coordinated benzenoid ligand is a characteristic reaction of 20-e complexes. $[Co(C_5H_5BPh)_2]^-$ in 59 is readily protonated at C-2, producing complex 79 (60) which alternatively can be obtained by hydride addition to cation 61 (69). Formal electrophilic addition is also observed in the reaction of $[Co(C_5H_5BMe)_2]^-$ in 58 with PhI which affords the two isomeric phenylation products 80 and 81 (60).

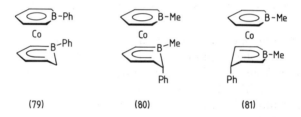

(79)   (80)   (81)

2. H/D Exchange

A number of the more robust 18-e borabenzene complexes undergo electrophilic substitution like their Cp analogs. This has first been observed for $Fe(C_5H_5BMe)_2$ (65) which exchanges the four protons at C-2 and C-6 when treated with CF_3CO_2D at 25°C; no further exchange was found even after 24 hours at reflux (58). $Fe(C_5H_5BH)_2$ (77) shows the same exchange but is ultimately destroyed by the acid (67). The much more reactive complex $Co(C_4Me_4)(C_5H_5BMe)$ (38), on treatment with CF_3CO_2D at 23°C, gives quantitative exchange at C-2 and C-6 within 5 minutes and at C-4 within 4 hours (71). It seems that electrophilic attack is charge controlled (39) and the observed regioselectivity reflects the charge distribution of the borabenzene ligand.

3. Acylation and Ring-Member Substitution

The ferrocene analog 65 is reported to give a 20% yield of the 2-acetyl derivative 82 when treated with $MeCOCl/AlCl_3$ (Scheme 14) (58). How-

SCHEME 14. i, MeCOCl/AlCl$_3$ (amount unspecified), in CH$_2$Cl$_2$, 0°C; ii, H$_2$O; iii, MeCOCl/AlCl$_3$ (fourfold excess), in CH$_2$Cl$_2$, 0°C.

ever, when a large excess of the Friedel–Crafts reagent is used, quantitative formation of the blood-red cation **83** is observed (Scheme 14) (66); this remarkable reaction is termed ring-member substitution (66). Similarly the mixed complexes FeCp(C$_5$H$_5$BR) **62** and **63** produce the cation [FeCp(PhMe)]$^+$ with a remarkable specificity (66).

The comparatively unreactive complex Mn(CO)$_3$(C$_5$H$_5$BMe) (**14**) with MeCOCl/AlCl$_3$ produces the 2-acetyl derivative **84** and small amounts of [Mn(CO)$_3$(PhMe)]$^+$ (27). The product ratio is rather insensitive to reaction conditions. It is reasonable to assume a common intermediate **85** (of unspecified stereochemistry at C-6) which under kinetic control may either irreversibly deprotonate to **84** or undergo a rearrangement ultimately leading to the ring-member substitution product (27).

<table>
<tr><td>

>=O
B—Me
Mn
(CO)$_3$

(84)

</td><td>

>=O]$^+$
B—Me
Mn
(CO)$_3$

(85)

</td></tr>
</table>

Further insight into the relationship between Friedel–Crafts acetylation and ring-member substitution has been gained from acetylation studies with the extremely reactive cobalt complex **38** (71). Four complex products **86–89** are found, and the reaction conditions are of great influence. With the mild catalyst AsCl$_3$ in CH$_2$Cl$_2$ at 40°C, the 2-acetyl derivative **86** is the main product (63%); the remarkable product **87** of a homoannular disubstitution is found in trace amounts (3%) (71). With SnCl$_4$ in CH$_2$Cl$_2$ at 35°C, **87** is now a main product (37%), and, in addition, the cations **88** and **89** are formed. With AlCl$_3$ only fast formation of the cations **88** and **89** is observed (71). Quite remarkably, both acetylation products **86** and **87** can be dissolved in superacids (CF$_3$SO$_3$H for **86**; CF$_3$CO$_2$H/BF$_3$ for **87**) to convert to the corresponding ring-member substitution products **88** and **89**, respectively (71). Obviously deprotonation of the intermediates **90** and **91**

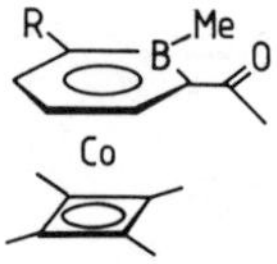

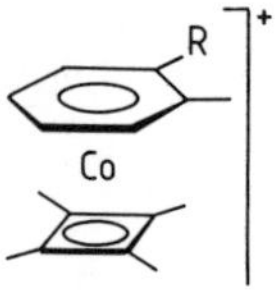

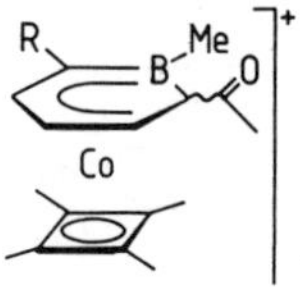

(86): R = H
(87): R = MeCO

(88): R = H
(89): R = MeCO

(90): R = H
(91): R = MeCO

becomes reversible under the most acidic conditions when $AlCl_3$ is the catalyst; the system then switches to thermodynamic control to give solely ring-member substitution products (71). A possible mechanism for the ring-member substitution has been proposed (71).

Vilsmeier formylation of **38** is possible with 65% yield (71). The B-phenyl analog **39** with $MeCOCl/AsCl_3$ also affords a 2-acetyl derivative; the B-phenyl group is not attacked under these conditions (71).

VIII

APPLICATIONS

Organocobalt complexes catalyze the cyclocotrimerization of acetylenes and nitriles, which affords pyridine and benzene derivatives (100). (Cyclopentadienyl)cobalt complexes such as CoCp(COD) favor pyridine formation (100), and modification of the Cp ligand has considerable influence on the activity of the catalyst and the chemo- and regioselectivity of the catalytic process (101).

Borabenzene complexes of cobalt such as $Co(C_5H_5BPh)(COD)$ (**51**) and its B-ethyl analog show the same type of catalysis but improved activity and chemoselectivity (77). Thus, **51** as the catalyst precursor gave the hitherto best results in the catalytic synthesis of the valuable 2-vinylpyridine from C_2H_2 and $CH_2{=}CHCN$ (120°C, 51 bar, 2 hours, turnover number 2164) (77,101). Furthermore, this catalyst for the first time allowed the synthesis of pyridine from C_2H_2 and HCN under mild conditions (110°C, 23 bar, 60 minutes, turnover number 103) (77).

IX

CONCLUDING COMMENTS

In the 16 years since the discovery of the first borabenzene derivative, the chemistry of borabenzene metal complexes has lost its pioneering character and gained maturity. Close similarities between bora-3,5-cyclohexadien-2-yl and cyclopentadienyl compounds have been established as well as distinct differences. The larger ring size is the origin of these differences: the more electron withdrawing character of the borabenzene ring, the stabilization of anions, the destabilization of cations. In addition, the boranediyl insertion, the various boranediyl extrusion reactions, and the ring-member substitution have added the fascination of the totally unexpected to the field.

The chemistry of borabenzene metal complexes has largely remained the domain of a single research group. We hope that this review will provide stimuli and orientation, especially to those who intend to contribute to the further development of this area.

X

ADDENDUM

Lithium 2-diisopropylamino-2-boratanaphthalene (cf. **28**) with Me_2-$NCH_2CH_2NMe_2$ (TMEDA) forms a crystalline derivative Li(TMEDA) ($C_9H_7BNPr^i_2$) which has been characterized by X-ray structure determination (102). The lithium cation is situated above the center of the boratabenzene moiety of the anion and, in addition, is chelated by one TMEDA ligand. Thus, the crystal consists of discrete tight ion-pair molecules.

^{59}Co NMR shifts of a series of complexes CoL(COD) (L = Cp, C_5Me_5, C_5H_4Ph, etc.) have been interpreted in terms of electron density at the cobalt atom (103). On this basis it is found once again that bora-3,5-cyclohexadien-2-yl ligands in Co(C_5H_5BR)(COD) [R = Ph (**51**), Et] are weaker donors and stronger acceptors than the Cp ligand (101).

Borabenzene–pyridine (**5**) and 2-boranaphthalene–pyridine readily add $M(CO)_3$ groups (M = Cr, Mo, W) to the borabenzene ring. X-Ray structural data are available for $5 \cdot Cr(CO)_3$, $5 \cdot Mo(CO)_3$, and the 2-boranaphthalene derivative ($C_9H_7B—NC_5H_5$)$Cr(CO)_3$ (104).

Transition metal complexes of 2-boratanaphthalene ligands begin to appear in the literature. The 2-ethoxy compound **28** has been used to synthesize Co[P(OPr)$_3$]$_2$(C_9H_7BOEt) and Co(COD)(C_9H_7BOEt) (101). $Rh(C_2H_4)_2$($C_9H_7BNPr^i_2$), Rh(COD)($C_9H_7BNPr^i_2$), and a complex $Fe(C_9H_7BPh)_2$ have also been made (102). The available NMR spectroscopic evidence shows that the metal is η^6-bonded to the boron-containing ring in all cases.

REFERENCES

1. G. E. Herberich, G. Greiss, and H. F. Heil, *Angew. Chem. Int. Ed. Engl.* **9**, 805 (1970).
2. G. E. Herberich, *in* "Comprehensive Organometallic Chemistry" (G. Wilkinson, F. G. A. Stone, and E. W. Abel, eds.), Vol. 1, p. 381. Pergamon, Oxford, 1982.
3. W. Siebert, *Adv. Organomet. Chem.* **18**, 301 (1980).
4. R. N. Grimes, *Coord. Chem. Rev.* **28**, 47 (1979).
5. C. W. Allen and D. E. Palmer, *J. Chem. Educ.* **55**, 497 (1978).
6. *Pure Appl. Chem.* **55**, 409 (1983).
7. G. E. Herberich and G. Greiss, *Chem. Ber.* **105**, 3413 (1972).
8. G. Schmid, *Angew. Chem. Int. Ed. Engl.* **9**, 819 (1970).
9. D. F. Shriver, *J. Am. Chem. Soc.* **85**, 3509 (1963).
10. H. Nöth and G. Schmid, *Angew. Chem. Int. Ed. Engl.* **2**, 623 (1963).
11. M. F. Hawthorne, D. C. Young, and P. A. Wegener, *J. Am. Chem. Soc.* **87**, 1818 (1965).
12. R. N. Grimes, *in* "Comprehensive Organometallic Chemistry" (G. Wilkinson, F. G. A. Stone, and E. W. Abel, eds.), Vol. 1, p. 459. Pergamon, Oxford, 1982.
13. K. P. Callahan and M. F. Hawthorne, *Adv. Organomet. Chem.* **14**, 145 (1976).
14. M. F. Hawthorne, *J. Organomet. Chem.* **100**, 97 (1975).
15. E. Wiberg and A. Bolz, *Ber. Dtsch. Chem. Ges.* **73**, 209 (1940).
16. M. J. S. Dewar, *in* "Progress in Boron Chemistry" (H. Steinberg and A. L. McCloeskey, eds.), Vol. 1, p. 235. Pergamon, Oxford, 1964.
17. R. Prinz and H. Werner, *Angew. Chem. Int. Ed. Engl.* **6**, 91 (1967).
18. G. Huttner and B. Krieg, *Chem. Ber.* **105**, 3437 (1972).
19. H. Nöth and W. Regnet, *Z. Anorg. Allg. Chem.* **352**, 1 (1967).
20. M. Scotti, H. Werner, D. L. S. Brown, S. Cavell, J. A. Connor, and H. A. Skinner, *Inorg. Chim. Acta* **25**, 261 (1977).
21. G. Huttner and B. Krieg, *Angew. Chem. Int. Ed. Engl.* **11**, 42 (1972).
22. G. Huttner, B. Krieg, and W. Gartzke, *Chem. Ber.* **105**, 3424 (1972).
23. A. J. Ashe, III and P. Shu, *J. Am. Chem. Soc.* **93**, 1804 (1971).
24. J. M. Schulman, R. L. Disch, and M. L. Sabio, *J. Am. Chem. Soc.* **104**, 3785 (1982).
25. G. Raabe, E. Heyne, W. Schleker, and J. Fleischhauer, *Z. Naturforsch., Teil A* **39**, 678 (1984).
26. R. Boese, N. Finke, J. Henckelmann, G. Maier, P. Paetzold, H. R. Reisenauer, and G. Schmid, *Chem. Ber.* **118**, 1644 (1985).
27. G. E. Herberich, W. Boveleth, B. Hessner, W. Koch, E. Raabe, and D. Schmitz, *J. Organomet. Chem.* **265**, 225 (1984).
28. G. Huttner and W. Gartzke, *Chem. Ber.* **107**, 3786 (1974).
29. E. Raabe, Dissertation, Rheinisch-Westfälische Technische Hochschule, Aachen (1984).
30. D. Rehder, I. Müller, and J. Kopf, *J. Inorg. Nucl. Chem.* **40**, 1013 (1978).
31. A. F. Berndt and R. E. Marsh, *Acta Crystallogr.* **16**, 118 (1963).
32. L. R. Byers and L. F. Dahl, *Inorg. Chem.* **19**, 277 (1980).
33. R. F. Bryan, P. T. Greene, M. J. Newlands, and D. S. Fields, *J. Chem. Soc. A* 3068 (1970).
34. W. Bunder and E. Weiss, *J. Organomet. Chem.* **92**, 65 (1975).
35. M. R. Churchill and B. G. DeBoer, *Inorg. Chem.* **13**, 1411 (1974).
36. T. A. Albright and R. Hoffman, *Chem. Ber.* **111**, 1578 (1978).
37. T. A. Albright, P. Hofmann, and R. Hoffman, *J. Am. Chem. Soc.* **99**, 7546 (1977).

38. M. C. Böhm, R. Gleiter, G. E. Herberich, and B. Hessner, *J. Chem. Phys.* **89**, 2129 (1985).
39. N. M. Kostić and R. F. Fenske, *Organometallics* **2**, 1319 (1983).
40. D. W. Clack and K. D. Warren, *Inorg. Chem.* **18**, 513 (1979).
41. D. W. Clack and K. D. Warren, *J. Organomet. Chem.* **208**, 183 (1981).
42. G. E. Herberich, T. Lund, and J. B. Raynor, *J. Chem. Soc. Dalton Trans.* 985 (1975).
43. R. Bucher, Dissertation, ETH Zürich, 1977.
44. R. Kühne, Dissertation, ETH Zürich, 1984.
45. G. E. Herberich and E. Raabe, unpublished observations.
46. G. E. Herberich, B. Hessner, and T. T. Kho, *J. Organomet. Chem.* **197**, 1 (1980).
47. G. E. Herberich, H. J. Becker, B. Hessner, and L. Zelenka, *J. Organomet. Chem.* **280**, 147 (1985).
48. G. E. Herberich and H. J. Becker, *Angew. Chem. Int. Ed. Engl.* **12**, 764 (1973).
49. G. E. Herberich and H. J. Becker, *Z. Naturforsch., Teil B* **29**, 439 (1974).
50. H. Nöth and B. Wrackmeyer, *in* "NMR Basic Principles and Progress" (P. Diehl, E. Fluck, and R. Kosfeld, eds.), Vol. 14. Springer-Verlag, Berlin, 1978.
51. G. E. Herberich, H. J. Becker, and C. Engelke, *J. Organomet. Chem.* **153**, 265 (1978).
52. G. E. Herberich, J. Hengesbach, G. Huttner, A. Frank, and U. Schubert, *J. Organomet. Chem.* **246**, 141 (1983).
53. E. O. Fischer and R. J. J. Schneider, *Chem. Ber.* **103**, 3684 (1970).
54. E. O. Fischer and R. D. Fischer, *Z. Naturforsch., Teil B* **16**, 556 (1961).
55. P. S. Braterman, "Metal Carbonyl Spectra," p. 169. Academic Press, New York, 1975.
56. J. K. Burdett, *J. Chem. Soc. A* 1195 (1971).
57. G. E. Herberich, G. Greiss, H. F. Heil, and J. Müller, *Chem. Commun.* 1328 (1971).
58. A. J. Ashe, III, E. Meyers, P. Shu, T. von Lehmann, and J. Bastide, *J. Am. Chem. Soc.* **97**, 6865 (1975).
59. D. L. Lichtenberger and R. F. Fenske, *J. Am. Chem. Soc.* **98**, 50 (1976).
60. G. E. Herberich, W. Koch, and H. Lueken, *J. Organomet. Chem.* **160**, 17 (1978).
61. G. E. Herberich, H. J. Becker, K. Carsten, C. Engelke, and W. Koch, *Chem. Ber.* **109**, 2382 (1976).
62. U. Koelle, *J. Organomet. Chem.* **157**, 327 (1978).
63. G. E. Herberich and D. Söhnen, *J. Organomet. Chem.* **254**, 143 (1983).
64. G. E. Herberich and W. Koch, *Chem. Ber.* **110**, 816 (1977).
65. G. E. Herberich and E. Bauer, *Chem. Ber.* **110**, 1167 (1977).
66. G. E. Herberich and K. Carsten, *J. Organomet. Chem.* **144**, C1 (1978).
67. A. J. Ashe, III, W. Butler and H. F. Sandford, *J. Am. Chem. Soc.* **101**, 7066 (1979).
68. G. E. Herberich, H. J. Becker, and G. Greiss, *Chem. Ber.* **107**, 3780 (1974).
69. G. E. Herberich, C. Engelke, and W. Pahlmann, *Chem. Ber.* **112**, 607 (1979).
70. R. N. Leyden and M. F. Hawthorne, *Inorg. Chem.* **14**, 2018 (1975).
71. G. E. Herberich and A. K. Naithani, *J. Organomet. Chem.* **241**, 1 (1983).
72. G. E. Herberich and H. J. Becker, *Z. Naturforsch., Teil B* **28**, 828 (1973).
73. U. Koelle, *J. Organomet. Chem.* **152**, 225 (1978).
74. K. H. Gustafsson, *Acta Chem. Scand., Ser. B* **32**, 765 (1978).
75. G. E. Herberich and B. Hessner, *Chem. Ber.* **115**, 3115 (1982).
76. G. E. Herberich and W. Pahlmann, *J. Organomet. Chem.* **97**, C51 (1975).
77. H. Bönnemann, W. Brijoux, R. Brinkmann, and W. Meurers, *Helv. Chim. Acta* **67**, 1616 (1984).
78. G. E. Herberich, and H. J. Becker, *Angew. Chem. Int. Ed. Engl.* **14**, 184 (1975).
79. H. F. Heil, Dissertation, Technische Universität, München (1971).
80. G. E. Herberich and J. Schwarzer, *Angew. Chem. Int. Ed. Engl.* **9**, 897 (1970).
81. J. Y. Corey, *Adv. Organomet. Chem.* **13**, 139 (1975).

82. H. D. Verkruijse and M. Hasselaar, *Synthesis* 292 (1979).
83. A. J. Ashe, III and W.-T. Chan, *J. Org. Chem.* **44**, 1409 (1979).
84. A. J. Ashe, III, W.-T. Chan, and E. Perozzi, *Tetrahedron Lett.* 1083 (1975).
85. H. F. Sandford, Ph. D. thesis, University of Michigan (1979).
86. R. van Veen and F. Bickelhaupt, *J. Organomet. Chem.* **77**, 153 (1974).
87. R. van Veen and F. Bickelhaupt, *J. Organomet. Chem.* **74**, 393 (1974).
88. P. Jutzi, *Angew. Chem. Int. Ed. Engl.* **11**, 53 (1972).
89. I. Ander, *in* "Comprehensive Heterocyclic Chemistry" (A. R. Katritzky and C. W. Rees, eds.), Vol. 1, p. 936. Pergamon, Oxford, 1984.
90. G. E. Herberich and H. Ohst, *Z. Naturforsch., Teil B* **38**, 1388 (1983).
91. N. Finke, Dissertation, Rheinisch–Westfälische Technische Hochschule, Aachen (1985).
92. G. E. Herberich, E. Bauer, J. Hengesbach, U. Kölle, G. Huttner, and H. Lorenz, *Chem. Ber.* **110**, 760 (1977).
93. G. E. Herberich, M. Thönnessen, and D. Schmitz, *J. Organomet. Chem.* **191**, 27 (1980), and references cited therein.
94. G. E. Herberich, J. Hengesbach, U. Kölle, and W. Oschmann, *Angew. Chem. Int. Ed. Engl.* **16**, 42 (1977).
95. U. Koelle, *Inorg. Chim. Acta* **47**, 13 (1981).
96. J. D. L.. Holloway, W. L. Bowden, and W. E. Geiger, Jr., *J. Am. Chem. Soc.* **99**, 7089 (1977).
97. Y. Mugnier, C. Moise, J. Tirouflet, and E. Laviron, *J. Organomet. Chem.* **186**, C49 (1980).
98. W. E. Geiger, Jr., *J. Am. Chem. Soc.* **96**, 2632 (1976).
99. W. Siebert, R. Full, J. Edwin, K. Kinberger, and C. Krüger, *J. Organomet. Chem.* **131**, 1 (1977).
100. H. Bönnemann, *Angew. Chem. Int. Ed. Engl.* **17**, 505 (1978).
101. H. Bönnemann, *Angew. Chem. Int. Ed. Engl.* **24**, 264 (1985).
102. P. Paetzold, N. Finke, P. Wennek, G. Schmid, and R. Boese, *Z. Naturforsch., Teil B.* **41** (1986). preparation.
103. H. Bönnemann, W. Brijoux, R. Brinkmann, W. Meurers, R. Mynott, W. von Philipsborn, and T. Egolf, *J. Organomet. Chem.* **272**, 231 (1984).
104. R. Boese, N. Finke, T. Keil, P. Paetzold, and G. Schmid, *Z. Naturforsch., Teil B* **40**, 1327 (1985).

ADVANCES IN ORGANOMETALLIC CHEMISTRY, VOL. 25

The Synthesis of Organometallics by Decarboxylation Reactions

GLEN B. DEACON, SUELLEN J. FAULKS,* and GEOFFREY N. PAIN*

Chemistry Department
Monash University
Clayton, Victoria 3168, Australia

I

INTRODUCTION

Decarboxylation reactions of metal carboxylates [Eq. (1)], are of increasing value in the synthesis of organometallic compounds (*1–5*). The reverse reaction, e.g., carbonation of Grignard and organolithium reagents (*6,7*), is a well-known source of carboxylic acids. Early reviews of the

$$—M(O_2CR) \rightarrow —MR + CO_2 \tag{1}$$

* Present address: Telecom Australia Research Laboratories, Clayton, Victoria 3168, Australia.

237

Copyright © 1986 by Academic Press, Inc.
All rights of reproduction in any form reserved.

synthesis of organometallics by thermal decarboxylation (*1*) and of the synthesis of mercurials by decarboxylation (*3,4*) are available. The substantial Russian monograph on the synthesis of organometallics by decarboxylation of metal acylates (*2*) is not readily accessible. A brief account of decarboxylation is included in a recent book on metal carboxylates (*5*), and a recent book on mercurials has a brief but thorough summary of their synthesis by decarboxylation (*5a*). This article provides a current appraisal of the use of decarboxylation of metal carboxylates in organometallic synthesis. For brevity, only reactions which produce isolable or stable characterizable organometallics have been included. Metal salt catalyzed, oxidative decarboxylations of carboxylic acids may have mechanisms involving organometallic intermediates, but they are not considered here. A recent review of these reactions is available (*8*). The synthesis of metal carboxylates has been thoroughly treated in a recent book (*5*) and hence does not need coverage in this review. Not all decarboxylation syntheses involve preformed metal carboxylates. These are sometimes prepared *in situ*, mainly by reaction of chloro complexes with silver or thallium(I) carboxylates [Eq. (2), M' = Ag or Tl]. Some cases of organometallic synthesis by decarboxylation of organo–oxycarbonyl complexes [Eq. (3)], have been included.

$$—M—Cl + M'(O_2CR) \rightarrow —M(O_2CR) + M'Cl \downarrow \qquad (2)$$

$$—MC(=O)OR \rightarrow —MR + CO_2 \qquad (3)$$

Decarboxylation can be achieved by

1. Heating the metal carboxylate in the solid state, in solution, or as a suspension
2. Radical initiation, which may involve irradiation, electrochemistry, or addition of a radical source.

This division is maintained in this article. However, it is not necessarily a mechanistic division, since thermally induced reactions may also proceed by a radical mechanism.

II

MECHANISM

A. Preamble

The mechanisms of thermally initiated decarboxylations giving organometallic compounds have generally not been studied in detail, and mechanistic conclusions have largely been derived from substituent or solvent effects. In the few cases where both polar and nonpolar solvents have been examined, e.g., $(2,6\text{-}F_2C_6H_3CO_2)_2Hg$ (9) or $PhHgO_2CC_6F_5$ (10), reaction usually proceeds readily in the former but not in the latter. This, together with substituent effects (see below), points to a heterolytic mechanism. However, successful thermal decarboxylations in nonpolar solvents are certainly known (1), and it is possible that homolytic mechanisms contribute in some cases. Initially, an account of heterolytic decarboxylation mechanisms is given to provide a basis for discussion of reactions for which mechanistic proposals have been made. Homolytic decarboxylation mechanisms are then considered.

B. *Possible Mechanisms for Heterolytic Decarboxylation*

A number of possible unimolecular mechanisms can be envisaged for decarboxylation ranging from one involving a free carbanion intermediate [Eq. (4)], to one involving a free carbonium ion intermediate [Eq. (5)]. In

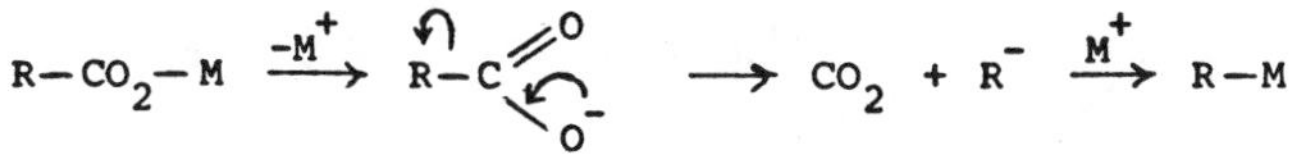

$$R-CO_2-M \xrightarrow{-M^+} R-C \overset{O}{\underset{O^-}{\diagup}} \longrightarrow CO_2 + R^- \xrightarrow{M^+} R-M \qquad (4)$$

$$R-C \overset{O}{\underset{O-M}{\diagup}} \longrightarrow CO_2 + R^+ + M^- \longrightarrow R-M \qquad (5)$$

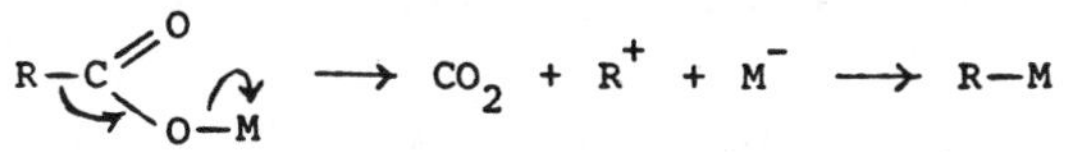

intermediate $S_E i$ mechanisms, the transition state has either carbanion [Eq. (6)] or carbonium ion [Eq. (7)] character, depending on whether R—C bond cleavage is slightly ahead [Eq. (6)] or slightly behind [Eq. (7)] M—C bond formation. Bimolecular mechanisms [e.g., Eq. (8)], where the

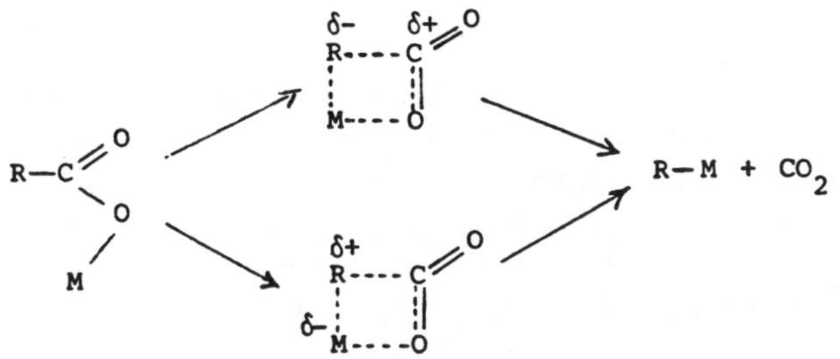

$$\tag{6}$$

$$R-M + CO_2$$

$$\tag{7}$$

$$\longrightarrow 2R-M + 2CO_2$$

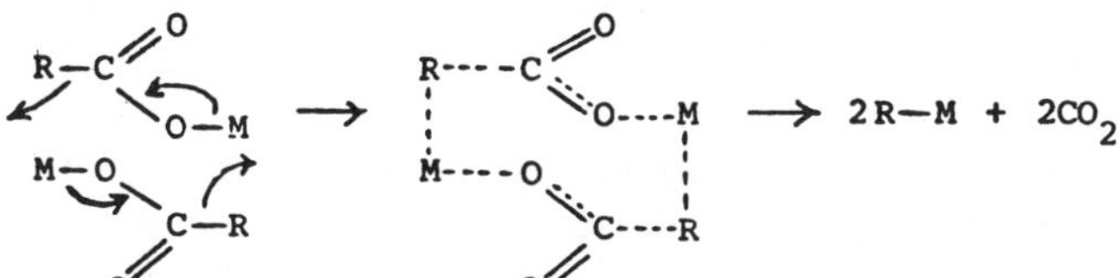

$$\tag{8}$$

transition state can have carbonium or carbanion character depending on the sequence of bond breaking and formation, are also possible.

Decarboxylation by the carbanion mechanism [Eq. (4)], would be aided by electron-withdrawing substituents in the organic group and would have a rate comparable with that of the free carboxylate ion. The mechanism is similar to that for the S_E1 decarboxylation of carboxylic acids [e.g., Eq. (4), M = H] (11,11a). Free carbanions are unlikely in syntheses giving good yields at elevated temperatures. Thus, preparations of $(C_6F_5)_2Hg$ and $(C_6F_5)_2Hgbpy$ (bpy = 2,2'-bipyridyl) at 150–200° C (10) in 50–80% yield would be impossible with the thermally unstable $C_6F_5^-$ [e.g., C_6F_5Li decomposes above 0° C (12)] as an intermediate. Coordination of the metal to an adjacent π bond may stabilize the carbanion (13,14). Donation of π electron density to the metal should aid decarboxylation by increasing the electron-withdrawing character of the organic group. Rapid $\pi \rightarrow \sigma$

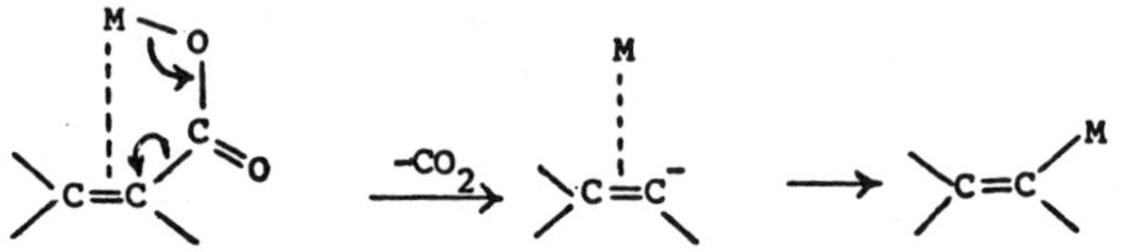

$$(9)$$

complex rearrangement is needed [see Eq. (9)] to maintain the stereochemistry of the double bond (14). The facile decarboxylation of metal cyanoacetates [e.g., Eq. (10)] (15) could conceivably involve a similar

$$\text{CuO}_2\text{CCH}_2\text{CN} \xrightarrow[50°\,C]{\text{HCONMe}_2} \text{CuCH}_2\text{CN} + \text{CO}_2 \qquad (10)$$

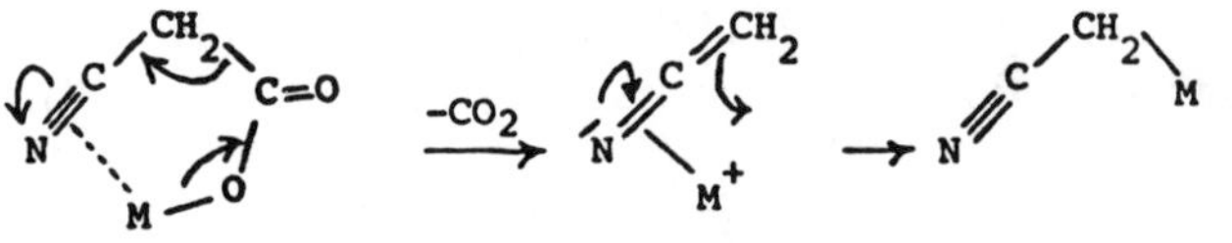

$$(11)$$

mechanism [Eq. (11)]. The mechanisms shown in Eqs. (9) and (11) are similar to those proposed for the decarboxylation of β-keto-, 2-cyano-, α-nitro-, and β–γ-unsaturated carboxylic acids (16).

A specific case of the carbonium ion mechanism [Eq. (5)] with reasonable plausibility is decarboxylation of metal arenoates by classic electrophilic aromatic substitution [Eq. (12)]. This mechanism would be favored by electron-donating substituents and has been invoked to explain the relative ease of decarboxylation of p-methoxybenzoic acid in molten mercuric trifluoroacetate (17) as well as the very facile decarboxylation on reaction of polymethoxybenzoic acids with mercuric acetate (18) (see below).

The transition states for the S_Ei mechanisms [Eqs. (6) and (7)] [and for the electronically similar versions of the transition state for the bimolecular

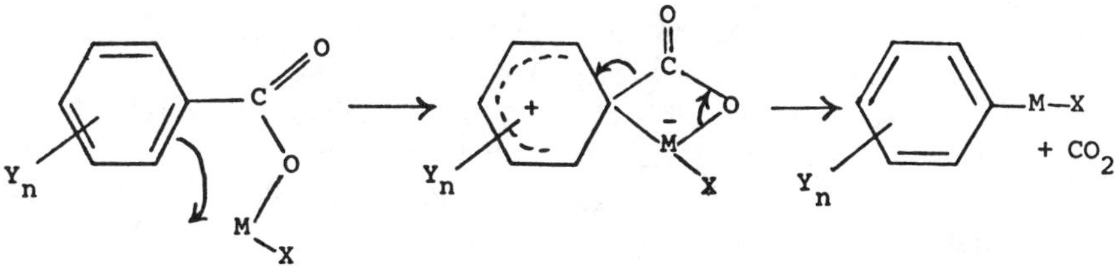

$$(12)$$

reaction, Eq. (8)] are stabilized by electron-withdrawing and electron-donating substituents, respectively. A more detailed picture of the transition state for the unimolecular $S_E i$ decomposition of an aromatic carboxylate [Eq. (13)] reveals a three center $(O{=})CCM({-}O)$ bond with

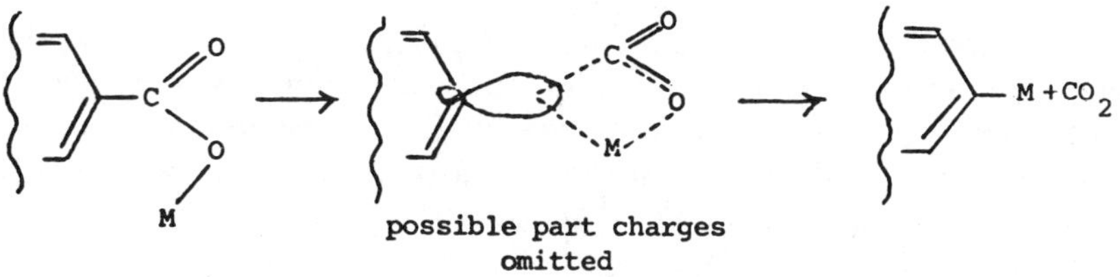

possible part charges
omitted

$$(13)$$

the two electrons originally from the $C{-}CO_2$ bond as the bonding electrons, and with a ring carbon sp^2 hybrid orbital involved in bonding. Any charge (δ^+ or δ^-) carried by the ring carbon in this orbital in the transition state would not be substantially affected by the resonance effects of substituents in the aromatic ring, since the sp^2 hybrid is orthogonal to the π orbitals. However, the partial carbanionic or carbonium transition states would be stabilized by inductive electron-withdrawing and -donating substituents, respectively. A wide range of decarboxylations are known to be facilitated by inductive electron-withdrawing substituents. On the other hand, decarboxylation by classic electrophilic aromatic substitution [Eq. (12)] would be aided by both resonance and inductive electron-donating substituents.

Where a metal has an established oxidation state two higher than the initial state, an oxidative addition (of $R{-}CO_2{}^-$) decarboxylation mechanism can be envisaged [Eq. (14)]. Decomposition of the oxidative addition

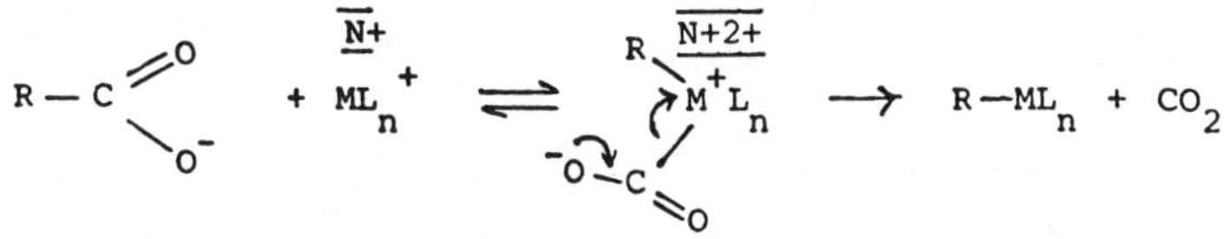

$$(14)$$

intermediate would be facilitated by electron-withdrawing substituents in the organic group. This mechanism is considered possible for the copper–quinoline catalyzed decarboxylation of carboxylic acids (*14*). A refinement of this mechanism [Eq. (15)] has been proposed for the decarboxylation of rhodium(I) polyhalogenobenzoates (*19*). The transition state would be stabilized by electron-withdrawing substituents.

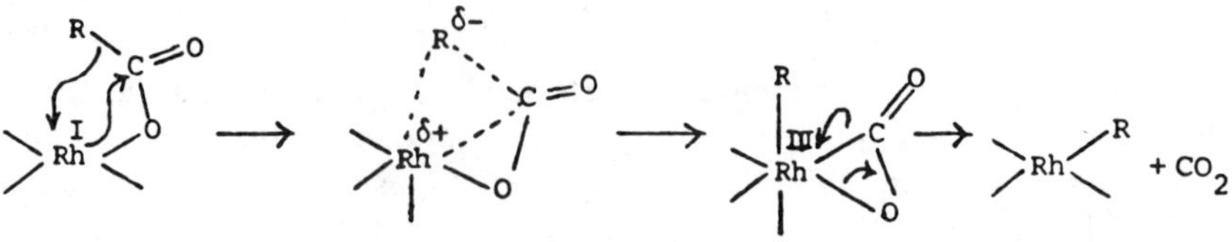

$$\tag{15}$$

Enhanced decarboxylation in polar solvents may be due to stabilization of polar transition states and/or solvent coordination to the metal (*20*). Coordination of solvent or ligands may aid decarboxylation by weakening metal–oxygen bonding (*10*). It also reduces the electrophilicity of the metal, the consequences of which are considered later.

Involvement of radical pathways in thermal decarboxylation is possible where no solvent or a nonpolar solvent is used. The pyrolysis of silver carboxylates in the absence of solvent under nitrogen gives radicals but no organometallics (*21*).

C. Homolytic Decarboxylation

Formation of organometallics by radical initiated decarboxylation is largely restricted to preparations of monoorganomercurials from mercuric carboxylates (see Section IV). These reactions are used as examples in the following discussion.

Decarboxylation on irradiation of mercuric carboxylates can be envisaged as proceeding by a chain mechanism:

Initiation $\qquad Hg(O_2CR)_2 \xrightarrow{h\nu} RCO_2^{\cdot} + {}^{\cdot}HgO_2CR$ $\qquad (16)$

$\qquad\qquad {}^{\cdot}HgO_2CR \rightarrow Hg + RCO_2^{\cdot}$ $\qquad (17)$

Propagation $\qquad RCO_2^{\cdot} \rightarrow R^{\cdot} + CO_2$ $\qquad (18)$

$\qquad R^{\cdot} + Hg(O_2CR)_2 \rightarrow RHgO_2CR + RCO_2^{\cdot}$ $\qquad (19)$

Termination $\qquad 2\,R^{\cdot} \rightarrow R_2$ $\qquad (20)$

When peroxide initiation is used, a chain mechanism may still obtain. The initiation reactions are shown in Eqs. (21)–(23). Propagation steps (18)

$$(R'CO_2)_2 \rightarrow 2R'CO_2^{\textbf{·}} \tag{21}$$

$$R'CO_2^{\textbf{·}} \rightarrow R'^{\textbf{·}} + CO_2 \tag{22}$$

$$R'^{\textbf{·}} + Hg(O_2CR)_2 \rightarrow R'Hg(O_2CR) + RCO_2^{\textbf{·}} \tag{23}$$

and (19) and termination by Eq. (20) follow. However, radical displacement giving $R'Hg(O_2CR)$ [Eqs. (21)–(23)] may predominate, and the peroxide becomes a stoichiometric reagent rather than an initiator. When $R' \neq R$, the occurrence of the chain mechanism is established by isolation of $RHg(O_2CR)$ rather than $R'Hg(O_2CR)$ as the main product. Alternatively, the paths may be distinguished by use of isotopic labeling. Clearly, peroxide initiated decarboxylation is likely to lead to product mixtures. Formation of mercurous salts [Eq. (24)], formation of mercury [Eq. (25)], and mercuration of aromatic solvents [Eq. (26)] can also complicate radical decarboxylation.

$$2\,^{\textbf{·}}HgO_2CR \rightarrow Hg_2(O_2CR)_2 \tag{24}$$

$$Hg_2(O_2CR)_2 \rightarrow Hg + Hg(O_2CR)_2 \tag{25}$$

$$ArH + Hg(O_2CR)_2 \rightarrow ArHg(O_2CR) + RCO_2H \tag{26}$$

In electrochemical decarboxylation, radicals are generated by anodic oxidation [Eqs. (27) and (28)]. As with reactions initiated by irradiation, propagation by Eqs. (18) and (19) and termination by Eq. (20) follow.

$$Hg(O_2CR)_2 \rightarrow HgO_2CR^+ + RCO_2 \tag{27}$$

$$RCO_2^- \rightarrow RCO_2^{\textbf{·}} + e^- \tag{28}$$

III

THERMAL DECARBOXYLATION

A. *Decarboxylation of Metal Polyhalogenoalkanoates*

Thermal decarboxylation has proved useful in the preparation of a variety of trihalogenomethyl- and a few polyfluoroalkylorganometallics. The thermal stability of the required organometallic limits the application of this technique (*1*,22).

Silver 2*H*-hexafluoroisobutyrate, $(CF_3)_2CHCO_2Ag$, in pyridine solution at room temperature, or in acetonitrile or dimethylformamide (dmf) at 40–50° C, loses carbon dioxide to form 1*H*-hexafluoroisopropylsilver, $(CF_3)_2CHAg$. The alkylsilver compound was characterized spectroscopically without isolation (*23*).

Fluoroalkylmercuric fluorocarboxylates, $RHgO_2CR$ (R = CF_3,C_2F_5, HF_2CCF_2, or C_5F_{11}), were prepared by heating the mercuric carboxylate, $(RCO_2)_2Hg$, to 300° C (24). Removal of the second carbon dioxide to give the dialkylmercurial, R_2Hg, was facilitated if a complex of the mercuric carboxylate was used (25) [Eq. (29), R = CF_3,C_2F_5, or C_3F_7; L =

$$(RCO_2)_2HgL \xrightarrow{\ 180-220°\ C\ } R_2HgL + 2\ CO_2 \qquad (29)$$

2,2'-bipyridyl (bpy) or 1,10-phenanthroline (phen)]. Ligand-free bis-(trifluoromethyl)mercury was obtained when mercuric trifluoroacetate was heated in the presence of anhydrous potassium carbonate $(26,27)$. The asymmetric mercurial $PhHgCF_3$ was also prepared by this method from phenylmercury trifluoroacetate. Reported syntheses of aryl(trifluoro-methyl)mercurials by heating arylmercuric trifluoroacetates, $RHgO_2CCF_3$ (R = m-FC_6H_4 or p-FC_6H_4), in dimethoxyethane (dme) (28) were found to be unrepeatable (29).

Mercuric $2H$-hexafluoroisobutyrate, $[(CF_3)_2CHCO_2]_2Hg$, when heated in refluxing pyridine (30) or to 40–50° C in dmf (31) gave bis($1H$-hexafluoroisopropyl)mercury, $[(CF_3)_2CH]_2Hg$. Competition between mer-curation and decarboxylation was studied by repeating the reaction in dmf in the presence of mercuric acetate (31). The main path was found to be mercuration at the α carbon of the hexafluoroisobutyrate group followed by decarboxylation. The low yield of bis($1H$-hexafluoroisopropyl)mercury showed that carbon dioxide elimination is slower than mercuration under these conditions.

Sodium trihalogenoacetates when reacted with mercuric salts in dme rapidly lost carbon dioxide and formed bis(trihalogenomethyl)mercurials [Eq. (30), X = Cl or Br, Y = Cl or OAc $(32,33)$] and/or tri-halogenomethylmercuric salts [Eq. (31), X = Cl, Y = Cl or Br (32) or X = Br, Y = Br (33)]. Since the evolution of carbon dioxide was more

$$2\ NaO_2CCX_3 + HgY_2 \rightarrow (CX_3)_2Hg + 2NaY + 2\ CO_2 \qquad (30)$$

$$NaO_2CCX_3 + HgY_2 \rightarrow CX_3HgY + NaY + CO_2 \qquad (31)$$

rapid in the presence of mercuric salts, intermediate trihalogenoaceta-tomercury(II) species were proposed. However mercuric trichloroacetate cannot be an intermediate, since the authentic compound, prepared from mercuric butoxide and trichloroacetic acid, underwent decomposition into trichloromethylmercuric chloride on heating in heptane, decane, or dme (34). A plausible path to the bis(trihalogenomethyl)mercurials may involve successive formation of CX_3CO_2HgY, CX_3HgY, and $CX_3HgO_2CCX_3$ intermediates.

Polyhalogenomethyltin and -lead compounds are of interest as possible sources of dihalogenocarbenes, $:CX_2$. Trimethyltin trihalogenoacetates,

$Me_3SnO_2CCX_3$ ($CX_3 = CF_3$, CCl_3, CBr_3 or CF_2Cl) were heated in cy-clooctene where any carbenes generated would add to the double bond

$$Me_3SnO_2CCX_3 \; + \quad \bigcirc\!\!\!| \longrightarrow \bigcirc\!\!\!<\!CX_2 \; + Me_3SnX + CO_2 \qquad (32)$$

[Eq. (32)] (*35*). Carbene addition was observed only for $CX_3 = CCl_3$ and CBr_3 (*35*). Spectroscopic studies of the decomposition of trimethyltin tri-chloroacetate demonstrated that decarboxylation to form Me_3SnCCl_3 was followed by formation of $:CCl_2$ and Me_3SnCl. The latter catalyzed the decarboxylation step, possibly by forming a complex with $Me_3SnO_2CCl_3$ (*36*).

Chloromethyltriphenyllead compounds, Ph_3PbR ($R = CCl_3$ or $CHCl_2$), have been proposed as products of the decomposition of the triphenyllead chloroacetates in refluxing pyridine on the basis of equivocal spectroscopic data (*37*). Trichloromethyltriphenyllead was isolated for use as a source of dichlorocarbene from the reaction of triphenyllead chloride and sodium trichloroacetate in dme at 85° C (*38*) [Eq. (33)].

$$Ph_3PbCl + NaO_2CCCl_3 \; \rightarrow \; Ph_3PbCCl_3 + NaCl + CO_2 \qquad (33)$$

Attempted syntheses of trifluoromethyl derivatives of germanium, tin, and lead by thermal decarboxylation either resulted in decomposition of the trifluoroacetate without forming carbon dioxide (*22,39,40*) or gave carbon dioxide but no trifluoromethyl organometallic (*22*). In the latter case, the metal fluoride was detected. This suggests that the trifluoromethyl com-pound is thermally unstable and decomposes by fluoride abstraction.

Heating dimethyltrifluoroacetoxyarsine to 205° C in a vacuum produced dimethyltrifluoromethylarsine (*41*). This contrasts with the thermal insta-bility of trifluoroacetoxyphosphines which eliminate carbon monoxide and form mixtures of trifluoromethylphosphine oxide compounds (*42*).

When colloidal selenium was heated with mercuric trifluoroacetate or silver trifluoroacetate, bis(trifluoromethyl)diselenide was formed (*43*). Later work with selenium/silver carboxylate, RCO_2Ag ($R = CF_3$, C_2F_5, or C_3F_7), mixtures at 280° C in a vacuum produced a mixture of the bis(perfluoroalkyl)selenide and the bis(perfluoroalkyl)diselenide (*44*). Formation of a polyselenium trifluoroacetate, which decarboxylates to produce the trifluoromethylselenides, was the proposed mechanism for $R = CF_3$ (*44*). However, silver trifluoroacetate is a source of trifluoro-methyl radicals when heated above 260° C (*21*), hence the trifluoromethyl-selenides may be formed by reaction of trifluoromethyl radicals with selenium, as in the reaction of CF_3I with selenium [Eq. (34)] (*45*).

$$\text{Se} \xrightarrow[260-285°\text{ C}]{\text{CF}_3\text{I}} (\text{CF}_3)_2\text{Se} + (\text{CF}_3)_2\text{Se}_2 \qquad (34)$$

The complexes $MCCl_3(CO)(PPh_3)_2(M = Rh$ or Ir) may be formed on reaction of $MH(CO)(PPh_3)_3$ with trichloroacetic acid (46).

B. Decarboxylation of Metal Cyanoacetates

Applications of decarboxylation of metal cyanoacetates are currently restricted to derivatives of tin, lead, and copper. Both copper(I) and copper(II) cyanoacetate lost carbon dioxide in dmf at 50° C and formed cyanomethylcopper(I) (15). In the presence of tertiary phosphines, the copper(I) cyanoacetate reaction was reversible [Eq. (35), R = Ph, Et, Bu,

$$\text{NCCH}_2\text{CO}_2\text{Cu(PR}_3)_x \xrightleftharpoons[50°\text{ C}]{\text{RT}} \text{NCCH}_2\text{Cu(PR}_3)_x + \text{CO}_2 \qquad (35)$$

OPh, or Ome; $x = 1, 2$, or 3] (47). The carboxylation reaction was favored when the ratio of tertiary phosphine to copper salt was increased to 3:1. This suggests that an intermediate $Cu—(CO_2)$ complex is not formed. At this ratio the copper complexes are considered coordinately saturated. Carboxylation was also favored when phosphines of high σ donor ability were used.

The reversible complexing of carbon dioxide by bis[bis(1,2-diphenylphosphino)ethane]iridium(I) chloride, $[Ir(dpe)_2]Cl$, in acetonitrile [Eq. (36)] (48) appears not to involve carboxylation of a cyanomethyliridium(III) complex or its formation by decarboxylation of the cyanoacetate

$$[Ir(dpe)_2]Cl + CH_3CN + CO_2 \xrightleftharpoons[\text{vac, }80°\text{ C}]{1 \text{ atm, RT}} [IrH(dpe)_2(O_2CCH_2CN)]Cl \qquad (36)$$

complex (48). Work with tetrakis(trimethylphosphine)iridium(I) chloride showed the corresponding cyanoacetate complex, $[IrH(PMe_3)_4-(O_2CCH_2CN)]Cl$, was not formed by carbon dioxide insertion into $[IrH(PMe_3)_4CH_2CN]Cl$ (48).

Loss of carbon dioxide from triorganotin and -lead cyanoacetates when heated *in vacuo* produced cyanomethyltriorganotin and -lead compounds [Eq. (37), M = Sn, R = Ph or Bu (49,50); M = Pb, R = Ph (51)].

$$R_3MO_2CCH_2CN \rightarrow R_3MCH_2CN + CO_2 \qquad (37)$$

Disproportionation to form tetraethyltin was observed when triethyltin cyanoacetate was heated (50). The fate of the cyanoacetate group was not established.

C. *Decarboxylation of Metal Alkynoates and Alkenoates*

Studies of the thermal decarboxylation of phenylpropiolates of copper, tin, and lead parallel those of the corresponding cyanoacetate compounds. Copper phenylpropiolate decarboxylated irreversibly in dmf at 35°C and reversibly in the presence of tertiary phosphines in dmf (*52*). Triorgano(phenylethynyl)tin and -lead compounds, $R_3MC\equiv CPh$ [M=Sn, R = Ph or Bu (*53*); M = Pb, R = Ph (*51*)], were isolated when the triorganotin or -lead phenylpropiolates, $R_3MO_2CC\equiv CPh$, were heated *in vacuo*. Triphenyllead phenylpropiolate also decarboxylated in refluxing toluene (*54*).

Decarboxylations of organotin and -lead propiolates do not yield the expected ethynyl compounds, $R_3MC\equiv CH$, but the dimetallated acetylenes $R_3MC\equiv CMR_3$ (M = Sn or Pb, R = Ph) are obtained (*53,54*). The disproportionation reaction [Eq. (39)] is thought to follow decarboxylation [Eq. (38)]. Independently synthesized ethynyltriphenyltin has been observed to undergo disproportionation [Eq. (39), R = Ph, M = Sn] at room temperature (*55*).

$$R_3MO_2CC\equiv CH \ \rightarrow \ R_3MC\equiv CH + CO_2 \tag{38}$$

$$2 \ R_3MC\equiv CH \ \rightarrow \ R_3MC\equiv CMR_3 + HC\equiv CH \tag{39}$$

Bis(triorganostannyl)acetylenes are also isolated after heating either bis(triorganostannyl) acetylenedicarboxylates, $R_3SnO_2CC\equiv CCO_2SnR_3$ (R = Et, Pr, or Bu), at 160–180°C [Eq. (40)] (*53*) or *O*-triethylstannyl triethylstannylpropiolate above 100°C [Eq. (41)] (*56*). Bis(triphenylplumbyl) acetylenedicarboxylate was also reported to decarboxylate on

$$R_3SnO_2CC\equiv CCO_2SnR_3 \ \rightarrow \ R_3SnC\equiv CSnR_3 + 2 \ CO_2 \tag{40}$$

$$Et_3SnO_2CC\equiv CSnEt_3 \ \rightarrow \ Et_3SnC\equiv CSnEt_3 + CO_2 \tag{41}$$

melting, but the product was not identified (*54*). Chlorodiethyltin propiolate undergoes decarboxylation and disproportionation when heated and forms diethyldiethynyltin [Eq. (42)] (*57*).

$$2 \ ClEt_2SnO_2CC\equiv CH \ \rightarrow \ Et_2Sn(C\equiv CH)_2 + Et_2SnCl_2 + 2 \ CO_2 \tag{42}$$

Syntheses of alkenyl organometallics by thermal decarboxylation are limited to one group of organotin compounds [Eq. (43), R = Ph, Me, Et, Pr, or Bu] (*58*).

$$Ph_2C=C(CN)CO_2SnR_3 \ \rightarrow \ Ph_2C=C(CN)SnR_3 + CO_2 \tag{43}$$

However, some binuclear π-allyl palladium complexes have been obtained by facile decarboxylations from palladium chloride and some

$$2 \underset{CH_2}{\overset{Me}{\underset{|}{\overset{|}{C}}}} \underset{CO_2H}{\overset{R}{\underset{|}{\overset{|}{C}}}} - X + 2PdCl_2 \xrightarrow{25°C} 2CO_2 + \left[Me - \overset{CH_2}{\underset{C}{\overset{|}{C}}} - Pd \underset{Cl}{\overset{}{\diagdown}} \right]_2 + 2HCl \qquad (44)$$

alkenoic acids [Eq. (44), R = Me, X = CN, CO_2H, CO_2Et, CO_2NH_2; R = Et; X = CO_2Et] (59). These reactions predated the synthesis of σ-bonded transition metal organometallics by decarboxylation (1,2).

D. Decarboxylation of Metal Polyhalogenoarenecarboxylates

Thermal decarboxylation has been widely used for the synthesis of polyhalogenoarylorganometallics, i.e., those of d^8 (Rh^I, Ir^I, Ni^{II}, Pd^{II}, Pt^{II}) and d^{10} (Cu^I, Ag^I, Au^I) transition metals, and Group 2B–4B main group metals. The success of the method is partly due to the thermal stability of the organometallics, which do not readily undergo halogen abstraction reactions. In addition, polyhalogenoaryl groups are strongly electron-withdrawing (12,60) thus promoting thermal decarboxylation by mechanisms shown in Eqs. (4), (6), (14), and (15).

1. Transition Metal Derivatives

The polyhalogenophenylrhodium(I) compounds, trans-RhR(CO)-$(PPh_3)_2$ (R = C_6F_5, C_6Cl_5, p-HC_6F_4, m-HC_6F_4, p-$MeOC_6F_4$, $4,5$-$H_2C_6F_3$, or $3,5$-$H_2C_6F_3$), have been prepared in pyridine either by decarboxylation of the corresponding trans-$RhO_2CR(CO)(PPh_3)_2$ complexes [Eq. (45), M = Rh] or by reactions between trans-$RhCl(CO)(PPh_3)_2$ and the corresponding thallous carboxylates (61,62). The latter reactions proceed by formation [Eq. (46), M = Rh] and decarboxylation [Eq. (45),

$$trans\text{-}MO_2CR(CO)(PPh_3)_2 \rightarrow trans\text{-}MR(CO)(PPh_3)_2 + CO_2 \qquad (45)$$

$$trans\text{-}MCl(CO)(PPh_3)_2 + TlO_2CR \rightarrow trans\text{-}MO_2CR(CO)(PPh_3)_2 + TlCl \qquad (46)$$

M = Rh] of intermediate carboxylatorhodium(I) complexes (62). No reaction was observed when R = o-HC_6F_4 or $2,6$-$F_2C_6H_3$, hence two ortho-fluorines and one other fluorine substituent are necessary for decarboxylation to occur under the conditions examined. The ease of decarboxylation decreased in the order R = C_6F_5 > p-$MeOC_6F_4$ > p-HC_6F_4 > m-HC_6F_4 > $4,5$-$H_2C_6F_3$ > $3,5$-$H_2C_6F_3$, consistent with the S_Ei (carbanion character) mechanism [Eq. (6)] or oxidative addition

[Eq. (15)] (*19,61,62*). Analogous organoiridium(I) complexes, *trans*-IrR(CO)(PPh$_3$)$_2$ (R = C$_6$F$_5$, *p*-HC$_6$F$_4$, or *m*-HC$_6$F$_4$), have been prepared by Eqs. (46) and (45) (M = Ir) (*19*).

Nickel(II) carboxylates, Ni(O$_2$CR)$_2$L·xH$_2$O (L = bpy, x = 2, R = C$_6$F$_5$, *p*-MeOC$_6$F$_4$, or *p*-EtOC$_6$F$_4$, or L = phen, x = 1, R = *p*-MeOC$_6$F$_4$ or *p*-EtOC$_6$F$_4$), decarboxylated in refluxing toluene to give polyfluoroarylnickel complexes, NiR$_2$L [Eq. (47)] (*63*). The presence of

$$Ni(O_2CR)_2L \cdot xH_2O \rightarrow 2 CO_2 + NiR_2L + x H_2O \qquad (47)$$

electron-withdrawing substituents in the organic groups is consistent with an heterolytic mechanism [e.g., Eq. (6)], but the use of toluene as medium and the need for a radical initiator to induce decarboxylation of Ni(O$_2$CC$_6$F$_5$)$_2$phen·H$_2$O (Section IV) raises the possibility of a homolytic mechanism. When L = phen and R = *p*-MeOC$_6$F$_4$ or *p*-EtOC$_6$F$_4$, the preparation of NiR$_2$L was accompanied by formation of NiR$_2$L′ [L′ = 2-(*p*-alkoxytetrafluorophenyl)-1,10-phenanthroline or 2,9-bis(*p*-alkoxy-tetrafluorophenyl)-1,10-phenanthroline] complexes [Eq. (48), R = *p*-MeOC$_6$F$_4$ or *p*-EtOC$_6$F$_4$] (*63*). Specific *ortho*-substitution was attributed

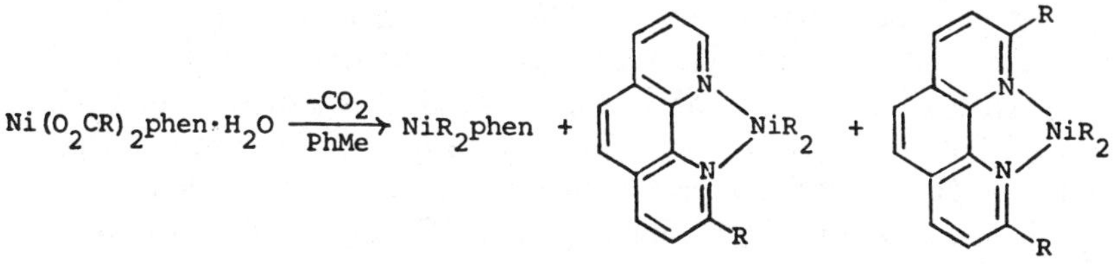

$$(48)$$

to nucleophilic attack by a highly carbanionic polyfluoroaryl group, derived from decarboxylation of a coordinated carboxylate, on the ortho position.

Bis(polyfluorophenyl)platinum(II) complexes, *trans*-PtR$_2$py$_2$ and PtR$_2$-bpy (R = C$_6$F$_5$, *p*-HC$_6$F$_4$, or *m*-HC$_6$F$_4$), were formed when either the corresponding carboxylates, *trans*-Pt(O$_2$CR)$_2$py$_2$ and Pt(O$_2$CR)$_2$bpy, or the platinum(II) halide complexes, *trans*-PtX$_2$py$_2$ and PtX$_2$bpy (X = Cl or Br), and the thallium(I) carboxylates, TlO$_2$CR, were heated in refluxing pyridine [Eqs. (49) and (50), M = Pt, L$_2$ = *trans*-(py)$_2$ or bpy] (*64*). The failure of the analogous 2,3,4,5-tetrafluorobenzoates to decarboxylate suggests that the elimination of carbon dioxide involves a transition state with a high degree of carbanionic character on the polyfluoroaryl group.

$$Pt(O_2CR)_2L_2 \rightarrow PtR_2L_2 + 2 CO_2 \qquad (49)$$

$$MX_2L_2 + 2 TlO_2CR \rightarrow MR_2L_2 + 2 TlX + 2 CO_2 \qquad (50)$$

Monopolyfluorophenylplatinum(II) complexes, $PtXRL_2$ ($L_2 = $ *trans*-$(py)_2$ or bpy, R $= C_6F_5$, X $= $ Cl or Br, R $= m$-HC_6F_4, X $= $ Cl; $L_2 = $ bpy, R $= p$-HC_6F_4, X $= $ Cl], have also been prepared from platinum(II) halide complexes and thallium(I) carboxylates [Eq. (51), M $= $ Pt] (*64*). The

$$MX_2L_2 + TlO_2CR \rightarrow MXRL_2 + TlX + CO_2 \qquad (51)$$

thallous salt reactions [Eqs. (50) and (51)] are thought to involve the intermediate complexes $[PtL_2py_2](O_2CR)_2$ or $[PtXL_2py]O_2CR$. In support, the dicarboxylates have been isolated for R $= o$-HC_6F_4 and $L_2 = $ bpy or $(py)_2$ and for R $= C_6F_5$ and $L_2 = (py)_2$; $[Pt(py)_4](O_2CC_6F_5)_2$ decarboxylated in pyridine to give *trans*-$Pt(C_6F_5)_2(py)_2$ [Eq. (52)].

$$PtCl_2L_2 \xrightarrow[py]{TlO_2CR} [PtL_2py_2](O_2CR)_2 \xrightarrow[py]{R\ =\ C_6F_5} trans\text{-}Pt(C_6F_5)_2(py)_2 \qquad (52)$$

Syntheses of PdR_2L_2 [$L_2 = $ *trans*-$(py)_2$, bpy, or phen, R $= C_6F_5$, p-HC_6F_4, or m-HC_6F_4; $L_2 = $ *cis*-$(py)_2$, R $= C_6F_5$] and $PdXRL_2$ [$L_2 = $ *trans*-$(py)_2$, bpy, or phen, R $= C_6F_5$, X $= $ Cl or Br; $L_2 = $ phen, R $= p$-HC_6F_4, X $= $ Cl; $L_2 = $ bpy, R $= m$-HC_6F_4, X $= $ Cl] have been accomplished by reactions (50) and (51) (M $= $ Pd) (*65*). The reaction of *trans*-$Pdpy_2Cl_2$ with thallous pentafluorobenzoate [Eq. (50)] differs from its platinum analog. The intermediate carboxylate appears to be $Pd(O_2CC_6F_5)_2py_2$ [cf. $[Pt(py)_4](O_2CC_6F_5)_2$, Eq. (52)], and the product is a mixture of *cis*- and *trans*-$Pd(C_6F_5)_2py_2$ [cf. *trans*-$Pt(C_6F_5)_2py_2$, Eq. (52)]. The loss of stereochemistry in the palladium reaction probably occurs in the second arylation step. Monoarylation gives only *trans*-$PdCl(C_6F_5)py_2$ which reacts with $TlO_2CC_6F_5$ in pyridine to give *cis*- and *trans*-$Pd(C_6F_5)_2py_2$ (*65*). Other differences between the palladium and platinum decarboxylation reactions are the more extensive displacement of 2,2'-bipyridyl by pyridine in the palladium reactions and the formation of isolable PdR_2phen and $PdXRphen$ complexes (see above) (*65*), in contrast to the complexity of the $PtCl_2phen/TlO_2CC_6F_5$ reaction (*64*).

Attempts to prepare the primary amine complex $Pt(C_6F_5)_2en$ (en $= $ ethane-1,2-diamine) by decarboxylation from $PtCl_2en$ and thallous pentafluorobenzoate in pyridine unexpectedly gave the organoamidoplatinum complex, $[N,N'$-bis(2,3,5,6-tetrafluorophenyl)ethane-1,2-diaminato(2-)] dipyridineplatinum(II) [Eq. (53)] (*66*). The proposed reaction sequence

$$PtCl_2en + 2TlO_2CC_6F_5 + 4\ py \rightarrow$$

$$Pt[N(p\text{-}HC_6F_4)CH_2]_2(py)_2 + 2\ TlCl + 2\ CO_2 + 2\ pyHF \qquad (53)$$

is given in Eqs. (54)–(56). Similar reactions between $PtCl_2en$ and thallous 2,3,5,6-tetrafluorobenzoate in pyridine in the presence of pentafluorobenzene or hexafluorobenzene also yield organoamidoplatinum compounds

$$PtCl_2en + 2\ TlO_2CC_6F_5 + 2\ py \rightarrow [Pten(py)_2](O_2CC_6F_5)_2 + 2\ TlCl \quad (54)$$

$$[Pten(py)_2](O_2CC_6F_5)_2 \rightarrow Pt[N(H)CH_2]_2(py)_2 + 2\ C_6F_5H + 2\ CO_2 \quad (55)$$

$$Pt[N(H)CH_2]_2(py)_2 + 2\ C_6F_5H + 2\ py \rightarrow$$
$$Pt[N(p\text{-}HC_6F_4)CH_2]_2(py)_2 + 2\ pyHF \quad (56)$$

$$PtCl_2en + 2\ TlO_2CC_6F_4H\text{-}p + 2\ RF + 4\ py \rightarrow$$
$$Pt[N(R)CH_2]_2(py)_2 + 2\ p\text{-}H_2C_6F_4 + 2\ TlCl + 2\ CO_2 + 2\ pyHF \quad (57)$$

[Eq. (57), R = p-HC$_6$F$_4$ or C$_6$F$_5$) (66). The reactions provide a novel route to a new class of air-stable organoamidoplatinum(II) complexes.

Decarboxylation has also been used to prepare organopalladium(II) and -platinum(II) complexes with phosphine ligands. Thus, MR$_2$L$_2$ [M = Pd, L$_2$ = (Ph$_2$PCH$_2$)$_2$, R = C$_6$F$_5$; M = Pt, L$_2$ = (Ph$_2$PCH$_2$)$_2$ or (Ph$_2$P)$_2$-(CH$_2$)$_3$, R = C$_6$F$_5$ or p-HC$_6$F$_4$] have been obtained from MCl$_2$L$_2$ and TlO$_2$ CR (mole ratio 1:2) in boiling pyridine in high yield [Eq. (50)] (65,67). The corresponding reactions of MCl$_2$(PPh$_3$)$_2$ complexes gave mixtures of products. From trans-PdCl$_2$(PPh$_3$)$_2$, cis- and trans-Pd(C$_6$F$_5$)$_2$(PPh$_3$)$_2$, trans-PdCl(C$_6$F$_5$)(PPh$_3$)$_2$, and cis-Pd(C$_6$F$_5$)$_2$(py)$_2$ were obtained (65), whilst for cis-PtCl$_2$(PPh$_3$)$_2$, the products were trans-Pt(C$_6$F$_5$)$_2$(PPh$_3$)$_2$, trans-PtCl(C$_6$F$_5$)(PPh$_3$)$_2$, and the unexpected cis-Pt(C$_6$F$_5$)$_2$py(PPh$_3$) (67). The last compound was also obtained together with trans-Pt(C$_6$F$_5$)$_2$(PPh$_3$)$_2$ on reaction of trans-PtCl(C$_6$F$_5$)(PPh$_3$)$_2$ with an equimolar amount of thallous pentafluorobenzoate. From PtCl$_2$L$_2$ [L$_2$ = (Ph$_2$PCH$_2$)$_2$ or (Ph$_2$P)$_2$(CH$_2$)$_3$] and TlO$_2$CC$_6$F$_5$ (mole ratio 1:1), the complexes PtCl(C$_6$F$_5$)L$_2$ were obtained, whilst a similar reaction of cis-PtCl$_2$(PPh$_3$)$_2$ gave trans-PtCl(C$_6$F$_5$)(PPh$_3$)$_2$ and trans-Pt(C$_6$F$_5$)$_2$py(PPh$_3$) (67).

Heating solid copper(I) and silver(I) pentafluorobenzoates to 200–240°C in vacuo (68) or the latter to 300°C under nitrogen (21) gave decafluoro-biphenyl and no organometallic products. However, Group 1B penta-fluorophenyl compounds can be isolated when milder conditions are used and the organometallic product is stabilized by complex formation. Thus, carbon dioxide was evolved from pyridine or acetonitrile solutions of silver pentafluorobenzoate at 25°C to give pentafluorophenylsilver pyridine or acetonitrile complexes (69), and pentafluorophenyltriphenylphos-phinegold(I) was formed by heating Ph$_3$PAuO$_2$CC$_6$F$_5$ at 130°C (70). The isolation of a quinoline complex of pentafluorophenylcopper(I) from heating copper(I) pentafluorobenzoate in quinoline (71) provides support-ing evidence for the proposed [e.g., (20)] formation of organocopper(I) intermediates in copperquinoline-catalyzed decarboxylations of carboxylic acids.

2. Main Group Element Compounds

Heating zinc and cadmium pentahalogenobenzoates at 200–300°C under vacuum resulted in decomposition giving bis(pentahalogenophenyl)-metallics [Eq. (58), R = C_6F_5, M = Zn (68) or Cd (72); R = C_6Cl_5, M = Cd (73)] or thermally unstable pentachlorophenylzinc chloride [Eq. (59)] (73). Mercuric pentafluorobenzoate readily decarboxylated at

$$M(O_2CR)_2 \rightarrow R_2M + 2\ CO_2 \qquad (58)$$

$$Zn(O_2CC_6Cl_5)_2 \xrightarrow{-CO_2} [(C_6Cl_5)_2Zn] \xrightarrow{-C_6Cl_4} C_6Cl_5ZnCl \xrightarrow{-C_6Cl_4} ZnCl_2 \qquad (59)$$

210°C in air [Eq. (58), R = C_6F_5, M = Hg] (10) [cf. hemidecarboxylation of mercuric perfluoroalkanoates at 300°C (24) (Section III, A)]. Complexes of mercuric pentafluorobenzoate gave the corresponding complexes of bis(pentafluorophenyl)mercury [Eq. (60), L = bpy at mp; L = phen or $(Ph_2PCH_2)_2$ in pyridine] (10).

$$(RCO_2)_2HgL \rightarrow (C_6F_5)_2HgL + 2\ CO_2 \qquad (60)$$

A wide range of bis(polyhalogenophenyl)mercurials have been prepared by decarboxylation of the corresponding mercuric polyhalogenobenzoates in boiling pyridine [Eq. (58), M = Hg, R = C_6Cl_5 (74), C_6Br_5, p-FC_6Br_4, p-ClC_6Br_4, p-$MeOC_6Br_4$ (75), or 2,6,-$F_2C_6H_3$ (9)]. In the last case, 2,6-difluorophenylmercuric 2,6-difluorobenzoate was also obtained [Eq. (61), R = 2,6-$F_2C_6H_3$]. The carboxylates $Hg(O_2CR)_2$ [R = p-

$$Hg(O_2CR)_2 \rightarrow RHgO_2CR + CO_2 \qquad (61)$$

MeC_6Br_4, o-MeC_6Br_4, m-$MeOC_6Br_4$, 2,6-$Me_2C_6Br_3$ (75), or 2,6-$Cl_2C_6H_3$ (9)] failed to react under these conditions. The substituent effects are consistent with an S_Ei mechanism in which the transition state develops carbanionic character. Decarboxylation of mercuric 2,6-difluorobenzoate in boiling pyridine was not inhibited by the addition of the free radical scavenger, p-dinitrobenzene and did not occur in the higher boiling, less polar xylene, thereby providing strong evidence against a homolytic mechanism (9).

Some mercuric carboxylates which failed to eliminate carbon dioxide in boiling pyridine have been found to decarboxylate in higher boiling polar solvents. Thus, $Hg(O_2CR)_2$ (R = p-MeC_6Br_4) gave the corresponding diarylmercurial in refluxing pyridine/nitrobenzene [Eq. (58), M = Hg], but the method could not be extended to R = o-MeC_6Br_4 or 2,6-$Me_2C_6Br_3$ (75). Mercuric 2,6-dichlorobenzoate was converted into bis(2,6-dichloro-phenyl)mercury in boiling dimethyl sulfoxide [Eq. (58), R = 2,6-$Cl_2C_6H_3$, M = Hg] (9). In the same medium, the 2,6-difluorobenzoate gave only bis(2,6-difluorophenyl)mercury (cf. pyridine, above), and mercuric 2,6-

dibromobenzoate underwent hemidecarboxylation [Eq. (61), R = 2,6-$Br_2C_6H_3$]. The rate of decarboxylation, R = 2,6-$F_2C_6H_3$ > 2,6-$Cl_2C_6H_3$ > 2,6-$Br_2C_6H_3$, in this solvent is consistent with the S_Ei (carbanionic) mechanism [Eq. (6)]. For the slower reacting $Hg(O_2CR)_2$ (R = 2,6-$Cl_2C_6H_3$ or 2,6-$Br_2C_6H_3$) derivatives, competition from reaction of the mercuric salt with the solvent considerably lowered the yields of the mercurials (9).

Although $Hg(O_2CR)_2$ (R = m-$MeOC_6Br_4$ or o-MeC_6Br_4) could not be decarboxylated in boiling pyridine and pyridine/nitrobenzene, respectively, the corresponding acids were decarboxylated in molten mercuric trifluoroacetate [Eq. (62)] (75). The products were characterized as the

$$RCO_2H + Hg(O_2CCF_3)_2 \xrightarrow{180-220°C} RHgO_2CCF_3 + CO_2 + CF_3CO_2H \qquad (62)$$

polybromophenylmercuric chlorides following work up with sodium chloride.

Unsymmetrical di(organo)mercurials, $RHgR'$ [R = Ph, $R' = C_6F_5$ (10), p-$MeOC_6F_4$, p-$EtOC_6F_4$ (76), C_6Cl_5 (74), or C_6Br_5 (75); R = p-MeC_6H_4 or p-$MeOC_6H_4$, $R' = C_6F_5$ (10) or C_6Cl_5 (74); R = Me, $R' = C_6F_5$ (10)] have been prepared from the organomercuric carboxylates in hot pyridine [Eq. (63)]. In addition, $RHgC_6Cl_5$ (R = Ph, p-

$$RHg(O_2CR') \rightarrow RHgR' + CO_2 \qquad (63)$$

MeC_6H_4, or Cl) compounds have been prepared from the appropriate chloromercury(II) compound and thallous pentachlorobenzoate in boiling pyridine [Eq. (64)] (74).

$$RHgCl + TlO_2CC_6Cl_5 \rightarrow RHgC_6Cl_5 + TlCl + CO_2 \qquad (64)$$

Syntheses of Group 3B organometallics by thermal decarboxylation are restricted to tris(polyfluorophenyl)thallium(III) compounds. Di(organo)thallium(III) carboxylates, R_2TlO_2CR (R = C_6F_5, p-$MeOC_6F_4$, p-HC_6F_4, or m-HC_6F_4), decarboxylated in boiling pyridine or at 310°C under nitrogen [Eq. (65)] (77). The resulting tri(organo)thallium(III) com-

$$R_2TlO_2CR \rightarrow R_3Tl + CO_2 \qquad (65)$$

pounds were isolated as air stable 1,4-dioxan or 4,4'-bipyridyl complexes. Higher yields were obtained for reactions in pyridine than for those without a solvent, where only $(C_6F_5)_3Tl$ was obtained in preparatively useful yield. Considerable competition from other decomposition paths was observed in the absence of a solvent.

The rate of decarboxylation in pyridine was found to decrease in the series $C_6F_5 > p$-$MeOC_6F_4 > p$-$HC_6F_4 > m$-$HC_6F_4 >> o$-HC_6F_4 (no reaction). The increase in rate with the increase in inductive electron-withdrawing character of the organic group indicates a transition state which has considerable carbanionic character, as in the $S_E i$ mechanism [Eq. (6)] or even a free carbanion mechanism [Eq. (4)] (77). If ion pairing is extensive in pyridine, the transition states for the two mechanisms would be similar.

Group 4B organometal carboxylates, $Ph_2Pb(O_2CR)_2$ ($R = C_6F_5$ or $p\text{-}EtOC_6F_4$) (78) and Ph_3MO_2CR [M = Pb, R = C_6F_5, $p\text{-}MeOC_6F_4$, or $p\text{-}EtOC_6F_4$ (76); M = Sn, R = C_6F_5, $p\text{-}MeOC_6F_4$, or $p\text{-}EtOC_6F_4$ (79); M = Ge, R = C_6F_5 or $p\text{-}EtOC_6F_4$ (79)], decarboxylated in refluxing pyridine and formed Ph_2PbR_2 and Ph_3MR compounds, respectively [Eqs. (66) and (67).] The germanium carboxylates were generated *in situ*

$$Ph_2Pb(O_2CR)_2 \rightarrow Ph_2PbR_2 + 2\ CO_2 \tag{66}$$

$$Ph_3MO_2CR \rightarrow Ph_3MR + CO_2 \tag{67}$$

from triphenylgermanium bromide and the appropriate silver carboxylate (79). Triphenylpolyfluorophenyllead compounds were also isolated in small amounts from the decarboxylation of diphenyllead dicarboxylates, $Ph_2Pb(O_2CR)_2$, and indeed Ph_3PbR was the only organolead compound identified for R = $p\text{-}MeOC_6F_4$ (78). The Ph_3PbR compounds were probably formed by decarboxylation of Ph_3PbO_2CR derivatives which were generated by the rearrangement given in Eq. (68). Heating under vacuum was a much less satisfactory decarboxylation method. Thus,

$$2\ Ph_2Pb(O_2CR)_2 \rightarrow Ph_3PbO_2CR + PhPb(O_2CR)_3 \tag{68}$$

diphenylbis(pentafluorophenyl)lead was isolated from the pyrolysis of diphenyllead bis(pentafluorobenzoate) *in vacuo*, but other diphenyllead polyfluorobenzoates, $Ph_2Pb(O_2CR)_2$ (R = $p\text{-}MeOC_6F_4$ or $p\text{-}EtOC_6F_4$), gave tetraphenyllead and no detectable polyfluoroaryllead compounds (78).

E. *Decarboxylation of Metal Arenecarboxylates*

Syntheses of aryl organometallics other than polyhalogenoaryls by thermal decarboxylation are comparatively rare. There are several reasons for this. For transition elements, the thermal stability of simple aryls is often low, especially by comparison with polyhalogenoaryl derivatives, thereby excluding syntheses at elevated temperatures. Electron-withdrawing substituents frequently aid thermal decarboxylation (Section III,A–D), and their absence inhibits major mechanistic paths to both transition metal and main group element derivatives, e.g., $S_E i$ (carbanionic) and oxidative addition (Section II). In thermal decomposition of mercuric arenoates, electron-withdrawing substituents deactivate (or block in the case of perhalogenoarenoates) the aromatic ring toward competition from mercuration, which is a classic aromatic substitution reaction and which is favored by electron-donating substituents (80). In contrast to the thermal decomposition of mercuric pentafluorobenzoate [Eq. (58), $M = Hg$, $R = C_6F_5$], pyrolysis of mercuric benzoate at 175°C gave benzoic acid and mercurated benzoate groups [Eq. (69)] (81,82). Although the

$$n \ Hg(O_2CC_6F_4H\text{-}o)_2 \rightarrow n \ o\text{-}HC_6F_4CO_2H + (\ o\text{-}HgC_6F_4CO_2)_n \qquad (69)$$

ortho mercurated species was predominant (82), mercuration was not regiospecific as claimed in an early report (81). At 190°C under vacuum, mercuric 2,3,4,5-tetrafluorobenzoate underwent ortho mercuration [Eq. (70)] (83).

$$n \ Hg(O_2CC_6F_4H\text{-}o)_2 \rightarrow n \ o\text{-}HC_6F_4CO_2H + (o\text{-}HgC_6F_4CO_2)_n \qquad (70)$$

The decomposition conditions also influence the competition between mercuration and decarboxylation. The mercuric carboxylates, $Hg(O_2CR)_2$ ($R = 2,6\text{-}F_2C_6H_3$, $2,6\text{-}Cl_2C_6H_3$, or $2,6\text{-}Br_2C_6H_3$), which decarboxylate in solution (Section III,D), gave carboxylic acids and mercuration products, predominantly based on 3-mercurated-2,6-dihalogenobenzoate groups, when heated under vacuum (9). On heating under nitrogen at atmospheric pressure, mercuric 2,6-dichlorobenzoate underwent both mercuration and decarboxylation giving m-dichlorobenzene and a complex mercurial with mainly 1,3-dimercurated-2,6-dichlorobenzene repeat units and 2,6-dichlorophenyl terminal groups (9). Suppression of mercuration in polar solvents can be attributed to reduced electrophilicity of the metal owing to coordination of the solvents to mercury. On the other hand, lowered electrophilicity should not substantially affect decarboxylation by the S_Ei (carbanion character) mechanism [Eq. (6)] since $R—CO_2$ bond breaking occurs slightly in advance of R—Hg bond formation.

Successful thermal decarboxylation of metal arenoates other than polyhalogenoarenoates are restricted to mercury compounds and fall into three categories, namely (i) those where electron-withdrawing substituents other than halogens are present in the organic groups, (ii) those where substituents and/or conditions are used which favor a different mechanism, e.g., classic electrophilic aromatic substitution, or (iii) those where the conditions are sufficiently forcing for both mercuration and decarboxylation to occur.

Other electron-withdrawing substituents have been little used apart from the synthesis of nitrophenylmercurials in early decarboxylation studies [e.g., Eq. (58), $M = Hg$, $R = 2,4,6\text{-}(O_2N)_3C_6H_2$]. These investigations have been adequately reviewed (1–4). Anomalies in reported decompositions of mercuric *mono*nitrobenzoates [see discussion in (1)] have not been resolved.

Successful decarboxylations where the substituents favor classic electrophilic aromatic substitution are known. Reaction of 2-hydroxy-1-naphthalenecarboxylic acid with mercuric acetate in cold acetic acid has been reported to give 2-hydroxynaphthalen-1-ylmercuric acetate [Eq. (71)] (84). Although the result probably requires reinvestigation,

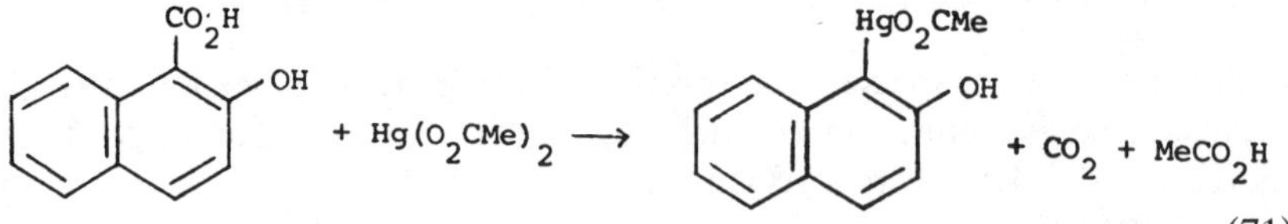

$$(71)$$

especially in view of the ease of decarboxylation, the carboxyl group occupies the position most activated toward classic electrophilic aromatic attack and the occurrence of decarboxylation can be rationalized in terms of a mechanism similar to that in Eq. (12).

At first sight, use of several strongly electron-donating groups in the aromatic ring might be expected to favor mercuration over decarboxylation. However, some polymethoxybenzoic acids gave facile and specific decarboxylation products on reaction with mercuric acetate in aqueous methanol at room temperature and lower [Eq. (72), R = 2,6-$(MeO)_2C_6H_3$, 2,3,4-, 2,4,5-, or 2,4,6-$(MeO)_3C_6H_2$] (18,85). In the reaction with 2,4,6-trimethoxybenzoic acid, use of a longer reaction time gave

$$2 \, RCO_2H + Hg(O_2CMe)_2 \rightarrow RHgO_2CR + CO_2 + 2 \, MeCO_2H \qquad (72)$$

$$2 \, RCO_2H + Hg(O_2CMe)_2 \rightarrow R_2Hg + 2 \, CO_2 + 2 \, MeCO_2H \qquad (73)$$

bis(2,4,6-trimethoxyphenyl)mercury [Eq. (73), R = 2,4,6-$(MeO)_3C_6H_2$] (18). Most other polymethoxybenzoic acids gave either no reaction [e.g., R = 3,4,5-$(MeO)_3C_6H_2$] or the mercuric salt, $Hg(O_2CR)_2$ [e.g., R = 2,3-$(MeO)_2C_6H_3$], at room temperature, and the mercuration product $+Hg(MeO)_xC_6H_{4-x}CO_2+_n$ [e.g., R = 3,5-$(MeO)_2C_6H_3$] on heating (18,85). Both mercuration and decarboxylation occurred only in the case of 2,4-dimethoxybenzoic acid [Eq. (74)] (85). Decarboxylation of 2,6-

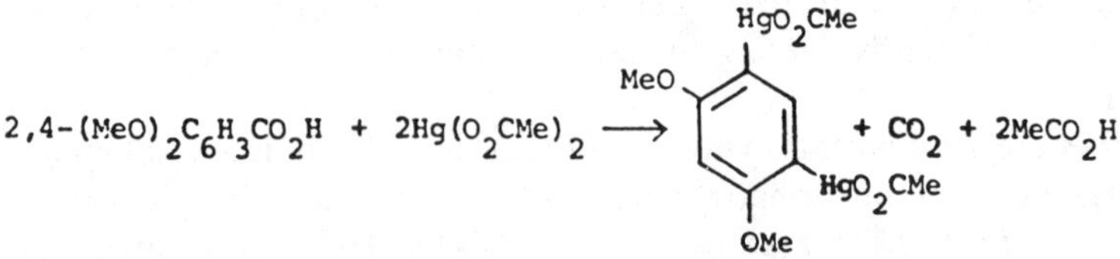

$$(74)$$

dimethoxybenzoic acid [Eq. (72), R = 2,6-(MeO)$_2$C$_6$H$_3$] is unaffected by radical inhibitors or the absence of light, hence a radical mechanism is excluded. The substituent effects indicate decarboxylation by classic electrophilic aromatic substitution. Two or more methoxy substituents ortho or para to the carboxyl group favor decarboxylation over mercuration except in Eq. (74). In this case, if decarboxylation precedes mercuration, then removal of the deactivating carboxyl group would facilitate subsequent mercuration.

In the third category, both mercuration and decarboxylation were induced by pyrolysis of mercuric 2,6-dichlorobenzoate at 210–230°C under nitrogen (9) (see above) and of phenylmercuric benzoate at 300°C [Eq. (75)], where the main product was a polymeric ortho dimercurated

$$n \text{ PhHgO}_2\text{CPh} \rightarrow n \text{ PhH} + n \text{ CO}_2 + -(\text{C}_6\text{H}_4\text{Hg})_n \qquad (75)$$

benzene (86). Benzoic acid and mono- and disubstituted benzoic acids underwent permercuration and partial decarboxylation on reaction with an excess of molten mercuric trifluoroacetate at 180–240°C [Eqs. (76) and (77), e.g., X = H, o-, m-, or p-Me, MeO, NO$_2$, CF$_3$, or F] (17). The ease

$$\text{XC}_6\text{H}_4\text{CO}_2\text{H} + 4 \text{ Hg(O}_2\text{CCF}_3)_2 \rightarrow \text{XC}_6(\text{HgO}_2\text{CCF}_3)_4\text{CO}_2\text{H} + 4 \text{ CF}_3\text{CO}_2\text{H} \qquad (76)$$

$$\text{XC}_6\text{H}_4\text{CO}_2\text{H} + 5 \text{ Hg(O}_2\text{CCF}_3)_2 \rightarrow \text{XC}_6(\text{HgO}_2\text{CCF}_3)_5 + \text{CO}_2 + 5 \text{ CF}_3\text{CO}_2\text{H} \qquad (77)$$

of decarboxylation during permercuration increased in the order p-XC$_6$H$_4$CO$_2$H $\leq$ m-XC$_6$H$_4$CO$_2$H $<$ o-XC$_6$H$_4$CO$_2$H for both electron-withdrawing (e.g., X = F, CF$_3$, NO$_2$) and electron-donating (X = Me) substituents, with the exception of X = OMe. Since steric assistance of decarboxylation could be eliminated from comparison of the behavior of ortho mono- and disubstituted acids, an S$_E$i mechanism was proposed with either slightly carbanionic or slightly carbonium ion character for the transition state, depending on whether the substituents have inductive electron-withdrawing or donating character [see Eqs. (6) and (7) and Section II]. For X = OMe, where ease of decarboxylation increased m-X $<$ o-X $\simeq$ p-X, classic electrophilic aromatic ipso substitution was proposed (17).

F. Decarboxylation of Metal Heteroarenecarboxylates

Studies are restricted to syntheses of Group 2B·organometallics and mainly to mercurials. Early studies of reactions of mercuric salts with furan- and thiophencarboxylic acids or their salts have been thoroughly reviewed (1–4). More recently several tetrafluoropyridyl derivatives have been prepared by decarboxylation [Eqs. (78) (87), (79) (88), or (80),

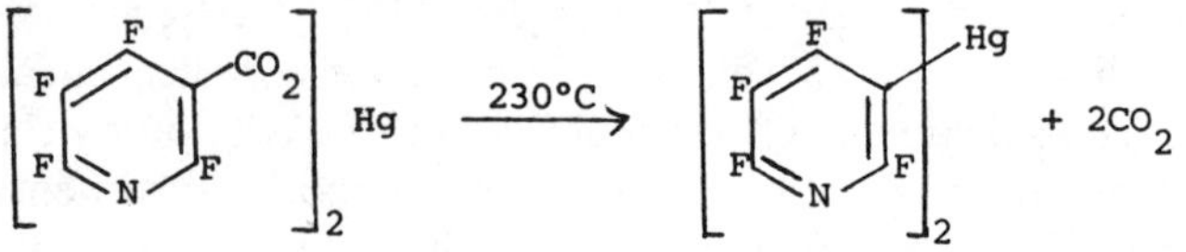

$$\left[\begin{array}{c} F \\ F \\ \text{-}CO_2 \\ F \quad N \quad F \end{array}\right]_2 Hg \xrightarrow{230°C} \left[\begin{array}{c} F \\ F \quad Hg \\ F \quad N \quad F \end{array}\right]_2 + 2CO_2 \qquad (78)$$

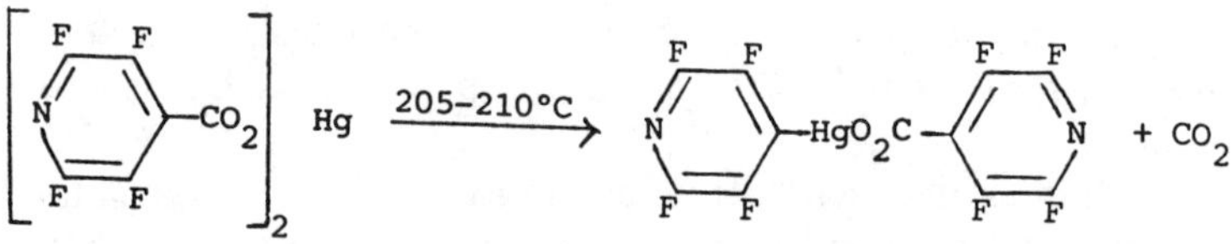

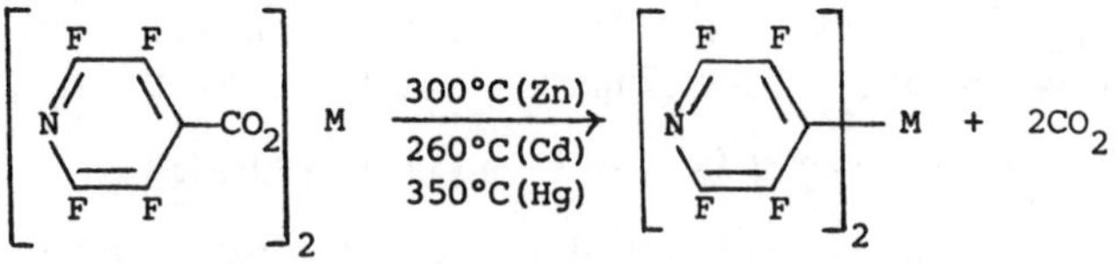

$$\left[\begin{array}{c} F \quad F \\ N \quad \text{-}CO_2 \\ F \quad F \end{array}\right]_2 Hg \xrightarrow{205-210°C} \begin{array}{c} F \quad F \\ N \quad \text{-}HgO_2C\text{-} \quad N \\ F \quad F \qquad F \quad F \end{array} + CO_2 \quad (79)$$

$$\left[\begin{array}{c} F \quad F \\ N \quad \text{-}CO_2 \\ F \quad F \end{array}\right]_2 M \xrightarrow[\substack{260°C\,(Cd) \\ 350°C\,(Hg)}]{300°C\,(Zn)} \left[\begin{array}{c} F \quad F \\ N \\ F \quad F \end{array}\right]_2 M + 2CO_2 \qquad (80)$$

$$\left(\begin{array}{c} N \\ \text{-}CO_2 \\ N \\ Me \end{array}\right)_2 Hg \xrightarrow{240-260°C} \left(\begin{array}{c} N \\ N \\ Me \end{array}\right)_2 Hg + 2CO_2 \qquad (81)$$

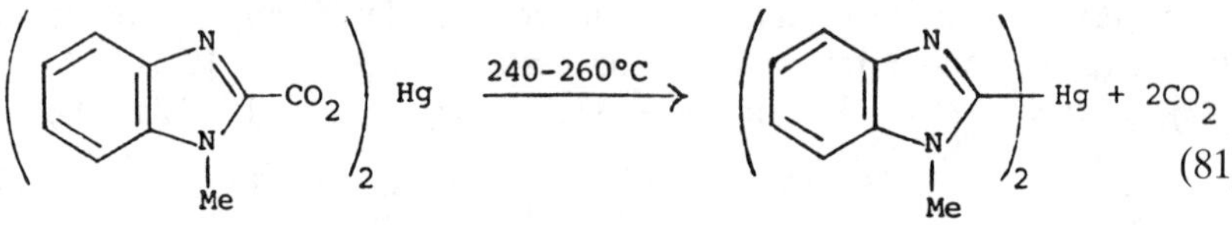

M = Zn, Cd, or Hg (*88*)]. Bis(3-methylbenzimidazol-2-yl) mercury has also been prepared by this method [Eq. (81)] (*89*).

G. Decarboxylation of Metal Arene- and Heteroarenedicarboxylates

The hemidecarboxylation of sodium phthalate on reaction with mercuric acetate in boiling water [Eq. (82), X = H] (*90*) was the first reported thermal decarboxylation. The reaction has been observed for a number of arenes with two adjacent carboxylate groups (*1–4,91*) and has been named the Pesci reaction (*91*). Studies of 3-substituted sodium phthalates or of preformed mercuric 3-substituted phthalates have shown that the sterically hindered carboxyl group (the 2-carboxyl) is preferentially eliminated whether X is electron-donating or electron-withdrawing [Eq. (82), X = Me (*91*), Cl, NO$_2$ (*91,93*), Br (*93*), or CO$_2$H (*94*)]. A similar conclusion was drawn from the decomposition of mercuric 1,2-naphthalenedicarboxylate and 3,4-phenanthrenedicarboxylate (*91*).

The decarboxylation of mercuric 3- and 4-nitro-1,8-naphthalenedicarboxylates has been studied to determine substituent electronic effects [Eq. (83)] (*95,96*), since significant steric effects are unlikely. There is disagreement over the products. In an early study, 3- and 4-nitro-1,8-naphthalenedicarboxylates were found to undergo decarboxylation predominantly in the ring without the nitro group (*95*), consistent

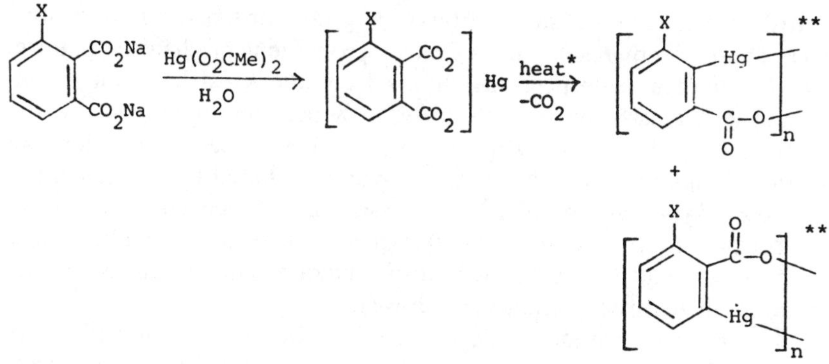

(82)

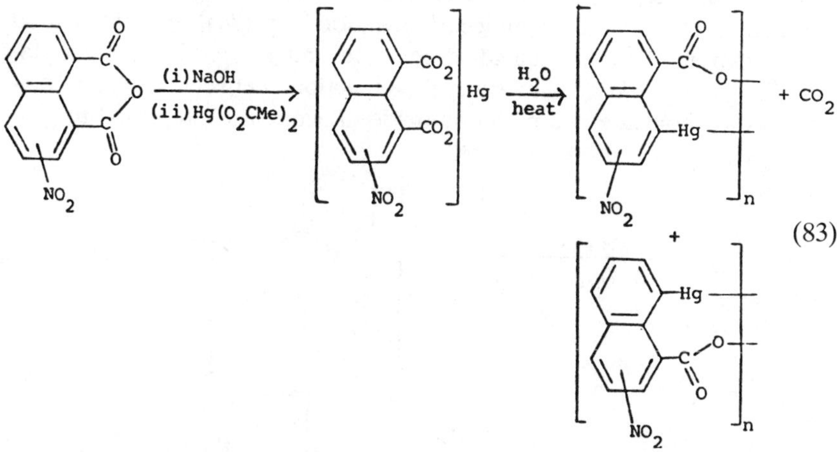

(83)

with either an S_Ei (carbonium character) mechanism or classic electrophilic aromatic ipso substitution. However, Takahashi has claimed that decarboxylation of the 3-, and 4-nitro-1,8-naphthalenedicarboxylates proceeded regiospecifically with replacement of the 8-carboxylate group for the

---

* Where the mercuric salt was formed *in situ*, the decomposition was effected in boiling water (e.g., Ref. *90*). Where the preformed mercuric salt was used, the solvent was hexamethylphosphoramide (*91*).

** These and related compounds are represented as having associated structures since the more familiar (e.g., Refs. *2,3,91*) internal salt or anhydrohydroxomercuriocarboxylate structure, e.g.,

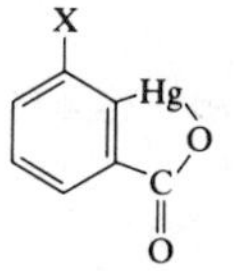

is inconsistent both with the insolubility in common solvents and the strongly preferred linear stereochemistry of mercury in organomercurials (*80,92*).

3-nitro derivative and of the 1-carboxylate group for the 4-nitro derivative (96). These recent results do not have a consistent mechanistic pattern, and the claimed regiospecificity differs from the usual behavior of unsymmetrical dicarboxylates (91,93–95), hence reexamination of the reactions is desirable. An analogous Pesci decarboxylation of · 2-nitro-1,8-naphthalenedicarboxylic anhydride resulted in replacement of the 1-carboxylate group (96). In this case, steric effects could well favor loss of this group, but the claimed regiospecificity seems unlikely since steric strain would be less than for substituted phthalates which give isomer mixtures on decarboxylation (see above).

Other decarboxylations reported by Takahashi have proved irreproducible. The reaction between 2,2'-biphenyldicarboxylic acid anhydride in alkali and mercuric oxide in acetic acid was claimed to give 2'-mercurio-2-biphenylcarboxylate by hemidecarboxylation [Eq. (84)] (97) but yielded instead mercuric 2,2'-biphenyldicarboxylate and negligible decarboxylation [Eq. (85)] (98). Similarly, the reaction between sodium 2,3-pyridinedicarboxylate and mercuric oxide in acetic acid, reported to give

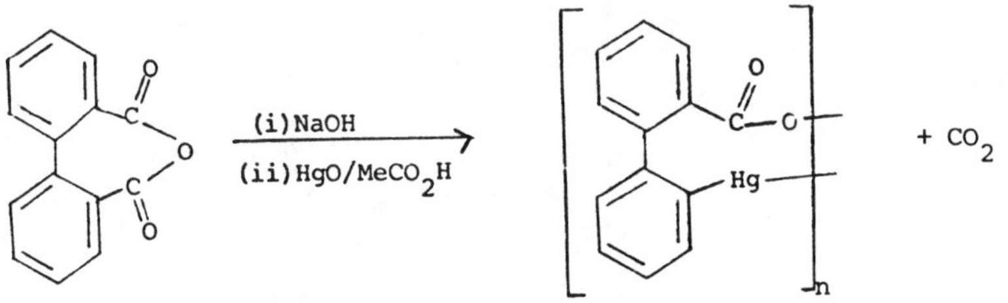

$$+ \; CO_2 \quad (84)$$

$$Hg \quad (85)$$

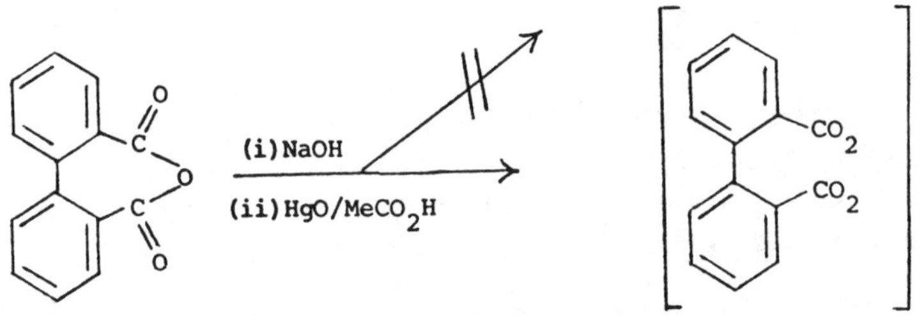

3-carboxylatopyridin-2-ylmercury(II) by a Pesci decarboxylation [Eq. (86)] (99), gave little decarboxylation, and mercurated acetic acid was the sole organometallic product [Eq. (87) (100). However, authentic mercuric 2,3-

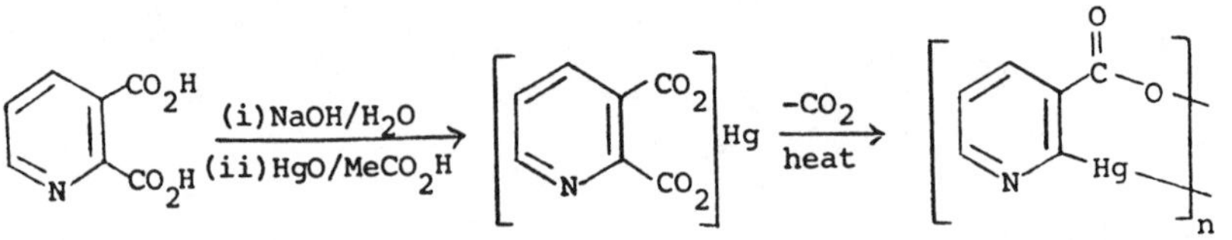

$$(86)$$

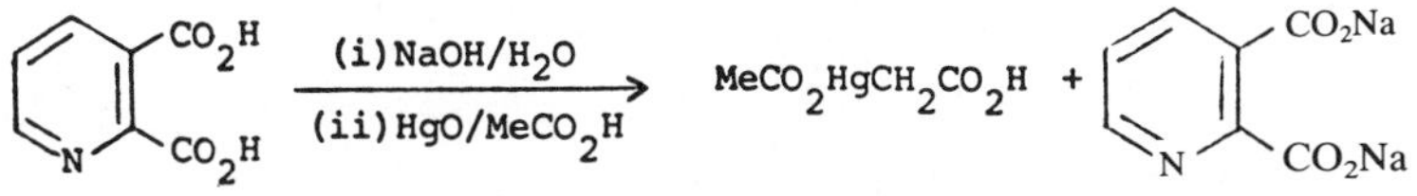

$$\underset{\text{(ii)}\,HgO/MeCO_2H}{\overset{\text{(i)}\,NaOH/H_2O}{\longrightarrow}} \quad MeCO_2HgCH_2CO_2H \;+\; \tag{87}$$

pyridinedicarboxylate underwent decarboxylation in dimethyl sulfoxide to give a mixture of 2-carboxylatopyridin-3-ylmercury(II) (major isomer) [cf. Eq. (86)] and 3-carboxylatopyridin-2-ylmercury(II) (minor isomer) [Eq. (88)] (*100*). Increased regioselectivity in favor of the major isomer

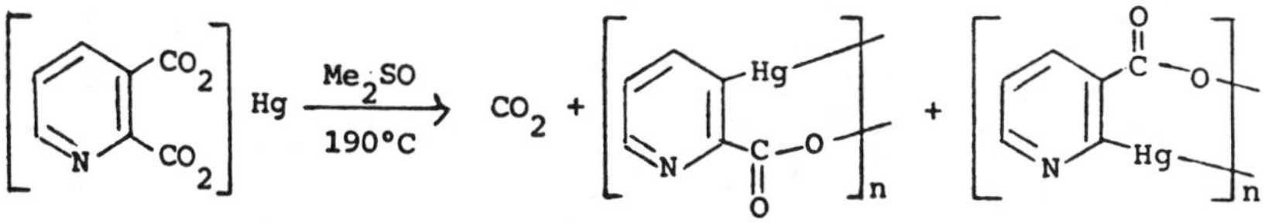

$$\xrightarrow[190^{\circ}C]{Me_2SO} \quad CO_2 \;+ \tag{88}$$

was observed when the solvent purity was increased. The preferred substitution position suggests classic electrophilic ipso attack.

The irreproducibility (*98,100*) of the Pesci decarboxylations [Eqs. (84) and (86)] (*97,99*) has further implications since the decarboxylation products were claimed to undergo novel nucleophilic displacements of mercury (*96,97,99*). Some reported nucleophilic demercurations of an authentic mercurial (*96*) could not be repeated (*98*). A Pesci type hemidecarboxylation of 5-norbornene-2,3-dicarboxylic acid has been reported by Takahashi (*101*), but this has also been found to be irreproducible (*101a*).

Prior to 1985, decarboxylation of metal (hydrocarbon) arenedicarboxylates was restricted to mercury and to acids with two adjacent carboxylate groups.

However, a recent study has shown that

$$(Ph_3P)_2\overline{PtO(O)CC_6H_4\text{-}o\text{-}C(O)O}$$

and *cis*-$(Ph_3P)_2Pt(O(O)CCH{=}CHCO_2H)_2$ at 210°C and 130°C give $(Ph_3P)_2\overline{PtC_6H_4\text{-}o\text{-}C(O)O}$ and $(Ph_3P)_2\overline{PtCH{=}CHC(O)O}$, respectively (*101b*). Much wider scope has been observed in decarboxylation of perfluoroarenedicarboxylates. Decarboxylation of mercuric tetrafluorophthalate is a two-step process [Eq. (89)], giving, after loss of both molecules of carbon dioxide, perfluorotribenzo[*b*,*e*,*h*][1,4,7]trimercuronin (*102*). The intermediate ortho-mercurated carboxylate was also obtained (but with unknown degree of association) by ortho mercuration of mercuric 2,3,4,5-tetrafluorobenzoate [Eq. (70)] and decomposed at 300°C into $(C_6F_4Hg)_3$ (*83*). Bis(phenylmercuric) tetrafluorophthalate when refluxed in pyridine lost one carbon dioxide per molecule [Eq. (90), M = PhHg] (*79*). The analogous triphenyltin compound (M = Ph$_3$Sn)

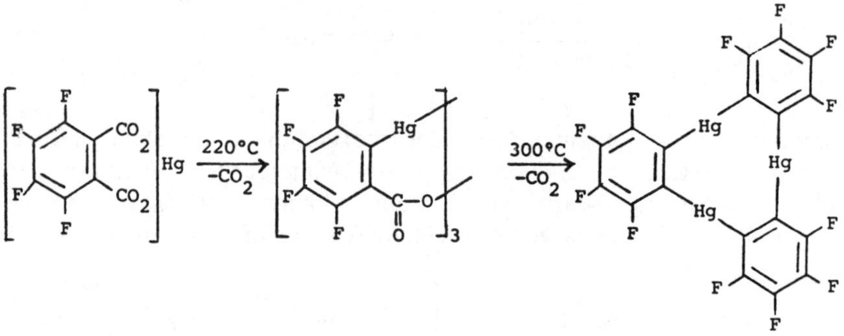

(89)

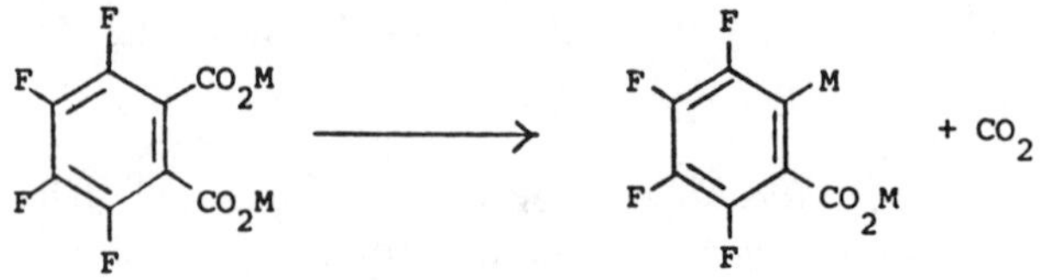

$+ CO_2$ (90)

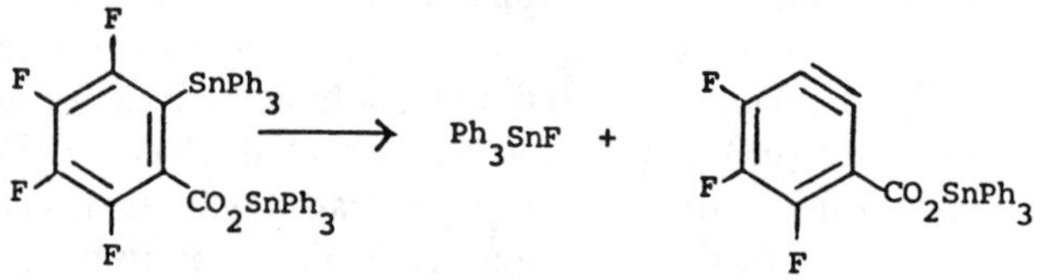

Ph_3SnF +   (91)

gave Ph_3SnF which was probably formed by the thermal decomposition given in Eq. (91).

Main group tetrafluoroterephthalates lose both carboxyl groups. Mercury and cadmium tetrafluoroterephthalates when heated *in vacuo* gave polymeric tetrafluorophenylenemercury and -cadmium compounds [Eq. (92), M = Hg or Cd] (*103*). The same organomercurial and mercury

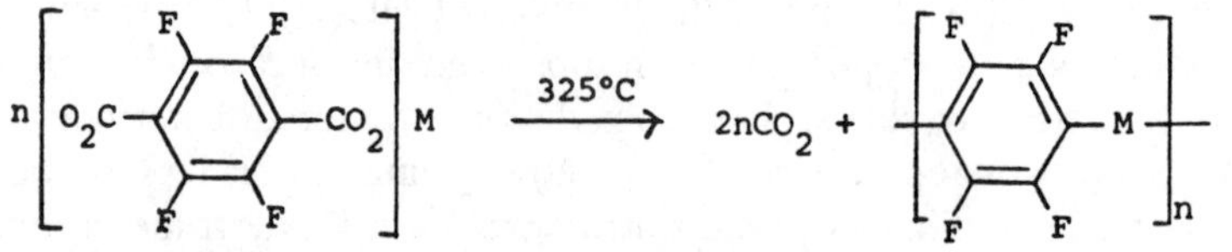

$2nCO_2$ +   (92)

metal were obtained from pyrolysis of $Hg_2[p\text{-}(O_2C)_2C_6F_4]$. Decomposition of zinc tetrafluoroterephthalate did not give a recognizable zinc polymer (*103*). Bis(organomercuric) and bis(chloromercuric) tetrafluoroterephthalates also lost two carbon dioxide molecules on heating in pyridine and formed μ-2,3,5,6-tetrafluoro-1,4-phenylenedimercury(II)   compounds [Eq. (93), R = Ph (79), Me (104), or Cl (105)].

$$p\text{-}(RHgO_2C)_2C_6F_4 \rightarrow p\text{-}(RHg)_2C_6F_4 + 2\ CO_2 \tag{93}$$

A similar reaction was observed for Group 4B triorganometal tetrafluoroterephthalates in refluxing pyridine [Eq. (94), M = Ge, Sn, or Pb] (79).

$$p\text{-}(Ph_3MO_2C)_2C_6F_4 \rightarrow p\text{-}(Ph_3M)_2C_6F_4 + 2\ CO_2 \tag{94}$$

Dimethyl   (μ-2,4,5,6-tetrafluoro-1,3-phenylene)dimercury(II)   was formed when bis(methylmercuric) tetrafluoroisophthalate was heated to 100°C in pyridine [Eq. (95)] (104).

$$m\text{-}(MeHgO_2C)_2C_6F_4 \rightarrow m\text{-}(MeHg)_2C_6F_4 + 2\ CO_2 \tag{95}$$

Mercuric octafluoro-2,2'-biphenyldicarboxylate at 300°C gave a tetrameric mercurial [Eq. (96)] (106).

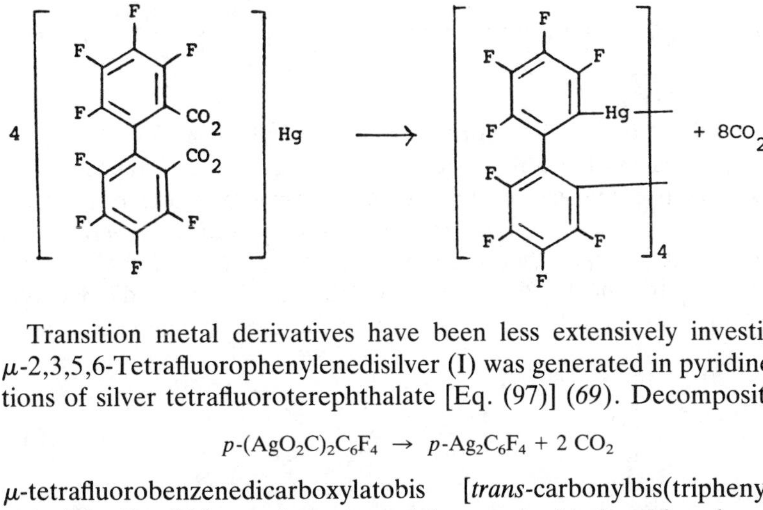

$$\tag{96}$$

Transition metal derivatives have been less extensively investigated. μ-2,3,5,6-Tetrafluorophenylenedisilver (I) was generated in pyridine solutions of silver tetrafluoroterephthalate [Eq. (97)] (69). Decomposition of

$$p\text{-}(AgO_2C)_2C_6F_4 \rightarrow p\text{-}Ag_2C_6F_4 + 2\ CO_2 \tag{97}$$

μ-tetrafluorobenzenedicarboxylatobis   [trans-carbonylbis(triphenylphosphine)rhodium(I)] complexes in pyridine resulted in hemidecarboxylation in each case [Eq. (98)] (19). Prolonged heating in an attempt to

$$o\text{-}m\text{-},\ \text{or}\ p[trans\text{-}(Ph_3P)_2(OC)RhO_2C]_2C_6F_4 \rightarrow$$
$$CO_2 + o\text{-},\ m\text{-},\ \text{or}\ p\text{-}[trans\text{-}(Ph_3P)_2(OC)Rh][trans\text{-}(Ph_3P)_2(OC)RhO_2C]C_6F_4 \tag{98}$$

remove the second carboxyl group resulted in decomposition, giving trans-Rh(O$_2$CR)CO(PPh$_3$)$_2$ (R = o-HC$_6$F$_4$) or trans-RhR(CO)(PPh$_3$)$_2$ (R = m-HC$_6$F$_4$ or p-HC$_6$F$_4$). The occurrence of Eq. (98) for the tetrafluorophthalate parallels the historic Pesci reaction [e.g., Eqs. (82) and (83)].

H. *Decarboxylation of Metal Carborane Carboxylates*

2,2'-Bipyridylnickel(II) *o*-carborane-1,2-dicarboxylate eliminated two carbon dioxide molecules in refluxing decane and formed a complex with a C_2Ni three-membered ring [Eq. (99), M = Ni] (*107*). Both carbon dioxide

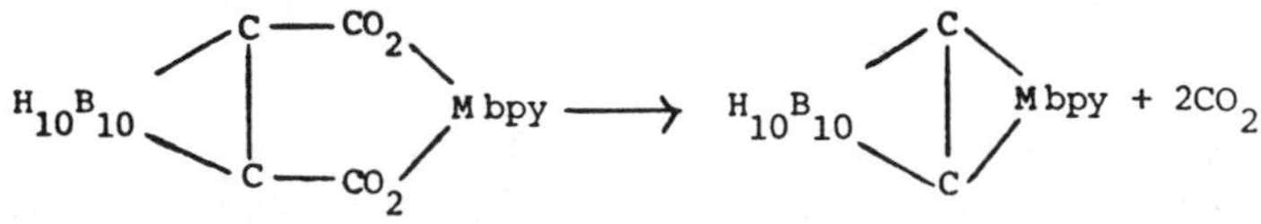

$$H_{10}B_{10} \underset{C-CO_2}{\overset{C-CO_2}{\big<}} \Big| M\,bpy \longrightarrow H_{10}B_{10} \underset{C}{\overset{C}{\big<}} \Big| M\,bpy + 2CO_2 \quad (99)$$

molecules were lost simultaneously as the intermediate for consecutive loss, 1-nickel(II) *o*-carborane-2-carboxylate, when synthesized independently, did not decarboxylate significantly in refluxing decane but required heating in an argon atmosphere [Eq. (100)] (*107*). In the decomposition of

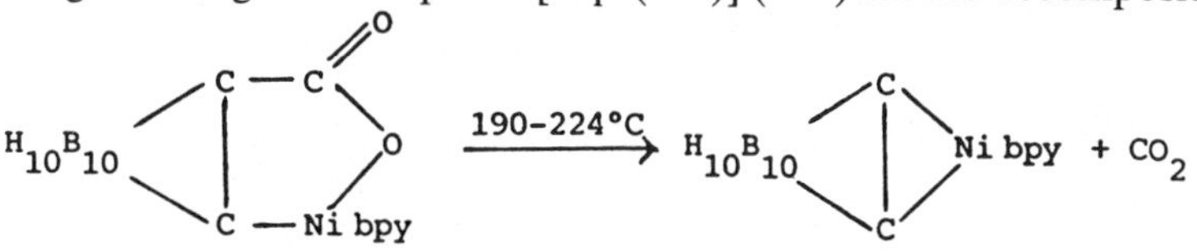

$$H_{10}B_{10} \underset{C-Ni\,bpy}{\overset{C-C\overset{O}{\underset{O}{\big<}}}{\big<}} \xrightarrow{190-224\,^\circ C} H_{10}B_{10} \underset{C}{\overset{C}{\big<}} \Big| Ni\,bpy + CO_2 \quad (100)$$

2,2'-bipyridylcopper(II) *o*-carborane-1,2-dicarboxylate under argon at 115–125° C, the loss of two molecules of carbon dioxide was established by thermogravimetric analysis. Formation of the carboranecopper(II) complex [Eq. (99), M = Cu], was inferred by comparison of the reactivity of the product with that of the authentic compound (*108*). Complexes of copper(I) *o*-carborane-1-carboxylate also eliminated carbon dioxide on heating [Eq. (101)] (*108*). For L = bpy, an unstable copper carborane

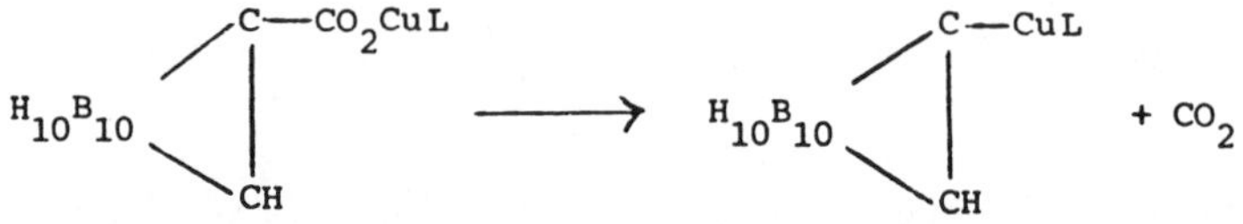

$$H_{10}B_{10} \underset{CH}{\overset{C-CO_2CuL}{\big<}} \Big| \longrightarrow H_{10}B_{10} \underset{CH}{\overset{C-CuL}{\big<}} \Big| + CO_2 \quad (101)$$

complex was isolated when the carboxylate was heated at 100° C under argon. The carboxylate (L = thf) decarboxylated in quinoline to give a quinolinecarboranecopper(I) complex.

Mercuric *o*-carborane-1-carboxylate complexes decarboxylated when heated above their melting points or when refluxed in decane or benzene solution [Eq. (102), L = phen, bpy, or (py)$_2$] (*109,110*). Asymmetric organomercury carboranes and their 1,10-phenanthroline complexes were also formed by thermal decarboxylation [e.g., Eq. (103), R = Me or Ph] (*111*).

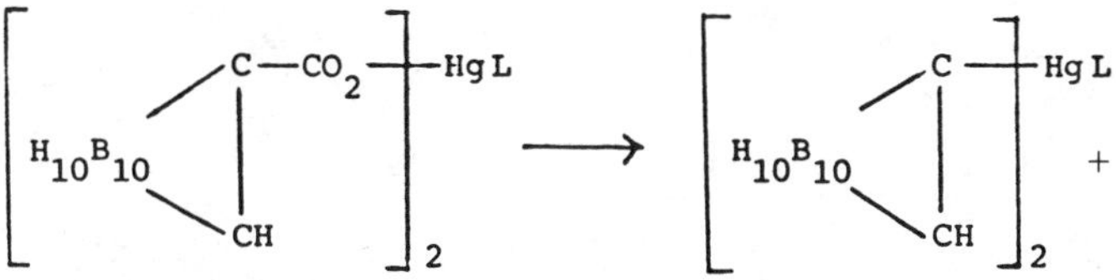

$$\left[H_{10}B_{10} \diagup\diagdown \begin{array}{c} C-CO_2 \\ | \\ CH \end{array} -HgL \right]_2 \longrightarrow \left[H_{10}B_{10} \diagup\diagdown \begin{array}{c} C \\ | \\ CH \end{array} -HgL \right]_2 + 2CO_2 \quad (102)$$

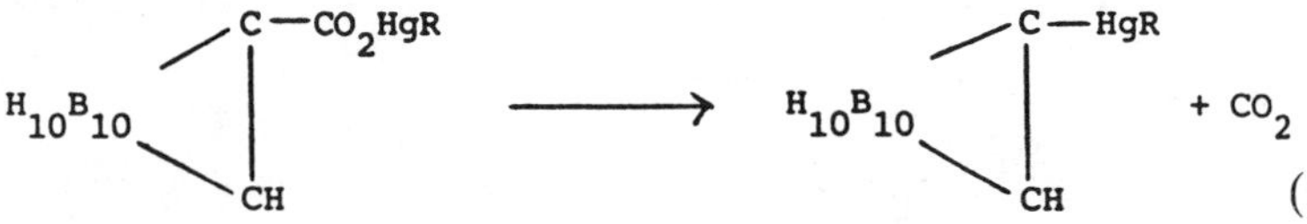

$$H_{10}B_{10} \diagup\diagdown \begin{array}{c} C-CO_2HgR \\ | \\ CH \end{array} \longrightarrow H_{10}B_{10} \diagup\diagdown \begin{array}{c} C-HgR \\ | \\ CH \end{array} + CO_2 \qquad (103)$$

I. Decarboxylation of Alkoxycarbonylmetal Compounds

In this section, the discussion is extended from metal carboxylates (MO_2CR) to alkoxycarbonylmetallics $[MC(=O)OR]$, in which the ligand is carbon–metal bonded. Decarboxylation has been observed for an alkoxycarbonyliron complex in boiling tetrahydrofuran or benzene [Eq.(104)] (*112*) and for several alkoxycarbonylplatinum(II) complexes in boiling benzene [Eq. (105), L = Ph_3P or Ph_2PMe, R^1, R^2, R^3 = H; R^1, R^3 = H, R^2 = Me; R^1, R^2 = H, R^3 = Me; R^2, R^3 = H, R^1 = Me] (*113*).

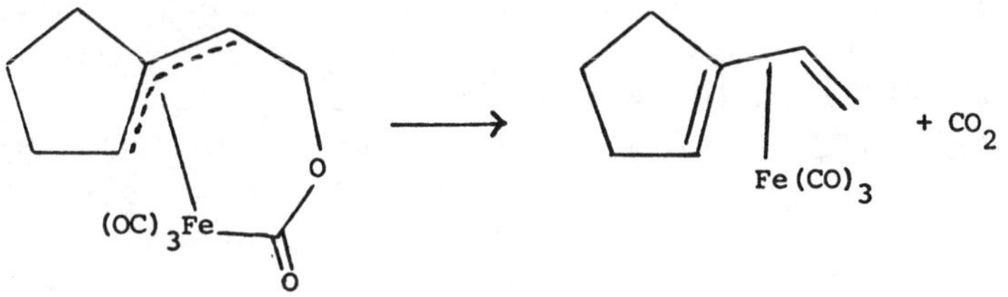

$$\longrightarrow \quad \text{Fe(CO)}_3 \quad + CO_2 \qquad (104)$$

$$Pt(\overset{O}{\overset{\|}{C}}-O-CHR^1CR^2=CHR^3)ClL_2 \longrightarrow \left[\begin{array}{c} L \\ \diagdown \\ Pt \\ \diagup \\ L \end{array} \diagdown \begin{array}{c} R^1 \\ -R^2 \\ R^3 \end{array} \right] Cl + CO_2$$

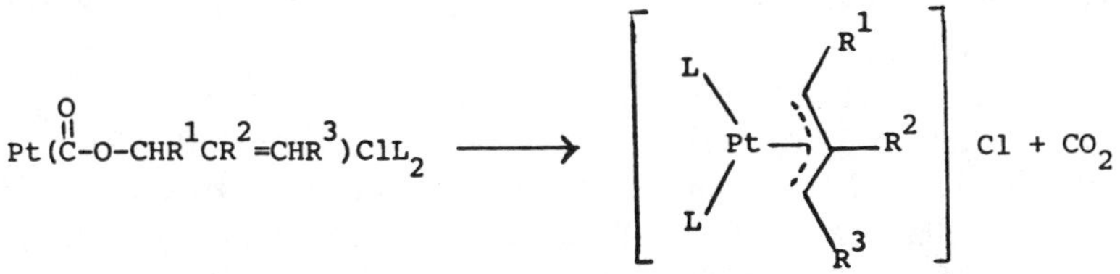

$$(105)$$

Reaction (105) could also be induced in benzene at room temperature by addition of silver perchlorate. Of related interest is the decarboxylation of diketene either by an osmium cluster to give an η^3-allylosmium complex [Eq. (106)] (*114*) or by $(\eta\text{-}C_5H_5)Mn(CO)_2(thf)$ to give $(\eta\text{-}C_5H_5)$-$Mn(CO)_2(allene)$ (*115*).

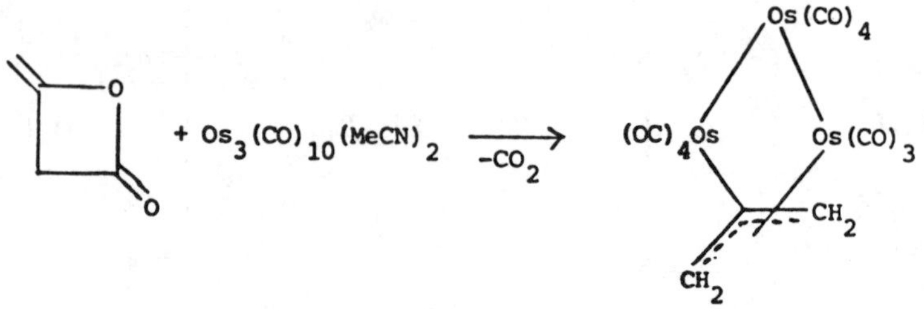

$$+ \ Os_3(CO)_{10}(MeCN)_2 \ \xrightarrow[-CO_2]{} \ (OC)_4Os \cdots Os(CO)_3 \quad (106)$$

J. Summary and Conclusions

Thermal decarboxylation syntheses of organometallics from metal carboxylates have been achieved for Rh(I), Ir(I), Ni(II), Pd(II), Pt(II), Group 1B [M(I) compounds], Group 2B, Tl, Sn, and Pb, and there are isolated examples for Cu(II), Ge, As, and Se (Sections III,A–H). Transition metal examples are currently limited to d^8 and d^{10} configurations, apart from one exception [Eq. (99), M = Cu]. In addition, a few Fe, Os, and Mn organometallics have been prepared from reagents other than carboxylates (Section III,I). Reversible decarboxylations are limited to a few copper(I) carboxylates at this point (Sections III,B and C) (47,52). Organic groups which can be attached by thermal decarboxylation are predominantly those with electron-withdrawing substituents (Sections III,A–D, and F), but there are recent syntheses (18,85) of polymethoxy-phenylmercurials (Section III,E).

The limitations on organometallic synthesis by thermal decarboxylation can arise from (i) a thermodynamic preference for the reverse reaction, (ii) mechanistic limitations, (iii) thermal instability of the target organometallics, and (iv) a preference for alternative decomposition paths.

The reverse reaction, CO_2 insertion, is well known for Group 1A and 2A organometallics (7). There are also examples for reactive organometallics of other elements (116), e.g., vanadium (117) and ytterbium (118).

The predominance of organometallics with electron-withdrawing substituents (Sections III,A–D, and F) can partly be attributed to the promotion of some mechanisms by these substituents [e.g., Eqs. (4), (6), and (15)] (Section II,B). This imposes limitations, as a number of polyhalogenobenzoates, notably 2,3,4,5-tetrafluorobenzoates, have insufficient electron-withdrawing capacity for decarboxylation to occur (Section III,D). Although mechanisms promoted by electron-donating substituents can be formulated (Section II,B), there is little evidence yet for their operation apart from classic electrophilic aromatic decarboxylation [Eq. (12)] (see also Section III,E).

Some effects of thermal instability have been noted [Eqs. (32) and (59)]. Formation of coupling products, e.g., biaryls, has sometimes been attributed to thermally induced self-coupling of organometallics formed by decarboxylation (*1,20*).

Alternative paths for decomposition of the metal carboxylate can lead to ketones, acid anhydrides, esters, acid fluorides (*1,11,22,68,77,78*), and various coupling products (*21,77,78*), and aspects of these reactions have been reviewed (*1,11*). Competition from these routes is often substantial when thermal decomposition is carried out in the absence of a solvent (Section III,D), and their formation is attributable to homolytic pathways (*11,21,77,78*). Other alternative paths are reductive elimination rather than metal–carbon bond formation [Eq. (36)] (Section III,B) and formation of metal–oxygen rather than metal–carbon bonded compounds [e.g., Eqs. (107) (*119*) and (108) (*120*)]. Reactions (36) and (108) are reversible, and CO_2 activation (*116*) is involved in the reverse reactions (*48,120*).

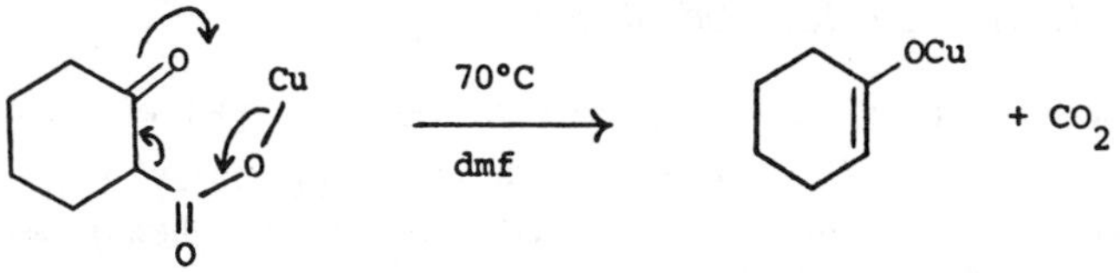

$$+ CO_2 \qquad (107)$$

$$+ CO_2$$

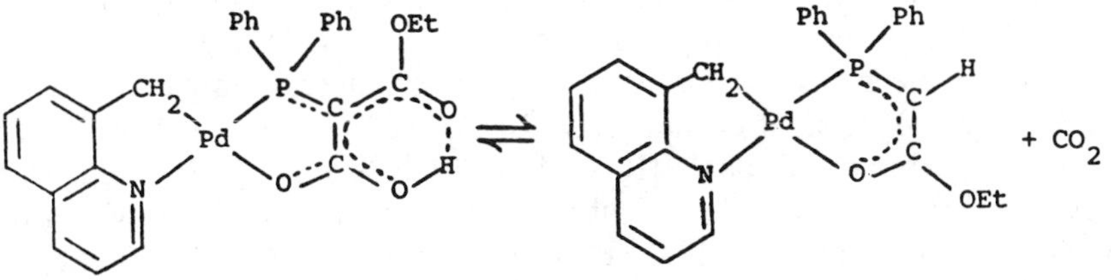

$$(108)$$

Competition between decarboxylation and mercuration on thermal decomposition of mercuric carboxylates has already been considered (Section III,E).

The limits of thermal decarboxylation in terms of elements, oxidation states, and attachment of organic groups have not yet been reached. Promising elements for investigation include Co, Ru, In, Te, and Bi, and promising electronic configurations d^6 and d^7. Greater scope for reversible reactions may be anticipated among d^8–d^{10} metals. Within electron-withdrawing organic groups, thermal decarboxylation can be extended. For example, aryl organometallics with carboxymethyl and trifluoromethyl substituents should be accessible by thermal decarboxylation. For transition metal aryls, use of electron-withdrawing ortho substituents that are also bulky, e.g., CO_2Me or NO_2, should enhance thermal stability (*121*).

Successful syntheses by classical electrophilic aromatic decarboxylation (Section III,E) offer promise that a range of aryl organometallics containing strongly electron-donating groups could be prepared for electrophilic metals (e.g., Tl^{III}, Pb^{IV}, Au^{III}).

IV

RADICAL INITIATED DECARBOXYLATION

Successful syntheses involving radical initiated decarboxylation are mainly limited to monoorganomercurials, where the method has extensive uses. Homolytic decarboxylation mechanisms have been discussed in Section II,C.

A. Peroxide Initiated Decarboxylation

Organometallic formation may result from a chain mechanism [Eqs. (21)–(23) and (18)–(20)] and/or radical displacement [Eqs. (21)–(23), alone]. The reaction of ^{13}C-labeled mercuric cyclohexanoate with cyclohexylcarbonyl peroxide (1:1) gave mainly unlabeled organomercurial, which was derived from radical displacement (122). Decarboxylation by a chain mechanism was reported for the syntheses of organomercuric carboxylates of straight chain alkyls [R = $Me(CH_2)_n$, n = 0–8, 10, or 15 (123–131)], branched alkyls [R = $Me_2CH(CH_2)_n$, n = 0 or 2 (132) or $Me_3C(CH_2)_n$, n = 0–2 (133)], substituted alkyls [R = cyclopentylmethyl (134), cyclohexylmethyl (134), $Ph(CH_2)_n$, n = 2 or 3 (135), $Cl(CH_2)_5$ (135), $MeO_2C(CH_2)_n$, n = 2, 4, or 7 (136), $Me(CH_2)_nO(CH_2)_2$, n = 0–4 (137), $Me_3Si(CH_2)_2$ (138)], and cycloaliphatics [R = cyclo-C_nH_{2n-1}, n = 3–7, 11, or 12 (122,130,139–144), 4-methylcyclohexyl (144) bornyl, 7,7-dimethyl-2-oxobicyclo[2,2,1]heptyl, 1-adamantyl (146,147)].

Syntheses are limited to mercuric salts of weak acids (2,110). Generally, increasing the length of the straight alkyl chain decreases the extent of decarboxylation (e.g., Ref. 133). Electron-withdrawing substituents suppress decarboxylation. For example, mercurials are not formed with MeO_2C, Cl, and $Me(CH_2)_nO$ substituents on the α carbon (137,148,149), but some decarboxylation occurs with these on the β carbon (135–137). Chain decarboxylation predominated in reactions in benzene, butyric acid [R = $Me(CH_2)_2$] (150), or acetic acid (R = Me) (124). The chain reaction was also observed for R = $Me(CH_2)_2$ in the absence of solvent and in ethylacetate or heptane solution, but in these media the radical displacement reaction was dominant (2,150). When benzene was used as solvent some mercuration of the solvent produced phenylmercuric carboxylates [Eq. (26), Ar = Ph].

Although peroxide initiated reactions can produce preparative yields of organomercury salts, a mixture of products was usually obtained due to chain mechanism/radical displacement competition and to the prevalence of side reactions [Eqs. (24), (25), and (109)]. An alternative to the

$$R^{\cdot} + Hg(O_2CR)_2 \rightarrow RCO_2R + {}^{\cdot}HgO_2CR \tag{109}$$

propagation steps (18) and (19) in which the carboxyl radical is complexed by the mercuric carboxylate prior to a concerted decarboxylation has been proposed [Eqs. (110) and (111)] (143). Complex formation [Eq. (110)] is

$$RCO_2^{\cdot} + Hg(O_2CR)_2 \rightarrow [Hg(O_2CR)_3]^{\cdot} \tag{110}$$

$$[Hg(O_2CR)_3]^{\cdot} \xrightarrow{-CO_2} [RHg(O_2CR)_2]^{\cdot} \rightarrow RHgO_2CR + RCO_2^{\cdot} \tag{111}$$

supported by spectroscopic evidence (151) and by inhibition of decarboxylation by alkali metal acetates (152). However, since there is not retention of configuration when the α carbon of the radical is asymmetric [R = 4-methylcyclohexyl (145) or bornyl (146)], a free organic radical, $R^{\cdot}$, is formed [Eq. (18)].

In the one example not involving mercurials, $Ni(O_2CC_6F_5)_2phen \cdot H_2O$, which failed to decarboxylate in boiling toluene [c.f. Eq. (47)] (Section III,D), eliminated carbon dioxide in benzene in the presence of benzoyl peroxide (63).

B. *Photochemically Initiated Decarboxylation*

Mercuric carboxylates, which decarboxylate by a chain mechanism when initiated by peroxides, also decarboxylate under UV irradiation ($123,128,129,131-140,142,144-146,153-155$). In addition, decarboxylation was observed for mercuric benzoate and mercuric α-naphthoate (123). Side reactions [Eqs. (24), (25), (109)] observed in peroxide initiated reactions also occurred on UV irradiation, and mercurous salt formation [Eq.(24)] was more extensive under the latter conditions. Decarboxylation giving methylmercuric acetate occurred on irradiation of mercuric acetate in aqueous solution and is considered to be of environmental significance ($156,157$). Stepwise decarboxylation giving $(CF_3)_2Hg$ occurred on irradiation of solid mercuric trifluoroacetate at $-196°$ C (158), but, at $20°$ C, trifluoromethyl radicals diffused from the solid and dimerized (158). No other diorganomercurial has been formed by radical decarboxylation, and the reaction is not preparatively competitive with the thermal decarboxylation synthesis of $(CF_3)_2Hg$ ($26,27$) (Section III,A).

Ethylenediamine-, 2,2'-bipyridyl-, and 1,10-phenanthrolinecobalt(III) complexes which contain glycinate rings decarboxylated on photolysis to form three-membered $Co{-}N{<}$ rings ($159{-}163$). When the glycinate ring was part of an aminopolycarboxylate such as ethylenediamine-N,N'-diacetate, the organocobalt(III) compound was unstable and was characterized spectroscopically ($159{-}161$). However the aminomethylcobalt(III) complexes formed in Eq. (112) (L = bpy or phen) were isolated

$$[L_2CoO_2CCH_2NH_2](ClO_4)_2 \xrightarrow{h\nu} [L_2CoCH_2NH_2](ClO_4)_2 + CO_2 \qquad (112)$$

and analyzed ($162,163$), and a crystal structure for L = bpy confirmed the presence of the $Co{-}C{-}N$ ring (164). The complex $[(bpy)_2CoCH_2NH_2]^{2+}$ has also been obtained from other glycinato-cobalt(III) complexes in reactions which combine decarboxylation and ligand exchange ($164a$). Similar aminoalkylcopper compounds have been proposed as intermediates in the photochemical decomposition of copper–amino acid complexes (165). The platinum complexes

$$(Ph_3P)_2PtC_6H_4\text{-}o\text{-}C(O)O$$

and $(Ph_3P)_2PtCH = CHC(O)O$ prepared recently by thermal decarboxylation (Section III,G) have also been obtained by photo-induced decarboxylation ($101b$).

C. Electrochemically Initiated Decarboxylation

In a recent innovation, the decarboxylation of the mercuric carboxylates, $(RCO_2)_2Hg$ [R = $Me(CH_2)_n$, $n = 0{-}4$, CF_3, or Me_3CCH_2], has been initiated electrochemically ($166{-}168$). Mercurous carboxylates, $(RCO_2)_2Hg_2$ [R = $Me(CH_2)_n$, $n = 0{-}4$, 8, Me_2CH, CF_3, C_6H_5, $PhCH_2$, cyclohexyl, 1-adamantyl], were generated by electrochemical oxidation of the carboxylic acid at a mercury anode ($166,169$). Attempted electrochemical initiation of decarboxylation of these salts resulted in mercury metal and $RHgO_2CR$ (168). The mercurial was obtained by decarboxylation of $(RCO_2)_2Hg$, which was formed by rearrangement [Eq. (25)].

D. Conclusion

Radical initiated decarboxylation appears to have considerable scope in organometallic synthesis, especially where the incoming radical is part of a coordinated ligand.

ACKNOWLEDGMENT

We are grateful to the Australian Research Grants Scheme for support, and to Monash University for a Post Doctoral Fellowship to G.N.P.

REFERENCES

1. G. B. Deacon, *Organometal. Chem. Rev. A* **5**, 355 (1970).
2. Yu. A. Ol'dekop and N. A. Maier, "Synthesis of Organometallic Compounds by Decarboxylation of Metal Acylates." Science and Technology, Minsk, 1976.
3. L. G. Makarova and A. N. Nesmeyanov, "The Organic Compounds of Mercury," *in* "Methods of Elemento-Organic Chemistry" (A. N. Nesmeyanov and K. A. Kocheshkov, eds.), Vol. 4, pp. 259–264, 276–284. North-Holland Publ., Amsterdam, 1967.
4. H. Straub, K. P. Zeller, and H. Leditschke, "Organoquecksilber Verbindungen" *in* Houben-Weyl, "Methoden der Organischen Chemie" (E. Muller, ed.), Vol. 13, Part 2b, pp. 114–128. Thieme, Stuttgart, 1974.
5. R. C. Mehrotra and R. Bohra, "Metal Carboxylates." Academic Press, London, 1983.
5a. R. C. Larock, "Organomercury Compounds in Organic Synthesis," p. 101. Springer-Verlag, Berlin and New York, 1985.
6. M. S. Kharasch and O. Reinmuth, "Grignard Reactions of Non-metallic Substances." Constable, London, 1954.
7. B. J. Wakefield, "Compounds of the Alkali and Alkaline Earth Metals in Organic Synthesis" *in* "Comprehensive Organometallic Chemistry" (G. Wilkinson, F. G. A. Stone, and E. W. Abel, eds.), Vol. 7, Ch. 44, pp. 38, 39, and references therein. Pergamon, Oxford, 1982.
8. Yu. A. Serguchev and I. P. Beletskaya, *Russ. Chem. Rev.* **49**, 1119 (1980).
9. G. B. Deacon and G. N. Stretton, *J. Organomet. Chem.* **218**, 123 (1981).
10. J. E. Connett, A. G. Davies, G. B. Deacon, and J. H. S. Green, *J. Chem. Soc. C* 106 (1966).
11. B. R. Brown, *Q. Rev. Chem. Soc.* **5**, 131 (1951).
11a. P. Segura, J. F. Bunnett, and L. Villanova, *J. Org. Chem.* **50**, 1041 (1985).
12. S. C. Cohen and A. G. Massey, *Adv. Fluorine Chem.* **6**, 83 (1970).
13. T. Cohen and R. A. Schambach, *J. Am. Chem. Soc.* **92**, 3189 (1970).
14. T. Cohen, R. W. Berninger, and J. T. Wood, *J. Org. Chem.* **43**, 837 (1978).
15. T. Tsuda, T. Nakatsuka, T. Hirayama, and T. Saegusa, *Chem. Commun.* 557 (1974).
16. J. March, "Advanced Organic Chemistry. Reactions Mechanisms and Structure," 2nd Ed., p. 571. McGraw-Hill, New York, 1977.
17. G. B. Deacon and G. J. Farquharson, *Aust. J. Chem.* **30**, 293 (1977).
18. G. B. Deacon, M. F. O'Donoghue, G. N. Stretton, and J. M. Miller, *J. Organomet. Chem.* **233**, Cl (1982).
19. S. J. Faulks, Ph. D. thesis, Monash University (1983).
20. J. Chodowska-Palicka and M. Nilsson, *Acta Chem. Scand.* **25**, 3451 (1971).
21. E. K. Fields and S. Meyerson, *J. Org. Chem.* **41**, 916 (1976).
22. P. Sartori and M. Weidenbruch, *Chem. Ber.* **100**, 2049 (1967).
23. V. R. Polishchuk, L. A. Fedorov, P. O. Okulevich, L. S. German, and I. L. Knunyants, *Tetrahedron Lett.* 3933 (1970).
24. P. E. Aldrich, *U. S. Patent* 30043859 (1962), *Chem. Abstr.* **57**, 149441 (1962).
25. J. E. Connett and G. B. Deacon, *J. Chem. Soc. C* 1058 (1966).
26. I. L. Knunyants, Ya. F. Komissarov, B. L. Dyatkin, and L. T. Lantseva, *Bull. Acad. Sci. USSR Div. Chem. Sci.* **22**, 912 (1973).
27. R. J. Lagow, R. Eujen, L. L. Gerchman, and J. A. Morrison, *J. Amer. Chem. Soc.* **100**, 1722 (1978).
28. D. N. Kravtsov, B. A. Kvasov, L. S. Golovchenko, and E. I. Fedin, *J. Organomet. Chem.* **36**, 227 (1972).
29. D. Seyferth, S. P. Hopper, and G. J. Murphy, *J. Organomet. Chem.* **46**, 201 (1972).
30. B. L. Dyatkin, E. P. Mochalina, L. T. Lantseva, and I. L. Knunyants, *Zh. Vses. Khim. Obshch.* **10**, 469 (1965); *Chem. Abstr.* **63**, 14691b (1965).

31. V. R. Polishchuk, L. S. German, and I. L. Knunyants, *Bull. Acad. Sci. USSR, Div. Chem. Sci.* **20,** 547 (1971).
32. T. J. Logan, *J. Org. Chem.* **28,** 1129 (1963).
33. R. Robson and I. E. Dickson, *J. Organomet. Chem.* **15,** 7 (1968).
34. Yu. A. Ol'dekop, N. A. Maier, A. A. Erdman, and Z. P. Zubreichuk, *J. Gen. Chem. USSR* **49,** 359 (1979).
35. D. Seyferth, F. M. Armbrecht, Jr., B. Prokai, and R. J. Cross, *J. Organomet. Chem.* **6,** 573 (1966).
36. F. M. Armbrecht, Jr., W. Tronich, and D. Seyferth, *J. Am. Chem. Soc.* **91,** 3218 (1969).
37. I. Balǎzs, V. Fǎrçǎsan, and I. Haiduc, *Rev. Roum. Chim.* **22,** 379 (1977).
38. D. Seyferth, G. J. Murphy, R. L. Lambert, Jr., and R. E. Mammarella, *J. Organomet. Chem.* **90,** 173 (1975).
39. N. K. Hota and C. J. Willis, *Can. J. Chem.* **46,** 3921–24 (1968).
40. R. S. Dickson, Ph. D. thesis, University of Adelaide (1962).
41. W. R. Cullen and L. G. Walker, *Can. J. Chem.* **38,** 472 (1960).
42. P. Sartori and M. Thomzik, *Z. Anorg. Allg. Chem.* **394,** 157 (1972); P. Sartori and R. Hochleitner, *Z. Anorg. Allg. Chem.* **404,** 164 (1974); P. Sartori, R. H. Hochleitner, and G. Hägele, *Z. Naturforsch. Teil. B* **31,** 76 (1976).
43. N. N. Yarovenko, V. N. Shemanina, and G. B. Gazieva, *J. Gen. Chem. USSR* **29,** 924 (1959).
44. H. J. Emeléus and M. J. Dunn, *J. Inorg. Nucl. Chem.* **27,** 752 (1965).
45. R. E. Banks, "Fluorocarbons and their Derivatives," 2nd Edn. Macdonald, London, 1970.
46. E. B. Boyar, W. G. Higgins, and S. D. Robinson, *Inorg. Chim. Acta* **76,** L293 (1983).
47. T. Tsuda, Y. Chujo, and T. Saegusa, *J. Am. Chem. Soc.* **100,** 630 (1978).
48. A. D. English and T. Herskovitz, *J. Am. Chem. Soc.* **99,** 1648 (1977).
49. J. G. A. Luijten and G. J. M. van der Kerk, "Investigations in the Field of Organotin Chemistry." Tin Research Institute, Middlesex, 1955.
50. J. G. A. Luijten and G. J.M. van der Kerk, *J. Appl. Chem. London* **6,** 93 (1956).
51. L. C. Willemsens and G. J.M. van der Kerk, "Investigations in the Field of Organolead Chemistry." Int. Lead Zinc Research Org., New York, 1965.
52. T. Tsuda, Y. Chujo, and T. Saegusa, *Chem. Commun.* 963 (1975).
53. J. G. A. Luijten and G. J. M. van der Kerk, *Recl. Trav. Chim. Pays-Bas. Belg.* **83,** 295 (1964).
54. A. G. Davies and R. J. Puddephatt, *J. Chem. Soc. C* 317 (1968).
55. C. Beermann and H. Hartmann, *Z. Anorg. Allg. Chem.* **276,** 20 (1954).
56. M. G. Voronkov, R. G. Mirskov, V. G. Chernova, and L. V. Skochilova, *Dokl. Vses. Konf. Khim. Atsetilena, 4th* **2,** 193 (1972); *Chem. Abstr.* **79,** 78903x (1973).
57. N. V. Komarov and A. A. Andreev, *Zh. Obshch. Khim.* **50,** 693 (1980); *Chem. Abstr.* **94,** 103490e (1981).
58. R. A. Cummins, P. Dunn, and D. Oldfield, *Aust. J. Chem.* **24,** 2257 (1971).
59. R. Hüttel and H. Schmid, *Chem. Ber.* **101,** 252 (1968).
60. T. Chivers, *Organometal. Chem. Rev. A* **6,** 1 (1970).
61. G. B. Deacon, S. J. Faulks, and I. L. Grayson, *Transition Met. Chem.* **3,** 317 (1978).
62. G. B. Deacon, S. J. Faulks, and J. M. Miller, *Transition Met. Chem.* **5,** 305 (1980).
63. P. G. Cookson and G. B. Deacon, *J. Organomet. Chem.* **33,** C38 (1971); *Aust. J. Chem.* **25,** 2095 (1972).
64. G. B. Deacon and I. L. Grayson, *Transition Met. Chem.* **7,** 97 (1982).
65. G. B. Deacon and I. L. Grayson, *Transition Met. Chem.* **8,** 131 (1983).
66. G. B. Deacon, B. M. Gatehouse, I. L. Grayson, and M. C. Nesbit, *Polyhedron* **3,** 753 (1984).

67. G. B. Deacon and P. W. Elliot, unpublished results (1983); P. W. Elliot, B. Sc. (Hons) Report, Monash University (1983).
68. P. Sartori and M. Weidenbruch, Chem. Ber. **100**, 3016 (1967).
69. P. Sartori and H. Kuhn, J. Fluorine Chem. **16**, 528 (1980).
70. C. M. Mitchell and F. G. A. Stone, J. Chem. Soc. Dalton Trans. 102, (1972).
71. A. Cairncross, J. R. Roland, R. M. Henderson, and W. A. Sheppard, J. Am. Chem. Soc. **92**, 3187 (1970).
72. M. Schmeisser and M. Weidenbruch, Chem. Ber. **100**, 2306 (1967).
73. M. Weidenbruch and S. Böke, Chem. Ber. **103**, 510 (1970).
74. G. B. Deacon and P. W. Felder, J. Chem. Soc. C 2313 (1967).
75. G. B. Deacon, G. J. Farquharson, and J. M. Miller, Aust. J. Chem. **30**, 1013 (1977).
76. G. B. Deacon and P. W. Felder, Aust. J. Chem. **23**, 1359 (1970).
77. G. B. Deacon and R. J. Phillips, Aust. J. Chem. **31**, 1709 (1978).
78. P. G. Cookson, G. B. Deacon, P. W. Felder, and G. J. Farquharson, Aust. J. Chem. **27**, 1895 (1974).
79. G. B. Deacon and G. J. Farquharson, J. Organomet. Chem. **135**, 73 (1977).
80. J. L. Wardell, "Mercury" in "Comprehensive Organometallic Chemistry" (G. Wilkinson, F. G. A. Stone, and E. W. Abel, eds.), Ch. 17. Pergamon, Oxford, 1982.
81. O. Dimroth, Ber. Dtsch. Chem. Ges. **35**, 2853 (1902).
82. D. Tunaley, Ph. D. thesis, Monash University (1979).
83. G. B. Deacon and L. C. Turner, unpublished results; L. C. Turner, B. Sc. (Hons) Report, Monash University (1974).
84. I. G. Kerkhof, Rec. Trav. Chim. **51**, 755 (1932).
85. G. B. Deacon and M. F. O'Donoghue, unpublished results; M. F. O'Donoghue, Ph. D. Thesis, Monash, 1984.
86. W. T. Reichle, J. Organomet. Chem. **18**, 105 (1969).
87. R. D. Chambers, F. G. Drakesmith, J. Hutchinson, and W. K. R. Musgrave, Tetrahedron Lett. 1705 (1967).
88. P. Sartori and H. Adelt, J. Fluorine Chem. **3**, 275 (1973/1974).
89. B. A. Tertov, A. V. Koblik, and P. P. Onishchenko, J. Gen. Chem. USSR **44**, 599 (1974).
90. L. Pesci, Atti. Real. Accad. Lincei [V], **10**, i, 362 (1901); see also J. Chem. Soc. Abstr. **80**, 576 (1901).
91. M. S. Newman and M. C. Vander Zwan, J. Org. Chem. **38**, 319 (1973).
92. L. G. Kuz'mina and Y. T. Struchkov, Croat. Chem. Acta **57**, 701 (1984).
93. F. C. Whitmore and P. J. Culhane, J. Amer. Chem. Soc. **51**, 602 (1929).
94. F. C. Whitmore and R. P. Perkins, J. Amer. Chem. Soc. **51**, 3352 (1929).
95. G. J. Leuck, R. P. Perkins, and F. C. Whitmore, J. Amer. Chem. Soc. **51**, 1831 (1929).
96. T. Takahashi, Yakugaku Zasshi **98**, 358 (1978).
97. T. Takahashi, S. Togashi, M. Morishita, and S. Takeda, Chem. Pharm. Bull. **30**, 3020 (1982).
98. G. B. Deacon, G. N. Stretton, and M. J. O'Connor, Syn. Comm. **13**, 1041 (1983).
99. T. Takahashi, Chem. Pharm. Bull. **27**, 2473 (1979).
100. G. B. Deacon and G. N. Stretton, Aust. J. Chem. **38**, 419 (1985).
101. T. Takahashi, Chem. Pharm. Bull. **27**, 870 (1979).
101a. G. B. Deacon and G. N. Stretton, Unpublished results (1985).
101b. O. J. Scherer, K. Hussong, and G. Wolmershäuser, J. Organomet. Chem. **289**, 215 (1985).
102. P. Sartori and A. Golloch, Chem. Ber. **101**, 2004 (1968).
103. P. Sartori and H. J. Frohn, Chem. Ber. **107**, 1195 (1974).

104. M. W. Buxton, R. H. Mobbs, and D. E. M. Wotton, *J. Fluorine Chem.* **1,** 179 (1971/1972).

105. H. B. Albrecht and G. B. Deacon, *Aust. J. Chem.* **25,** 57 (1972).

106. S. C. Cohen and A. G. Massey, *J. Organomet. Chem.* **10,** 471 (1967).

107. Yu. A. Ol'dekop, N. A. Maier, A. A. Erdman, and V. P. Prokopovich, *Dokl. Chem.* **257,** 118 (1981).

108. Yu. A. Ol'dekop, N. A. Maier, A. A. Erdman, and V. P. Prokopovich, *Vestsi. Akad. Navuk BSSR, Ser. Khim. Navuk* 71 (1981); *Chem. Abstr.* **96,** 20149r (1982).

109. Yu. A. Ol'dekop, N. A. Maier, A. A. Erdman, and V. P. Prokopovich, *J. Gen. Chem. USSR* **47,** 1533 (1977).

110. Yu. A. Ol'dekop, N. A. Maier, A. A. Erdman, and V. P. Prokopovich, *Dokl. Chem.* **243,** 570 (1978).

111. Yu. A. Ol'dekop, N. A. Maier, A. A. Erdman, and V. P. Prokopovich, *Vestsi Akad. Navuk BSSR, Ser. Khim. Navuk* 85 (1979); *Chem. Abstr.* **92,** 129003g (1980).

112. G. D. Annis, S. V. Ley, C. R. Self, and R. Sivaramakrishnan, *Chem. Commun.* 299 (1980).

113. H. Kurosawa, *Inorg. Chem.* **14,** 2148 (1975).

114. A. J. Arce and A. J. Deeming, *Chem. Commun.* 364 (1982).

115. W. A. Herrmann, J. Weichmann, M. L. Ziegler, and H. Pfisterer, *Angew. Chem. Int. Ed.* **21,** 551 (1982).

116. D. J. Darensbourg and R. A. Kudaroski, *Adv. Organomet. Chem.* **22,** 129 (1983), and references therein.

117. G. A. Razuvaev, L. I. Vyshinskaya, V. V. Drobotenko, G. Ya. Mal'kova, and N. N. Vyshinsky, *J. Organomet. Chem.* **239,** 335 (1982).

118. G. B. Deacon, P. I. Mackinnon, and T. D. Tuong, *Aust. J. Chem.* **36,** 43 (1983).

119. T. Tsuda, Y. Chujo, S. Takahashi, and T. Saegusa, *J. Org. Chem.* **46,** 4980 (1981).

120. P. Braunstein, D. Matt, Y. Dusausoy, J. Fischer, A. Mitschler, and L. Ricard, *J. Amer. Chem. Soc.* **103,** 5115 (1981).

121. J. Chatt, *Adv. Organomet. Chem.* **12,** 1 (1974).

122. T. N. Shatkina, A. N. Lovtsova, T. I. Pekhk, E. T. Lippmaa, and O. A. Reutov, *Dokl. Chem.* **220,** 44 (1975).

123. Yu. A. Ol'dekop and N. A. Maier, *Bull. Acad. Sci. USSR Div. Chem. Sci.* **15,** 1127 (1966).

124. Yu. A. Ol'dekop and N. A. Maier, *J. Gen. Chem. USSR* **30,** 295 (1960).

125. Yu. A. Ol'dekop and N. A. Maier, *J. Gen. Chem. USSR* **30,** 320 (1960).

126. Yu. A. Ol'dekop and N. A. Maier, *J. Gen. Chem. USSR* **30,** 639 (1960).

127. Yu. A. Ol'dekop, N. A. Maier, A. A. Erdman, and Yu. A. Dzhomidava, *J. Gen. Chem. USSR* **40,** 270 (1970).

128. Yu. A. Ol'dekop, N. A. Maier; A. A. Erdman, and Yu. A. Dzhomidava, *J. Gen. Chem. USSR* **40,** 607 (1970).

129. Yu. A. Ol'dekop, N. A. Maier, A. A. Erdman, and S. S. Stanovaya, *J. Gen. Chem. USSR* **40,** 275 (1970).

130. Yu. A. Ol'dekop, N. A. Maier, and Yu. A. Dzhomidava, *Vestsi Akad. Navuk BSSR, Ser. Khim. Navuk* 115 (1971); *Chem. Abstr.* **75,** 36280d (1971).

131. Yu. A. Ol'dekop, N. A. Maier, and A. L. Isakhanyan, *Vestsi Akad. Navuk BSSR, Ser. Khim. Navuk* 63 (1970); *Chem. Abstr.* **74,** 42453h (1971).

132. Yu. A. Ol'dekop, N. A. Maier, A. A. Erdman, and T. I. Pryamushko, *Vestsi Akad. Navuk BSSR, Ser. Khim. Navuk* **65,** (1970); *Chem. Abstr.* **74,** 13242h (1971).

133. Yu. A. Ol'dekop, N. A. Maier, and A. L. Isakhanyan, *Vestsi Akad. Navuk, BSSR Ser. Khim. Navuk* 102 (1971); *Chem. Abstr.* **76,** 45449p (1972).

134. Yu. A. Ol'dekop, N. A. Maier, Yu. D. But'ko, and M. S. Mindel', *J. Gen. Chem. USSR* **41**, 1073 (1971).
135. Yu. A. Ol'dekop, N. A. Maier, and A. L. Isakhanyan, *Vestsi Akad. Navuk BSSR, Ser. Khim. Navuk* 116 (1973); *Chem. Abstr.* **78**, 97776d (1973).
136. Yu. A. Ol'dekop, N. A. Maier, A. A. Erdman, and S. S. Gusev, *J. Gen. Chem. USSR* **39**, 1080 (1969).
137. Yu. A. Ol'dekop, N. A. Maier, A. L. Isakhanyan, and V. N. Pshenichnyi, *Vestsi Akad. Navuk BSSR, Ser. Khim. Navuk* 101 (1971); *Chem. Abstr.* **77**, 34660v (1972).
138. Yu. A. Ol'dekop, N. A. Maier, A. A. Erdman, Z. P. Zubreichuk, and I. A. Shingel, *Vestsi Akad. Navuk BSSR, Ser. Khim. Navuk* 102 (1982); *Chem. Abstr.* **97**, 72416r (1982).
139. Yu. A. Ol'dekop, N. A. Maier, and Yu. D. But'ko, *J. Gen. Chem. USSR* **41**, 2279 (1971).
140. Yu. A. Ol'dekop, N. A. Maier, Yu. D. But'ko, and M. S. Mindel', *J. Gen. Chem. USSR* **41**, 835 (1971).
141. T. N. Shatkina, K. S. Mazel', and O. A. Reutov, *Dokl. Chem.* **219**, 911 (1974).
142. Yu. A. Ol'dekop, N. A. Maier, and Yu. D. But'ko, *J. Gen. Chem. USSR* **40**, 612 (1970).
143. A. N. Lovtsova, T. N. Shatkina, and O. A. Reutov, *Dokl. Chem.* **220**, 76 (1975).
144. Yu. A. Ol'dekop, N. A. Maier, Yu. D. But'ko, and I. A. Shingel, *Vestsi Akad. Navuk BSSR, Ser. Khim. Navuk* 63 (1974). *Chem. Abstr.* **82**, 112148d (1975).
145. Yu. A. Ol'dekop, N. A. Maier, and Yu. D. But'ko, *J. Gen. Chem. USSR* **41**, 2067 (1971).
146. Yu. A. Ol'dekop, N. A. Maier, A. A. Erdman, and Yu. D. But'ko, *J. Organomet. Chem.* **201**, 39 (1980).
147. Yu. A. Ol'dekop, N. A. Maier, A. A. Erdman. T. A. Rubakha, and I. A. Shingel, *Dokl. Chem.* **235**, 445 (1977).
148. Yu. A. Ol'dekop and N. A. Maier, *J. Gen. Chem. USSR.* **30**, 3440 (1960).
149. Yu. A. Ol'dekop, N. A. Maier, A. A. Erdman, and E. N. Kalinichenko, *Vestsi Akad. Navuk BSSR, Ser. Khim. Navuk* 89 (1972); *Chem. Abstr.* **77**, 19127h (1972).
150. Yu. A. Ol'dekop, N. A. Maier, and A. A. Erdman, *Vestsi Akad. Navuk BSSR, Ser. Khim. Navuk* 74 (1971); *Chem. Abstr.* **74**, 124585s (1971).
151. I. P. Zyat'kov, M. M. Zubareva, I. A. Shingel, Yu. A. Ol'dekop, N. A. Maier, and A. A. Erdman, *Zh. Prikl. Spektrosk.* **15**, 283 (1971).
152. Yu. A. Ol'dekop, N. A. Maier, and V. L. Shirokii, *Vestsi Akad. Navuk BSSR, Ser. Khim. Navuk* 99 (1978); *Chem. Abstr.* **89**, 23720m (1978).
153. Yu. A. Ol'dekop and N. A. Maier, *J. Gen. Chem. USSR* **30**, 324 (1960).
154. Yu. A. Ol'dekop, N. A. Maier, and V. I. Gesel'berg, *J. Gen. Chem. USSR* **30**, 2550 (1960).
155. Yu. A. Ol'dekop and N. A. Maier, *J. Gen. Chem. USSR* **32**, 1428 (1962).
156. H. Akagi and E. Takabatake, *Chemosphere* **2**, 131 (1973).
157. G. B. Deacon, "Some Fundamental and Environmental Aspects of Mercury Chemistry" *in* "Mercury in the Environment" (G. B. Deacon and J. D. Smith, eds.), pp. 96–126. R.A.C.I., Melbourne, 1984.
158. N. D. Kagramanov, A. K. Mal'tsev, and O. M. Nefedov, *Bull. Acad. Sci. USSR Div. Chem. Sci.* **26**, 1697 (1977).
159. A. L. Poznyak and V. E. Stel'mashok, *Koord. Khim.* **5**, 1670 (1979); *Chem. Abstr.* **93**, 177086d (1980).
160. V. I. Pavlovski and A. L. Poznyak, *Dokl. Akad. Nauk BSSR* **24**, 1103 (1980); *Chem. Abstr.* **94**, 112411g (1981).
161. V. E. Stel'mashok and A. L. Poznyak, *Russ. J. Inorg. Chem.* **26**, 1324 (1981).

162. A. L. Poznyak and V. I. Pavlovski, *Z. Chem.* **21**, 74 (1981).
163. A. L. Poznyak and V. I. Pavlovski, *Russ. J. Inorg. Chem.* **26**, 292 (1981).
164. A. L. Poznyak, V. I. Pavlovski, E. B. Chuklanova, T. N. Polynova, and M. A. Porai-Koshits, *Mh. Chem.* **113**, 561 (1982).
164a. A. L. Poznyak and V. I. Pavlovski, *Z. Anorg. Allg. Chem.* **485**, 225 (1985).
165. P. Natarajan and G. Ferraudi, *Inorg. Chem.* **20**, 3708 (1981).
166. Yu. A. Ol'dekop, N. A. Maier, and V. L. Shirokii, *Dokl. Chem.* **229**, 500 (1976).
167. Yu. A. Ol'dekop, N. A. Maier, and V. L. Shirokii, *Vestsi Akad. Navuk BSSR, Ser. Khim. Navuk* 74 (1978); *Chem. Abstr.* **89**, 6389t (1978).
168. Yu. A. Ol'dekop, N. A. Maier, and V. L. Shirokii, *Vestsi Akad. Navuk BSSR, Ser. Khim. Navuk* 82 (1979); *Chem. Abstr.* **90**, 152317e (1979).
169. Yu. A. Ol'dekop, N. A. Maier, and V. L. Shirokii, *J. Gen. Chem. USSR* **48**, 381 (1978).

ADVANCES IN ORGANOMETALLIC CHEMISTRY, VOL. 25

Detection of Transient Organometallic Species by Fast Time-Resolved IR Spectroscopy

MARTYN POLIAKOFF

Department of Chemistry
University of Nottingham
Nottingham NG7 2RD, England

and

ERIC WEITZ

Department of Chemistry
Northwestern University
Evanston, Illinois 60201

I

INTRODUCTION

In his 1981 Nobel Lecture, Professor R. Hoffmann (1) described transition metal fragments as "the building blocks" of organometallic chemistry. Complicated molecular structures become understandable when broken down into a collection of transition metal fragments, such as $Fe(CO)_4$, $Cr(CO)_5$, or $Fe(\eta^5\text{-}C_5H_5)$, bonded through their frontier orbitals like some molecular construction kit. Similarly one can view many reactions of organometallic compounds in terms of two processes, (i) the generation of transition metal fragments and (ii) the reaction or recombination of these fragments.

Copyright © 1986 by Academic Press, Inc.
All rights of reproduction in any form reserved.

There is one important difference between the roles of fragments in the understanding of structure and in the analysis of reactions. The fragments used to build up structures are largely conceptual and need never have been observed as free entities. The fragments generated in reactions must necessarily have at least a transient existence and can, in principle, be detected and characterized. Since conventional diffraction techniques are largely inappropriate for characterizing transient species, the fragments must be detected spectroscopically. Although uv–vis spectroscopy is now routinely used for detecting intermediates generated by flash photolysis, it has severe limitations when applied to organometallic species. These limitations arise from the general broadness and lack of resolvable fine structure in the electronic absorptions of most organometallic species (2). Thus, uv–vis spectra rarely provide much structural information about transition metal fragments. Such information *can* be provided by vibrational spectroscopy, particularly infrared (IR) which, in the case of some metal carbonyl species, can even provide accurate estimates of bond angles (3).

Until recently, fast time-resolved IR spectroscopy has been a technique fraught with difficulty. Generally it has been easier to use low temperature techniques, particularly matrix isolation (2,4), to prolong the lifetime of the fragments so that conventional spectrometers can be used. In the last 5 years, however, there have been major advances in fast IR spectroscopy. It is now posssible to detect metal carbonyl intermediates at room temperature in both solution and gas phase reactions. In Section II of this article, we explain the principles of these new IR techniques and describe the apparatus involved in some detail. In Section III we give a self-contained summary of the organometallic chemistry that has already been unravelled by time-resolved IR spectroscopy.

Any new technique relies heavily on what has gone before. In the remainder of this introduction, first we outline briefly the role of matrix isolation in characterizing transition metal fragments and then consider what conventional flash photolysis with uv–vis detection has revealed about the reactivity of these fragments. It is the timescale of these reactions which dictates the speed of the IR spectroscopy required to detect the intermediates.

A. *Low Temperature Techniques*

1. *Matrix Isolation*

The basic principles of matrix isolation are relatively well known, and its application to organometallic chemistry has been recently reviewed (4). Of relevance here are low temperature experiments, in which a stable metal carbonyl species is isolated in a large excess of an inert solid, the matrix, and is then photolyzed to generate unstable fragments. The most common matrix materials are solid gases (e.g., noble gases or CH_4 at 10–30 K) or frozen hydrocarbons (5,6) or more recently cast polymer films (7), both of which have rather wider temperature ranges than solid gases. The great strength of matrix isolation is the wide range of spectroscopic techniques which can be brought to bear on a particular isolated metal fragment. IR and uv–vis are the most important of these techniques (2–4), but Raman (8), electron spin resonance (esr) (9), magnetic circular dichroism (mcd) (10), fluorescence (11), and even Mössbauer (12) spectroscopy have all been used successfully on organometallic species.

Most of the possible binary $M(CO)_x$ species have been characterized for the first row transition metals, Cr to Cu, as well as many binary carbonyls of second and third row metals.[1] Two surprising features emerged from these studies. The first was the low symmetry of many of the fragments [e.g., $Cr(CO)_4$ (14) and $Fe(CO)_4$ (15) have C_{2v} structures in the matrix], and the second was the enormous reactivity of some of the fragments [i.e., $Cr(CO)_5$ appeared to coordinate CH_4 or even Xe into its vacant site (16)]. Although both the structure and the reactivity have since been elegantly rationalized on theoretical grounds (17–19), many people originally thought that these were merely artifacts induced by the solid matrix. We show below how flash photolysis, particularly with IR detection, has demonstrated that the low symmetry structures and high reactivity are real and important features of these systems.

The success of matrix isolation with mononuclear metal carbonyls is largely due to their unique spectroscopic properties. The C—O stretching vibrations, ν_{C-O}, are largely uncoupled to other vibrations of the molecules (i.e., M—C bending and stretching vibrations). The ν_{C-O} give rise to very intense IR absorption bands (10^3 to 5×10^4 dm^3 mol^{-1} cm^{-1}) usually in the region between 2150 and 1750 cm^{-1}, and, most importantly, the band pattern in this region can be reproduced both in frequency and relative intensity, using very simple force field models (20). For most mononuclear carbonyl fragments, it has been possible to deduce a unique structure by combining the results of partial ^{13}CO enrichment with C—O factored force field calculations (3).

Matrix isolation has many limitations, even for metal carbonyls: (i) it cannot easily be used for charged species[2]; (ii) very little kinetic information can be obtained because of the restricted temperature range and limited diffusion; (iii) the solid matrix cage can effectively block some

---

[1] Some of the carbonyls with a high degree of coordinative unsaturation were produced by co-condensation of metal atoms and CO rather than photolytically (13).

[2] Some metal carbonyl radical ions have been generated from neutral carbonyls, *in situ* in the matrix (21,23).

pathways in photochemical reactions. For example, photolysis of matrix-isolated $Mn_2(CO)_{10}$ provides no evidence for formation of $Mn(CO)_5$ radicals (5), even though these radicals are produced in solution (see Section III,C). Presumably the radicals, once formed, immediately recombine without escaping from the matrix cage [Eq. (1)].

$$Mn_2(CO)_{10} \overset{h\nu, 20K}{\underset{\Delta}{\rightleftarrows}} 2\, Mn(CO)_5 \qquad (1)$$

Often, however, this cage effect can be circumvented and the fragment can be generated by a different route, as in the case of $Mn(CO)_5$ (9,22) [Eq. (2)].

$$HMn(CO)_5 + CO \xrightarrow{h\nu, 20K} Mn(CO)_5 + HCO \qquad (2)$$

Despite the limitations, matrix isolation has been used to generate a large number of transition metal fragments containing carbonyl groups. The frequencies of their C—O bands have been measured and these data form a "spectral library" which has played a central role in the interpretation of time-resolved IR experiments.

2. Low Temperature Solutions

Many of the species generated in low temperature matrices are coordinatively saturated species, e.g., $Ni(CO)_3N_2$ (24) or $Cr(CO)_5(H_2)$ (25). Unlike the unsaturated transition metal fragments, these species may have significant activation barriers for reaction or decomposition and can be stabilized by merely lowering the temperature. Liquefied noble gases have recently proved to be very useful solvents for stabilizing unstable metal carbonyl compounds (26). So far, over 30 compounds containing olefin, dinitrogen, and dihydrogen ligands have been characterized, and another "spectral library" is accumulating. A particular advantage of these solvents is their total lack of IR bands which, when combined with Fourier transform interferometry (FT–IR), allows weak absorptions due to coordinated ligands to be detected. Thus the H—H stretching vibration of coordinated dihydrogen in $Cr(CO)_5(H_2)$ can be detected (27) in liquid Xe, although it is too weak to be observed in matrices (25).

Of course, even in low temperature solutions, unstable compounds may not be very long-lived. Modern fast-scanning FT–IR interferometers can produce high signal-to-noise spectra in a single scan. This means that metal carbonyl compounds with half-lives as short as 2 seconds can be easily detected using an unmodified interferometer (28,29). With improved interferometers, we anticipate that such studies will soon be extended to compounds with lifetimes $\sim$100 mseconds. However, detection of shorter lived species, such as reaction intermediates, requires much faster and more sensitive techniques.

B. *Flash Photolysis of Metal Carbonyls*

It is not our intention in this section to provide a comprehensive review of flash photolysis of organometallic species; rather, we summarize some key experiments which establish the timescales of different types of reactions. Understandably, much more work has been done on the flash photolysis of metal carbonyls in solution than in the gas phase, and so we begin with solution experiments.

1. *Metal Carbonyls in Solution*

Flash photolysis of metal carbonyls in solution was pioneered by Kelly and Koerner von Gustorf (*30*). The most complete of these studies has been carried out on the photolysis of $Cr(CO)_6$ (*30*). The historical development of these experiments, which forms an intriguing story in its own right, has been recently retold (*2*). The salient features are as follows:

1. A broad transient absorption in the visible region is observed when a solution of $Cr(CO)_6$ is photolyzed with a uv flash. The position of the absorption and its decay rate are very sensitive to trace impurities in the solution (*30,31*).

2. The pseudo-first order decay kinetics of this absorption are consistent with the decay of $Cr(CO)_5$. Thus, in cyclohexane solution under 1 atm pressure CO, the half-life of decay is 25 μseconds ($k = 2.8 \times 10^{-4}$ second^{-1}) (*30*).

3. The assignment of the band to $Cr(CO)_5$ is not conclusive. Direct comparison with visible absorption bands of matrix-isolated $Cr(CO)_5$ is difficult because of wavelength shifts produced by $Cr(CO)_5$ matrix interactions (*16*).

4. $Cr(CO)_5$ interacts with solvent molecules and in solution cannot be considered as "naked." The interaction is much weaker with fluorocarbon solvents than hydrocarbon (*33*). Using a pulsed laser photolysis source (frequency tripled NdYAG) and C_7F_{14} as a solvent, Kelly and Bonneau (*33*) measured the rate constants for the reaction of $Cr(CO)_5$ with C_6H_{12}, CO, and other ligands [Eq. (3)].

$$Cr(CO)_5 + Q \xrightarrow{k} Cr(CO)_5Q \qquad Q = C_6H_{12}, \ CO, \ etc. \qquad (3)$$

These reactions had similar rate constants, $\sim 4 \times 10^9$ dm^3 mol^{-1} second^{-1}, which approached the diffusion-controlled limit. Thus, for 10^{-2} M concentration of added ligand the half-life of $Cr(CO)_5$ would be ~ 17 nseconds. Interest in these experiments has been reawakened by the recent reports of photoactivation of alkanes by metal carbonyl species (*34*).

5. In the absence of added CO, $Cr(CO)_5$ reacts in C_7F_{14} solution with excess unphotolyzed $Cr(CO)_6$ to form a new transient species. This was identified as $Cr_2(CO)_{11}$ on the basis of reaction kinetics but its structure is unknown (33).

6. Picosecond flash photolysis measurements on $Cr(CO)_6$ in pure C_6H_{12} show that $Cr(CO)_5(C_6H_{12})$ is formed within 25 pseconds of the uv flash (35). Thus the lifetime of "naked" $Cr(CO)_5$ in solution must be even shorter. This means that even if at some future date sub-picosecond flash photolysis apparatus with IR detection were available, an IR spectrum of "naked" $Cr(CO)_5$ in solution might be uninformative. The Heisenberg uncertainty principle dictates that a light pulse, 0.1 psecond long, would be ~ 50 cm^{-1} wide, too wide to give an IR spectrum with useful structural information. Thus "naked" $Cr(CO)_5$ must be sought in the gas phase where, in the absence of any solvent, its lifetime should be longer. (See Sections I,C and III,A,2.)

The photolysis of $Cr(CO)_6$ illustrates many of the important features of metal carbonyl photochemistry in solution. Generally, metal carbonyl intermediates are quite easy to detect; quantum yields for their formation are high and their uv–visible absorptions are intense. Other organometallic intermediates may be harder to observe because of lower quantum yields as, for example, with the recent observation of $(\eta^5\text{-}C_5H_5)_2Mo$ in THF solution (36). Reaction rates are fast, and therefore good time resolution is required. However, most solvents have no absorptions in the uv–visible region, and reactions (e.g., bimolecular radical recombination) can often be slowed down by dilution and a corresponding pathlength increase without adversely affecting the signal-to-noise ratio of the spectrum or the intensity of the transient absorption. By contrast, solvent absorption is a serious problem for IR measurements, even for the ν_{C-O} region of the spectrum. Typically IR solution cells are limited to <1 mm pathlength for most applications. These short pathlengths make it harder to build time-resolved IR apparatus which are optically as efficient as the equivalent uv–visible equipment.

Flash photolysis has now been applied to a wide range of metal carbonyl species in solution, including $Mn_2(CO)_{10}$ (37), $[CpFe(CO)_2]_2$ (38), and $[CpMo(CO)_3]_2$ (39). In almost every case, interesting data have emerged, but, as with $Cr(CO)_5$, the structural information is usually minimal. Thus, the radical $Mn(CO)_5$ has been generated in solution by flash photolysis (37), the rate constant for its bimolecular recombination has been measured, but the experiments did not show whether it had D_{3h} or C_{4v} symmetry. Some experiments have been unsuccessful. Although the fragment $Fe(CO)_4$ is well known in matrices (15), it has never been observed by flash photolysis in solution, probably because it has no absorption maxima in the visible or near-uv. Despite this, flash photolysis of $Fe(CO)_5$ in cyclohexane solution suggests that $Fe(CO)_4$ must be very reactive because the trimer $Fe_3(CO)_{12}$ is formed within 100 μ seconds of the flash (40).

Thus, overall, it is clear that flash photolysis with uv–visible detection is effective in establishing the broad outlines of the photochemistry of a particular metal carbonyl. Intermediates can be identified from their reaction kinetics, and sometimes, with the help of uv–vis data from matrix isolation experiments. Structural information from uv–vis flash photolysis is at best sketchy. Many questions remain unanswered. Time-resolved IR measurements can fill in some of these answers.

2. Metal Carbonyls in the Gas Phase

Relatively little work has been done on the flash photolysis of gas phase metal carbonyls, partly because of the low volatility of many of the compounds. Early work by Callear (41,42) provided some evidence for $Ni(CO)_3$ generated from $Ni(CO)_4$ in the gas phase (41) and Fe atoms produced from $Fe(CO)_5$ (42). This latter process has even been used as the basis of an Fe atom laser (43). More recently Breckenridge and Sinai (44) studied the flash photolysis of $Cr(CO)_6$. Their results, interpreted largely on the basis of data from matrix isolation experiments, were in broad agreement with Kelly and Bonneau's solution work (33). In particular, they found no evidence for loss of more than one CO group [Eqs. (4) and (5)].

$$Cr(CO)_6 \xrightarrow{\text{355 nm}} Cr(CO)_5 + CO \tag{4}$$

$$Cr(CO)_6 \xrightarrow{\text{355 nm}} Cr(CO)_4 + 2\,CO \tag{5}$$

This contrasts with the results obtained by Yardley who, in gas phase trapping experiments with $Cr(CO)_6$ (45) and $Fe(CO)_5$ (46), showed that uv light of wavelength shorter than 355 nm could promote loss of two or more CO groups.

As with solution experiments, flash photolysis in the gas phase has produced evidence for the existence of intermediates but no information about their structure. In principle gas phase IR spectra can provide much more information, although the small rotational B value of gaseous carbonyls and low lying vibrational excited states preclude the observation of rotational fine structure. As described in Section II, time-resolved IR experiments in the gas phase do not suffer from problems of solvent absorption, but they do require very fast detection systems. This requirement arises because gas-kinetic reactions in the gas phase are usually one or two orders of magnitude faster than the corresponding diffusion controlled reactions in solution, with the same concentrations of reactants.

3. Unimolecular Processes

Until now, we have only described bimolecular reactions of photochemically generated fragments. These bimolecular reactions, involving either dimerization of fragments or reaction with a ligand (e.g., CO), can always be slowed down, in principle at least, by diluting the solution or reducing the gas pressure. There are, however, interesting unimolecular processes which cannot necessarily be slowed down in this way. The most important of these processes are (i) isomerization and (ii) relaxation of electronically excited states.

a. Isomerization. Quite frequently, the formation of matrix-isolated fragments, produced by photolysis of metal carbonyls, involves not only loss of CO but also intramolecular isomerization. Perhaps the simplest case involved the photolysis of *trans*-$(^{13}CO)W(CO)_4CS$ which yielded *cis*-$(^{13}CO)W(CO)_3CS$ as the principal product (*47*), while the best example is the photolysis of $Fe_2(CO)_9$ (*48*) [Eq. (6)].

$$Fe_2(CO)_9 \xrightarrow[\text{Ar matrix}]{h\nu, 20K} Fe_2(CO)_8 \xrightarrow{\Delta, 30K} Fe_2(CO)_8 \qquad (6)$$
$$\text{(CO bridged)} \qquad \text{(no CO bridge)}$$

Both examples clearly involve isomerization. For $Fe_2(CO)_9$ the processes of photolysis and isomerization can be observed separately. The first product is stabilized in an Ar matrix and can be isomerized by gently annealing the matrix (*17*). In the case of *trans*-$(^{13}CO)W(CO)_4CS$, it was possible to deduce that isomerization occurred *after* CO loss, presumably in some excited state of the initial photoproduct, *trans*-$(^{13}CO)W(CO)_3CS$, but this product could not be frozen out before isomerization took place (*48*). The activation barriers to intramolecular isomerization are generally small (*49*), and although the temperatures of solid matrices are low enough to freeze out thermal isomerization in the *ground state* of metal carbonyls (*50*), matrices cannot usually prevent isomerization in the course of photochemical reactions. Valuable mechanistic information can be lost. A recent case is the photolysis of matrix-isolated $[CpFe(CO)_2]_2$ (*6,7*) [Eq. (7)].

$$CpFe(CO)(\mu\text{-}CO)_2Fe(CO)Cp \xrightarrow[\text{matrix}]{h\nu} CpFe(\mu\text{-}CO)_3FeCp \qquad (7)$$
$$(Cp = \eta^5\text{-}C_5H_5)$$

Does isomerization to form three CO-bridges occur before or after the loss of CO from the parent carbonyl? This reaction is discussed further in Section III,C.

In general, intramolecular isomerization in coordinatively unsaturated species would be expected to occur much faster than bimolecular processes. Some isomerizations, like those occurring with $W(CO)_4CS$ (*47*) are anticipated to be very fast, because they are associated with electronic relaxation. Assuming reasonable values for activation energies and *A*-factors, one predicts that, in solution, many isomerizations will have half-lives at room temperature in the range 10^{-7} to 10^{-6} seconds. The principal means of identifying transients in uv–visible flash photolysis is decay kinetics and their variation with reaction conditions. Such identification will be difficult if not impossible with unimolecular isomerization, particularly since uv–visible absorptions are not very sensitive to structural changes (see Section I,B). These restrictions do not apply to time-resolved IR measurements, which should have wide applications in this area.

b. Electronic Excited States of Organometallic Species. The photochemistry of most organometallic compounds differs from that of inorganic coordination compounds because, in general, the quantum yields for dissociation of organometallic species are high while those for coordination compounds are low. Despite the limitations outlined above, a considerable amount of information is available about organometallic intermediates, but little is known about the electronically excited states from which the intermediates are formed. The reverse is true for coordination compounds. The result has been a considerable difference in emphasis in the interpretation of photochemical and photophysical processes in organometallic and coordination compounds (*51*).

A knowledge of excited states is important for an understanding of why particular intermediates are generated, and of the wavelength dependence of photochemical pathways. Adamson (*52*) has coined the term *thexi* (a contraction of thermally equilibrated excited) for these states. So far, little is known about thexi states in organometallic photochemistry. Thexi states are inherently different from intermediates in that they cannot be stabilized by trapping techniques such as matrix isolation, and their lifetimes are generally much shorter. Lifetimes of thexi states can, of course, be somewhat lengthened by lowering the temperature (*53*).

Those organometallic thexi states which have been detected have involved compounds where the quantum yield for photodissociation is very low. Time-resolved uv–visible absorption and emission studies have been made on $W(CO)_5L$ and $W(CO)_4L'$ species (L = acetylpyridine, L' = *o*-phenanthroline) (*54*), but, as in the case of intermediates, these studies provided lifetimes but no structural information.

A technique which does provide structural information is time-resolved resonance Raman spectroscopy (RR) (55). Since this technique involves uv–visible excitation and detection, it suffers from none of the limitations associated with time-resolved IR measurements (solvent absorptions, slow detection, etc.; see Section II). RR has been successfully applied to thexi states of bipyridyl complexes (56,57), which are of importance in electron transfer processes. The most organometallic of these is (bipy)Re(CO)$_3$Cl, which Wrighton and co-workers (57) have shown to have a relatively long-lived excited state ($t_{1/2}$ 200 nseconds) and a structure which involves localization of an electron on the bipy ligand, yielding a coordinated radical anion bipy$^{\cdot-}$. This experiment indicates the limitation of RR: The structural information is good but localized. No information is obtained about the metal center or the coordinated carbonyl groups. In principle, time-resolved IR experiments, using equipment currently available (see Section II), should be able to detect such thexi states.

Clearly RR is attractive as a technique for detecting intermediates. It has, for example, been very successful with intermediates in organic and bioinorganic reactions (55). Unfortunately the high laser intensities that are needed in RR may be incompatible with the photosensitivity of most organometallic intermediates. For example, in an attempt to detect Cr(CO)$_5$ in solution by RR, the only detectable transient signal was emission from excited Cr atoms (58).

C. Timescales of Reactions

As discussed above, the solution environment provides for a set of time scales different from the gas phase environment. In solution, there are typically 10^{13} collisions second^{-1} of a solute molecule with solvent molecules. Thus, if a photolytically generated species is expected to have a large cross section for reaction with solvent and it is desired to monitor that reaction, both generation and monitoring must be done on a picosecond (psecond) or even sub-psecond timescale. That monitoring this rapid is necessary has been confirmed in an experiment on Cr(CO)$_6$ in cyclohexane solution where psecond photolysis and monitoring was not rapid enough to detect the "naked" Cr(CO)$_5$ that existed before coordination with cyclohexane (35).

However, there is another operative timescale in solution. This is that timescale for reaction with other photolytically generated species or with added reactants. This reaction cannot take place faster than the diffusion-limited reaction rate which is concentration dependent (59). Typical diffusion-controlled reaction rate constants are $\sim 10^9$–10^{10} dm^3 mol^{-1} second^{-1}. By comparison, a typical gas-kinetic rate constant, the fastest rate of reaction for neutral gas phase species, is $\sim 10^{11}$ dm^3 mol^{-1} sec^{-1}. Thus it is very interesting to note that for identical concentrations, the fastest gas phase reactions will be one to two orders of magnitude faster than the fastest possible solution phase reaction. This puts a much less stringent requirement on the time resolution necessary in a solution phase versus gas phase transient absorption apparatus. For example, systems which use uv flash lamps to generate transients have proved successful in measuring rates of bimolecular reactions in solution (60,61). Such flashlamps typically have pulse widths in the μsecond range but if concentrations are kept low ($\sim 10^{-5}$ molar), even diffusion controlled bimolecular reaction half-lives in solution will be 10^{-4} to 10^{-5} second, well within the range of what can be measured with a flash lamp pumped system.

From the above discussion, it is obvious that response time of a system and its sensitivity are intrinsically linked. In fact they have a reciprocal relationship. As sensitivity increases it is possible to look at bimolecular reactions of species at lower and lower concentrations. In these circumstances the requirements on the response time for a system will get less and less. Of course, there are limits to how far this can be pushed, particularly with time resolved IR measurements in solutions, where absorption by the solvent is significant. Also, as indicated previously, coordination of a nascent photofragment by solvent molecules can occur on an exceedingly rapid timescale (35). Additionally, as the concentration of added reactant is diminished, reactions with impurities in the solvent or with small concentrations of atmospheric gases become a problem. Nevertheless, over a wide range of concentration there is a trade-off between minimum detectable signal and timescale.

The timescales of the processes mentioned above are summarized in Fig. 1. In the next section we look at the technical problems of recording IR spectra on these timescales.

II

TIME-RESOLVED IR MEASUREMENTS: TECHNICAL CONSIDERATIONS

A.. Historical Development

Transient absorption techniques now have a venerable history. The development of "flash kinetic spectroscopy" was the work of Norrish and Porter (62). This technique typically employed a flash lamp to produce

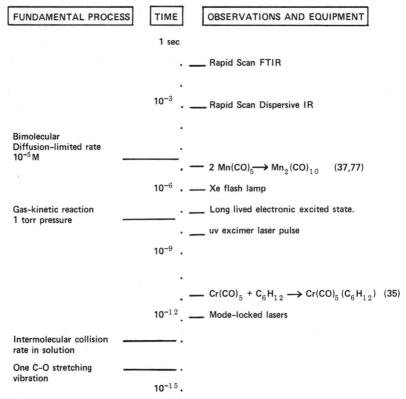

FIG. 1. Timescales of fundamental processes in solution and the gas phase, compared to observed reaction rates and equipment performance. Note that the scale is logarithmic.

transient species in the gas phase via photodissociation and a second smaller flash lamp to provide the probe radiation. In its early manifestations, the probe radiation was essentially limited to the visible and uv regions of the electromagnetic spectrum, principally due to the low detectivity and lack of availability of infrared detectors. Nevertheless, the technique allowed for the first real-time studies of a wide variety of radicals and other transient species. With appropriately modernized equipment, it is in wide use today. For their pioneering efforts, Norrish and Porter were awarded the Nobel Prize, and it can be seen from the Introduction what an important role flash photolysis has already played in understanding the mechanisms of organometallic reactions.

It has been long recognized that vibrational spectra often have more easily accessible structural information than similar spectra in the visible or uv regions. This is due to the finite IR spectral shifts associated with even minor structural changes in molecules. Pimentel (*63*) was the first to develop a viable technique for the measurement of IR spectra of transient species. The technique, so-called rapid scan spectroscopy, was a modification of the Norrish and Porter technique. Rapid scan spectroscopy used a uv flash lamp to produce transient species in the gas phase which were probed by an IR spectrometer modified to allow rapid rotation of the grating. The grating rotation was synchronized with the flash lamp pulse. Further modifications replaced the original IR source with a higher intensity source and provided for multipassing of the sample cell, with an effective pathlength up to 40 m. With this arrangement, IR spectral data could be obtained over a few 100 cm^{-1} on a timescale of 1 cm^{-1} μsecond^{-1} with a wavelength resolution of $\sim 1 \text{ cm}^{-1}$. Even with this relatively limited time resolution, this apparatus was able to provide a number of breakthroughs in the detection of transients. The gas phase IR spectra of CF_3, CH_3, and CD_3 were all first observed with this apparatus (*63*). A similar apparatus built in the Soviet Union (*64*) could scan $\sim 400 \text{ cm}^{-1}$ in 1.4 mseconds with a repetition rate of 300 Hz. Since that time, there has been considerable progress in the area of detection of transient species both in the infrared and visible spectral regions.

A modern variation on the rapid scan spectrometer, which is under development, uses a laser-generated plasma as a high intensity broad-band IR source (*65*). This method has been used to probe the ν_{C-O} absorption of $W(CO)_6$. Another technique TRISP (time-resolved IR spectral photography), which involves up-conversion of IR radiation to the visible, has also been used to probe transients (*66*). This method has the enormous advantage that efficient phototubes and photodiodes can be used as detectors. However, it is a technically challenging procedure with limitations on the frequency range which depend on the optical material used as an up-converter.

B. *IR Kinetic Spectroscopy*

Both Porter's original flash photolysis apparatus and Pimentel's rapid scan spectrometer recorded the whole spectral region in a time which was short compared to the decay of the transient species. Kinetic information was obtained by repeatedly firing the photolytic flash lamp and making each spectroscopic measurement at a different time delay after each flash. The decay rate could then be extracted from this series of delayed spectra. Such a process clearly has limitations, particularly for IR measurements, where the decay must be slow compared to the scan rate of the spectrum.

Modern time-resolved IR measurements typically use a different strategy. For each uv flash, kinetic measurements are made at one IR wavelength. The monitoring wavelength is changed and the "lamp" is flashed again. Thus, using a number of flashes, data are accumulated for wavelengths across the spectral region of interest. These data can then be used to construct "point-by-point" spectra corresponding to any particular time delay after the flash.[3] Thus the essential requirements for such measurements are (i) a monochromatic IR beam, (ii) a pulsed uv light source, and (iii) an IR detector and associated electronics (Fig. 2).

Four different laboratories have built IR kinetic spectrometers for use with organometallic compounds. A fundamental feature of all these spectrometers is that the detector is AC coupled. This means that the spectrometers only measure *changes* in IR absorption. Thus, in the time-resolved IR spectrum, bands due to parent compounds destroyed by the flash appear as *negative* absorptions, bands due to photoproducts appear as *positive* absorptions, and static IR absorptions, due to solvents, for example, do not register at all. The important features of these spectrometers are listed in Fig. 2. Since three spectrometers have a line-tunable CO laser as the monochromatic light source, we begin with the CO laser. Then we look in more detail at spectrometers designed for gas phase and solution experiments.

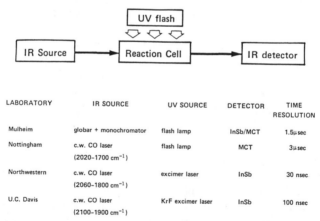

LABORATORY	IR SOURCE	UV SOURCE	DETECTOR	TIME RESOLUTION
Mulheim	globar + monochromator	flash lamp	InSb/MCT	1.5µsec
Nottingham	c.w. CO laser (2020–1700 cm⁻¹)	flash lamp	MCT	3µsec
Northwestern	c.w. CO laser (2060–1800 cm⁻¹)	excimer laser	InSb	30 nsec
U.C. Davis	c.w. CO laser (2100–1900 cm⁻¹)	KrF excimer laser	InSb	100 nsec

FIG. 2. Schematic diagram of an IR kinetic spectrometer and details of the four sets of equipment which have been used successfully for organometallic photochemistry. The spectrometers at the Max Planck Institut für Strahlenchemie in Mülheim (*60*) and the University of Nottingham (*61*) are for use with solutions, while those at Northwestern University (*68*) and the University of California (Davis) (*69*) are for gas phase samples.

[3] A similar technique has been used to obtain time-resolved IR spectra of carbonyl intermediates in "stopped-low" experiments on a msecond timescale (*67*).

1. *The CO Laser*

Over the past decade, c.w. CO lasers have frequently been used for monitoring transient species, most often vibrationally excited CO, in the gas phase (*70*). The CO laser is particularly suitable for detecting organometallic transients because its tunable output spans the ν_{C-O} absorptions of nearly all transition metal carbonyl compounds. The laser typically lases on one line out of a series of lines corresponding to specific vibration–rotation transitions of CO (*71*). The particular line is selected by a diffraction grating at one end of the laser cavity. Rotation of the grating changes the laser transition and hence the IR frequency of the output. These transitions are spaced at roughly 4 cm^{-1} increments since the rotational B value for CO is 1.9 cm^{-1}. In the CO laser, vibrationally excited CO is produced up to very high quantum numbers. Laser action can occur whenever there is a population inversion on a specific vibration–rotation transition. Laser action is accompanied by population transfer of excited molecules from vibrational level V and rotational level J to state $V' = V - 1, J' = J + 1$ (P branch transitions) [Eq. (8)].

$$CO(V, J) \rightarrow CO(V - 1, J + 1) + h\nu(IR) \tag{8}$$

This process can, and does, continue as long as there is a sufficient population inversion between state V', J' and state $V' - 1, J' + 1$. This results in a cascade of population and correspondingly the CO laser is described as a "cascade laser". Output of the laser can be made to span the region from the CO $V = 1 \rightarrow V = 0$ transition (at ~2140 cm^{-1}) to very high vibrational states of CO (in the 1600 cm^{-1} region or even further to lower frequencies). These wavelengths correspond to CO laser transitions higher than $V = 20$. Operating on low vibrational states requires care in laser design and operation. Typically, if it is desired to operate the laser on a transition lower than $V = 5 \rightarrow 4$, the CO laser tube must be cooled below room temperature. Cooling lowers the rotational temperature of the laser medium. This lowered temperature facilitates the attainment of a population inversion. Particular care is necessary to obtain lasing on the CO $V = 1 \rightarrow 0$ transition. One must avoid the buildup of unexcited CO anywhere in the laser cavity. Unexcited CO will act as an intra-cavity absorber and quench laser action. Additional laser frequencies can be obtained by using ^{13}CO laser gas, but this is only practicable with a sealed laser. In the past, nitrogen-cooled lasers, essential for laser action at high frequency, have been destroyed by explosions caused by ozone, and great care must be taken to avoid any build up of solid ozone in the laser tube.

2. Time-Resolved IR in the Gas Phase

An apparatus at Northwestern is shown schematically in Fig. 3. It has been used to obtain both IR spectral and kinetic data on coordinatively unsaturated compounds in the gas phase. It uses a uv excimer laser which has a significant advantage over a flash lamp in that a high intensity pulse of uv radiation can be generated on a much shorter timescale. A typical excimer laser produces 10–300 mJ of uv radiation in a ~10 nsecond pulse at a number of wavelengths according to the lasing gas (ArF 193 nm, KrF 249 nm, XeF 355 nm, and XeCl 308 nm). This radiation is fairly well collimated and highly monochromatic compared to a flash lamp source, which typically has much longer pulse-lengths, in the μsecond range. A μsecond pulse can be a severe detriment to obtaining time-resolved information in the gas phase for very reactive species (see Fig. 1). Faster flashlamps can only be obtained by sacrificing considerable intensity in the photolysis pulse.

The IR probe radiation, provided by a liquid N_2-cooled line-tunable CO laser, is passed through the cell collinearly or almost collinearly with the uv photolysis radiation. The probe beam is then directed onto an IR detector using a filter for either selectively blocking the photolysis beam or

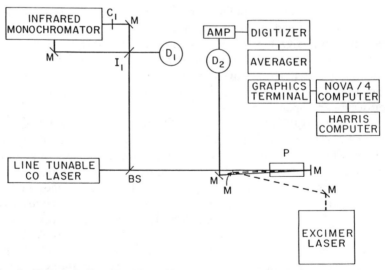

FIG. 3.   Schematic diagram of the Northwestern apparatus for IR laser kinetic measurements in the gas phase. D_1 and D_2 are InSb detectors with D_2 being a high speed photovoltaic detector. M = Mirror, I = iris, C = chopper, BS = beam splitter, P = photolysis cell. [Reproduced with permission from Ouderkirk et al. (75).]

selectively reflecting the IR beam, or both. The most used IR detector is based on an InSb element. Generally, a detector of this type has a D^* of 5×10^{10} $Hz^{1/2}$ W^{-1} with a useful detection range from about ~10,000 to 1780 cm^{-1}. D^* is a measure of detector sensitivity, but a more relevant parameter for this type of experiment is noise equivalent power (NEP): the amount of incident power falling on the detector which is necessary to produce a signal-to-noise ratio (S/N) of unity. Since the background noise of any electrical circuit depends on the bandwidth, the NEP has units of W $(Hz)^{-1/2}$. A typical InSb detector has a NEP of 10^{-10} W $(Hz)^{-1/2}$. Thus for the 30 MHz bandwidth used in the Northwestern apparatus, the NEP is ~0.5 μW. By comparison, a typical c.w. CO laser operates in the range 10 mW to 1 W. This illustrates an important principle of detection via a c.w. probe. It takes very little c.w. power to push a detector into a region where an experiment is *not* detector noise limited. This is a significant advantage over an IR emission experiment, which normally operates in the detector noise limited regime (72). To realize this advantage over emission experiments, it is important to utilize a probe source with enough photons per unit bandwidth to exceed the NEP of the detector. In initial experiments with the apparatus in Fig. 3, a globar was used as the IR probe source. When the globar was replaced by the CO laser a three order of magnitude improvement in S/N took place (73)! In retrospect this result is not surprising because a CO laser probe has far more photons over a few cm^{-1} bandwidth than does a globar and the laser beam is more easily collimated.

Although an InSb detector is a very good infrared detector, it is limited to operation at frequencies higher than ~1780 cm^{-1}. (Lower frequency operation can be achieved at temperatures higher than 77° K but with degradation of D^* and increase in NEP.) For operation at frequencies lower than ~1780 cm^{-1} other detectors should be used. Obvious choices are cooled Ge detectors, doped with Au, Cu, or Hg. The Cu:Ge and Hg:Ge detectors require liquid helium for cooling, which extends the viable range to $\geqslant 15$ μm, while Au Ge operates at 77 K but is viable to only ~10 μm. Another choice is HgCdTe (MCT) which can be used to at least 20 μm at 77 K, but when used in the longer wavelength region of its detection range it has significantly lower D^* than typical Ge doped detectors (74). Following detection, the signal is fed to appropriate amplifiers, a transient digitizer and, if necessary, a signal averager. With any detector, care must be taken to provide a system that is not only sensitive but has a linear response over the large dynamic range required for a c.w. probe experiment. Normally with photoconductive detectors, this will require some modification of the standard bias circuit. With both photovoltaic and photoconductive detectors, care must be taken to ensure

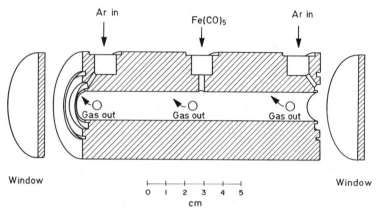

FIG. 4.   Schematic cross section of the Northwestern flow gas cell. Fe(CO)₅ enters through the center port, Ar purge gas through ports by the windows. [Reproduced with permission from Ouderkirk *et al.* (*75*).]

that the amplifiers have a linear response over the expected range of signal amplitudes.

After the data are acquired, transient signals taken at individual wavelengths can be analyzed for kinetic information or an entire time-resolved spectrum can be synthesized. This is achieved by instructing the computer to assemble a point-by-point spectrum, corresponding to a particular delay time after the photolysis pulse. The spectrum is constructed out of points that are 4 cm^{-1} apart and is similar to a spectrum produced by a normal IR spectrometer but with 30 nsecond time resolution!

Another important aspect of a gas phase apparatus is the photolysis cell (*73,75*). A poorly designed cell can produce spurious signals. In any system in which energy is deposited inhomogeneously, shock waves will be generated due to the expansion of the inhomogeneously heated gas. The shock waves will modulate the radial density in the cell and thus the composition of the gas. This modulation will appear as a spurious absorption. This phenomenon, sometimes known as "acoustic waves", is obviously undesirable. Shock waves have been eliminated in the Northwestern apparatus by making the uv laser beam as uniform as possible and by adding a suitable pressure of rare gas to the cell. The rare gas increases the heat capacity of the contents of the cell. This is turn diminishes the overall temperature rise and reduces the amplitude of any potential shock waves. For chemical systems in which involatile products are not generated and the parent carbonyl is regenerated, experiments can be carried out in a sealed cell. However, a flow cell is necessary for experiments with parent gases that produce involatile photoproducts, either directly or following reaction (75). In particular, the $Fe(CO)_5$ system (see Section III,A,1) produces large quantities of involatile species which, in a conventional static or flow cell, render the windows opaque within a few laser pulses. This problem is eliminated in the cell, shown in Fig. 4, where the windows are purged by a "curtain" of rare gas. Purge rates and pressures can be balanced so there is little or no mixing of purge gas with the sample mixture.

The equipment at the University of California (Davis) (59,76) is broadly similar to the Northwestern apparatus but photolysis cells up to 1 m long are used. A full description has been published recently (76).

3. Time-Resolved IR of Solutions

In principle a time-resolved IR spectrometer (60, 61) for use with solutions is similar to the Northwestern spectrometer described above. For many experiments the time-resolution requirements are fewer (see Fig. 1), and successful results can be achieved with a Xe flash lamp. The first successful time-resolved IR spectrometer (60) for organometallic photochemistry was built at the Max Planck Institut für Strahlenchemie in Mülheim. This spectrometer uses a globar and monochromator for the IR probe and Xe flash lamps ($t_{1/2} = 1.5$ μseconds) for the photolysis source. Although much weaker than a laser source, the globar has the advantage that its wavenumber range is unlimited. At Mülheim, transient signals have been recorded over the range 2260–1740 cm^{-1} (see Section III). A key to the success of this equipment has been the design of the flash lamp/IR cell assembly, shown in Fig. 5, with the lamps very close to the cell, which has an evacuable flow-system controlled by a solenoid.

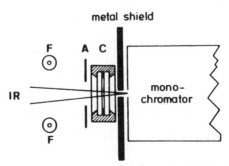

FIG. 5. Diagram of part of the Mülheim apparatus for IR kinetic measurements on solutions showing the arrangement of flash tubes (F), sample cell (C), aperture (A), and IR monitoring beam. [Reproduced with permission from Hermann *et al.* (60); copyright 1982, American Chemical Society.]

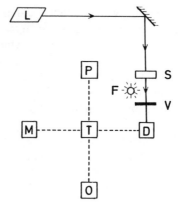

FIG. 6. Schematic diagram of the Nottingham apparatus for IR kinetic measurements on solutions. Solid lines represent the light path, broken lines the electrical connections. L = Line tunable CO laser, S = sample cell, F = flash lamp, P = photodiode, D = fast MCT IR detector, T = transient digitizer, O = oscilloscope, and M = microcomputer. Nonfocussing optics were used throughout, and the IR laser beam was heavily attenuated by a variable path cell V, filled with liquid methanol, placed immediately in front of the detector. [Reproduced with permission from Moore *et al.* (*61*).]

The Nottingham spectrometer (*61*) also has the cell very close to the flash lamp but uses a CO laser as the IR probe (Fig. 6) with a corresponding increase of sensitivity. Most solvents (apart from liquefied noble gases) absorb IR. This means that as the IR probe wavelength is changed the IR power falling on the detector can change dramatically. In the Nottingham apparatus (*61*), this problem is overcome by running the laser with much more IR power than is required for the detector and then attenuating the beam with a cell filled with methanol before it reaches the detector. The amount of attenuation can then be reduced or increased to compensate for changes in absorption by the sample. In Mülheim, the slits of the monochromator can be opened wide when the solvent is strongly absorbing, or, in extreme cases, the solvent can be changed for one that is less absorbing (*77*)!

In all these spectrometers the IR signals are recorded using transient digitizers, which convert the continuous analog signal into digital form. An important point is the choice of digitization rate. Thus for a spectrometer designed to detect species with lifetimes of ~10 μseconds, one requires a digitization rate substantially faster than 1 μsecond (1 MHz), which would give only 10 data points within a 10 μsecond lifetime. Finally, unimolecular processes normally cannot be monitored using a flash lamp, because even in solution they are usually faster than the duration of the flash (see Fig. 1).

4. *Improved IR Light Sources*

Now that the parameters of typical prototype systems have been considered, it seems worthwhile to consider further variations that could be used to enhance the power and/or versatility of the basic apparatus. An obvious improvement would be to use a completely tunable IR probe source. As mentioned above, a globar source is completely tunable but with a lower than desired number of photons per unit bandwidth (*60,73*). There are a number of other means of producing completely tunable IR probe radiation with enough intensity to pull the detector out of the blackbody-noise regime.

One possibility is an IR diode laser system (*78*) which can produce tunable radiation in virtually any region of the IR from about 350 to 4000 cm^{-1}. Sufficient power can be produced over this entire region to get above background noise with a good semiconductor detector, but tunable single-mode power does fall off into the μW range at low wavenumbers. In addition to the more obvious advantages, the complete tunability of a diode laser would allow for construction of a "point-by-point" spectrum with virtually any desired resolution. However, diode laser systems are not without potential problems. Although continuous single mode tuning ranges have been increased, they are still limited to a few cm^{-1}. Overall, tuning ranges for individual diodes have constantly been improved, but they are currently limited to, at best, ~200 cm^{-1} per diode. This necessitates the purchase of several diodes for true versatility. Amplitude and frequency stability and determining the precise laser frequency are also problems. Nevertheless, the potential advantages certainly outweigh the problems. IR diode lasers have already been used to produce high resolution spectra of molecules cooled in supersonic jets (*79*), to measure homogeneous linewidths in matrices (*80*), and to obtain gas phase kinetics and dynamical information (*81*). Their use in the detection of transient species is certain to provide both the expected advantages and unanticipated benefits.

Another method of producing tunable IR radiation involves difference frequency generation (*82*). In this method, the beams from two dye lasers operating at frequencies v_x and v_y in the visible region are mixed in a nonlinear crystal with the concomitant generation of the difference frequency $(v_x - v_y)$ in the IR region. Recently, this method (*82*) has been elegantly applied to obtain the gas phase IR spectrum of the v_1 and v_3 modes of singlet CH_2. An advantage over the diode laser is that it is easier to determine the absolute frequency of the IR beam because the IR frequency is generated via the difference between two visible frequencies and the determination of frequencies is easier in the visible than in the IR.

Stability may not be as much of a problem as with a diode source. However, there are problems with this method as well. The range of tunability is limited by the absorption properties of the nonlinear crystal which generates the difference frequency. At present, tunability is limited to wavenumbers $\geqslant 2500$ cm^{-1} and conversion efficiencies are low. Typical laser powers in the CH_2 experiments (82) were ~20 μW (compared to the power of the CO lasers, 10 mW–1 W). This produces a situation where IR detectors, particularly fast ones, may be close to or background noise limited. However, it is clear that more applications of this technique will appear in the future.

C. *Alternative Experimental Methods for Transient Detection*

Before beginning a discussion of alternative experimental detection techniques, it is worthwhile considering the fundamental differences in data collection that can result from different experimental techniques. IR kinetic spectroscopy involves uv flash generation of transients and monitoring of transients at a finite number of IR wavelengths. The data that are obtained are continuous in time at a given wavelength but are fairly coarse-grained in the spectral domain. We call this "dense in time and coarse in frequency." Obviously the situation can be changed by the use of an IR diode laser, or a nonlinear mixing device because, with these devices, data become relatively dense in each domain. This is indeed the ideal situation as far as information content is concerned. The first flash photolysis apparatus developed by Norrish and Porter (62) works in the opposite mode: it is dense in the frequency domain but coarse in time. These early flash photolysis experiments involved generation of transients with one flash and probing of transients with another flash which was delayed relative to the first. A photographic plate was used to detect absorption of the dispersed emission of the second flash. This relatively dense frequency information was obtainable from the photograph, while each experiment produced only one time point.

Normally, time-resolved FT–IR spectroscopy (TRS FT–IR) possesses the same data characteristics. In a typical TRS FT–IR experiment, interferograms are assembled for a specific delay time after the photolysis pulse, and the data produced are normally finer-grained in frequency than in time. This type of experiment is complementary to experiments with fine-grained time information. It is particularly useful where a wide spectral range is necessary and works reasonably well for highly reproducible events which occur on relatively long timescales (fractions of seconds) (83). It is also an appealing system for use on shorter timescales, and it has recently been used for the first time to detect an organometallic intermediate on a 50 μsecond timescale (*84*). However, the technique is highly susceptible to artifacts, and changes in commercial TRS software will be necessary before the method is likely to be adopted for general use.

Another useful technique is time-resolved Raman spectroscopy of molecules (*55*). Raman spectroscopy has the advantage that, in principle, fine-grained time *and* fine-grained frequency information can be obtained in a single experiment. This could be accomplished by recording spectral data with a gated optical multichannel analyzer detector (OMA) which uses a diode array to acquire spectoscopic data. In principle the output of each diode can also be followed to obtain time-resolved data at each frequency. In practice, however, the number of time points obtainable is limited by the read-out, digitization, and processing rates of the controlling computer. Typically these rates limit the time between successive spectra to the order of 10 mseconds. Thus, in effect, this system can provide fine-grained spectral data with coarser time data.

An additional problem with Raman spectroscopy relates to sensitivity. Raman scattering is an intrinsically low cross-section scattering process. Cross sections can be dramatically increased by tuning the excitation source, normally a laser, to coincide with an electronic absorption of the transient. This process is known as resonance Raman spectroscopy. However, a problem in the application of this process to organometallics involves the generally high photosensitivity of organometallic species which can lead to sample decomposition. A flow cell can be of aid in this regard. Another potential problem involves the variable enhancement of modes via resonance Raman spectroscopy. The modes that contain the most structurally significant information may not be enhanced. Nevertheless, Raman spectroscopy and, in particular, resonance Raman are powerful and important tools for the study of transients (*55–57*). Although it has not been extensively applied to organometallic species it has been used in a number of studies of other transient species and its application in the organometallic area is likely to grow.

III

IR KINETIC SPECTROSCOPY OF ORGANOMETALLIC TRANSIENTS

This section must necessarily be regarded as a preliminary review. The first paper (*60*) on fast time-resolved IR spectroscopy of an organometallic species was published in mid-1982. By the end of 1984, a further 10 papers had been published, accepted, or submitted to journals. In order to present a better summary, we have also included some unpublished work. The published work can be conveniently divided into three categories: (A) gas phase photochemistry; (B) photochemistry of $Cr(CO)_6$ in solution; and (C) photolysis of dinuclear metal carbonyls.

A. Gas Phase Photochemistry

There have been three primary motives behind the study of metal carbonyl photochemistry in the gas phase: first, to discover the shapes of metal carbonyl fragments in the absence of perturbing solvents or matrices; second, to probe the effect of uv photolysis wavelength on product distribution; and third, to measure the reaction kinetics of carbonyl fragments. All three areas have already proved fruitful. The photochemistry of two molecules, $Fe(CO)_5$ and $Cr(CO)_6$, has been studied in detail.

1. Fe(CO)₅

Figure 7 shows a series of time-resolved IR spectra recorded after uv photolysis of $Fe(CO)_5$ in the gas phase (75). One can see clearly that $Fe(CO)_5$ was destroyed (bands labeled V), also that new bands labeled II–IV appeared. These bands decayed at different rates indicating that there must have been at least three different photoproducts, which were identified as $Fe(CO)_2$, $Fe(CO)_3$, and $Fe(CO)_4$ by comparison with matrix isolation data (85). This assignment was further supported by photolysis in the presence of added CO, when $Fe(CO)_2$ and $Fe(CO)_3$ have very short lifetimes (68).

The bands due to $Fe(CO)_4$ are shown in Fig. 8. This spectrum (68) was particularly important because it showed that in the gas phase $Fe(CO)_4$ had at least two ν_{C-O} vibrations. Although metal carbonyls have broad ν_{C-O} absorptions in the gas phase, much more overlapped than in solution or in a matrix, the presence of the two ν_{C-O} bands of $Fe(CO)_4$ was clear. These two bands show that in the gas phase $Fe(CO)_4$ has a distorted non-tetrahedral structure. The frequencies of these bands were close to those of $Fe(CO)_4$ isolated in a Ne matrix at 4 K (86). Previous matrix, isolation experiments (15) (see Section I,A) has shown that $Fe(CO)_4$ in the matrix had a distorted C_{2v} structure (Scheme 1) and a paramagnetic ground state. This conclusion has since been supported by both approximate (17,18) and ab initio (19) molecular orbital calculations for $Fe(CO)_4$ with a $^3B_2$ ground state. The observation of a distorted structure for $Fe(CO)_4$ in the gas phase proved that the distortion of matrix-isolated $Fe(CO)_4$ was not an artifact introduced by the solid state.

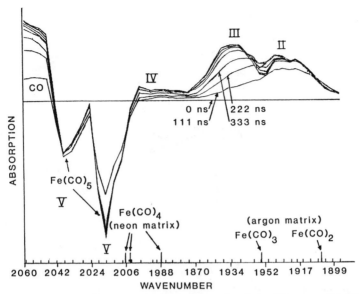

FIG. 7. Portion of the IR spectra obtained after uv photolysis of 30 mtorr of $Fe(CO)_5$ in 5 torr of Ar buffer gas, using a KrF (249 nm) excimer laser pulse. Adjacent traces were taken at intervals of 111 nseconds. Features are assigned as follows: V, $Fe(CO)_5$; IV, $Fe(CO)_4$; III, $Fe(CO)_3$; and II, $Fe(CO)_2$. Also shown at the high frequency end is an absorption due to free CO. [Reproduced with permission from Ouderkirk *et al.* (*75*), but note that the peaks have been renumbered.]

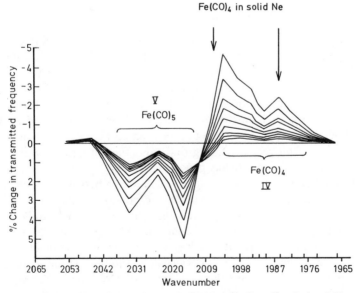

FIG. 8. Portion of the infrared spectrum shown following photolysis of 30 mtorr of $Fe(CO)_5$ in 100 torr of CO with a KrF excimer laser pulse. Adjacent traces are taken at 3-μsecond intervals. The first trace is the one with the largest excursion from the abscissa. Nine traces are shown with the first being 3 μseconds after the excimer laser pulse photolyzes the sample. Bands are labelled V, $Fe(CO)_5$; IV, $Fe(CO)_4$. [Reproduced from Ouderkirk *et al.* (*68*) with permission; copyright 1983, American Chemical Society.]

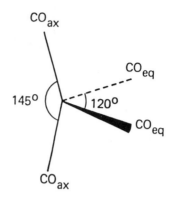

SCHEME 1.

Since $Fe(CO)_4$ has a triplet ground state its reaction with CO to form $Fe(CO)_5$ should be spin forbidden [Eq. (9)].

$$Fe(CO)_4(^3B_2) + CO \rightarrow Fe(CO)_5(^1A_1') \qquad (9)$$

The forbidden nature of this reaction is supported by the rate constant for this reaction ($3.5 \pm 0.9 \times 10^7$ dm^3 mol^{-1} second^{-1}), as measured from the IR kinetics experiment. The rate was more than two orders of magnitude less than the rate of the spin allowed reaction (85) [Eq. (10)].

$$Fe(CO)_3(^3A_1) + CO \rightarrow Fe(CO)_4(^3B_2) \qquad (10)$$

The relative yields of the photoproducts $Fe(CO)_4$, $Fe(CO)_3$, and $Fe(CO)_2$ were strongly wavelength dependent, and the relative amount of $Fe(CO)_2$ increased as shorter wavelength uv light was used. The formation of all of these photoproducts appeared to involve absorption of only a single uv photon by a molecule of $Fe(CO)_5$ (46,68,75). This is quite different from photochemistry in a matrix. When $Fe(CO)_3$ and $Fe(CO)_2$ are produced photolytically from matrix-isolated $Fe(CO)_5$, one uv photon is required to remove *each* CO group, and the formation of the lower carbonyl species is a multistep process (87,88).

There may be several reasons for the difference between gas phase and matrix photochemistry, and we outline one possible explanation. Even at 355 nm (XeF laser), a uv photon has more energy (equivalent to 335 kJ mol^{-1}) than is needed to break one M—CO bond (89,90). In a matrix, the isolated $Fe(CO)_5$ molecule is in intimate contact with the matrix material, and any excess energy can be rapidly lost to the matrix. In the gas phase, collisions are the principal pathway for loss of this excess energy. Under the conditions used in the gas phase photolysis, the mean time between collisions was relatively long and the excess energy could not be lost before the loss of further CO groups occurred (75). Another interesting feature of these gas phase reactions was that both the $Fe(CO)_x$ and CO were produced in "hot" rotational–vibrational states (see Section III,A,2).

The loss of more than one CO group from $Fe(CO)_5$ (75) is consistent with the trapping results obtained by Yardley and co-workers (46). However, the IR laser kinetic measurements (75) differ from the trapping experiments in that they provided *direct* evidence for the formation of $Fe(CO)_x$ ($x = 4, 3$, and 2). Although the ultimate fate of these fragments was not established in detail, polynuclear species were seen to be formed (73,75). These experiments produced the first IR spectra of completely uncoordinated metal carbonyl fragments (68,73,75), which were in gratifying agreement with the results of matrix isolation studies (15).

2. $Cr(CO)_6$

Less work has been done on the time-resolved IR spectroscopy of $Cr(CO)_6$ in the gas phase (69,91,92). Broadly, the results were similar to those with $Fe(CO)_5$. Ultraviolet photolysis (249 nm) provides evidence for several CO loss products (92) tentatively assigned as $Cr(CO)_5$ and $Cr(CO)_4$.

The most complete spectroscopic study has been performed using 355 nm light (XeF laser) (91). At this wavelength, $Cr(CO)_5$ was found to be the predominant photoproduct, in broad agreement with the results of Breckenridge (44) (see Section I,B,2). The IR absorptions of gas phase $Cr(CO)_5$ (91) are close in frequency to those of $Cr(CO)_5$ isolated in a Ne matrix (16). These gas phase bands are assigned to the e (1980 cm^{-1}) and a_1 (1948 cm^{-1}) modes of square pyramidal $Cr(CO)_5$. Thus, in the gas phase, away from any solid state perturbations, "naked" $Cr(CO)_5$ appears to have the same C_{4v} geometry as already found in the matrix (93), in agreement with molecular orbital predictions (17,18). The controversy surrounding possible D_{3h} ground-state geometries of $Cr(CO)_5$ now appears to be over (94).

There have been two independent determinations of the rate constant for the recombination of $Cr(CO)_5$ with CO [Eq. (11)].

$$Cr(CO)_5 + CO \xrightarrow{k} Cr(CO)_6 \qquad (11)$$

The two values, $2.3 \pm 0.4 \times 10^{10}$ dm^3 mol^{-1} second^{-1} (92) and $1.5 \pm 0.3 \times 10^{10}$ dm^3 mol^{-1} second^{-1} (91), are both much larger than the rate constant for the reaction of $Fe(CO)_4$ with CO, $3.5 \pm 0.9 \times 10^7$ dm^3 mol^{-1} second^{-1} (85). This difference may reflect the spin forbidden nature of the reactions of $Fe(CO)_4$. The results fully support the proposition (95), made 10 years ago on the basis of matrix isolation experiments, that the apparent reactivity of $Fe(CO)_4$ in solution (40) is illusory. $Fe(CO)_4$ appears more reactive than $Cr(CO)_5$ in solution because $Fe(CO)_4$ reacts with the solvent much more slowly than $Cr(CO)_5$ does.

The photolysis of $Cr(CO)_6$ also provides evidence for the formation of both CO (69) and $Cr(CO)_x$ species (91,92) in vibrationally excited states. Since CO lasers operate on vibrational transitions of CO, they are particularly sensitive method for detecting vibrationally excited CO. It is still not clear in detail how these vibrationally excited molecules are formed during uv photolysis. For $Cr(CO)_6$ (69,92), more CO appeared to be formed in the ground state than in the first vibrational excited state, and excited CO continued to be formed after the end of the uv laser pulse. Similarly, $Fe(CO)_x$ and $Cr(CO)_x$ fragments were initially generated with IR absorptions that were shifted to long wavelength (75,91). This shift was apparently due to rotationally–vibrationally excited molecules which relaxed at a rate dependent on the pressure of added buffer gas.

These IR kinetic experiments (75) were the first examples of vibrationally excited metal carbonyls to be observed. More detailed studies on the behavior of "hot" carbonyls should provide an intriguing insight into the photophysics of these molecules. We now look at metal carbonyl photochemistry in solution.

B. Photochemistry of $Cr(CO)_6$ in Solution

There were substantial problems with impurities and degassing in the initial time-resolved IR measurements on $Cr(CO)_6$ in solution. The spectroscopic results were very encouraging, but an unidentified $Cr(CO)_5X$ species (X = impurity) was observed (60). The system has now been examined in detail by the Mülheim group (96), who overcame the earlier experimental problems.

1. $Cr(CO)_5(solvent)$

It had already been established by uv–vis flash photolysis (35) that $Cr(CO)_5(solvent)$ was the first observable intermediate in the photolysis of $Cr(CO)_6$. Figure 9 shows the IR spectrum (96) of the photoproduct $Cr(CO)_5(C_6H_{12})$ in cyclohexane solution. The spectra were obtained using $Cr(CO)_5(^{13}CO)$ (96). The extra spectroscopic information provided by the ^{13}CO group was sufficient to show that the spectrum was consistent

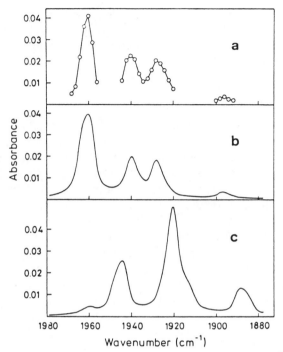

FIG. 9. (a) Transient IR spectrum of the photoproducts of a $6 \times 10^{-4} \, M$ solution of $Cr(CO)_5(^{13}CO)$ in CO-saturated C_6H_{12} solution. The spectrum is a superposition of the IR bands of $Cr(CO)_{5-x}(^{13}CO)_x$, $x = 0, 1$. Fresh solution had to be used for each measured point since isotopic scrambling occurs after each flash. (b) Simulated IR spectrum for $Cr(CO)_{5-x}(^{13}CO)_x$, $x = 0, 1$, calculated assuming a C_{4v} structure, axial–equatorial bond angle 93°, and Lorentzian band shapes 7 cm^{-1} FWHM. (c) Simulated IR spectrum for $Cr(CO)_{5-x}(^{13}CO)_x$, $x = 0, 1$, calculated assuming a D_{3h} structure. Note the excellent agreement between the observed spectrum trace (a) and the simulated spectrum trace (b). [Reproduced with permission from Church *et al.* (*96*); copyright 1985, American Chemical Society.]

with that predicted for a C_{4v} molecule. The figure illustrates the high quality of structural information which is obtainable with time-resolved IR measurements.

The IR kinetic measurements (*96*) of the rate constants for reaction of $Cr(CO)_5(C_6H_{12})$ with CO were very similar to those measured using uv–vis flash photolysis (*30,33*). In the presence of added ligands, $Cr(CO)_5(C_6H_{12})$ decayed to give $Cr(CO)_5L$ products. For both $L = CO$ and $L = H_2O$ the activation energy was 22 ± 5 kJ mol^{-1} (*96*), but surprisingly the rate of addition of H_2O was much faster than that of CO. Similar spectra were obtained in *n*-heptane solution (*97*), but the *n*-heptane complex decayed somewhat faster than the cyclohexane one, suggesting that the Cr—(*n*-heptane) interaction was slightly weaker than that of Cr—(cyclohexane) (*97*).

2. Cr(CO)₅(H₂)

In the presence of H_2, the decay of $Cr(CO)_5(C_6H_{12})$ yielded a product (*98*) with ν_{C-O} bands at frequencies very close to those of $Cr(CO)_5(H_2)$ in liquid Xe solution (*27*). The IR kinetic traces in Fig. 10 show the decay of $Cr(CO)_5(C_6H_{12})$ and the corresponding formation of $Cr(CO)_5(H_2)$ (*98*). This compound, which in liquid xenon solution appears to contain coordinated dihydrogen (*27*), is suprisingly long-lived in hydrocarbon solution at room temperature. At 25°C, $Cr(CO)_5(H_2)$ had a pseudo-first order decay (*98*) with a significant isotope effect [k_{H_2} ~2.7 second^{-1} (1.3 atm H_2) and k_{D_2} ~0.5 second^{-1} (1.3 atm D_2)], which presumably reflects differences between the Cr—(H_2) and Cr—(D_2) bond energies (*98*). A still unexplained facet of these experiments is the widely differing rates of reaction of H_2 (19000 second^{-1}) and D_2 (10000 second^{-1}) with $Cr(CO)_5(C_6H_{12})$.

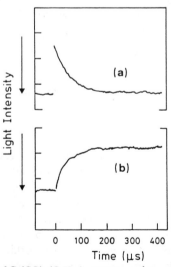

FIG. 10. (a) The decay of $Cr(CO)_5(C_6H_{12})$ at 1960 cm^{-1} together with (b) the concurrent grow-in of $Cr(CO)_5(H_2)$ at 1974 cm^{-1} in H_2-saturated cyclohexane solution. [Reproduced with permission from Church *et al.* (*98*).]

3. $Cr(CO)_5(N_2)$

The reaction of $Cr(CO)_5(C_6H_{12})$ with dinitrogen is similar to that with dihydrogen (99) [Eq. (12)].

$$Cr(CO)_5(C_6H_{12}) + N_2 \rightarrow Cr(CO)_5(N_2) + C_6H_{12} \qquad (12)$$

IR kinetic measurements on $Cr(CO)_5(N_2)$ were a particular technological triumph (99) because not only were the strong ν_{C-O} bands observed but also the very weak $\nu_{N\equiv N}$ (2240 cm^{-1}) and natural abundance $\nu_{^{13}CO}$ bands were detected. The compound $Cr(CO)_5(N_2)$ decayed at 25°C with a pseudo-first order rate constant of 1.7 second^{-1}. Thus, $Cr(CO)_5(H_2)$ and $Cr(CO)_5(N_2)$ have similar thermal stabilities, and it has been one of the great surprises of the Mülheim work (96–99) to find how long-lived "unstable" molecules can be.

4. $Cr(CO)_5(H_2O)$

After much difficulty the mysterious $Cr(CO)_5$ (impurity) species was identified as the complex $Cr(CO)_5(H_2O)$ (96). It could be formed by saturating the hydrocarbon solvents with water. Surprisingly, the water could be largely removed again by careful pumping under vacuum. The time-resolved IR spectrum of $Cr(CO)_5(H_2O)$ in cyclohexane (96), was similar to that previously reported for $Cr(CO)_5(H_2O)$ at low temperatures (100). One of the most interesting aspects of these experiments (96) was the kinetic analysis of the decay of $Cr(CO)_5(H_2O)$ that was consistent with a dissociative pathway with an activation energy of 75 $\pm$ 15 kJ mol^{-1}, which may well represent the $Cr-H_2O$ bond dissociation energy [Eq. (13)].

$$Cr(CO)_5(H_2O) \rightarrow Cr(CO)_5 + H_2O \qquad \Delta H \sim 75 \text{ kJ mol}^{-1} \qquad (13)$$

As might be expected, O_2 impurities were very effective scavengers of $Cr(CO)_5$, but more unexpectedly CO_2 had no effect. Reactions carried out under pressures of CO_2 (97) were indistinguishable from those carried out under Ar. This contrasts sharply with matrix studies (101) where CO_2 was found to photooxidize metal carbonyls.

C. Dinuclear Metal Carbonyl Species

There is still controversy over the primary steps in the photolysis of dinuclear metal carbonyls, such as $Mn_2(CO)_{10}$ or $[CpFe(CO)_2]_2$, and the field has been recently reviewed (102,103). The controversy has centered on the number of primary photoproducts, their identity, the effects of photolysis wavelength, and the possibility of heterophotolysis (21,103).

Both $Mn_2(CO)_{10}$ (77) and $[CpFe(CO)_2]_2$ (61) have now been studied by IR kinetic spectroscopy.

1. $Mn_2(CO)_{10}$

In general, uv–vis flash photolysis has shown that most dinuclear carbonyls have two primary photoproducts in solution: a radical product and a CO-loss product (37–39), both of which could be identified through their reaction kinetics (i.e., bimolecular recombination for radicals, and bimolecular or pseudo-first order CO-dependent decay for products with CO loss). In the case of $Mn_2(CO)_{10}$, there was evidence for wavelength dependence of photolysis products (104); the yield of $Mn(CO)_5$ increased relative to $Mn_2(CO)_9$ as the photolysis wavelength increased from uv to visible. Quantum yield measurements (105) suggested that there might well be a third photoreactive species, possibly a CO-bridged isomer of $Mn_2(CO)_{10}$. Matrix isolation only provided evidence for formation of $Mn_2(CO)_9$ (5), presumably because of the cage effect (see Section I,A,1). Elegant matrix experiments (106), involving photolysis and spectroscopy with plane polarized lights, suggested that $Mn_2(CO)_9$ had an asymmetric π-bonded CO bridge,

Church and co-workers (77) have obtained time-resolved IR spectra of both $Mn(CO)_5$ and $Mn_2(CO)_9$ by flash photolysis of $Mn_2(CO)_{10}$ in solution. The spectra (Fig. 11) were in close agreement with the spectra of matrix isolated $Mn(CO)_5$ (22) and $Mn_2(CO)_9$ (5,106). There was a bridging ν_{C-O} band for $Mn_2(CO)_9$ showing that it has a CO-bridged structure in solution as well as in the matrix. Structural information of this type could not have been obtained from uv–vis spectroscopy. Similarly, the IR spectra indicated that $Mn(CO)_5$ had the same C_{4v} structure in solution (77) as in the matrix (22). In solution (77), the yield of $Mn_2(CO)_9$ was approximately equal to that of $Mn(CO)_5$. Bearing in mind that one molecule of $Mn_2(CO)_{10}$ produces two molecules of $Mn(CO)_5$ [Eq. (14)], CO loss from $Mn_2(CO)_{10}$ [Eq. (15)], must be the major process at these photolysis wavelengths (37,77).

$$Mn_2(CO)_{10} \xrightarrow{uv} 2\ Mn(CO)_5 \tag{14}$$

$$Mn_2(CO)_{10} \xrightarrow{uv} Mn_2(CO)_9 + CO \tag{15}$$

$Mn_2(CO)_9$ reacted with CO at a rate well below the diffusion-controlled limit (77), and the bimolecular rate constant was solvent dependent [$k =$

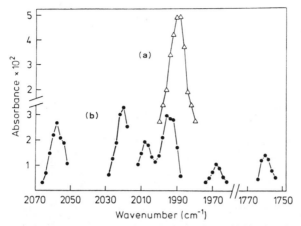

FIG. 11.   Transient IR spectra of the photoproducts of $Mn_2(CO)_{10}$ in *n*-heptane solution immediately after the flash. (a) $Mn(CO)_5$ (upper absorbance scale). (b) $Mn_2(CO)_9$ (lower absorbance scale); note the prominent bridging CO band at 1760 cm^{-1}. The spectrum of $Mn(CO)_5$ could be separated from that of $Mn_2(CO)_9$, which it overlaps, because of its faster decay rate. [Reproduced with permission from Church *et al.* (*77*).]

2.7×10^6 dm^3 mol^{-1} $second^{-1}$ (*n*-heptane) and 1.2×10^6 dm^3 mol^{-1} $second^{-1}$ (cyclohexane)]. This solvent dependence is reminiscent of the behavior of $Cr(CO)_5$ (see Section II,B) and may well indicate an interaction between the solvent and $Mn_2(CO)_9$ (*77*). By contrast, $Mn(CO)_5$ decayed at a rate close to the diffusion-controlled limit with less solvent dependence of the rate constant (*77*) ($k \sim 10^9 dm^3$ mol^{-1} $second^{-1}$). More recent IR studies (*107*) using shorter timescale measurements confirmed that $Mn_2(CO)_9$ is formed in solution within 100 nseconds of the uv flash, in agreement with the results of uv–vis flash photolysis (*37*).

2.   [CpFe(CO)₂]₂ (Cp = η⁵-C₅H₅)

Ultraviolet–visible flash photolysis (*38*) of $[CpFe(CO)_2]_2$ provided evidence for formation of both $CpFe(CO)_2$ and $Cp_2Fe_2(CO)_3$ as primary photolysis products. Matrix isolation studies (*6,7*) found $Cp_2Fe_2(CO)_3$ as the sole product. Rest and co-workers (*7*) showed, using ^{13}CO enrichment, that $Cp_2Fe_2(CO)_3$ had an unusual structure, $CpFe(\mu\text{-}CO)_3FeCp$, and it was unexpectedly stable. The C_5Me_5 analog ($\eta^5\text{-}C_5Me_5$)Fe(μ-CO)₃Fe(η^5-C_5Me_5) was stable up to room temperature (*108,109*). A surprising feature of the matrix photochemistry (*6,7*) was that only the trans isomer of $[CpFe(CO)_2]_2$ was photolyzed, and it may be that this was the consequence of concerted CO loss from the trans isomer and isomerization to form the triply bridged product (*6*).

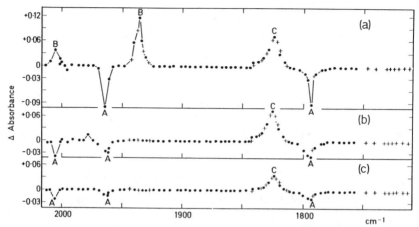

FIG. 12. Transient IR difference spectra showing changes in absorbance: (a) 5 μseconds, (b) 25 μseconds, and (c) 1.25 mseconds after the UV flash photolysis of [CpFe(CO)$_2$]$_2$ in cyclohexane solution under 1 atm pressure of CO. Bands pointing upward represent an increase in absorbance (i.e., formation of a compound) and those pointing downward a decrease [i.e., depletion of starting material, (A)]. The bands are assigned as follows: A, [CpFe(CO)$_2$]$_2$; B, CpFe(CO)$_2$; and C, CpFe(μ-CO)$_3$Fe(Cp). Points marked ● were recorded with a ^{12}CO laser and those marked + with a ^{13}CO laser. [Reproduced with permission from Moore et al. (61).]

Time-resolved IR measurements by Moore, Simpson, and co-workers (61) showed that both CpFe(CO)$_2$ and CpFe(μ-CO)$_3$FeCp were formed within 5 μseconds of photolysis of [CpFe(CO)$_2$]$_2$ in cyclohexane solution. The spectra are shown in Fig. 12. CpFe(μ-CO)$_3$FeCp has similar IR absorption frequencies in the matrix (6,7) and in solution (61). Interestingly, CpFe(CO)$_2$ was the first unsaturated species to be identified by time-resolved IR *without* previous matrix isolation data being available. CpFe(μ-CO)$_3$FeCp reacts with CO [Eq. (16)] much more slowly (k ~4.5 × 10^4 dm^3 mol^{-1} second^{-1}) than Mn$_2$(CO)$_9$ reacts (77)

$$CpFe(\mu\text{-}CO)_3FeCp + CO \xrightarrow{k} [CpFe(CO)_2]_2 \qquad (16)$$

(k ~1.2 × 10^6 dm^3 mol^{-1} second^{-1}), and the rate constant for CpFe(μ-CO)$_3$FeCp showed less solvent dependence. The lifetime of CpFe(μ-CO)$_3$FeCp had a considerable dependence on CO pressure; the half-life in *n*-heptane was reduced from 1.4 mseconds (~1 atm CO) to 25 μseconds (~100 atm CO) (110). The relatively slow rate of reaction of CpFe(μ-CO)$_3$FeCp with CO may well be a consequence of a triplet ground state (109,111) which would render the process in Eq. (16) spin forbidden. Nevertheless, the photolysis of [CpFe(CO)$_2$]$_2$ under 1 atm CO is completely reversible even for many thousands of flashes (84).

The relative yields of $CpFe(CO)_2$ and $CpFe(\mu\text{-}CO)_3FeCp$ were wavelength dependent. At wavelengths >450 nm, little of the bridged product was formed (61). Very recent measurements on a much shorter timescale (107) have shown that formation of $CpFe(CO)_2$ occurred within 30 nseconds, of the uv laser flash but that $CpFe(\mu\text{-}CO)_3FeCp$ was formed more slowly (~1 μsecond). This delay is not predicted by the proposed mechanism (6), concerted elimination of CO and isomerization. Possibly $CpFe(\mu\text{-}CO)_3FeCp$ is formed in solution by isomerization of a photoproduct (107) with a terminally bonded CO group [Eq. (17)]. A transient IR absorption ~1951 cm^{-1} (107) might be due to this first photoproduct.

$$CpFe(CO)(\mu\text{-}CO)_2Fe(CO)Cp \xrightarrow{uv}$$

$$CpFe(CO)(\mu\text{-}CO)_2FeCp \xrightarrow[t_{1/2} \sim 500 \text{ nseconds}]{\text{isomerize}} CpFe(\mu\text{-}CO)_3FeCp \qquad (17)$$

Unfortunately, it has not yet been possible to decide which photoproducts are formed from which isomer (cis or trans) of the parent $[CpFe(CO)_2]_2$.

Time-resolved IR measurements appear to be a general method for studying the photochemistry of $[CpM(CO)_x]_2$ compounds in solution. Thus, photolysis of $[CpMo(CO)_3]_2$ in cyclohexane solution produced two products (110) of which one is $CpMo(CO)_3$, with IR absorptions close to those reported for matrix-isolated $CpMo(CO)_3$ (112).

A particularly interesting question which remains unanswered is whether dinuclear photoproducts are produced directly from the photoexcited parent molecule or whether they are formed by reaction of free radicals within the solvent cage. In principle this question can be answered by making time-resolved IR measurements on the molecules in the gas phase, where no solvent cage can interfere. Thus, it may transpire that a full understanding of the photolysis of these dinuclear compounds will require complementary experiments in solution and in the gas phase.

IV

CONCLUDING REMARKS

Already a considerable number of transient organometallic species have been characterized by IR kinetic spectroscopy (see Table I). Like most other "sporting" techniques for structure determination, IR kinetic spectroscopy will not always provide a complete solution to every problem. What it can do is to provide more *structural* information, about metal carbonyl species at least, than conventional uv–visible flash photolysis. This structural information is obtained without loss of kinetic data, which can even be more precise than data from the corresponding uv–visible

TABLE I

METAL CARBONYL INTERMEDIATES DETECTED BY IR
KINETIC SPECTROSCOPY

Species	Precursor(s)[a]	Ref.
Gas Phase		
$Fe(CO)_4$	$Fe(CO)_5$	
$Fe(CO)_3$	$Fe(CO)_5$	*68,73,75*
$Fe(CO)_2$	$Fe(CO)_5$	
$Cr(CO)_5$	$Cr(CO)_6$	*91,92*
$Cr(CO)_4$	$Cr(CO)_6$	*92*
$Mn(CO)_5$	$Mn_2(CO)_{10}$	*107*
Solution		
$Cr(CO)_5(C_6H_{12})$	$Cr(CO)_6/C_6H_{12}$	*60,96*
$Cr(CO)_5(n\text{-}C_7H_{16})$	$Cr(CO)_6/(n\text{-}C_7H_{16})$	*97*
$Cr(CO)_5(H_2)$	$Cr(CO)_6/H_2$	*98*
$Cr(CO)_5(N_2)$	$Cr(CO)_6/N_2$	*99*
$Cr(CO)_5(H_2O)$	$Cr(CO)_6/H_2O$	*96*
$Mn(CO)_5$	$Mn_2(CO)_{10}$	*77*
$CpFe(CO)_2$[b]	$[CpFe(CO)_2]_2$[b]	*61*
$CpMo(CO)_3$[b]	$[CpMo(CO)_3]_2$[b]	*110*
$Mn_2(CO)_9$	$Mn_2(CO)_{10}$	*77*
$CpFe(\mu\text{-}CO)_3FeCp$[b]	$[CpFe(CO)_2]_2$[b]	*61*

[a] More than one fragment may be produced simultaneously
from a particular precursor.
[b] $Cp = (\eta^5\text{-}C_5H_5)$.

experiments. Time-resolved IR measurements become particularly powerful when combined with matrix isolation and studies in low temperature
solvents. This provides a "triad" of IR techniques with which to probe
organometallic intermediates and excited states (Scheme 2).

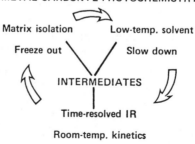

METAL CARBONYL PHOTOCHEMISTRY

Matrix isolation   Low-temp. solvent

Freeze out   Slow down

INTERMEDIATES

Time-resolved IR

Room-temp. kinetics

SCHEME 2.

The development of IR kinetic spectroscopy has been challenging. Organometallic chemists have had to learn about lasers and electronics, while chemical physicists have learned about organometallic chemistry. However the final apparatus has turned out to be relatively uncomplicated and not difficult to use. We therefore anticipate that such equipment will become more widely available in the near future.

ACKNOWLEDGMENTS

We would like to thank all those who have given us data in advance of publication. We are particularly grateful to our colleagues and co-workers at Nottingham and Northwestern Universities for their help, advice, and criticism. We wish to thank NATO for a collaborative grant (No. 591/83) which enabled us to write this review. We gratefully acknowledge the generous support of the following organizations for our time-resolved work: *Nottingham*: the Science & Engineering Research Council, the Paul Instrument Fund of the Royal Society, Nicolet Instruments Ltd., J. K. Lasers Ltd., and Perkin-Elmer Ltd.; *Northwestern*: the National Science Foundation (CHE 82-06979), the Air Force Office of Scientific Research (83-0372), and the Donors of the Petroleum Research Fund administered by the American Chemical Society.

REFERENCES

1. R. Hoffmann, *Angew. Chem. Intl. Ed.* **21,** 711 (1982).
2. J. J. Turner and M. Poliakoff *in* "Inorganic Chemistry: Toward the 21st Century" (M. H. Chisholm, ed.), ACS Symposium Series, No. 211, p. 35 (1983).
3. J. K. Burdett, M. Poliakoff, J. J. Turner, and H. Dubost, *in* "Advances in Infrared and Raman Spectroscopy" (R. J. H. Clark and R. E. Hester, eds.) Vol. 2, p. 1. Heyden, London, (1976).
4. J. K. Burdett. *Coord. Chem. Rev.* **27,** 1 (1978); R. B. Hitam, K. A. Mahmoud, and A. J. Rest, *Coord. Chem. Rev.* **55,** 1 (1984).
5. A. F. Hepp and M. S. Wrighton, *J. Am. Chem. Soc.* **105,** 6249 (1983).
6. A. F. Hepp, J. P. Blaha, C. Lewis, and M. S. Wrighton, *Organometallics* **3,** 174 (1984).
7. R. H. Hooker, K. A. Mahmoud, and A. J. Rest, *J. Chem. Soc. Chem. Commun.* 1022 (1983).
8. J. K. Burdett, A. J. Downs, G. P. Gaskill, M. A. Graham, J. J. Turner, and R. F. Turner, *Inorg. Chem.* **17,** 523 (1978); J. A. Crayston, M. J. Almond, A. J. Downs, M. Poliakoff, and J. J. Turner, *Inorg. Chem.* **23,** 3051 (1984).
9. R. L. Sweany and M. C. R. Symons, *Organometallics* **1,** 834 (1982).
10. T. J. Barton, R. Grinter, A. J. Thomson, B. Davies, and M. Poliakoff, *J. Chem. Soc. Chem. Commun.* 841 (1977); P. A. Cox, P. Grebenik, R. N. Perutz, M. D. Robinson, R. G. Grinter, and D. R. Stern, *Inorg. Chem.* **22,** 3614 (1983).
11. J. Chetwynd-Talbot, P. Grebenik, R. N. Perutz, and M. A. Powell, *Inorg. Chem.* **22,** 1675 (1983).
12. C. H. F. Peden, S. F. Parker, P. H. Barrett, and R. G. Pearson, *J. Phys. Chem.* **87,** 2329 (1983).
13. For a general review, see "Cryochemistry" (M. Moskovits and G. A. Ozin, eds.). Wiley-(Interscience), New York, (1976).

14. R. N. Perutz and J. J. Turner, *J. Am. Chem. Soc.* **97**, 4800 (1975).
15. M. Poliakoff, *Chem. Soc. Rev.* **7**, 527 (1978).
16. R. N. Perutz and J. J. Turner, *J. Am. Chem. Soc.* **97**, 4791 (1975).
17. J. K. Burdett, *J. Chem. Soc. Faraday Trans. 2*, **70**, 1599 (1974).
18. M. Elian and R. Hoffmann, *Inorg. Chem.* **14**, 1058 (1975).
19. A. Veillard, *Nouv. J. Chim.* **5**, 599 (1981); C. Daniel, M. Bénard, A. Dedieu, R. Wiest, and A. Viellard, *J. Phys. Chem.* **88**, 4825 (1984).
20. P. S. Braterman, "Metal Carbonyl Spectra." Academic Press, London, (1975).
21. S. C. Fletcher, M. Poliakoff, and J. J. Turner, *J. Organomet. Chem.* **268**, 259 (1984).
22. S. P. Church, M. Poliakoff, J. A. Timney, and J. J. Turner, *J. Am. Chem. Soc.* **103**, 7515 (1981).
23. P. A. Breeze, J. K. Burdett, and J. J. Turner, *Inorg. Chem.* **20**, 3369 (1981).
24. A. J. Rest, *J. Organomet. Chem.* **40**, C76 (1972).
25. R. N. Perutz, unpublished results; R. L. Sweany, *J. Am. Chem. Soc.* **107**, 2374. (1985).
26. See, for example, J. J. Turner, M. B. Simpson, M. Poliakoff, and W. B. Maier, II, *J. Am. Chem. Soc.* **105**, 3898 (1983).
27. R. K. Upmacis, G. E. Gadd, M. Poliakoff, M. B. Simpson, J. J. Turner, R. Whyman, and A. F. Simpson, *J. Chem. Soc. Chem. Commun.* 27 (1985).
28. M. B. Simpson, M. Poliakoff, J. J. Turner, W. B. Maier, II, and J. G. McLaughlin, *J. Chem. Soc. Chem. Commun.* 1355 (1983).
29. R. J. Kazlauskas and M. S. Wrighton, *J. Am. Chem. Soc.* **104**, 5784 (1982).
30. J. M. Kelly, D. V. Bent, H. Hermann, D. Schulte-Frohlinde, and E. Koerner von Gustorf, *J. Organomet. Chem.* **69**, 259 (1974).
31. J. Nasielski, unpublished results.
32. M. Wyart-Remy, thesis, Université Libre de Bruxelles (1976).
33. R. Bonneau and J. M. Kelly, *J. Am. Chem. Soc.* **102**, 1220 (1980); J. M. Kelly, C. Long, and R. Bonneau, *J. Phys. Chem.* **87**, 3344 (1983).
34. A. H. Janowicz and R. G. Bergman, *J. Am. Chem. Soc.* **104**, 352 (1982); A. J. Rest, I. Whitewell, W. A. G. Graham, J. K. Hoyano, and A. D. McMaster, *J. Chem. Soc. Chem. Commun.* 624 (1984).
35. J. A. Welch, K. S. Peters, and V. Vaida, *J. Phys. Chem.* **86**, 1941 (1982).
36. R. N. Perutz and J. C. Scaiano, *J. Chem. Soc. Chem. Commun.* 457 (1984).
37. L. J. Rothberg, N. J. Cooper, K. S. Peters, and V. Vaida, *J. Am. Chem. Soc.* **104**, 3536 (1982); H. Yesaka, T. Kobayashi, K. Yasafuku, and S. Nagakura, *J. Am. Chem. Soc.* **105**, 6249 (1983).
38. J. V. Caspar and T. J. Meyer, *J. Am. Chem. Soc.* **102**, 7794 (1980).
39. C. R. Bock, J. L. Hughey, IV, and T. J. Meyer, *J. Am. Chem. Soc.* **97**, 4440 (1975).
40. N. Harritt, J. M. Kelly, and E. Koerner von Gustorf, unpublished results.
41. A. B. Callear and R. J. Oldman, *Nature (London)* **210**, 730 (1966).
42. A. B. Callear and R. J. Oldman, *Trans. Faraday Soc.* **63**, 2888 (1967).
43. D. W. Trainor and A. S. Mani, *Appl. Phys. Lett.* **33**, 81 (1978); D. W. Trainor and A. S. Mani, *J. Chem. Phys.* **68**, 5481 (1978).
44. W. H. Breckenridge and N. Sinai, *J. Phys. Chem.* **85**, 3557 (1981).
45. W. Tumas, B. Gitlin, A. M. Rosan, and J. T. Yardley, *J. Am. Chem. Soc.* **104**, 55 (1982).
46. G. Nathanson, B. Gitlin, A. M. Rosan, and J. T. Yardley, *J. Chem. Phys.* **74**, 361 (1981).
47. M. Poliakoff, *Inorg. Chem.* **15**, 2892 (1976).
48. M. Poliakoff and J. J. Turner, *J. Chem. Soc. A* 2939 (1971).
49. J. G. Bullitt, F. A. Cotton, and T. J. Marks, *J. Am. Chem. Soc.* **92**, 2155 (1970).

50. J. K. Burdett, and J. M. Grzybowski, M. Poliakoff, and J. J. Turner, *J. Am. Chem. Soc.* **98,** 5728 (1976).
51. For an excellent summary of Inorganic Photochemistry, see the series of papers in *J. Chem. Educ.* **60,** 784–881 (1983).
52. A. W. Adamson, *J. Chem. Educ.* **60,** 797 (1983).
53. E.g., G. Boxhoorn, A. Oskam, E. P. Gibson, R. Narayanaswamy, and A. J. Rest, *Inorg. Chem.* **20,** 783 (1981).
54. A. J. Lees and A. W. Adamson, *J. Am. Chem. Soc.* **104,** 3840 (1982); D. M. Manuta and A. J. Lees, *Inorg. Chem.* **22,** 572 (1983), and references therein.
55. R. E. Hester, *Spex Speaker* **27** 1 (1982).
56. P. G. Bradley, N. Kress, B. A. Homberger, R. F. Dallinger, and W. H. Woodruff, *J. Am. Chem. Soc.* **103,** 7441 (1981); M. Forster and R. E. Hester, *Chem. Phys. Lett.* **81,** 42 (1981); *Chem. Phys. Lett.* **85,** 287 (1982).
57. W. K. Smothers and M. S. Wrighton, *J. Am. Chem. Soc.* **105,** 1067 (1983).
58. L. Brus, personal communication.
59. J. W. Moore and R. G. Pearson, "Kinetics and Mechanism." Wiley, New York, 1981.
60. H. Hermann, F. -W. Grevels, A. Henne, and K. Schaffner, *J. Phys. Chem.* **86,** 5151 (1982).
61. B. D. Moore, M. B. Simpson, M. Poliakoff, and J. J. Turner, *J. Chem. Soc. Chem. Commun.* 972 (1984).
62. R. G. W. Norrish and G. Porter, *Nature (London)* **164,** 658 (1949).
63. K. C. Herr and G. C. Pimentel, *J. Chim. Phys.* **61,** 1509 (1964); K. C. Herr and G. C. Pimentel, *Appl. Opt.* **4,** 25 (1965); A. S. Lefohn and G. C. Pimentel, *J. Chem. Phys.* **55,** 1213 (1971); L. Y. Tan, A. M. Winer, and G. C. Pimentel, *J. Chem. Phys.* **57,** 4028 (1972).
64. G. S. Denisov, O. D. Dmitrievskii, K. G. Tochadze, and Yu. V. Ulashkevich, *Pribori Tech. Eksp.* **5,** 197 (1972).
65. A. W. Adamson and M. C. Cimolino, *J. Phys. Chem.* **88,** 488 (1984).
66. D. S. Bethune, J. R. Lankard, and P. P. Sorokin, *Opt. Lett.* **4,** 103 (1979); Ph. Avouris, D. S. Bethune, J. R. Lankard, J. A. Ors, and P. P. Sorokin, *J. Chem. Phys.* **74,** 2304 (1981).
67. S. E. Brady, J. P. Maher, J. Bromfield, K. Stewart, and M. Ford, *J. Phys. E: Sci. Instrum.* **9,** 19 (1976); P. K. Baker, N. G. Connelly, B. M. R. Jones, J. P. Maher, and K. R. Somers, *J. Chem. Soc. Dalton Trans.* 579 (1980).
68. A. J. Ouderkirk, P. Wermer, N. L. Schultz, and E. Weitz, *J. Am. Chem. Soc.* **105,** 3354 (1983).
69. T. R. Fletcher and R. N. Rosenfeld, *J. Am. Chem. Soc.* **105,** 6358 (1983).
70. M. L. Lin, in "Chemiluminescence and Bioluminescence" (D. Hercules, J. Lee, and M. L. Cormier, eds.), p. 61. Plenum, New York, (1973); D. S. Y. Hsu and M. C. Lin, *Chem. Phys. Lett.* **56,** 79 (1978).
71. M. L. Bhaumik, in "High Power Gas Lasers" (E. R. Pake, ed.), p. 243. Institute of Physics, London, (1976).
72. S. R. Leone, *Acct. Chem. Res.* **16,** 88 (1983).
73. A. J. Ouderkirk, Ph.D. thesis, Northwestern University (1983).
74. R. A. Smith, F. E. Jones, and R. P. Chasmar, "The Detection and Measurement of Infrared Radiation." Clarendon, Oxford (1957).
75. A. J. Ouderkirk, T. A. Seder, and E. Weitz, *Proc. SPIE* **458,** 148 (1984).
76. B. I. Sonobe, T. R. Fletcher, and R. N. Rosenfeld, *J. Am. Chem. Soc.* **106,** 4352 (1984).
77. S. P. Church, H. Hermann, F. -W. Grevels, and K. Schaffner, *J. Chem. Soc. Chem. Commun.* 785 (1984).

78. R. S. Eng, J. F. Butler, and K. J. Linden, *Opt. Eng.* **19,** 945 (1980).
79. P. K. Chakraborti, V. B. Kartha, R. K. Tulukdar, P. N. Bajaj, and A. Joshi, *Chem. Phys. Lett.* **101,** 397 (1983).
80. M. Dubs and Hs. H. Günthard, *Chem. Phys. Lett.* **64,** 105 (1979).
81. J. O. Chu, C. F. Wood, G. W. Flynn, and R. E. Weston, *J. Chem. Phys.* **80,** 1703 (1984).
82. H. Petek, D. J. Nesbitt, P. R. Ogilby, and C. B. Moore, *J. Phys. Chem.* **87,** 5367 (1983).
83. J. A. Graham, W. M. Grim, III, and W. G. Fateley, *J. Mol. Struct.* **113,** 311 (1984).
84. B. D. Moore, M. Poliakoff, M. B. Simpson, and J. J. Turner, *J. Phys. Chem.* **89,** 850 (1985).
85. A. J. Ouderkirk and E. Weitz, *J. Chem. Phys.* **79,** 1089 (1983).
86. M. Poliakoff and J. J. Turner, *J. Chem. Soc. Dalton Trans.* 1351 (1973).
87. M. Poliakoff, *J. Chem. Soc. Dalton Trans.* 210 (1974).
88. M. Poliakoff and J. J. Turner, *J. Chem. Soc. Faraday Trans. II* **70,** 93 (1974).
89. M. Bernstein, J. D. Simon, and K. S. Peters, *Chem. Phys. Lett.* **100,** 241 (1983).
90. K. E. Lewis, D. M. Golden, and G. P. Smith *J. Am. Chem. Soc.* **106,** 3905 (1984).
91. T. A. Seder, S. P. Church, A. J. Ouderkirk, and E. Weitz, *J. Am. Chem. Soc.* **107,** 1432 (1985).
92. T. R. Fletcher and R. N. Rosenfeld, *J. Am. Chem. Soc.* **107,** 2203 (1985).
93. R. N. Perutz and J. J. Turner, *Inorg. Chem.* **14,** 262 (1975).
94. E. P. Kundig and G. A. Ozin, *J. Am. Chem. Soc.* **96,** 3820 (1974); J. K. Burdett, M. A. Graham, R. N. Perutz, M. Poliakoff, A. J. Rest, J. J. Turner, and R. F. Turner, *J. Am. Chem. Soc.* **97,** 4805 (1975).
95. M. Poliakoff and J. J. Turner, *J. Chem. Soc. Dalton Trans.* 2276 (1974).
96. S. P. Church, H. Hermann, F. -W. Grevels, and K. Schaffner, *Inorg. Chem.* **24,** 418 (1985).
97. S. P. Church, personal communication.
98. S. P. Church, F. -W. Grevels, H. Hermann, and K. Schaffner, *J. Chem. Soc. Chem. Commun.* 30 (1985).
99. S. P. Church, F. -W. Grevels, H. Hermann, and K. Schaffner, *Inorg. Chem.* **23,** 3830 (1984).
100. J. M. Boylan, J. D. Black, and P. S. Braterman, *J. Chem. Soc. Dalton Trans.* 1646 (1980).
101. M. J. Almond, R. N. Perutz, and A. J. Downs, *Inorg. Chem.* **24,** 275 (1985).
102. G. L. Geoffroy and M. S. Wrighton, "Organometallic Photochemistry." Academic Press, New York, (1979).
103. A. E. Stiegman and D. R. Tyler, *Acc. Chem. Res.* **17,** 61 (1984).
104. K. Yasufuku, *J. Am. Chem. Soc.,* in press (1985).
105. A. Fox and A. Poë, *J. Am. Chem. Soc.* **102,** 2497 (1980); A. Poë, unpublished results.
106. I. R. Dunkin, P. Härter, and C. J. Shields, *J. Am. Chem. Soc.* **106,** 7248 (1984).
107. T. A. Seder, S. P. Church, B. D. Moore, and E. Weitz, unpublished results.
108. A. J. Rest, personal communication.
109. J. P. Blaha, B. E. Bursten, J. C. Dewan, R. B. Frankell, C. W. Randolph, B. A. Wilson, and M. S. Wrighton, *J. Am. Chem. Soc.* **107,** 4561 (1985).
110. B. D. Moore, M. B. Simpson, M. Poliakoff, and J. J. Turner, unpublished results.
111. J. W. Lauher, M. Elian, R. H. Summerville, and R. Hoffmann, *J. Am. Chem. Soc.* **98,** 3219 (1976).
112. K. A. Mahmoud, A. J. Rest, and H. G. Alt, *J. Organomet. Chem.* **246,** C37 (1983).

ADVANCES IN ORGANOMETALLIC CHEMISTRY, VOL. 25

Carbonyl Derivatives of Titanium, Zirconium, and Hafnium

DAVID J. SIKORA

Monsanto Industrial Chemicals Company
Monsanto Company
St. Louis, Missouri 63167

DAVID W. MACOMBER

Department of Chemistry
Kansas State University
Manhattan, Kansas 66506

MARVIN D. RAUSCH

Department of Chemistry
University of Massachusetts
Amherst, Massachusetts 01003

Copyright © 1986 by Academic Press, Inc.
All rights of reproduction in any form reserved.

I

INTRODUCTION

While the chemistry of metal carbonyl complexes has enjoyed a rather long and colorful history, being extensively studied and widely reviewed (1–3), the synthesis and reactivity of the group 4B (Ti, Zr, Hf) metal carbonyls have developed relatively slowly. Although the first well-characterized group 4B metal carbonyl complex, bis(η-cyclopentadienyl)-dicarbonyltitanium (**1**), was reported by Murray of Monsanto Co. in

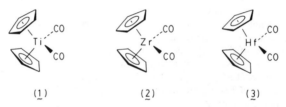

$$(\underline{1}) \qquad\qquad (\underline{2}) \qquad\qquad (\underline{3})$$

1959 ($4,5$), it has been within only the last 10 years that similar complexes of zirconium and hafnium have been synthesized (6–8). Thus, bis(η-cyclopentadienyl)dicarbonylzirconium (**2**) and bis(η-cyclopentadienyl) dicarbonylhafnium (**3**) represent the first well-characterized carbonyl complexes of these respective metals.

Since new carbonyl compounds of Ti, Zr, and Hf and their accompanying chemistry are appearing at a steady rate in the literature, it seems appropriate at this time to bring this rapidly growing area into perspective. In this review we will endeavor to categorize all reported group 4B metal carbonyls including their synthesis, formation,[1] structure, spectroscopic

---

[1] We differentiate between synthesis and formation because in many cases a substrate may be carbonylated resulting in the formation of a given carbonyl complex, however, this may not necessarily be considered as a viable synthetic route to the carbonyl complex in question. The original intent of the investigators may not have been to develop a new route to this carbonyl complex, but rather only to study the reaction of CO with the substrate.

properties, and, in most cases, their reactivity. For the sake of spatial requirements, we have restricted our discussion of reactivity to only those reactions which themselves produce group 4B metal carbonyl complexes.

II

BINARY CARBONYL COMPLEXES

In contrast to the vast number of mono- and multinuclear binary carbonyl complexes of the transition metals, no isolable binary carbonyls of titanium, zirconium, or hafnium have been reported.

Two U.S. patents issued to the Barium Steel Corporation in 1957 claim the formation of the heptacarbonyls $M(CO)_7$ (M = Ti, Zr, Hf) as intermediates for the purification of these metals (9,10). In this described refining process, the finely divided metal is treated with CO at 300–400°C and 4–8 atm. The resulting liquid heptacarbonyl compound is then thermally dissociated to the pure metal and CO. The alleged existence of these binary carbonyls seems highly unlikely without supporting evidence.

A report by Ozin et al. in 1977 describes the formation of $Ti(CO)_6$ via matrix cocondensation techniques (11). This green complex, while not isolated, was characterized by its infrared and ultraviolet–visible spectra. In a pure CO matrix, a color change from green to reddish-brown was observed on warming from 10 K to about 40–50 K. The infrared spectrum of the reddish-brown material showed no evidence for coordinated CO, thus suggesting the extreme thermal instability of $Ti(CO)_6$.

III

CARBONYL COMPLEXES NOT CONTAINING π-BONDED HYDROCARBON LIGANDS

While the majority of group 4B metal carbonyl complexes contain π-bonded hydrocarbon ligands, most notably η-cyclopentadienyl, recent studies by Wreford and co-workers have led to the identification and isolation of three novel phosphine-stabilized titanium carbonyl complexes (12,13).

The low temperature carbonylation of $(\eta\text{-}C_4H_6)_2Ti(dmpe)$ [dmpe = 1,2-bis(dimethylphosphino)ethane] resulted in the formation of $Ti(CO)_3(dmpe)_2$ (4) in 55% yield. Complex 4 along with its depe analog 5 [depe = 1,2-bis(diethylphosphino)ethane] could also be prepared via the

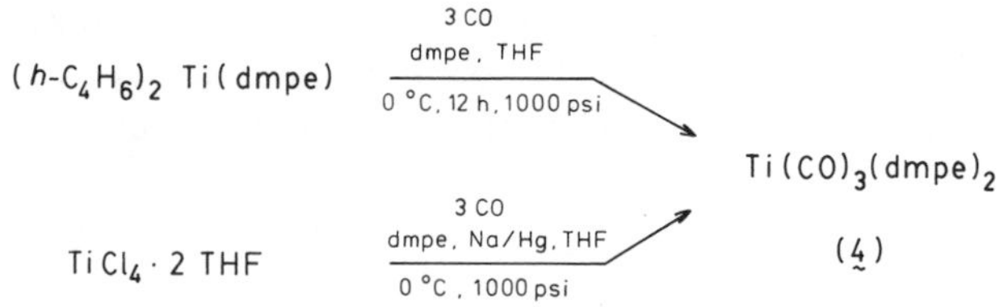

$$(h\text{-}C_4H_6)_2\,Ti(dmpe) \xrightarrow[\text{0 °C, 12 h, 1000 psi}]{\overset{\text{3 CO}}{\underset{\text{dmpe, THF}}{}}}$$

$$Ti(CO)_3(dmpe)_2$$

$$TiCl_4 \cdot 2\,THF \xrightarrow[\text{0 °C, 1000 psi}]{\overset{\text{3 CO}}{\underset{\text{dmpe, Na/Hg, THF}}{}}}$$

$$(4)$$

SCHEME 1

reductive carbonylation of $TiCl_4 \cdot 2THF$ in the presence of the appropriate bidentate phosphine. These synthetic methods are outlined in Scheme 1. The red complexes **4** and **5** exhibit metal carbonyl bands at 1810, 1720 cm^{-1} and at 1820, 1780 cm^{-1} (Nujol), respectively. X-ray crystallographic analysis of **4** revealed it to be monomeric, possessing a monocapped octahedral geometry.

Treatment of **4** with either PF_3 or ^{13}CO results in CO substitution believed to proceed via a dissociative process yielding $Ti(CO)_2(PF_3)$-$(dmpe)_2$ (**6**) and $Ti(^{13}CO)_3(dmpe)_2$. Structural characterization of **6** showed it also to be monomeric, but possessing a monocapped trigonal prismatic geometry. Complexes **4**, **5**, and **6** may be considered phosphine-substituted derivatives of the as yet unisolated $Ti(CO)_7$, thus representing the only isolable titanium carbonyl complexes where the titanium atom is in the zero oxidation state.

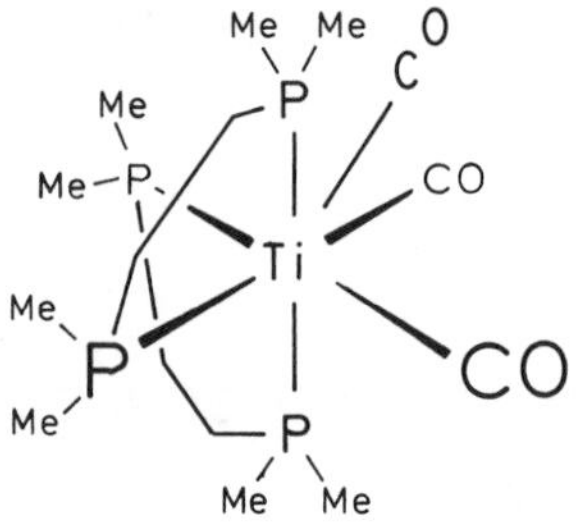

$$(4)$$

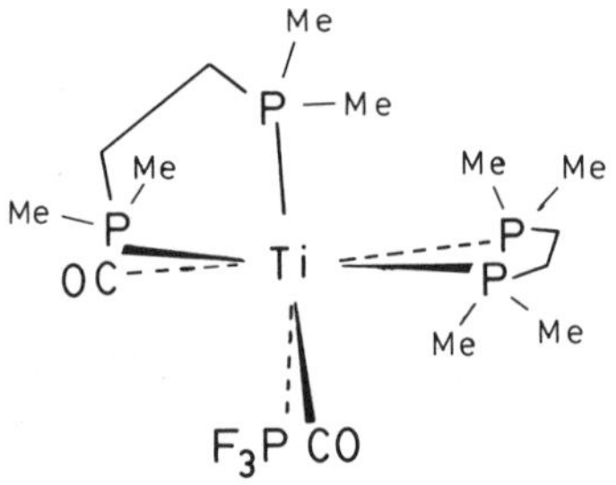

$$(6)$$

Through the use of high pressure infrared spectroscopy, Schmulbach and Ballintine were able to detect the formation of $TiCl_3(CO)(PEt_3)_2$ (**7**) on treatment of a benzene solution of $TiCl_3(PEt_3)_2$ with 800 psi of CO in a specially designed cell [$\nu(CO) = 1875$ cm^{-1}] (*14*). Under preparative conditions **7** could not be isolated free of the starting material, however; analysis of the electronic spectra (ambient pressure) of the reaction mixture was used to determine the yield of product (14%). At atmospheric pressure **7** readily decomposed at 40–42°C, releasing one equivalent of CO.

Two titanium carbonyl complexes derived from low temperature carbonylation of tetrabenzyltitanium and tetrabenzyltitanium(dicyclohexylamine) have been observed spectroscopically (IR), but not isolated (*15*). When a solution of $(PhCH_2)_4Ti$ in pentane was exposed to a stream of CO at −50° C, the original red solution changed to orange with uptake of two equivalents of CO per titanium atom and metal carbonyl bands appeared at 1952 and 1867 cm^{-1}, assignable to tetrabenzyldicarbonyltitanium (**8**). These bands then gradually decreased in intensity with concurrent formation of a yellow precipitate. This solid, which decarbonylated above −20°C, was identified as the CO insertion product, dibenzylbis-(phenylacetyl)titanium (**9**), as evidenced by its low temperature IR spectrum (KBr) which exhibited ketonic bands at 1640 and 1635 cm^{-1}. Elemental analysis of **9** together with the identification of the hydrolysis products, toluene and phenylacetaldehyde, confirmed its structural assignment. Scheme 2 illustrates the aforementioned reactions.

Similarly, when a pentane solution of tetrabenzyltitanium(dicyclohexylamine) (**10**) was treated with CO at −50°C, the red solution changed to yellow and a strong terminal metal carbonyl band at 1956 cm^{-1} could be observed, assignable to $(PhCH_2)_4Ti(CO)[NH(C_6H_{11})_2]$ (**11**). On standing for ~10 h at −50°C, a yellow solid crystallized from the pentane solution and the metal carbonyl band of **11** at 1956 cm^{-1} decreased in intensity as a new band at 1652 cm^{-1} appeared due to the CO insertion product, $(PhCH_2)_3(PhCH_2CO)Ti[(NH(C_6H_{11})_2]$ (**12**). Complex **12** was further

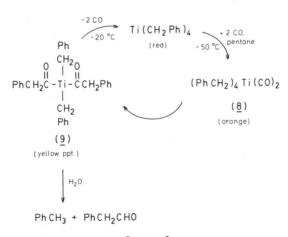

SCHEME 2

characterized by elemental analysis and the identification of its hydrolysis products, toluene, phenylacetaldehyde, and dicyclohexylamine. When a pentane solution of **12** was allowed to warm to $-10°C$, CO was evolved and complex **10** was obtained.

IV

FORMATION OF THE GROUP 4B METALLOCENE DICARBONYLS

A. Formation of Bis(η-cyclopentadienyl)dicarbonyltitanium

1. Preparation/Formation of $Cp_2Ti(CO)_2$ via the Reductive Carbonylation of $Cp_2TiCl_2/(Cp_2TiCl)_2$

Due to the commercial availability of Cp_2TiCl_2, its reduction and subsequent treatment with CO represents one of the most direct routes to $Cp_2Ti(CO)_2$ (**1**). An early example of this method was reported by Calderazzo et al. whereby Cp_2TiCl_2 was first reduced by sodium naphthalene at 25°C in THF for 24 hours (16). The resulting green titanocene, μ-$(\eta^5:\eta^5$-fulvalene)-di-μ-hydrido-bis(cyclopentadienyltitanium) (17–22), was then treated with 1 atm of CO in toluene for 4 hours at 20°C. However, this procedure proved inefficient for the production of **1** since only 10% of the required amount of CO for complete conversion was consumed.

In 1972 Van Tamelen and co-workers reported that the reduction of a toluene solution of Cp_2TiCl_2 by sodium under an argon atmosphere over a period of 6–10 days led to the formation of an "active titanocene" postulated as $(Cp_2Ti)_{1-2}$ based on its visible and infrared spectrum. Exposure of $(Cp_2Ti)_{1-2}$ in toluene to CO converted it to $Cp_2Ti(CO)_2$ (**1**) (23).

Bis(η-cyclopentadienyl)dicarbonyl has also been prepared electrochemically. When DME solutions of Cp_2TiCl_2 were electrolytically reduced in an atmosphere of CO, high yields of $Cp_2Ti(CO)_2$ (**1**) were obtained (24).

Recently, through the use of a specially designed cell, El Murr and Chaloyard have studied the electrochemical reduction of Cp_2TiX_2 (X = Cl, Br) under CO pressure (3 atm) (25). The addition of one electron to Cp_2TiCl_2 resulted in the absorption of CO, the appearance of an ESR signal ($g = 1.9755$), and the observation of a metal carbonyl band at 1950 cm^{-1} in the IR spectrum. When Cp_2TiCl_2 was reduced in an argon atmosphere an ESR signal corresponding to $(Cp_2TiCl_2)^-$ ($g = 1.9789$) was observed. The species postulated as corresponding to these data is $[Cp_2Ti(CO)Cl_2]^-$. Transfer of a second electron resulted in further uptake of CO, the absence of an ESR signal, and appearance of metal carbonyl bands assignable to $Cp_2Ti(CO)_2$ (1).

In 1975 Demerseman and co-workers reported two new preparations for $Cp_2Ti(CO)_2$ via the reductive carbonylation of Cp_2TiCl_2. The first of these involved the reaction of either Cp_2TiCl_2 or $(Cp_2TiCl)_2$ with $AlEt_3$ in a CO atmosphere. After these heptane suspensions or benzene solutions were stirred for 20 hours at 20°C, $Cp_2Ti(CO)_2$ (1) could be isolated in 30% yield (26). No speculation as to the mechanism of this reduction was discussed; however, alkylation and CO insertion steps are probably involved.

The second method, published later that year, represents probably the best laboratory preparation of $Cp_2Ti(CO)_2$ (1) to date (27). It involved the reduction of Cp_2TiCl_2 by means of aluminum filings activated with mercuric chloride. This reaction occurs via the green $(Cp_2TiCl)_2$ complex

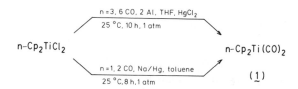

which is clearly visible throughout the early stages of the reduction. The method gives excellent yields of **1** (87%) and requires only atmospheric pressure of CO. Activated magnesium and zinc have also been employed as reducing agents.

A short time later, a similar synthesis was reported which used sodium amalgam as the reducing agent and reportedly gave a comparable yield of 81% under mild conditions (6). Also in 1976, Polish workers discovered that $Cp_2Ti(CO)_2$ (1) could be prepared by the magnesium reduction of Cp_2TiCl_2 under an argon atmosphere with subsequent treatment of the resulting black THF solution with carbon dioxide (28). No yields were

$$Cp_2TiCl_2 \ + \ Mg \ \xrightarrow[25\,°C,\,12\,h,\,1\,atm]{THF,\,argon} \ ``Cp_2Ti" \ + \ MgCl_2$$

$$``Cp_2Ti" \ + \ 2\,CO_2 \ + \ 2\,Mg \ \xrightarrow[25\,°C,\,1\,atm]{THF} \ Cp_2Ti(CO)_2 \ + \ 2\,MgO$$
$$(\underline{1})$$

given; however, this reaction was later studied by Demerseman *et al.* and found to proceed in ~25% yield (29). Further studies showed that aluminum and zinc were also effective for the reduction. When a THF solution of Cp_2TiCl_2 was reduced with aluminum, CO_2 was absorbed slowly over a 10-day period to give **1** in 80% yield. The addition of $AlCl_3$ was found to significantly enhance the rate of CO_2 uptake, but the yield of **1** remained 80%. In the absence of $AlCl_3$, with zinc as a reducing agent, the formation of **1** was similarly found to be slow, and gave **1** in only 18% yield. However, when $AlCl_3$ was added, an increase in both the rate of absorption of CO_2 and the yield of **1** was observed. Other Lewis acids were found to give similar results.

The reduction of Cp_2TiCl_2 by manganese in a CO atmosphere to give **1** has been reported independently by two research groups (29–31).

Most recently, the formation of $Cp_2Ti(CO)_2$ (**1**) by the CO-induced disproportionation of $(Cp_2TiCl)_2$ has been studied (31). Nardelli and co-workers found that exposure of a THF solution of $(Cp_2TiCl)_2$ to CO at 20°C over a 3-day period resulted in CO uptake corresponding to one-quarter the theoretical amount required for formation of **1**. A 1:1

$$(Cp_2TiCl)_2 + 2\ CO \xrightarrow[20\,°C,\,3\,days,\,1\,atm]{THF} Cp_2Ti(CO)_2 + Cp_2TiCl_2$$

(1)

stoichiometric mixture of $(Cp_2TiCl)_2$ and Cp_2TiCl_2 could be also isolated when the solvent was removed. The above observations were suggestive of adduct formation between $(Cp_2TiCl)_2$ and Cp_2TiCl_2 such that the chloride ligands of Cp_2TiCl_2 could occupy the two coordination sites of $(Cp_2TiCl)_2$ or each site of two molecules of Cp_2TiCl. The concept of adduct formation was corroborated by a molecular weight determination on a mixture of $(Cp_2TiCl)_2$ and Cp_2TiCl_2 which gave a value of 690 amu in good agreement with an actual value of 676 amu for $[(Cp_2TiCl)_2 \cdot Cp_2TiCl_2]$. Furthermore, the addition of 1 equivalent of Cp_2TiCl_2 or 2 equivalents of $Et_4N^+Cl^-$ to a solution of $(Cp_2TiCl)_2$ (1 equivalent) resulted in no CO absorption. The above data suggest that the disproportionation reaction proceeds by initial coordination of CO to $(Cp_2TiCl)_2$, disproportionation of this carbonyl species to **1** and Cp_2TiCl_2, and finally adduct formation between the product, Cp_2TiCl_2, and the starting material $(Cp_2TiCl)_2$.

2. Preparation/Formation of $Cp_2Ti(CO)_2$ via Titanocene Alkyl and Aryl Complexes

Bis(η-cyclopentadienyl)dicarbonyltitanium (**1**) was first prepared by Murray in 1958 via the high pressure carbonylation of $Cp_2Ti(\eta^1\text{-}C_5H_5)_2$ (32a) and $Cp_2Ti(n\text{-}C_4H_9)_2$ (32b), respectively (4,5). The former dialkyl complex was prepared via the addition of Cp_2TiCl_2 to a benzene solution containing slightly more than two equivalents of sodium cyclopentadienide or by the addition of $TiCl_4$ to a benzene solution containing slightly more than four equivalents of sodium cyclopentadienide. In either case, the mixture was allowed to react for 1.5 hours and then charged to an autoclave where it was heated to 100°C under 110 atm of CO for 8 hours. Workup resulted in an 18% yield of $Cp_2Ti(CO)_2$ (**1**).

$$Cp_2Ti(\eta^1\text{-}C_5H_5)_2 \ + \ 2\ CO \quad \xrightarrow[\text{100 °C, 8 h, 110 atm}]{C_6H_6} \quad Cp_2Ti\ (CO)_2$$
$$(\underline{1})$$

$$\text{``}Cp_2Ti(n\text{-}C_4H_9)_2\text{''} + \ 2\ CO \quad \xrightarrow[\text{150 °C, 8 h, 240 atm}]{Et_2O} \quad (\underline{1})$$

$$Cp_2Ti(\eta^1\text{-}C_5H_5) \ + \ 2\ CO \quad \xrightarrow[\text{80 °C, 150 atm}]{} \quad (\underline{1})$$

Bis(η-cyclopentadienyl)di-n-butyltitanium (*32b*), prepared in ethyl ether, was similarly charged to an autoclave, but was heated to higher temperature under higher CO pressure also for 8 hours. While the product displayed metal carbonyl bands for $Cp_2Ti(CO)_2$ in its infrared spectrum and evolved CO on treatment with iodine, no yield was given. Two years after Murray's initial preparation of $Cp_2Ti(CO)_2$ (**1**), Fischer and Löchner reported that the high pressure carbonylation of $Cp_2Ti(\eta^1\text{-}C_5H_5)$ (*32a*) also resulted in the formation of $Cp_2Ti(CO)_2$ (*33*).

In 1968, while investigating the efficacy of bis(η-cyclopentadienyl)-diphenyltitanium as a catalyst for the isomerization of 1,5-cyclooctadiene to 1,3-cyclooctadiene, Hagihara and co-workers found that the presence of CO inhibited the isomerization. Under the reaction conditions imposed, $Cp_2Ti(CO)_2$ (**1**) and benzophenone were isolated. Furthermore, **1** showed poor catalytic activity for this isomerization (*34*).

In the early 1970s Brintzinger and co-workers reported that various titanocene and titanocene hydride species (derived from Cp_2TiMe_2 and H_2) when treated with CO, led to the formation of $Cp_2Ti(CO)_2$ (**1**) in high yield. The reaction of solid Cp_2TiMe_2 with H_2 led to the formation of the violet dimeric titanocene complex **13** in 70% yield (*35*). Treatment of the solid violet hydride **13** with 1 atm of CO led to the evolution of hydrogen and subsequent formation of $Cp_2Ti(CO)_2$ (**1**) in 68% yield (*24*).

If **13** was allowed to stand at room temperature in an inert atmosphere over a prolonged period, or if a saturated hexane solution of Cp_2TiMe_2 was treated with excess hydrogen at low temperature with efficient stirring, a gray–green polymeric titanocene hydride **14** formed which exhibited reactivity similar to the violet hydride **13** (*24*). Thus, exposure of a toluene

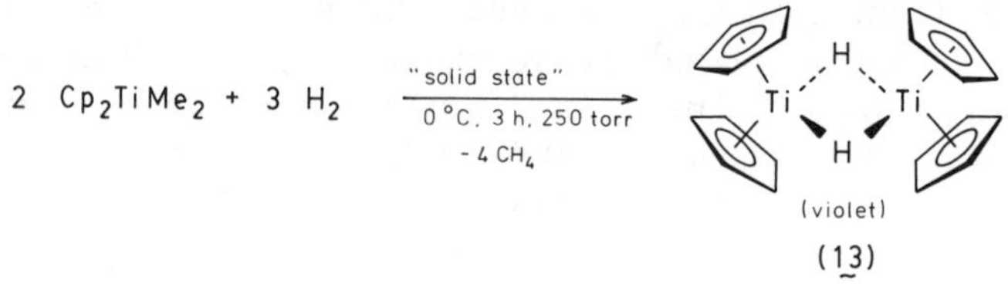

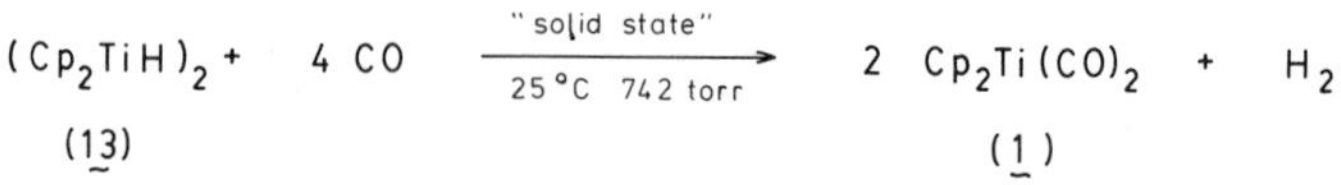

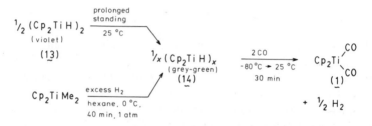

$$2 \ Cp_2TiMe_2 + 3 \ H_2 \quad \xrightarrow[\substack{0\,°C,\ 3\ h,\ 250\ torr \\ -4\ CH_4}]{\text{"solid state"}}$$

(violet)

(13)

$$(Cp_2TiH)_2 + 4 \ CO \quad \xrightarrow[25\,°C \quad 742\ torr]{\text{"solid state"}} \quad 2 \ Cp_2Ti(CO)_2 + H_2$$

(13)   (1)

slurry of **14** with CO at $-80°C$ with subsequent warming led to $Cp_2Ti(CO)_2$ (**1**) in good yield (*24*).

$$\frac{1}{2} (Cp_2TiH)_2 \quad \xrightarrow[25\,°C]{\substack{\text{prolonged} \\ \text{standing}}}$$

(violet)

(13)

$$\frac{1}{x} (Cp_2TiH)_x \quad \xrightarrow[\substack{-80\,°C \ \to\ 25\,°C \\ 30\ min}]{2\ CO} \quad Cp_2Ti \Big\langle \substack{CO \\ CO}$$

(grey-green)

(14)   (1)

$$Cp_2TiMe_2 \quad \xrightarrow[\substack{hexane,\ 0\,°C, \\ 40\ min,\ 1\ atm}]{excess\ H_2}$$

$$+ \ \frac{1}{2} H_2$$

A highly reactive form of titanocene could be obtained when a suspension of the gray–green hydride **14** was stirred in ethyl ether for 2 hours at room temperature. The solid gradually disappeared concurrent with the evolution of 0.5 equivalent of H_2 per equivalent of Ti. Molecular weight measurements showed this metastable form of titanocene (**15**) to be dimeric. Treatment of a cold ethereal solution of $(Cp_2Ti)_2$ (**15**) with CO resulted in a quantitative yield of $Cp_2Ti(CO)_2$ (**1**) (*24,36*).

Most recently, it has been shown that (**15**) can also react with CO_2 to yield $Cp_2Ti(CO)_2$ (**1**) (*37*).

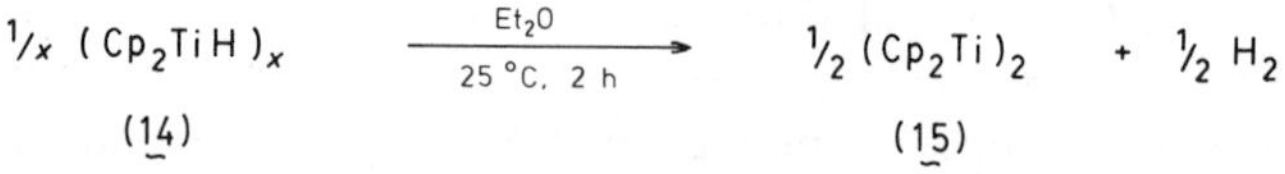

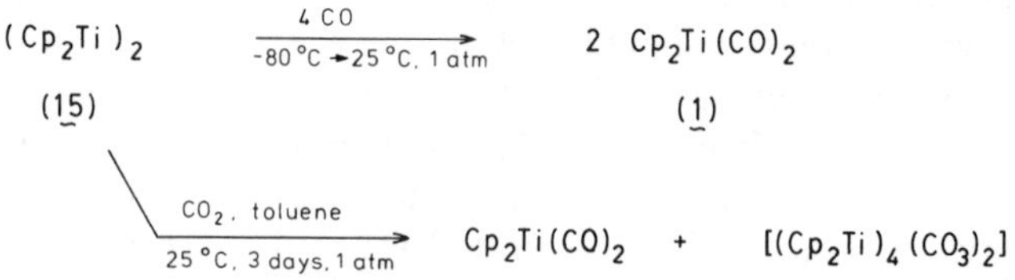

$$\frac{1}{x} (Cp_2TiH)_x \quad \xrightarrow[25\,°C,\ 2\ h]{Et_2O} \quad \frac{1}{2} (Cp_2Ti)_2 + \frac{1}{2} H_2$$

(14)   (15)

$$(Cp_2Ti)_2 \quad \xrightarrow[-80\,°C \ \to\ 25\,°C,\ 1\ atm]{4\ CO} \quad 2 \ Cp_2Ti(CO)_2$$

(15)   (1)

$$\xrightarrow[25\,°C,\ 3\ days,\ 1\ atm]{CO_2,\ toluene} \quad Cp_2Ti(CO)_2 + [(Cp_2Ti)_4(CO_3)_2]$$

Brintzinger and co-workers also found that the complex $(Cp_2TiEt)_2 \cdot 6MgCl_2 \cdot 7Et_2O$ (**16**), derived from Cp_2TiCl_2 and a six- to eightfold excess of EtMgCl, served as a source of the metastable titanocene **15** when treated with certain reagents (e.g., N_2, CO). When **16** was suspended and stirred in ethyl ether under 150 atm of CO at room temperature for 24 hours, $Cp_2Ti(CO)_2$ (**1**) could be isolated in about 70% yield. Both ethane and ethylene were identified in the autoclave vent gas. The reaction is believed to proceed via the intermediacy of **15** (*24*).

$$(Cp_2TiEt)_2 \; + \; 4\,CO \quad \xrightarrow[25°C.\,24h.\,150atm]{Et_2O} \quad 2\,Cp_2Ti\,(CO)_2 \; + \; C_2H_6 \; + \; C_2H_4$$

$$\text{(16)} \qquad\qquad\qquad\qquad\qquad \text{(1)}$$

In 1972 Floriani and Fachinetti reported that bis(η-cyclopentadienyl)-dibenzyltitanium (**17**) could be conveniently carbonylated to $Cp_2Ti(CO)_2$ (**1**) in heptane at only 1 atm of CO and at room temperature (*38*). However, the choice of solvent was crucial to the success of the reaction, since the use of benzene led predominantly to decomposition of **17** according to the reaction shown. When heptane was used as the solvent,

$$Cp_2Ti\,(CH_2Ph)_2 \; + \; CO \quad \xrightarrow[25°C,\,1\,atm]{C_6H_6} \quad \tfrac{1}{x}\,[\,Ti\,(C_5H_4)_2\,]_x \; + \; 2\;PhCH_3$$

$$\text{(17)} \qquad\qquad\qquad\qquad\qquad\qquad \text{Major}$$

$$+ \; Cp_2Ti(CO)_2 \; + \; (Ph\,CH_2)_2CO$$

$$\text{(1)}$$

$$\text{Minor}$$

the carbonylation reaction predominated, and $Cp_2Ti(CO)_2$ (**1**) could be isolated in 80% yield. Based on the amount of CO absorbed (2.6 equivalents CO per equivalent Ti) and the isolation of dibenzylketone, the reaction was postulated as proceeding via the acyl intermediate **18**.

$$Cp_2Ti\Big\langle{}^{CH_2Ph}_{CH_2Ph} \; + \; CO \quad \xrightarrow[\substack{25°C,\,3\,days.\\1\,atm}]{heptane} \quad \left[Cp_2Ti{-CO}\Big\langle{}^{CH_2Ph}_{CH_2Ph}\right] \longrightarrow$$

$$\text{(17)}$$

$$\left[Cp_2Ti\Big\langle{}^{CH_2Ph}_{\underset{O}{\overset{\shortparallel}{C}}CH_2Ph}\right] \; + \; 2\;CO \longrightarrow Cp_2Ti(CO)_2 \; +$$

$$\text{(18)} \qquad\qquad\qquad\qquad\qquad\qquad \text{(1)}$$

$$(PhCH_2)_2CO$$

Floriani and Fachinetti later observed that CO would readily insert into the titanium–alkyl bond of various titanocene alkyl halides resulting in the formation of novel bis(η-cyclopentadienyl)haloacyltitanium complexes (39). These complexes exhibited high thermal and oxidative stability and unlike intermediate **18**, did not react with CO to yield $Cp_2Ti(CO)_2$ (**1**). Floriani attributed this reactivity difference to the inherent instability of complexes in a CO atmosphere when an alkyl group and an acyl group are bonded to the same metal center. To support this proposal, bis(η-cyclopentadienyl)chloroacetyltitanium was treated with MeMgCl under 1 atm of CO. Good yields of $Cp_2Ti(CO)_2$ (**1**) and acetone were observed.

$$Cp_2Ti \underset{CH_3}{\overset{Cl}{<}} \; + \; CO \quad \xrightarrow[\substack{25°C, 2\,days, \\ 1\,atm}]{heptane} \quad Cp_2Ti \overset{Cl}{\underset{\underset{O}{\overset{\|}{C}}CH_3}{<}} \quad \xrightarrow[Et_2O]{CH_3MgCl/CO}$$

(yellow-orange crystalline solid)

$$\left[Cp_2Ti \overset{CH_3}{\underset{\underset{O}{\overset{\|}{C}}CH_3}{<}} \right] \xrightarrow{2\,CO} Cp_2Ti(CO)_2 \; + \; (CH_3)_2CO$$

$$(\underline{1})$$

Wailes *et al.* have briefly noted that Cp_2TiMe_2 (**19**) can be conveniently carbonylated to $Cp_2Ti(CO)_2$ at room temperature in an aliphatic solvent under 120 atm of CO pressure (40).

Another extremely reactive form of titanocene, namely black titanocene, was discovered by Rausch and Alt in 1974 as the product of the photolysis of Cp_2TiMe_2 (**19**) in either aliphatic or aromatic solvents (41). Irradiation of a hexane solution of **19** resulted in the deposition of a dark precipitate with concomitant evolution of essentially only methane. Benzene solutions of this photochemically generated titanocene reacted rapidly with CO to give red solutions from which $Cp_2Ti(CO)_2$ (**1**) could be isolated in 60% yield. Similarly, **1** could be prepared directly under photochemical conditions in similar yield if **19** was irradiated in a CO atmosphere (41,42).

$$Cp_2Ti(CH_3)_2 \xrightarrow[hexane]{h\nu} C_{10}H_{9-10}Ti \downarrow \; + \; 2\,CH_4$$

$$(\underline{19})$$

"black titanocene"

$$\xrightarrow[\substack{25°C,1\,atm, \, -2CH_4}]{h\nu, 2\,CO, \, pentane}$$

$$\underset{2\,h, 1\,atm}{\overset{2\,CO}{\underset{25°C}{\Big|}}}\,{C_6H_6}$$

$$Cp_2Ti(CO)_2$$

$$(\underline{1})$$

Along similar lines, Rausch *et al.* later found that the photolysis of bis(η-cyclopentadienyl)diphenyltitanium in benzene also resulted in the

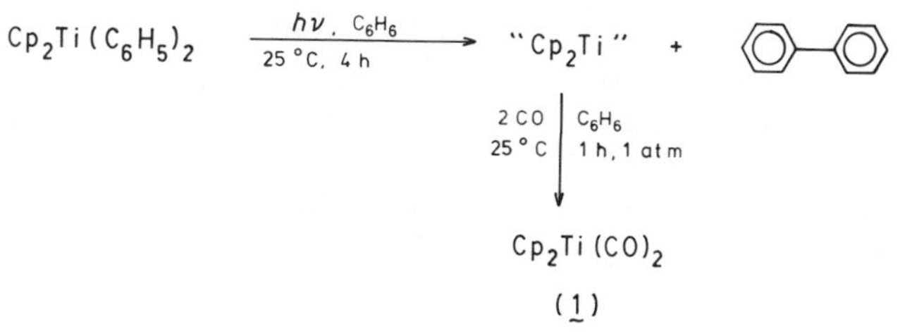

$$Cp_2Ti(C_6H_5)_2 \xrightarrow[25\,°C.\ 4\,h]{h\nu,\ C_6H_6} \text{``}Cp_2Ti\text{''} \ + \ \text{(Ph-Ph)}$$

$$\downarrow \begin{array}{c} 2\ CO \\ 25\,°C \end{array} \bigg| \begin{array}{c} C_6H_6 \\ 1\,h,\ 1\,atm \end{array}$$

$$Cp_2Ti(CO)_2$$

(1)

formation of a highly reactive black titanocene which upon exposure to CO gave $Cp_2Ti(CO)_2$ (1) in 32% yield (43). However, unlike the photolysis of Cp_2TiMe_2 where CH_4 was the predominant organic photoproduct, irradiation of Cp_2TiPh_2 in C_6D_6 revealed that significant amounts (~50%) of the reductive elimination product biphenyl-d_0 were formed.

In 1975 Demerseman and co-workers reinvestigated the carbonylation of $Cp_2Ti(CH_2Ph)_2$ (17) and studied the effect of the addition of AlEt₃ (26). In contrast to Floriani's observations, Demerseman et al. felt that the thermal decomposition of 17 in aromatic solvents proceeded via the bis(η-cyclopentadienyl) benzylphenylacetyltitanium intermediate 18, thus resulting in titanium alkoxide oligomers as opposed to the initially proposed $1/x[Ti(C_5H_4)_2]_x$ (38). By adding AlEt₃ to 17 in at least a molar ratio

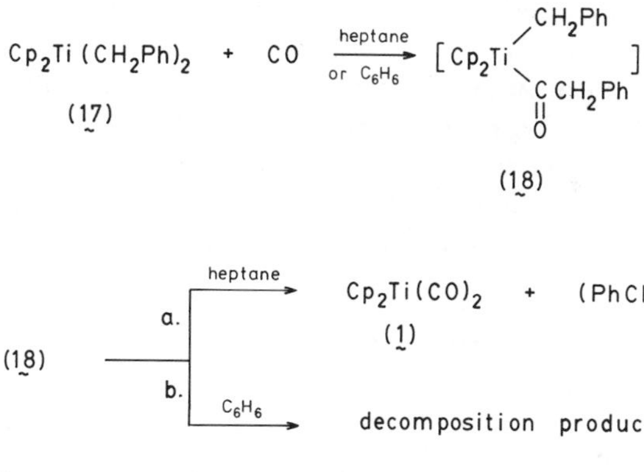

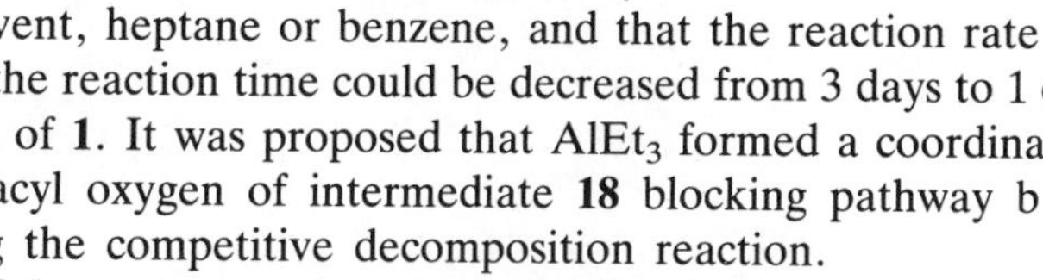

$$Cp_2Ti(CH_2Ph)_2 \ + \ CO \xrightarrow[\text{or } C_6H_6]{\text{heptane}} \left[Cp_2Ti \underset{\underset{O}{\overset{\|}{C}}CH_2Ph}{\overset{CH_2Ph}{<}} \right]$$

(17)

(18)

$$(18) \quad \begin{array}{l} \xrightarrow[a.]{\text{heptane}} \quad Cp_2Ti(CO)_2 \ + \ (PhCH_2)_2CO \\ \quad\quad\quad\quad (1) \\ \xrightarrow[b.]{\ C_6H_6\ } \quad \text{decomposition products} \end{array}$$

of 1 to 1, Demerseman found that pathway a became exclusive regardless of the solvent, heptane or benzene, and that the reaction rate increased such that the reaction time could be decreased from 3 days to 1 day for an 80% yield of 1. It was proposed that AlEt₃ formed a coordination bond with the acyl oxygen of intermediate 18 blocking pathway b and thus preventing the competitive decomposition reaction.

Some of the more recent work concerning the formation of 1 from alkyl complexes of titanocene has been reported by Teuben and co-workers. The reaction of Cp_2TiMe (20) with CO yields $Cp_2Ti(CO)_2$ (1) and acetone, along with the reduced organics isopropanol and ethanol as minor products (44). The reaction is believed to proceed via the monoacetyl complex $Cp_2Ti(COCH_3)$ (20a). Further reaction of 20a with unreacted 20 produces the products acetone and 1. The hydrogen for the organic reduction products is believed to be supplied by the cyclopentadienyl ligands, the methyl groups or the solvent.

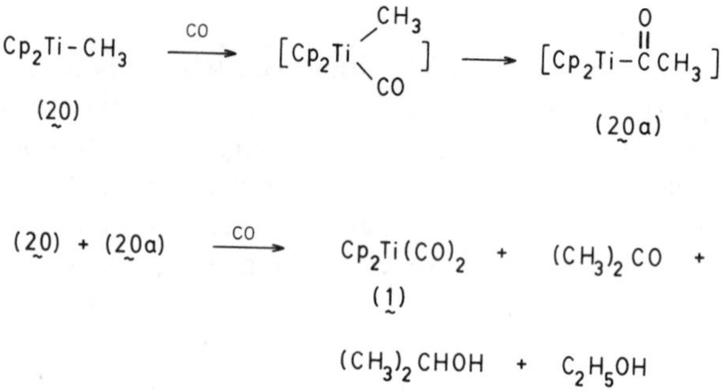

$$Cp_2Ti-CH_3 \xrightarrow{CO} [Cp_2Ti \underset{CO}{\overset{CH_3}{<}}] \longrightarrow [Cp_2Ti-\overset{O}{\overset{\|}{C}}CH_3]$$

(20)   (20a)

$$(20) + (20a) \xrightarrow{CO} Cp_2Ti(CO)_2 + (CH_3)_2CO +$$
$$(1)$$

$$(CH_3)_2CHOH + C_2H_5OH$$

Bis(η-cyclopentadienyl)neopentyltitanium reacted in a similar manner with CO to yield 1 (44). Similarly, Demerseman et al. have reported that bis(η-cyclopentadienyl)benzyltitanium reacts with CO in ethyl ether at atmospheric pressure over a 2-hour period to give 1 in 34% yield (27). In contrast, bis(η-cyclopentadienyl)phenyltitanium reacted with CO, but the formation of 1 was not observed (44).

The carbonylation of bis(η-cyclopentadienyl)-η^3-allyltitanium[2] (21) has also been reported to proceed under very mild conditions (45,46). Treatment of a pentane solution of 21 at room temperature under one atmosphere of CO for 1 hour resulted in the formation of 1 in 65% yield. While this reaction is perceived as proceeding in the same manner as the

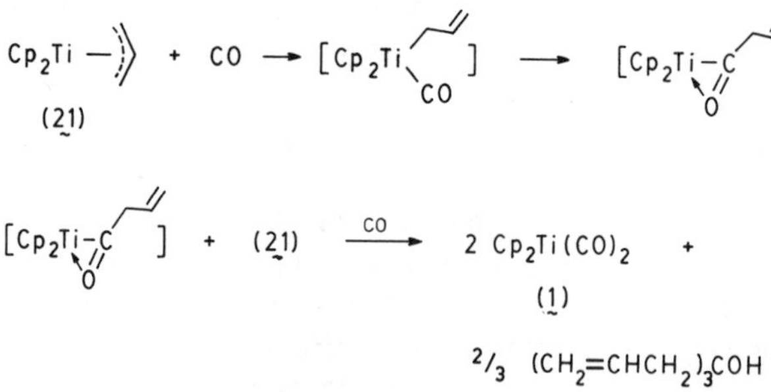

$$Cp_2Ti -\!\!\!\text{//} + CO \rightarrow [Cp_2Ti \underset{CO}{\overset{\text{//}}{<}}] \longrightarrow [Cp_2Ti-\overset{\text{//}}{\underset{O}{C}}]$$

(21)

$$[Cp_2Ti-\overset{\text{//}}{\underset{O}{C}}] + (21) \xrightarrow{CO} 2\,Cp_2Ti(CO)_2 +$$
$$(1)$$

$$\tfrac{2}{3}\,(CH_2\!\!=\!\!CHCH_2)_3COH$$

---

[2] Strictly speaking, η^3-allyl complexes are not considered as alkyl complexes; however, because certain ligands such as CO will induce the $\eta^3 \rightarrow \eta^1$ allyl rearrangement, η^3-allyls will be considered as such here for the sake of consistency.

carbonylation of Cp_2TiMe (**20**), the significant difference here is the formation of triallylmethanol as the major organic product (50%) instead of the diallylketone that might be expected based on the formation of acetone in the reaction of Cp_2TiMe with CO. A further difference can be recognized when comparing the reduced organic products since the isopropanol (the minor product from the carbonylation of Cp_2TiMe) is a secondary alcohol while triallylmethanol is a tertiary alcohol, thus suggesting alkylation of diallylketone. Surprisingly, the reduction of CO to the alcohol occurs in an aprotic solvent (i.e., without hydrolysis) such that the source of hydrogen is believed to have originated from the Cp rings.

Most recently, Lappert and co-workers have found that the carbonylation of bis(η-cyclopentadienyl)titanabenzocyclopentene resulted in the formation of $Cp_2Ti(CO)_2$ (**1**) and indan-2-one (*47a*).

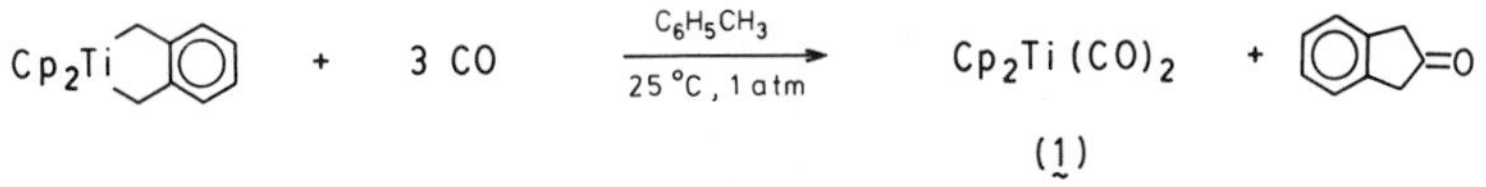

$$Cp_2Ti + 3\ CO \xrightarrow[25\,°C,\ 1\,atm]{C_6H_5CH_3} Cp_2Ti\,(CO)_2 + \text{(indanone)} =O$$

(**1**)

3. Preparation/Formation of $Cp_2Ti(CO)_2$ via other Titanium Complexes

Brintzinger and co-workers, while investigating the formation and chemical reactivity of various "titanocenes," prepared a novel triphenylphosphine–titanocene complex of empirical formula $[C_{10}H_{10}TiP(C_6H_5)_3]_2$ (**22**) whose structure to date has been undetermined. The purple complex **22** can be prepared by the four routes illustrated below (*24*).

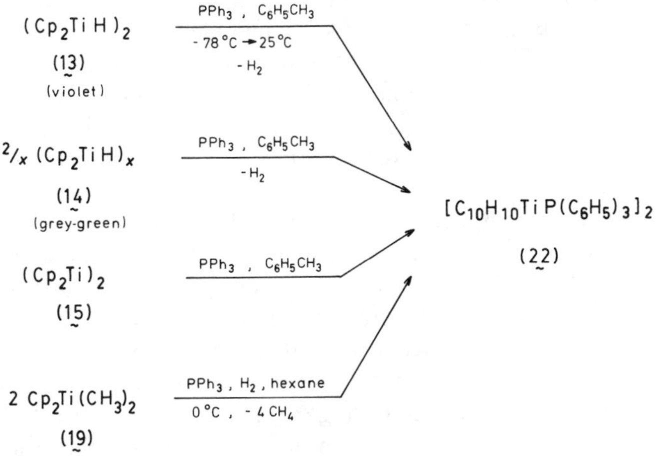

$(Cp_2Ti\,H)_2$ $\xrightarrow[\substack{-78\,°C\,\to\,25\,°C \\ -H_2}]{PPh_3,\ C_6H_5CH_3}$

(**13**)

(violet)

$^2/_x\ (Cp_2TiH)_x$ $\xrightarrow[-H_2]{PPh_3,\ C_6H_5CH_3}$

(**14**)

(grey-green)

$(Cp_2Ti)_2$ $\xrightarrow{PPh_3,\ C_6H_5CH_3}$

(**15**)

$2\ Cp_2Ti(CH_3)_2$ $\xrightarrow[0\,°C,\ -4\,CH_4]{PPh_3,\ H_2,\ hexane}$

(**19**)

$[C_{10}H_{10}Ti\,P(C_6H_5)_3]_2$

(**22**)

Allowing a toluene solution of **22** to stir under 1 atm of CO at room temperature for 2 hours led to a red–brown solution from which $Cp_2Ti(CO)_2$ (**1**) could be isolated in 33% yield (*24*).

$$[C_{10}H_{10}TiP(C_6H_5)_3]_2 \; + \; 4\;CO \xrightarrow[25\,°C,\,2\,h,\,1\,atm]{C_6H_5CH_3} \; 2\;Cp_2Ti(CO)_2 \; + \; 2\;PPh_3$$

(22)   (1)

The reaction of metastable titanocene **15** with dinitrogen oxide produced a highly reactive Ti(III) complex, $(Cp_2Ti)_2O$, and nitrogen. Treatment of this new complex with a large excess of CO resulted in a disproportionation reaction yielding $Cp_2Ti(CO)_2$ (**1**) and a yellow titanium(IV) polymer of empirical formula Cp_2TiO (*37*).

$$(Cp_2Ti)_2 \; + \; N_2O \xrightarrow[0\,°C,\,5\,min,\,1\,atm]{C_6H_5CH_3} \; (Cp_2Ti)_2O \; + \; N_2$$

(15)

$$(Cp_2Ti)_2O \; + 2CO \longrightarrow Cp_2Ti(CO)_2 \; + \; \text{``}Cp_2TiO\text{''}$$

(1)

$$Cp_2Ti(BH_4) \; + \; 2\;CO \xrightarrow[25\,°C,\,4\,days,\,1\,atm]{Et_3N} Cp_2Ti(CO)_2$$

(1)

Demerseman and co-workers have briefly noted that the reduction of $CpTiCl_3$ in THF with either aluminum or magnesium in the presence of CO gave $Cp_2Ti(CO)_2$ as the only titanium carbonyl species (*27*).

In 1976 Floriani reported that **1** could be prepared in 80% yield via the carbonylation of $Cp_2Ti(BH_4)$ in triethylamine (*8*). The facile displacement of labile ligands from various titanocene complexes by CO has led to the formation of $Cp_2Ti(CO)_2$ (**1**). The reduction of Cp_2TiCl_2 with sodium vapor in a rotating metal atom reactor in the presence of $P(OMe)_3$ at $-100°C$ gave $Cp_2Ti[P(OMe)_3]_2$ in 35% yield. When a pentane solution of $Cp_2Ti[P(OMe)_3]_2$ was exposed to 1 atm of CO at room temperature, **1** formed in high yield (*48*).

Similarly, the complexes $Cp_2Ti(CO)(PR_3)$ (R = Et, Ph, F) and $Cp_2Ti(PF_3)_2$ were found to readily react with CO under mild conditions to give compound **1** (see Section VI,A for their preparation) (*49,50*).

Unlike its zirconium and hafnium counterparts, $Cp_2Ti(CO)_2$ (**1**) appeared photochemically inactive in that no observable CO was evolved upon photolysis nor did the intensity of the cyclopentadienyl resonance in the ^1H-NMR spectrum change during the irradiation period (*49–51*). This apparent inertness prompted Rausch and Sikora to prepare the isotopically labeled complex $Cp_2Ti(C^{18}O)_2$ (**1a**). The reductive carbonylation of Cp_2TiCl_2 using $C^{18}O$ and aluminum as the reducing agent produced **1a** in 53% yield with metal carbonyl bands appearing at 1934 and 1856 cm^{-1}

(hexane) (50). The IR-monitored photolysis of a hexane solution of $Cp_2Ti(C^{18}O)_2$ (**1a**) in a $C^{16}O$ sparge led to complete exchange of $C^{18}O$ to $Cp_2Ti(C^{16}O)_2$ (**1**) [ν(CO) 1977 and 1899 cm^{-1}] in 20 minutes, with the observation of the intermediate $Cp_2Ti(C^{16}O)(C^{18}O)$ [ν(CO) 1959 and 1870 cm^{-1}]. All three dicarbonyl species were observable after 5 minutes of photolysis, while after 10 minutes only the bands due to $Cp_2Ti(C^{16}O)(C^{18}O)$ and $Cp_2Ti(C^{16}O)_2$ were observed ($50,51$).

When the same experiment was performed in the absence of light, exchange was found to occur, but at a much slower rate. After 30 minutes, bands assignable to $Cp_2Ti(C^{16}O)(C^{18}O)$ were barely detectable, while after 4.5 hours all three dicarbonyl species appeared in approximately equal concentrations. On terminating the reaction after 13 hours, the carbonyl bands due to $Cp_2Ti(C^{18}O)_2$ (**1a**) and $Cp_2Ti(C^{16}O)(C^{18}O)$ were clearly observable; however, $Cp_2Ti(C^{16}O)_2$ was the major species present ($50,51$).

These experiments confirmed the photolability of the carbonyl ligands of **1** and suggested that the initial photoproduct may be the coordinatively unsaturated species [$Cp_2Ti(CO)$]. This compound recombines with CO at a much faster rate than decomposing, thus appearing to show no net reaction upon photolysis.

B. *Formation of Bis(η-cyclopentadienyl)dicarbonylzirconium*

Bis(η-cyclopentadienyl)dicarbonylzirconium (**2**) was the first fully characterized zirconium carbonyl complex to appear in the scientific literature. In 1976 this complex was reported simultaneously and independently by American (6), French (7), and Italian (8) research groups. Previous to this, many of the methods which proved successful for the preparation of $Cp_2Ti(CO)_2$ (**1**) failed for the formation of $Cp_2Zr(CO)_2$ (**2**) ($5,26,38$).

The American researchers found that **2** could be obtained via the reduction of Cp_2ZrCl_2 in toluene under 1 atm of CO using excess sodium amalgam (6). While the mild conditions imposed here seemed to make this an attractive preparative route, **2** could be isolated in only 9% yield (21 mg). When the same reaction was performed under 200 atm of CO pressure, the yield of **2** was only 11%.

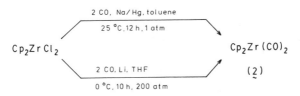

The French workers similarly reduced Cp_2ZrCl_2 under 200 atm of CO; however, lithium metal was used as the reducing agent and the solvent was THF (7). While an 80% yield of **2** was inferred by treatment of a portion of the reaction solution with I_2, the isolated yield of **2** was 46%. When the reaction was carried out at atmospheric pressure of CO, only a 3% yield of **2** was observed.

The Italian group reported that bis(η-cyclopentadienyl)bis(tetrahydroborato)zirconium could be converted to $Cp_2Zr(CO)_2$ (**2**) by treating the former with CO in triethylamine at 100 atm of pressure for 2 days at 50°C (8). Upon workup, **2** was obtained in 15% yield. Although treatment of

$$Cp_2Zr(BH_4)_2 \ + \ 2\ CO \ \xrightarrow[50\ °C,\ 2\ days,\ 100\ atm]{Et_3N} \ Cp_2Zr(CO)_2$$

$$(2)$$

$Cp_2Zr(BH_4)_2$ with triethylamine is known to give both $Cp_2Zr(H)(BH_4)$ and $(Cp_2ZrH_2)_x$, these workers did not speculate as to the intermediate species being carbonylated.

In 1978, Schwartz and Gell found that CO would induce reductive elimination of alkane in various zirconocene alkyl hydride complexes with concurrent formation of $Cp_2Zr(CO)_2$ (**2**) (52,53). It was postulated that CO initially coordinates to the 16-e^- complex **23** forming the coordinatively saturated species **24** which can then reductively eliminate alkane and/or rearrange to a zirconocene acyl hydride intermediate. When R = cyclohexylmethyl, methylcyclohexane reductively eliminated and $Cp_2Zr(CO)_2$ was isolated in 25% yield.

Along similar lines, Schwartz and Gell later reported that tertiary phosphines would also induce reductive elimination in bis(η-cyclopentadienyl)(cyclohexylmethyl)(hydrido)zirconium resulting in high yields of zirconocene bis(phosphine) complexes (53–55). Carbon monoxide was found to readily react with a benzene solution of $Cp_2Zr(PMePh_2)_2$

$$Cp_2Zr\begin{smallmatrix}H\\\\R\end{smallmatrix} \xrightarrow{+CO} [Cp_2Zr\begin{smallmatrix}H\\-CO\\R\end{smallmatrix}] \xrightarrow{-RH} [Cp_2Zr(CO)] \xrightarrow{+CO} Cp_2Zr(CO)_2$$

$$(23) \qquad\qquad (24) \qquad\qquad\qquad\qquad\qquad (2)$$

$$R = -CH_3, \ \text{—}\bigcirc, \ -CH_2\text{—}\bigcirc$$

$$Cp_2Zr\begin{smallmatrix}H\\\\CH_2\end{smallmatrix}\text{—}\bigcirc \quad + \quad \text{excess}\ \ PMeRPh \xrightarrow[-30\ °C\ \to\ 25\ °C,\ 1.25\ h]{toluene}$$

$$R = Me, Ph$$

$$Cp_2Zr(PMeRPh)_2 \quad + \quad \bigcirc\text{—}CH_3$$

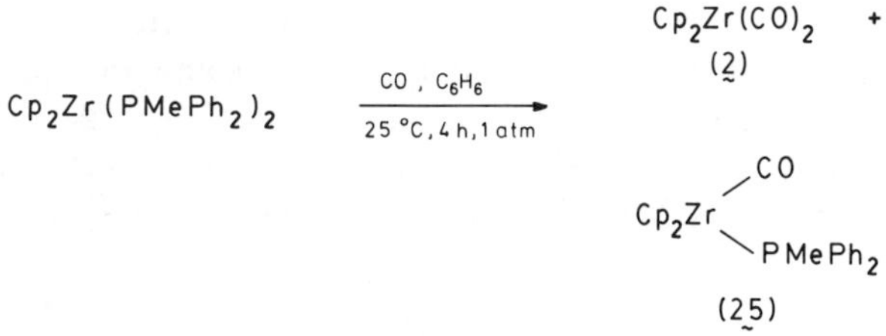

$$Cp_2Zr(PMePh_2)_2 \xrightarrow[25\,°C,\,4\,h,\,1\,atm]{CO\,,\,C_6H_6} \begin{array}{c} Cp_2Zr(CO)_2 \quad + \\ (2) \\[2em] Cp_2Zr{\Large\diagdown}^{CO}_{PMePh_2} \\ (25) \end{array}$$

containing 2 equivalents of $PMePh_2$ over a 4 hour period at room temperature and atmospheric pressure to give an equilibrium mixture of $Cp_2Zr(CO)_2$ (2) (40%) and $Cp_2Zr(CO)(PMePh_2)$ (25) (60%) (54,55).

Another example of such a ligand displacement reaction by CO was reported in 1980. Bis(η-cyclopentadienyl)bis(trimethylphosphite)-zirconium was prepared in a manner similar to that for $Cp_2Ti[P(OMe)_3]_2$ [reduction of Cp_2ZrCl_2 with sodium vapor in the presence of $P(OMe)_3$ in a rotating metal atom reactor]. Treatment of a pentane solution of $Cp_2Zr[P(OMe_3)]_2$ with CO resulted in the clean displacement of both $P(OMe)_3$ ligands and formation of $Cp_2Zr(CO)_2$ (2) in high yield (48).

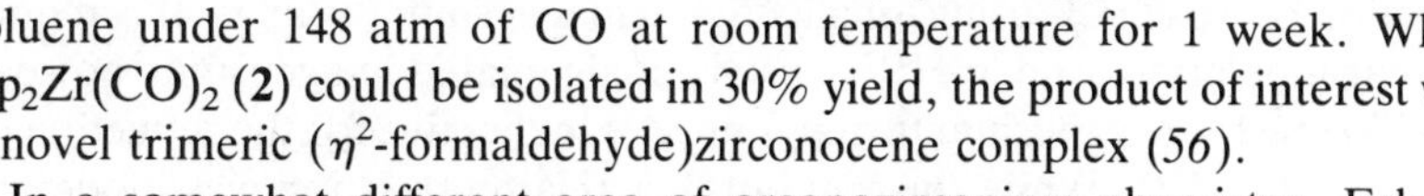

$$Cp_2Zr[P(OMe)_3]_2 + 2\,CO \xrightarrow[25\,°C,\,1\,atm]{pentane} \underset{(2)}{Cp_2Zr(CO)_2} + 2\,P(OMe)_3$$

Erker and co-workers in 1983 found that $Cp_2Zr(CO)_2$ (2) was formed, along with a mixture of other organozirconium products, when oligomeric bis(cyclopentadienyl)dihydridozirconium, $(Cp_2ZrH_2)_x$, was stirred in toluene under 148 atm of CO at room temperature for 1 week. While $Cp_2Zr(CO)_2$ (2) could be isolated in 30% yield, the product of interest was a novel trimeric (η^2-formaldehyde)zirconocene complex (56).

In a somewhat different area of organozirconium chemistry, Erker's group has recently reported the formation of $Cp_2Zr(CO)_2$ (2) in very good yield via the carbonylation of the zirconocene diene complexes 26a and 26b (57). In cases where the diene was structurally less complex (e.g.,

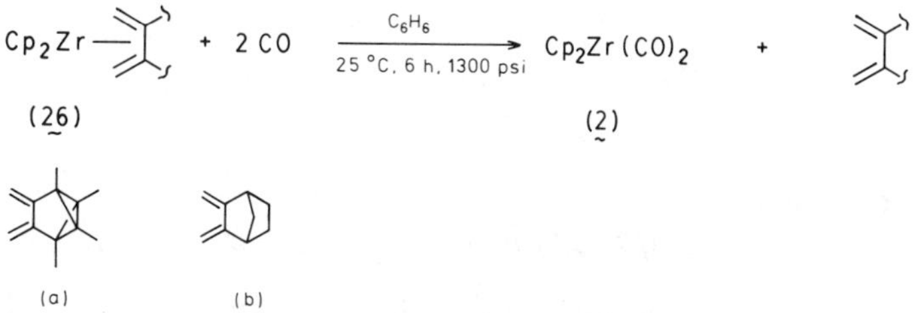

$$\underset{(26)}{Cp_2Zr-} + 2\,CO \xrightarrow[25\,°C,\,6\,h,\,1300\,psi]{C_6H_6} \underset{(2)}{Cp_2Zr(CO)_2} +$$

(a)   (b)

2,3-dimethyl-1,3-butadiene), reaction with CO and subsequent workup resulted in the formation of cyclopentenones.

The most recent preparation of **2** has been reported by Rausch and co-workers and holds many advantages over some of the previously described methods. Amalgamated magnesium (Grignard turnings) was found to smoothly reduce THF solutions of Cp_2ZrCl_2 under 1 atm of CO at 25°C and give $Cp_2Zr(CO)_2$ (**2**) in good yield (50%) and in ample amounts (1–2 g) (*58,59*).

$$Cp_2ZrCl_2 \xrightarrow[\text{25°C, 24 h, 1 atm}]{\text{2 CO, Mg/Hg, THF}} Cp_2Zr(CO)_2$$

(**2**)

C. Formation of Bis(η-cyclopentadienyl)dicarbonylhafnium

Like zirconium, the first fully characterized carbonyl complex of hafnium was reported in 1976 by Thomas and Brown (*6*). This complex, bis(η-cyclopentadienyl)dicarbonylhafnium (**3**) was prepared via the reductive carbonylation of Cp_2HfCl_2 using sodium amalgam. While the reaction proceeded to give a moderate yield of **3** (30%), this corresponded to only 60 mg of isolated product.

In a similar manner, Demerseman *et al.* attempted the high pressure reductive carbonylation of Cp_2HfCl_2 using lithium metal. However, after treatment of a THF solution with 200 atm of CO for 10 hours, no $Cp_2Hf(CO)_2$ (**3**) could be isolated (*7*).

Again, Rausch and co-workers have recently described what appears to be the best route to $Cp_2Hf(CO)_2$ (**3**). Analogous to their method for the reduction of Cp_2ZrCl_2 using amalgamated magnesium turnings, this reductive carbonylation of Cp_2HfCl_2 used amalgamated magnesium powder (50–100 mesh). Magnesium turnings proved ineffective for the reduction of Cp_2HfCl_2 (*58,59*). While the yield of **3** in this preparation was moderate (30%), it proceeded under mild conditions and gave adequate amounts of **3** (600 mg) for further chemical studies.

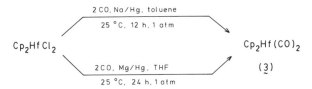

D. Formation of Bis(η-pentamethylcyclopentadienyl)dicarbonyltitanium

Bis(η-pentamethylcyclopentadienyl)dicarbonyltitanium (**27**) was first prepared by Brintzinger and Bercaw in 1971 (*24,60*). The thermolysis of $(\eta\text{-}C_5Me_5)TiMe_2$ in refluxing toluene gave a turquoise product which upon treatment with hydrogen yielded decamethyltitanocene (**28**) via the reductive elimination of H_2 from the intermediary $(\eta\text{-}C_5Me_5)_2TiH_2$. Decamethyltitanocene (**28**) was then observed to react readily with CO to give $(\eta\text{-}C_5Me_5)_2Ti(CO)_2$ (**27**) in ~68% yield.

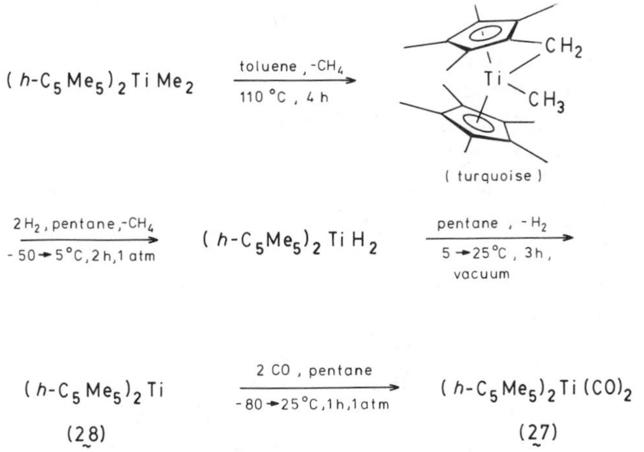

Some years later it was reported that $(\eta\text{-}C_5Me_5)_2TiCl_2$ could be reductively carbonylated under very mild conditions using either zinc (*61*) or amalgamated magnesium powder (*59,62*) as reducing agents. Yields of isolated $(\eta\text{-}C_5Me_5)_2Ti(CO)_2$ (**27**) could be obtained in ~70% by these methods.

Brubaker and Tung were able to prepare **27** photochemically by the irradiation of a toluene solution of bis(η-pentamethylcyclopentadienyl)diphenyltitanium in the presence of CO (*63*).

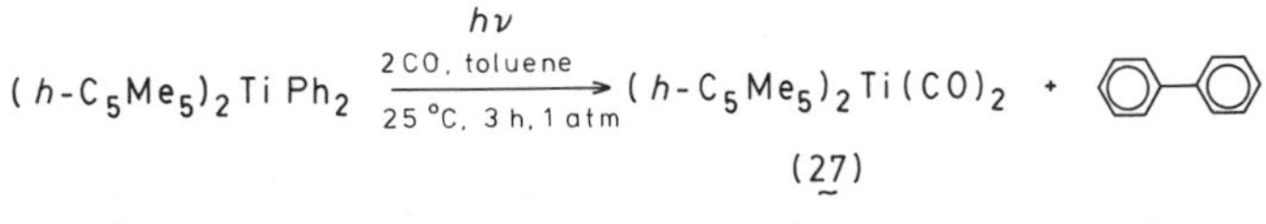

In order to study the photochemically induced lability of the carbonyl ligands of $(\eta\text{-}C_5Me_5)_2Ti(CO)_2$ (**27**), Rausch *et al.* prepared $(\eta\text{-}C_5Me_5)_2Ti(C^{18}O)_2$ (**27a**) via the reductive carbonylation of $(\eta\text{-}C_5Me_5)_2TiCl_2$ using magnesium metal as the reducing agent (*51*). The reaction produced the desired isotopically labeled complex **27a** in 59% yield with metal carbonyl bands appearing at 1900 and 1818 cm^{-1} (hexane). The IR-monitored photolysis of a hexane solution of $(\eta\text{-}C_5Me_5)_2Ti(C^{18}O)_2$ (**27a**) in a $C^{16}O$ sparge led to complete exchange of $C^{18}O$ to $(\eta\text{-}C_5Me_5)_2Ti(C^{16}O)_2$ (**27**)[$\nu(CO)$ 1940 and 1858 cm^{-1}] in 30 minutes, with the observation of the intermediate (η-

$C_5Me_5)_2Ti(C^{16}O)(C^{18}O)$ [$\nu(CO)$ 1926 and 1835 cm^{-1}]. The IR spectrum recorded after 5 minutes of irradiation showed the presence of all three dicarbonyl species, while after 15 minutes only carbonyl bands due to $(\eta\text{-}C_5Me_5)_2Ti(C^{16}O)(C^{18}O)$ and $(\eta\text{-}C_5Me_5)_2Ti(C^{16}O)_2$ (27) were observed (49,51).

In a similar experiment in the absence of light exchange was also found to occur, but at a much slower rate such that after 90 minutes the predominant species was still $(\eta\text{-}C_5Me_5)_2Ti(C^{18}O)_2$. While the carbonyl bands of $(\eta\text{-}C_5Me_5)_2Ti(C^{16}O)(C^{18}O)$ were clearly observable, bands assignable to $(\eta\text{-}C_5Me_5)_2Ti(C^{16}O)_2$ were absent (51).

Surprisingly, when a hexane solution of $(\eta\text{-}C_5Me_5)_2Ti(CO)_2$ (27) was photolyzed in the presence of PF_3, no evidence for the formation of $(\eta\text{-}C_5Me_5)_2Ti(CO)(PF_3)$ or $(\eta\text{-}C_5Me_5)_2Ti(PF_3)_2$ was observed. This result contrasts with the facile formation of $Cp_2Ti(CO)(PF_3)$ and $Cp_2Ti(PF_3)_2$ via the photolysis of $Cp_2Ti(CO)_2$ (1) with PF_3 (49,50).

Recently, Bercaw and co-workers have described the preparation of the first ethylene complex of titanium (64). The sodium amalgam reduction of a toluene solution of $(\eta\text{-}C_5Me_5)_2TiCl_2$ under an ethylene atmosphere afforded $(\eta\text{-}C_5Me_5)_2Ti(\eta\text{-}C_2H_4)$ in 80% yield. Treatment of this complex with CO at low temperature resulted in the displacement of C_2H_4 and quantitative formation of $(\eta\text{-}C_5Me_5)_2Ti(CO)_2$ (27).

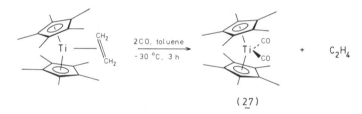

(27)

E. Formation of Bis(η-pentamethylcyclopentadienyl)dicarbonylzirconium

By the displacement of 3 equivalents of N_2 with CO from the novel permethylzirconocene dinitrogen complex 29, Bercaw *et al.* were able to prepare $(\eta\text{-}C_5Me_5)_2Zr(CO)_2$ (30) for the first time (65). Complex 29 was prepared by the reductive nitrogenation of $(\eta\text{-}C_5Me_5)_2ZrCl_2$ in toluene using sodium amalgam as the reducing agent (66).

Bis(η-pentamethylcyclopentadienyl)zirconium was proposed as an intermediate in the permethylzirconocene hydride promoted reduction of coordinated carbon monoxide by Bercaw and co-workers (65,67–69). In hopes of intercepting such an intermediate, CO was allowed to diffuse

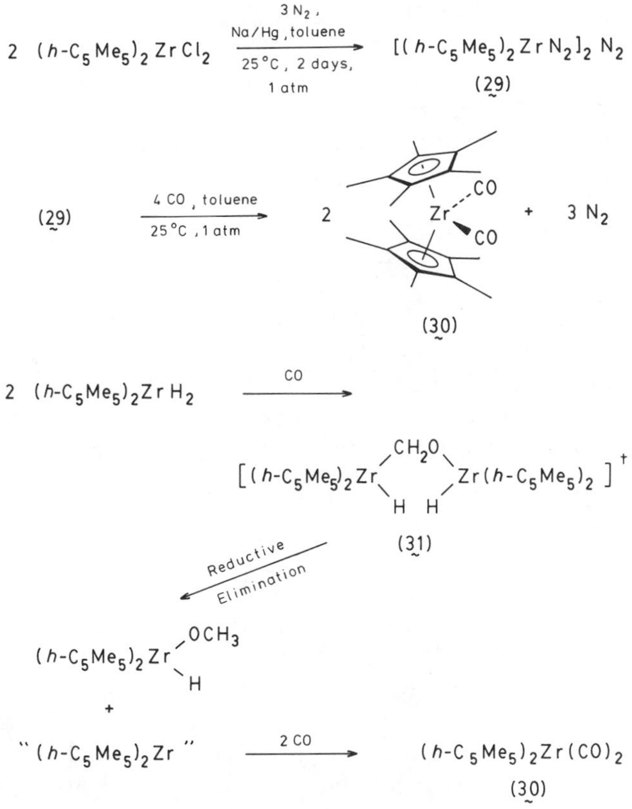

$$2 \ (h\text{-}C_5Me_5)_2 ZrCl_2 \xrightarrow[\substack{25°C, \ 2 \ days, \\ 1 \ atm}]{\substack{3 \ N_2, \\ Na/Hg, \ toluene}} [(h\text{-}C_5Me_5)_2 ZrN_2]_2 \ N_2$$
$$(29)$$

$$(29) \xrightarrow[25°C, \ 1 \ atm]{4 \ CO, \ toluene} 2 \quad \text{(30)} \quad + \quad 3 \ N_2$$

$$2 \ (h\text{-}C_5Me_5)_2 ZrH_2 \xrightarrow{CO}$$

$$\left[(h\text{-}C_5Me_5)_2 Zr \underset{H \quad H}{\overset{CH_2O}{\diagup \diagdown}} Zr(h\text{-}C_5Me_5)_2 \right]^{\dagger}$$
$$(31)$$

Reductive Elimination

$$(h\text{-}C_5Me_5)_2 Zr \underset{H}{\overset{OCH_3}{\diagup}}$$
+
$$\text{"}(h\text{-}C_5Me_5)_2 Zr\text{"} \xrightarrow{2 \ CO} (h\text{-}C_5Me_5)_2 Zr(CO)_2$$
$$(30)$$

† see ref. 69

slowly into a nitrogen blanketed toluene solution of $(\eta\text{-}C_5Me_5)_2 ZrH_2$. After several hours, the desired reduction product $(\eta\text{-}C_5Me_5)_2 Zr\text{-}$ $(H)(OCH_3)$ could be observed in a 1:1 ratio with $(\eta\text{-}C_5Me_5)_2 Zr(CO)_2$ (**30**).

Low temperature IR and ^{1}H-NMR studies have revealed that **30** can form via the loss of H_2 from $(\eta\text{-}C_5Me_5)_2 ZrH_2$; however, the ultimate

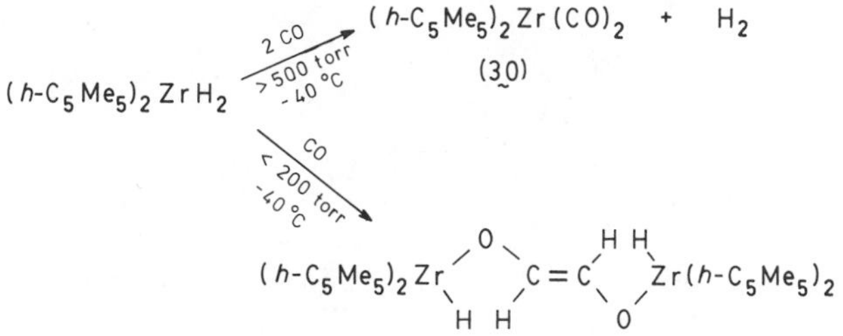

$$(h\text{-}C_5Me_5)_2 ZrH_2 \xrightarrow[\substack{> 500 \ torr \\ -40°C}]{2 \ CO} (h\text{-}C_5Me_5)_2 Zr(CO)_2 \ + \ H_2$$
$$(30)$$

$$\xrightarrow[\substack{< 200 \ torr \\ -40°C}]{CO} (h\text{-}C_5Me_5)_2 Zr \underset{H \quad H}{\overset{O}{\diagup \diagdown}} C = C \underset{O}{\overset{H \quad H}{\diagdown \diagup}} Zr(h\text{-}C_5Me_5)_2$$

organozirconium product distribution was found to be sensitive to the initial CO pressure (70).

In 1981 Brubaker and Tung reported that the photolysis of a toluene solution of bis(η-pentamethylcyclopentadienyl)diphenylzirconium in a CO atmosphere led to the formation (η-C$_5$Me$_5$)$_2$Zr(CO)$_2$ (30) in 25% yield (63).

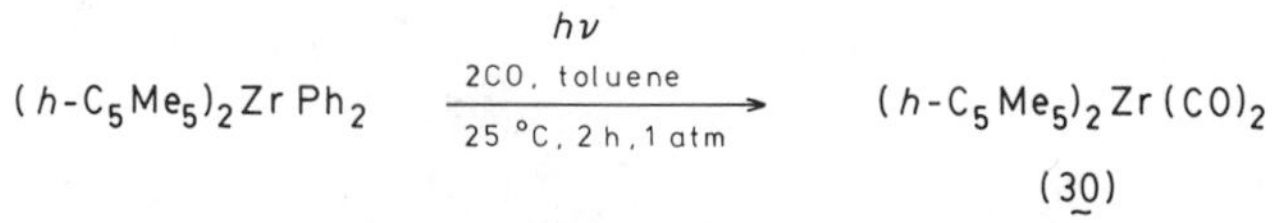

$$(h\text{-}C_5Me_5)_2ZrPh_2 \xrightarrow[\substack{25\,°C,\,2\,h,\,1\,atm}]{\substack{h\nu \\ 2CO,\,toluene}} (h\text{-}C_5Me_5)_2Zr(CO)_2$$
$$(30)$$

In the same year, Rausch and co-workers found that the reductive carbonylation of (η-C$_5$Me$_5$)$_2$ZrCl$_2$ in THF using amalgamated magnesium powder gave high yields of (η-C$_5$Me$_5$)$_2$Zr(CO)$_2$ (30) (80%) under mild reaction conditions (59,62). The low pressure employed, together with the

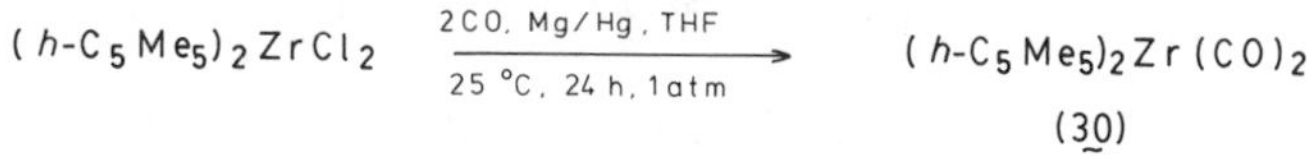

$$(h\text{-}C_5Me_5)_2ZrCl_2 \xrightarrow[\substack{25\,°C,\,24\,h,\,1\,atm}]{\substack{2CO,\,Mg/Hg,\,THF}} (h\text{-}C_5Me_5)_2Zr(CO)_2$$
$$(30)$$

high yield and direct conversion of (η-C$_5$Me$_5$)$_2$ZrCl$_2$ to 30, makes this perhaps the most attractive route for the preparation of 30.

In 1982, a most remarkable reaction was discovered by Bercaw et al. which involved the reductive coupling of the terminal carbonyl ligands of [CpFe(CO)$_2$]$_2$. When 1 equivalent of the permethylated zirconocene dinitrogen complex [(η-C$_5$Me$_5$)$_2$ZrN$_2$]$_2$N$_2$ (29) was treated with 2 equivalents of [CpFe(CO)$_2$]$_2$ in toluene above $-20°C$, 3 equivalents of N$_2$ evolved

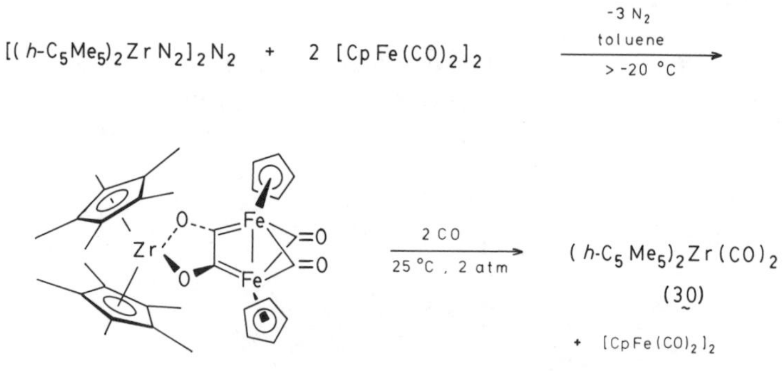

$$[(h\text{-}C_5Me_5)_2ZrN_2]_2N_2 \quad + \quad 2\,[CpFe(CO)_2]_2 \xrightarrow[\substack{> -20\,°C}]{\substack{-3\,N_2 \\ toluene}}$$

$$\xrightarrow[\substack{25\,°C,\,2\,atm}]{\substack{2\,CO}} (h\text{-}C_5Me_5)_2Zr(CO)_2$$
$$(30)$$
$$+ \quad [CpFe(CO)_2]_2$$

(32)

concomitant with the formation of a dark-red complex 32 which could be observed in 95% yield (71). The structure of complex 32 was obtained by X-ray diffraction studies and revealed that it contained a novel dioxozirconacyclopenta-3,4-diylidene diiron moiety.

The reaction of 32 with CO is significant in that over a period of several days at room temperature, $[Cp_2Fe(CO)_2]_2$ formed concurrently with $(\eta\text{-}C_5Me_5)_2Zr(CO)_2$ (**30**). When ^{13}CO was used, $(\eta\text{-}C_5Me_5)_2Zr(^{13}CO)_2$ resulted.

F. Formation of Bis(η-pentamethylcyclopentadienyl)dicarbonylhafnium

The existence of $(\eta\text{-}C_5Me_5)_2Hf(CO)_2$ (**33**) was first reported as the result of some low temperature IR spectroscopic studies (70). Treatment of a hexane solution of $(\eta\text{-}C_5Me_5)_2HfH_2$ with CO in an IR cell led to the formation of $(\eta\text{-}C_5Me_5)_2Hf(H)_2(CO)$ at $-35°C$ as observed by a high energy metal carbonyl band at 2036 cm^{-1}. As the solution was allowed to warm ($>-5°C$), bands assignable to $(\eta\text{-}C_5Me_5)_2Hf(CO)_2$ (**33**) appeared at 1941 and 1850 cm^{-1} at the expense of the band at 2036 cm^{-1}.

The first synthesis of $(\eta\text{-}C_5Me_5)_2Hf(CO)_2$ (**33**) on a preparative scale was reported by Rausch *et al.* in 1981 (62). Amalgamated magnesium powder (50–100 mesh), used as a reducing agent in the reductive carbonylation of Cp_2HfCl_2, $(\eta\text{-}C_5Me_5)_2TiCl_2$, and $(\eta\text{-}C_5Me_5)_2ZrCl_2$, proved ineffective for the reduction of $(\eta\text{-}C_5Me_5)_2HfCl_2$. Thus, a more active form of magnesium was required. The use of Rieke magnesium activated with mercuric chloride was found to be sufficient for the reduction of $(\eta\text{-}C_5Me_5)_2HfCl_2$, and $(\eta\text{-}C_5Me_5)_2Hf(CO)_2$ (**33**) could be prepared in 25% yield (360 mg) (59,62).

Complex **33** can be prepared via a method analogous to that used for the preparation of $(\eta\text{-}C_5Me_5)_2Zr(CO)_2$ (**30**) from $[(\eta\text{-}C_5Me_5)_2ZrN_2]_2N_2$ (65). The reduction of $(\eta\text{-}C_5Me_5)_2HfI_2$ in DME at $-41°C$ with Na–K alloy under 1 atm of N_2 gave $[(\eta\text{-}C_5Me_5)_2HfN_2]_2N_2$ (72), which upon treatment with CO yielded $(\eta\text{-}C_5Me_5)_2Hf(CO)_2$ (**33**) (73).

$$(h\text{-}C_5Me_5)_2HfCl_2 + 2\,CO \xrightarrow[25°C,\,24h,1\,atm]{Rieke\ Mg,\ THF} (h\text{-}C_5Me_5)_2Hf(CO)_2$$
$$(33)$$

$$(h\text{-}C_5Me_5)_2HfI_2 \xrightarrow[-41°C,\,4h,1\,atm]{3\,N_2,\,Na/K,DME} [(h\text{-}C_5Me_5)_2HfN_2]_2N_2$$
$$\downarrow 4\,CO$$
$$(h\text{-}C_5Me_5)_2Hf(CO)_2$$
$$(33)$$

Very recently, Bercaw and co-workers have reported a high yield preparation of **33** via the high pressure carbonylation of $(\eta\text{-}C_5Me_5)HfH_2$ in toluene (72).

$$(h\text{-}C_5Me_5)_2HfH_2 \xrightarrow[\text{25 °C, 1 h, 1500 psi}]{\text{2 CO, toluene}} (h\text{-}C_5Me_5)_2Hf(CO)_2$$

$$(33)$$

G. Formation of Metallocene Dicarbonyls with Other η^5-Bonded Rings

Many metallocene dicarbonyl complexes of group 4B have been synthesized containing cyclopentadienyl rings which have been substituted to varying degrees possessing various functionalities. Most of the synthetic routes to these compounds involve methods which have also been used for the preparation of $Cp_2M(CO)_2$ or $(\eta\text{-}C_5Me_5)_2M(CO)_2$ (M = Ti, Zr, Hf).

1. Preparation of $(\eta\text{-}C_5Me_5)CpM(CO)_2$ (M = Ti, Zr, Hf)

The mixed metallocene dicarbonyls of titanium, zirconium, and hafnium containing one cyclopentadienyl ring and one pentamethylcyclopentadienyl ring have been prepared by Rausch and co-workers in good yields via the reductive carbonylation of the corresponding metallocene dichlorides using amalgamated magnesium powder as the reducing agent (74). As in the reduction of $(\eta\text{-}C_5Me_5)_2HfCl_2$ (59,62), Rieke magnesium was used for the reduction of $(\eta\text{-}C_5Me_5)CpHfCl_2$.

$$(h\text{-}C_5Me_5)CpMCl_2 + 2\ CO \xrightarrow[\text{25 °C, 24 h, 1 atm}]{\text{Mg/Hg, THF}} (h\text{-}C_5Me_5)CpM\begin{matrix}CO\\\diagup\\\diagdown\\CO\end{matrix}$$

$$M = Ti\ 88\%$$
$$M = Zr\ 75\%$$
$$M = Hf\ 25\%$$

2. Preparation of Bis(η-indenyl)dicarbonyl Complexes of Titanium and Zirconium

Bis(η-indenyl)dicarbonyltitanium (34) was first reported in 1975 independently by two different research groups using different synthetic routes. Rausch and Alt discovered that the photolysis of a pentane solution of bis(η-indenyl)dimethyltitanium in a CO atmosphere resulted in the evolution of methane and the formation of 34 in 50% yield (75).

Demerseman et al. were able to prepare 34 via the carbonylation of bis(η-indenyl)dibenzyltitanium in the presence of $AlEt_3$ using a procedure

$$(h^5\text{-}C_9H_7)_2TiMe_2 \xrightarrow[\text{25 °C, 25 min, 1 atm}]{\overset{h\nu}{\text{2 CO, pentane}}} (h^5\text{-}C_9H_7)_2Ti(CO)_2$$

$$(34)$$

similar to that for the preparation of $Cp_2Ti(CO)_2$ (1) from $Cp_2Ti(CH_2Ph)_2$ (26). The related complex, bis(η-tetrahydroindenyl)dicarbonyltitanium, was also prepared in this manner (26). In neither case were the yields given.

More recently Rausch and Moriarty have been successful in preparing $(\eta^5\text{-}C_9H_7)_2Ti(CO)_2$ (34) as well as $(\eta^5\text{-}C_9H_7)_2Zr(CO)_2$ directly via the reductive carbonylation at atmospheric pressure of the corresponding bis(η-indenyl)dichlorides of titanium and zirconium, using amalgamated aluminum turnings and magnesium turnings, respectively, as the reducing agents (74). Both compounds could be isolated in 45% yield. This method represents the first and only report of bis(η-indenyl)dicarbonylzirconium.

3. Preparation of Ansa-Titanocene Dicarbonyl Derivatives

Ansa-metallocenes are those in which the cyclopentadienyl rings are linked by a molecular bridge that may contain various kinds and numbers of atoms. The chemistry of the ethylene-bridged complexes $(CH_2)_2(\eta\text{-}C_5H_4)_2TiCl_2$ (35) and $(CH_2)_2(\eta\text{-}C_5H_4)_2TiMe_2$ (36) have been studied in detail and were found to exhibit reactivities quite different from Cp_2TiCl_2 and Cp_2TiMe_2 (76). Most notable was the inaccessibility of the low valent titanocene species, $[(CH_2)_2(\eta\text{-}C_5H_4)_2Ti]_x$. Methods which proved successful for the reduction of Cp_2TiCl_2 (use of alkali metal reducing agent) or Cp_2TiMe_2 (treatment with H_2) were found to be ineffective for 35 and 36. In this regard, the preparations of $(CH_2)_2(\eta\text{-}C_5H_4)_2Ti(CO)_2$ (37) are for the most part unlike any of the procedures reported for the synthesis of $Cp_2Ti(CO)_2$ (1).

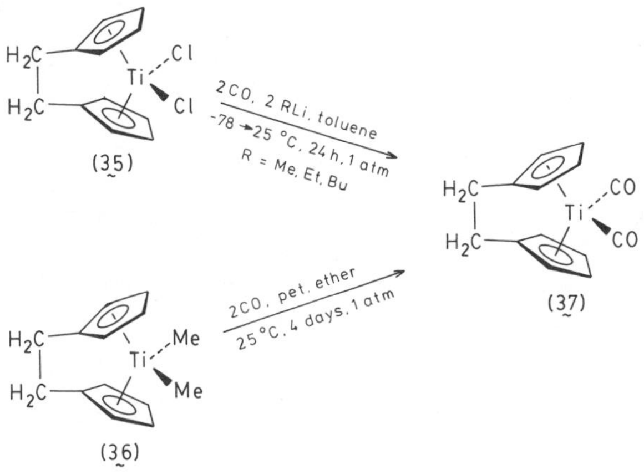

Treatment of a toluene solution of $(CH_2)_2(\eta\text{-}C_5H_4)_2TiCl_2$ (35) with 2 equivalents of methyl-, ethyl-, or butyllithium in a CO atmosphere resulted in the isolation of $(CH_2)_2(\eta\text{-}C_5H_4)_2Ti(CO)_2$ (**37**) in 78% yield upon workup (*76*). This preparation has similarities to Murray's original preparation of $Cp_2Ti(CO)_2$ (**1**) (*4,5*); however, the reaction conditions described by Murray were drastically different (150°C, 240 atm CO).

When a petroleum ether solution of $(CH_2)_2(\eta\text{-}C_5H_4)_2TiMe_2$ (**36**) was exposed to CO at atmospheric pressure and room temperature, the yellow solution gradually turned red–brown over a 24-hour period. After stirring for an additional 3 days at room temperature, workup of the solution gave $(CH_2)_2(\eta\text{-}C_5H_4)_2Ti(CO)_2$ in 78% yield along with acetone as the organic product (*76*). This preparation is similar to that of Wailes *et al.* by which Cp_2TiMe_2 was carbonylated to **1** at room temperature, but at a CO pressure of 120 atm (*40*). While the mild conditions used for the preparation of **37** have not been employed for the synthesis of $Cp_2Ti(CO)_2$ (**1**), it remains to be seen if such conditions would be capable of transforming Cp_2TiCl_2 and Cp_2TiMe_2 into **1**. The related ansa-titanocene dicarbonyls $(CH_2)(\eta\text{-}C_5H_4)_2Ti(CO)_2$ (*76*) and $(CH_2)_3(\eta\text{-}C_5H_4)_2Ti(CO)_2$ (*77*) are presumed to have been synthesized by the same methods as for **37**.

Assuming that $[(CH_2)_2(\eta\text{-}C_5H_4)_2Ti]_x$ is indeed an inaccessible reaction intermediate as reactivity evidence has suggested, then the formation of **37** must circumvent the intermediacy of such an intermediate. Thus, Brintzinger and co-workers have postulated that the Ti(II) center is most likely formed while in contact with one or two stabilizing CO ligands.

Recently an ansa-titanocene dicarbonyl complex containing a siloxy bridge has been reported by Curtis *et al.* (*78*). The reaction sequence is shown below.

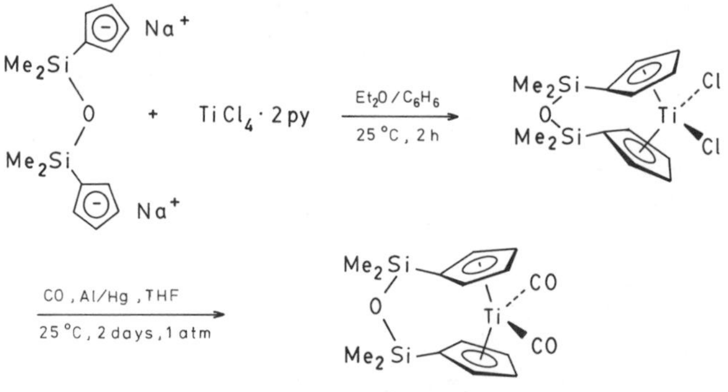

(**38**)

Reduction of $(Me_4Si_2O)(\eta\text{-}C_5H_4)_2TiCl_2$ with amalgamated aluminum in the presence of 1 atm of CO led to the corresponding dicarbonyl complex **38** in 80% yield.

4. Preparation of Functionally Substituted η-Cyclopentadienyl Dicarbonyl Complexes of Titanium and Zirconium

The availability of synthetic routes to cyclopentadienyl anions containing pendant tertiary phosphine moieties has provided the synthetic organometallic chemist with a means of preparing various bimetallic complexes. In the early 1980s two reports appeared in which bimetallic complexes of titanium and molybdenum (79) and titanium and manganese (80) were synthesized, whereby the molybdenum and manganese atoms were bonded to a phosphorus atom attached to one of the rings of $Cp_2Ti(CO)_2$ (1). The synthetic route to the titanium–molybdenum complex is illustrated below.

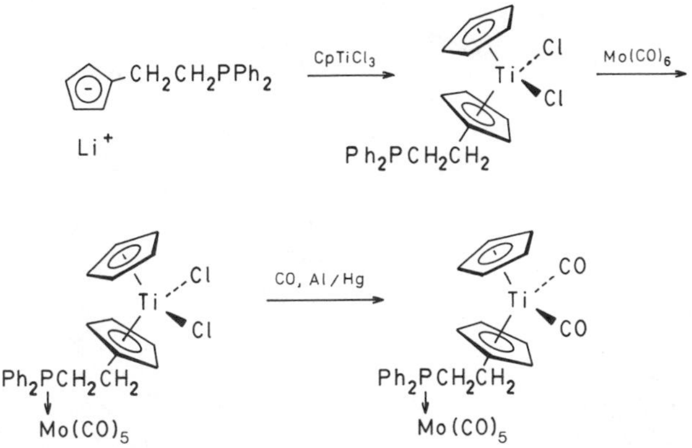

Along similar lines, the same researchers also prepared (η-$C_5H_4PPh_2)_2Ti(CO)_2$ via the reduction of (η-$C_5H_4PPh_2)_2TiCl_2$ in a CO atmosphere, however no bimetallic compounds involving this complex were noted (79). Rausch et al. have described the synthesis of (η-$C_5H_4PPh_2)CpTiCl_2$ via the reaction of [(diphenylphosphino)cyclopentadienyl]thallium with $CpTiCl_3$ in THF. Subsequent reduction of this dichloro derivative with aluminum amalgam under CO led to the formation of (η-$C_5H_4PPh_2)CpTi(CO)_2$ (39) in 25% yield (80). The reaction of 39 with $CpMn(CO)_2(THF)$ at room temperature over a 12-hour period gave the titanium–manganese bimetallic compound, $CpMn(CO)_2[(\eta$-$C_5H_4PPh_2)CpTi(CO)_2]$ (40) in 39% yield.

Two other functionally substituted η-cyclopentadienyl titanium dicarbonyl complexes prepared by Rausch and co-workers include the vinyl Cp compound (η-$C_5H_4CH{=}CH_2)CpTi(CO)_2$ (81) and the carbomethoxy Cp compound (η-$C_5H_4CO_2Me)CpTi(CO)_2$ (82). Both were synthesized via the aluminum-induced reductive carbonylation of the corresponding dichloride derivatives.

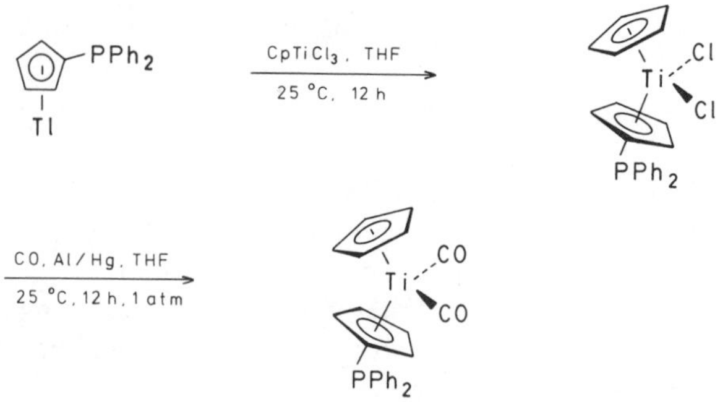

(39)

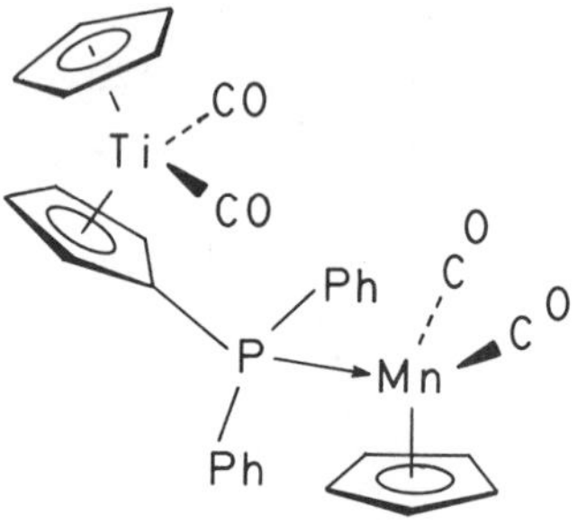

(40)

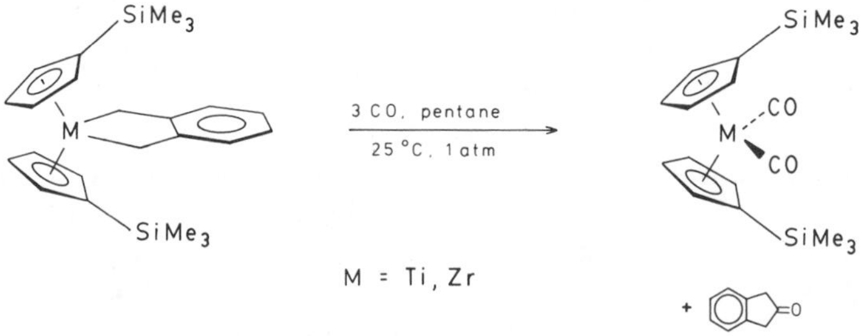

M = Ti, Zr

The trimethylsilylcyclopentadienyl complexes $(\eta\text{-}C_5H_4SiMe_3)_2M(CO)_2$ (M = Ti, Zr) have been recently prepared via the carbonylation of the metallabenzocyclopentene complexes $(\eta\text{-}C_5H_4SiMe_3)_2\overline{M(o\text{-}CH_2C_6H_4C\text{-}H_2)}$ (M = Ti, Zr) (47a). Most recently Lappert et al. have prepared $[\eta\text{-}C_5H_3(SiMe_3)_2]_2Zr(CO)_2$ via the reductive carbonylation of the corresponding dichloride using Na/Hg, Mg/HgCl$_2$, and Li(COT) as reducing agents. The hafnium dicarbonyl complex was prepared in a similar manner (47b).

V

PHYSICAL, SPECTROSCOPIC, AND STRUCTURAL PROPERTIES OF THE VARIOUS GROUP 4B METALLOCENE DICARBONYLS

A. Physical Properties

Because of the low oxidation state of the metal [M(II)] in the group 4B metallocene dicarbonyl compounds, all of them, perhaps with the exception of $(\eta—C_5Me_5)_2Ti(CO)_2$ (27), are very air sensitive and decompose rapidly on exposure to air, forming a yellow solid for the titanium compounds and cream-colored solids for the zirconium and hafnium analogs. While the dicarbonyl 27 is indeed air sensitive, its decomposition appears qualitatively to be much slower relative to the other related complexes.

The group 4B metallocene dicarbonyls are all brightly colored. The titanium complexes are various shades of red with the exception of bis(η-indenyl)dicarbonyltitanium which is green. The three zirconocene dicarbonyls $Cp_2Zr(CO)_2$ (2), $(\eta-C_5Me_5)_2Zr(CO)_2$ (30), and $(\eta-C_5Me_5)$-$CpZr(CO)_2$ appear black in crystalline form, but give dark red–green colored solutions. Bis(η-indenyl)dicarbonylzirconium is green both in the solid state and green when in solution. The analogous hafnium compounds are purple both as solids and in solution.

The group 4B metallocene dicarbonyls exhibit excellent solubility in aromatic and ethereal solvents and very good solubility in aliphatic solvents. In most cases the dicarbonyls can be purified by recrystallization from aliphatic solvents at about $-20°C$ (58,59) and/or by sublimation at 70–80°C and $10^{-2}–10^{-3}$ mm Hg (6–8,24).

B. Spectroscopic Properties

1. Infrared Spectra

All the group 4B metallocene dicarbonyl complexes exhibit two strong metal carbonyl stretching frequencies (A_1 and B_1) with the exception of bis(η-indenyl)dicarbonyltitanium (34) and bis(η-tetrahydroindenyl)dicarbonyltitanium; they exhibit three frequencies due to splitting of the B_1 band indicating the probable existence of two isomers. The metal carbonyl bands of all the dicarbonyl complexes are compiled in Table I. It is noteworthy to compare the difference in carbonyl stretching frequencies between $Cp_2M(CO)_2$ and $(\eta\text{-}C_5Me_5)_2M(CO)_2$ (M = Ti, Zr, Hf). Frequencies for the decamethylmetallocene compounds (27, 30, and 33) are lower in energy relative to the corresponding cyclopentadienyl analogs (1–3) by about 29–41 cm^{-1}. This corresponds to an average reduction in the stretching frequencies of 3.4 cm^{-1} for each substitution of a hydrogen atom by an "electron-releasing" methyl group. Such a decrease has also been observed between the metal carbonyl bands of $Cp_2Mo(CO)$ [$\nu(CO) = 1905$ cm^{-1}] (83) and $(\eta\text{-}C_5Me_5)_2Mo(CO)$ [$\nu(CO) = 1868$ cm^{-1}] (84) ($\Delta = 37$ cm^{-1}). This phenomenon is not uncommon nor unexpected and can be attributed to the enhanced π-backbonding of the electron-rich metal center of 27, 30, or 33 to the π^* orbitals of CO.

The metal carbonyl stretching frequencies of the metallocene dicarbonyls containing one cyclopentadienyl ligand and one pentamethylcyclopentadienyl ligand, $(\eta\text{-}C_5Me_5)CpM(CO)_2$ (M = Ti, Zr, Hf), are, as expected, intermediate to the values of the (bis)cyclopentadienyl and (bis)pentamethylcyclopentadienyl analogs. Also significant is the observed lowering of the metal carbonyl bands of $(Me_4Si_2O)\text{-}(\eta\text{-}C_5H_4)_2Ti(CO)_2$ (38) relative to $Cp_2Ti(CO)_2$ (1). This shift to lower frequency by ~15 cm^{-1} can be attributed to the "electron-releasing" effect of the silicon-substituted rings.

2. ^1H-NMR Spectra

The diamagnetic d^2 group 4B metallocene dicarbonyls possess fairly unambiguous proton NMR spectra. One sharp singlet is observed for $Cp_2M(CO)_2$ and $(\eta\text{-}C_5Me_5)_2M(CO)_2$ (M = Ti, Zr, Hf), indicating the equivalence of the 10 ring protons and 10 ring methyl groups, respectively. The hybrid compounds $(\eta\text{-}C_5Me_5)CpM(CO)_2$ (M = Ti, Zr, Hf) exhibit two sharp singlets in the ratio of 3:1 corresponding to the 15 equivalent methyl protons and the 5 equivalent cyclopentadienyl protons. The bis(η-indenyl)dicarbonyl complexes of titanium and zirconium show an A_2B pattern for the three protons of the C_5 ring, consistent with a symmetrical bonding of this ring to the metal. The four aromatic protons exhibit an expected A_2B_2 pattern.

The ansa-titanocene derivatives $(CH_2)_n(\eta\text{-}C_5H_4)_2Ti(CO)_2$ (n = 2, 3) and $(Me_4Si_2O)(\eta\text{-}C_5H_4)_2Ti(CO)_2$ display an A_2B_2 resonance pattern for the cyclopentadienyl protons of both rings, while the complexes possessing a monosubstituted cyclopentadienyl ring and an unsubstituted cyclopentadienyl ring exhibit both an A_2B_2 resonance pattern and a sharp singlet, respectively.

The chemical shifts and multiplicities of the various metallocene dicarbonyl compounds are listed in Table I.

TABLE I

SPECTROSCOPIC PROPERTIES OF THE GROUP 4B METALLOCENE DICARBONYLS

Compounds	IR $\nu(CO)$ (cm^{-1})	^{1}H NMR (ppm) $\delta(C_5H_{5-x})$ (x = 0, 1, 2)	M$^+$ (m/e)[h]	Ref.
$Cp_2Ti(CO)_2$	1977, 1899[a] 1965, 1883[b]	4.62 (s)[f]	234	5,6,27, 59,62
$Cp_2Zr(CO)_2$	1975, 1885[a] 1967, 1872[b]	4.95 (s)[f]	276	6,7,8 59,62
$Cp_2Hf(CO)_2$	1969, 1878[a] 1960, 1861[b]	4.81 (s)[f]	366	6,7,59, 62
$(\eta\text{-}C_5Me_5)_2Ti(CO)_2$	1940, 1858[a]	1.67 (s)[f]	374	59,62
$(\eta\text{-}C_5Me_5)_2Zr(CO)_2$	1945, 1852[a]	1.73 (s)[f]	416	59,62
$(\eta\text{-}C_5Me_5)_2Hf(CO)_2$	1940, 1844[a]	1.74 (s)[f]	506	59,62
$(\eta\text{-}C_5Me_5)CpTi(CO)_2$	1967, 1884[a]	4.80 (s), 1.84 (s)[f]	304	74
$(\eta\text{-}C_5Me_5)CpZr(CO)_2$	1965, 1875[c]	5.00 (s), 1.93 (s)[f]	346	74
$(\eta\text{-}C_5Me_5)_2CpHf(CO)_2$	1962, 1868[c]	4.85 (s), 1.95 (s)[f]	436	74
$(\eta^5\text{-}C_9H_7)_2Ti(CO)_2$	1981, 1913, 1903[c]	4.83 (t)[g] 5.05 (d) 6.80 A_2B_2	334	26,74,75
$(\eta^5\text{-}C_9H_7)_2Zr(CO)_2$	1985, 1899[c]	4.90 (d)[g] 5.12 (t) 6.55 A_2B_2	376	74
$(\eta^5\text{-}C_9H_{11})_2Ti(CO)_2$	1962, 1886, 1883[d]			26
$(CH_2)_2(\eta\text{-}C_5H_4)_2Ti(CO)_2$	1980, 1900[a]	4.98 (t)[f] 4.68 (t)	260	76
$(CH_2)_3(\eta\text{-}C_5H_4)_2Ti(CO)_2$		4.67 (t)[f] 4.41 (t)		77

(continued)

TABLE I (continued)

Compounds	IR ν(CO) (cm^{-1})	^{1}H NMR (ppm) δ(C$_5$H$_{5-x}$) (x = 0, 1, 2)	MS M$^+$ (m/e)[h]	Ref.
(Me$_4$Si$_2$O)(η-C$_5$H$_4$)$_2$Ti(CO)$_2$	1960, 1885[a]	4.88 (m) A_2B_2[f]	364	78
(η-C$_5$H$_4$PPh$_2$)$_2$Ti(CO)$_2$	1968, 1882[e]			79
Mo(CO)$_5$(η-C$_5$H$_4$CH$_2$CH$_2$PPh$_2$)CpTi(CO)$_2$	1966, 1883[e]	4.6[f]		79
(η-C$_5$H$_4$PPh$_2$)CpTi(CO)$_2$	1960, 1880[b]	5.12–5.30 (m)[e] 4.77–4.91 (m) 4.72 (s)		80
CpMn(CO)$_2$[(η-C$_5$H$_4$PPh$_2$)CpTi(CO)$_2$]	1960, 1880[f]	4.67–5.08 (m)[f] 4.46 (s)	594	80
(η-C$_5$H$_4$CH=CH$_2$)CpTi(CO)$_2$	1980, 1910[f]	5.58 (t)[f] 4.33 (t) 4.65 (s)		81
(η-C$_5$H$_4$CO$_2$Me)$_2$Ti(CO)$_2$	1990, 1910[b]		292	82
(η-C$_5$H$_4$SiMe$_3$)$_2$Zr(CO)$_2$	1970, 1880[i]	5.1 (m)[f]		47b
[η-C$_5$H$_3$(SiMe$_3$)$_2$]$_2$Zr(CO)$_2$	1962, 1875[f]	5.6[f]		47b
[η-C$_5$H$_3$(SiMe$_3$)$_2$]$_2$Hf(CO)$_2$	1950, 1855[i]	5.35[f]		47b

[a] Hexane.
[b] THF.
[c] Pentane.
[d] Heptane.
[e] CD$_2$Cl$_2$.
[f] C$_6$D$_6$(H$_6$).
[g] Acetone-d_6.
[h] Based on most abundant isotope.
[i] Nujol.

3. Mass Spectra

In all cases where the mass spectrum has been recorded (Table I), the metallocene dicarbonyl compounds exhibited a parent ion, M^+, together with peaks corresponding to the stepwise loss of the two CO ligands.

4. Photoelectron Spectra

Fragalia and co-workers have reported the details of the He(I) and He(II) excited photoelectron spectra of $Cp_2Ti(CO)_2$ and concluded that evidence exists for significant backbonding between the Ti $3d$ orbitals and empty carbonyl π^* orbitals. Further, there is no evidence of important overlap between Ti and Cp orbitals. A small electrostatic perturbation of the Cp ligands is caused by the titanium atom (85). Böhm has described an elaborate study of the low energy PE spectrum of $Cp_2Ti(CO)_2$ (1) by means of semiempirical MO calculations (86).

C. Structural Properties

Atwood and co-workers have elucidated by X-ray diffraction studies the structures of 11 of the group 4B metallocene dicarbonyl complexes, including $Cp_2M(CO)_2$ (87–90), $(\eta\text{-}C_5Me_5)_2M(CO)_2$ (62) $(\eta\text{-}C_5Me_5)$-$CpM(CO)_2$ (91) (M = Ti, Zr, Hf), and $(\eta^5\text{-}C_9H_7)_2M(CO)_2$ (91) (M = Ti, Zr). A salient feature of the bis(η-cyclopentadienyl) congeners is that the cyclopentadienyl rings of $Cp_2Ti(CO)_2$ (1) are eclipsed while those of $Cp_2M(CO)_2$ (M = Zr, Hf) are staggered. Thus, $Cp_2Zr(CO)_2$ (2) and $Cp_2Hf(CO)_2$ (3) are isostructural while (1) is not. The decamethylmetallocene analogs are similar to the bis(η-cyclopentadienyl) compounds in that $(\eta\text{-}C_5Me_5)_2Zr(CO)_2$ (30) and $(\eta\text{-}C_5Me_5)_2Hf(CO)_2$ (33) are isostructural. Table II contains pertinent structural data for the metallocene dicarbonyls, and ORTEP drawings of several of these complexes appear in Fig. 1–3.

VI

MONOCARBONYL DERIVATIVES OF TITANOCENE

A. Formation of Titanocene Monocarbonyl–Phosphine Complexes

While the replacement of carbon monoxide by a tertiary phosphine ligand represents one of the most fundamental substitution reactions in metal carbonyl chemistry, it was not until 1975, some 16 years after the

TABLE II

BOND DISTANCES (Å) AND ANGLES (°) FOR THE GROUP 4B METALLOCENE DICARBONYLS

Compound	M—C(σ)	M—C(η^5)$_{av}$	M—centroid$_{av}$	C(σ)—M—C(σ)	Centroid—M—centroid	Centroid—M—C(σ)$_{av}$	M—C—O	Ref.
Cp$_2$Ti(CO)$_2$	2.030(11)	2.347	2.025	87.9(6)	138.6	104.8	179.4(9)	87,88
(η-C$_5$Me$_5$)$_2$Ti(CO)$_2$	2.01(1)	2.384(12)	2.07	83.3(3)	147.9		176.3	62
Cp$_2$Zr(CO)$_2$	2.187(4)	2.48(2)	2.184	89.2(2)	143.4	102.9	178.6(4)	89
(η-C$_5$Me$_5$)$_2$Zr(CO)$_2$	2.145(9)	2.498(9)	2.20	86.3(5)	147.4		179.3(8)	62
Cp$_2$Hf(CO)$_2$	2.16(2)	2.45(5)	2.16	89.3(9)	141	104	178(1)	90
(η-C$_5$Me$_5$)$_2$Hf(CO)$_2$	2.14(2)		2.17	87(1)	148.2		178.2	62

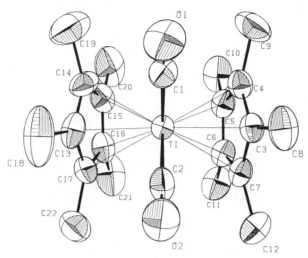

FIG. 1.   Molecular structure of $(\eta\text{-}C_5Me_5)_2Ti(CO)_2$ (**27**). [Ⓒ 1981 American Chemical Society (*62*).]

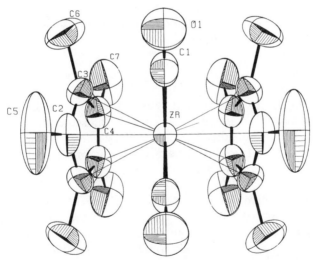

FIG. 2.   Molecular structure of $(\eta\text{-}C_5Me_5)_2Zr(CO)_2$ (**30**). [Ⓒ 1981 American Chemical Society (*62*).]

discovery of $Cp_2Ti(CO)_2$ (**1**), that a phosphine-substituted derivative of **1** was reported (*26*). Bis(η-cyclopentadienyl)carbonyltrimethylphos-phinetitanium was prepared via the thermolysis of $Cp_2Ti(CO)_2$ (**1**) and trimethylphosphine in refluxing hexane over a 2-hour period (*61*). It was isolated in 54% yield and was characterized by meticulous examination of

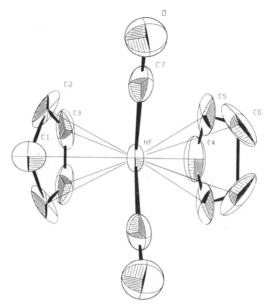

FIG. 3.   Molecular structure of $Cp_2Hf(CO)_2$ (**3**). [© 1979 American Chemical Society. (*90*).]

its IR spectrum which contained a diagnostic low energy metal carbonyl band at 1863 cm^{-1}.

Some years later, Rausch and Sikora prepared similar complexes, $Cp_2Ti(CO)(PR_3)$ (R = Et, Ph), which they subsequently found to be convenient sources of the 16-e$^-$, coordinatively unsaturated species "$Cp_2Ti(CO)$" (*49,50*). Refluxing hexane solutions of $Cp_2Ti(CO)_2$ (**1**) with excess triethylphosphine (8 hours) or triphenylphosphine (48 hours) led to the formation of $Cp_2Ti(CO)(PEt_3)$ (**41**) and $Cp_2Ti(CO)(PPh_3)$ (**42**) as extremely air-sensitive, brown–red solids in yields of 60 and 63%, respectively (*50*). Monocarbonyl complex **41** was also observed to form upon the photolysis of heptane solutions of **1** with excess PEt_3 over a 3-hour period at 15°C. However, the formation of substantial amounts of $Cp_2Ti(CO)(PEt_3)$ (**41**) could only be realized when an inert gas (Ar) was sparged through the solution during the photolysis to sweep out the dissociated CO. This phenomenon can be attributed to the equilibrium reaction shown below.

$$Cp_2Ti(CO)_2 \ + \ PEt_3 \ \underset{\text{no Ar Sparge}}{\overset{\substack{h\nu \\ \text{Ar Sparge}}}{\rightleftarrows}} \ Cp_2Ti(CO)(PEt_3)$$

$$(\underset{\sim}{1}) \qquad\qquad\qquad\qquad\qquad (\underset{\sim}{41})$$

$$+ \ CO$$

The PEt_3 ligand of $Cp_2Ti(CO)(PEt_3)$ (**41**) is extremely labile, and **41** reacts very rapidly with CO to reform $Cp_2Ti(CO)_2$. This process has been demonstrated independently with $Cp_2Ti(CO)(PEt_3)$ (**41**) prepared by the thermolysis reaction (*50*). Addition of CO to a solution of **41** in an aliphatic or aromatic solvent in the dark led to quantitative formation of **1** instantaneously. Thus, without the use of an argon sparge the equilibrium of the aforementioned photochemical reaction lies to the left, whereas removal of CO from the solution favors the formation of **41**.

Complexes **41** and **42** were characterized by their IR and ¹H-NMR spectra, and **41** also by elemental analysis. Table III contains the pertinent spectral data. Noteworthy are the very low energy terminal carbonyl bands for **41** and **42** at 1864 cm⁻¹ (hexane). The weak π-accepting abilities of PR_3 (R = Et, Ph) allow the lone CO ligand to π-backbond to the Ti(II) center to a much greater degree. The ¹H-NMR spectrum of **41** exhibited a doublet (J_{H-p} = 1.5 Hz) at δ 4.75 due to the coupling of the cyclopentadienyl protons with the ³¹P nucleus, while complex **42** exhibited a broad cyclopentadienyl singlet at δ 4.67.

Atwood and Rogers have determined the structure of **41** via an X-ray diffraction study (*50*). The Ti-P and Ti-CO bond distances were found to be 2.585(1) and 2.009(4) Å, respectively. An ORTEP drawing of $C_pTi(CO)(PEt_3)$ (**41**) appears in Fig. 4.

The thermal and photochemical reactions of $Cp_2Ti(CO)_2$ (**1**) with the strongly π-accepting phosphines $PF_2N(Me)PF_2$ and PF_3 were also investigated by Rausch and Sikora (*49–51*). Methylaminobis(difluorophosphine), $PF_2N(Me)PF_2$, has received much attention of late in organotransition

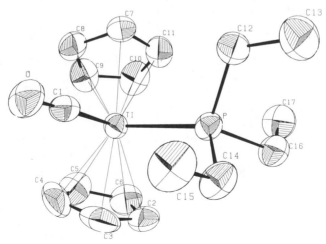

Fig. 4. Molecular structure of $Cp_2Ti(CO)(PEt_3)$ (**41**). [© 1983 American Chemical Society (*50*).]

$$Cp_2Ti(CO)_2 \ + \ excess \ PF_2N(Me)PF_2 \ \xrightarrow[\text{or pentane reflux}]{h\nu, \text{pentane}, 15\,°C}$$

(1)

$$Cp_2Ti \underset{PF_2}{\overset{CO}{<}}$$

(43)

metal chemistry, due primarily to the elegant work of R. B. King and co-workers (92). The reaction of 1 with $PF_2N(Me)PF_2$ under either photochemical or thermal conditions gave only the monosubstituted product, $Cp_2Ti(CO)[PF_2N(Me)PF_2]$ (43), in moderate yields. No evidence for the chelated complex, $Cp_2Ti[PF_2N(Me)PF_2]$, was found (51).

Complex 43 is a very air-sensitive, light orange solid which exhibited a doublet (J_{H-p} = 3.0 Hz) at δ 4.65 in its ^{1}H-NMR spectrum assignable to the cyclopentadienyl protons coupled with the ^{31}P nucleus. The assignment of $PF_2N(Me)PF_2$ as a monodentate ligand in 43 is confirmed by the 10:3 integration ratio of the cyclopentadienyl protons relative to the amino methyl protons and is corroborated by the appearance of the parent ion at $m/e = 373$ in the mass spectrum. The IR spectrum of 43 shows a strong metal carbonyl band at 1925 cm^{-1}. This shift to higher energy relative to the frequencies of $Cp_2Ti(CO)(PR_3)$ (R = Me, Et, Ph) at 1863 and 1864 cm^{-1} illustrates the stronger π-acidity of $PF_2N(Me)PF_2$.

All attempts to prepare $Cp_2Ti(CO)(PF_3)$ (44) via the thermolysis or photolysis of $Cp_2Ti(CO)_2$ (1) and PF_3 resulted in the contamination of 44 with significant amounts of $Cp_2Ti(CO)_2$ and/or $Cp_2Ti(PF_3)_2$ (45) (49,50). Furthermore, all attempts to separate mixtures of 1, 44, and 45 have proved unsuccessful. Since PF_3 is a gas (bp -101.8°C), its use in any thermal reaction almost always demands the use of a sealed vessel. Such a requirement is a distinct disadvantage when dealing with reversible metal carbonyl substitution reactions, since the liberated CO cannot be readily removed from the reaction mixture. When $Cp_2Ti(CO)_2$ (1) and excess PF_3 were allowed to react in a sealed tube at 60°C for 6 hours, an inseparable mixture of $Cp_2Ti(CO)_2$ (1) and $Cp_2Ti(CO)(PF_3)$ (44) was obtained (50).

The photolysis of a hexane solution of 1 with a PF_3 sparge led ultimately to pure $Cp_2Ti(PF_3)_2$ (45) as expected due to the PF_3 stream sweeping the photodissociated CO from the solution.[3] This reaction was conveniently monitored using IR spectroscopy since the intermediate $Cp_2Ti(CO)(PF_3)$

---

[3] The authors feel that a PF_3 sparge through a heated hexane solution of $Cp_2Ti(CO)_2$ (1) would also lead to $Cp_2Ti(PF_3)_2$ (45).

(44) could be identified by its carbonyl band at 1932 cm^{-1} (50). At time zero only the metal carbonyl bands due to 1 [ν(CO) = 1977 and 1899 cm^{-1}] were observed. After 10 minutes of photolysis, these two bands had decreased in intensity and a new strong carbonyl stretching frequency at 1932 cm^{-1}, assignable to 44, was observed. Further photolysis ($t = 30$ minutes) resulted in only the observation of 44 (substantial amounts of 45 would be expected to also have formed). Over the next hour the band at 1932 cm^{-1} decreased in intensity until it was no longer detectable at $t = 90$ minutes. The suggested pathway for the formation of Cp$_2$Ti(PF$_3$)$_2$ (45) is depicted below.

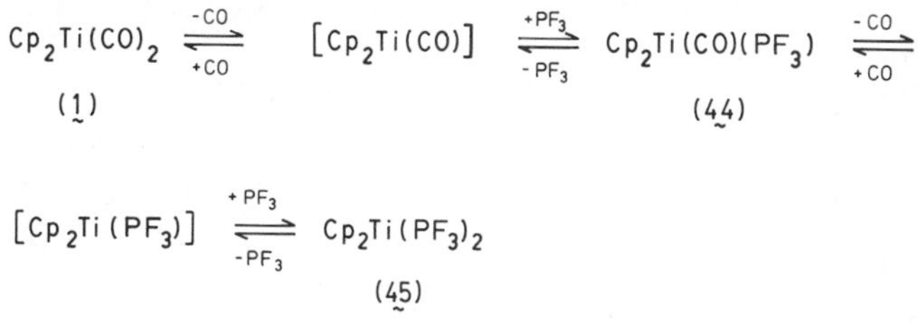

$$Cp_2Ti(CO)_2 \underset{+CO}{\overset{-CO}{\rightleftharpoons}} [Cp_2Ti(CO)] \underset{-PF_3}{\overset{+PF_3}{\rightleftharpoons}} Cp_2Ti(CO)(PF_3) \underset{+CO}{\overset{-CO}{\rightleftharpoons}}$$

(1)   (44)

$$[Cp_2Ti(PF_3)] \underset{-PF_3}{\overset{+PF_3}{\rightleftharpoons}} Cp_2Ti(PF_3)_2$$

(45)

When PF$_3$ was sparged through a hexane solution of Cp$_2$Ti(CO)$_2$ (1) in the absence of light for 90 minutes, a weak band at 1932 cm^{-1} assignable to Cp$_2$Ti(CO)(PF$_3$) (44) was observed in addition to strong bands due to Cp$_2$Ti(CO)$_2$ (1). In contrast, when CO was sparged through a hexane solution of Cp$_2$Ti(PF$_3$)$_2$ (45) in the absence of light, only the metal carbonyl bands of Cp$_2$Ti(CO)$_2$ could be observed at the end of the 90-minute reaction period (50).

A more facile route to Cp$_2$Ti(CO)(PF$_3$) (44), which circumvents the separation problem associated with either the thermal or photochemical reaction of 1 and PF$_3$, takes advantage of the extreme lability of PR$_3$ (R = Et, Ph) in Cp$_2$Ti(CO)(PR$_3$). The reaction of strong π-acceptor ligands (L) with Cp$_2$Ti(CO)(PR$_3$) leads to the rapid displacement of PR$_3$ and the formation of Cp$_2$Ti(CO)(L) in high yield.[4] Thus, merely bubbling PF$_3$ through a hexane solution of Cp$_2$Ti(CO)(PEt$_3$) (41) at room temperature for 45 seconds resulted in the quantitative formation of Cp$_2$Ti(CO)(PF$_3$) (44) as evidenced by IR and ^{1}H-NMR spectroscopy (50). A similar reaction was noted for 41 and PF$_2$N(Me)PF$_2$ (51).

Monocarbonyl 44 exhibited a doublet ($J_{H-P} = 3.5$ Hz) at $\delta 4.59$ in its ^{1}H-NMR spectrum, assignable to the cyclopentadienyl protons coupled with the ^{31}P nucleus. Its IR spectrum displayed a strong metal carbonyl band at 1932 cm^{-1}, while the mass spectrum showed a parent ion at

[4] While Cp$_2$Ti(CO)(PEt$_3$) (41) and Cp$_2$Ti(CO)(PPh$_3$) (42) show identical reactivity toward L, Cp$_2$Ti(CO)(PEt$_3$) (41) is preferred for preparative reactions since PEt$_3$ is more easily removed on workup than is PPh$_3$.

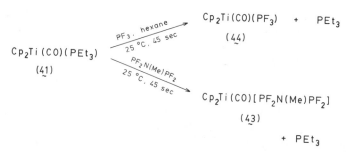

$m/e = 294$. The spectral properties of **43** were identical to those obtained when **43** was prepared thermally or photochemically from $Cp_2Ti(CO)_2$ (**1**) and $PF_2N(Me)PF_2$.

The titanocene monocarbonyl–triphenylphosphite complex, $Cp_2Ti-(CO)[P(OPh)_3]$ (**46**) has also been prepared by this method (50). The cyclopentadienyl protons of **46** appeared as a doublet ($J_{H-P} = 2.5$ Hz) **46**, while the parent ion ($m/e = 516$) was found in the mass spectrum.

$$Cp_2Ti(CO)(PEt_3) \ + \ P(OPh)_3 \ \xrightarrow[\substack{25\ °C,\ 30\ min \\ -PEt_3}]{hexane} \ Cp_2Ti(CO)[P(OPh)_3]$$
$$\qquad\qquad (41) \qquad\qquad\qquad\qquad\qquad\qquad\qquad\qquad (46)$$

The IR and ¹H-NMR spectral data for the various titanocene mono-carbonyl–phosphine complexes are compiled in Table III. Examination of the carbonyl stretching frequencies (Table III) nicely demonstrates the enhanced π-backbonding of the titanium center to CO as the π-accepting ability of the phosphine ligand decreases.

B. *Formation of Titanocene Monocarbonyl η^2-Acetylene Complexes*

The existence of π-bonded acetylene complexes for the group 5B (93) and 6B (84) metals is not uncommon; however, until 1974 no such complexes had been reported for the group 4B metals. It was at this time that Floriani and Fachinetti described the reaction of $Cp_2Ti(CO)_2$ (**1**) with diphenylacetylene in heptane solution under reduced pressure which led to the formation of the first group 4B η^2-acetylene complex, $Cp_2Ti(CO)(\eta^2-PhC\equiv CPh)$ (**47**) (94). Over a 3-hour period, the evolved CO was removed *in vacuo* and **47** precipitated as a yellow solid. The product was character-ized by its ¹H-NMR and IR spectra, molecular weight measurement, and elemental analysis. The cyclopentadienyl protons appeared as a singlet at δ 5.0 in the ¹H-NMR spectrum, while the IR spectrum showed a strong metal carbonyl band at 1995 cm⁻¹ (Nujol) and a metal π-acetylene band at 1780 cm⁻¹ (Nujol).

TABLE III

IR AND ^{1}H-NMR SPECTRAL PROPERTIES FOR TITANOCENE MONOCARBONYL–PHOSPHINE COMPLEXES

Complex	ν(CO) (cm^{-1}) (hexane)	ν(CO) (cm^{-1}) (THF)	$\delta(\eta^5\text{-}C_5H_5)$ (ppm) (C_6D_6)	J_{H-P}(Hz) ($\eta^5\text{-}C_5H_5$)	Ref.
CpTi(CO)(PF$_3$)	1932	1920	4.59 (d)	3.5	49,50
Cp$_2$Ti(CO)[PF$_2$N(Me)PF$_2$]	1925	1907	4.65 (d)	3.0	51
Cp$_2$Ti(CO)[P(OPh)$_3$]	1900	1887	4.72 (d)	2.5	50
Cp$_2$Ti(CO)(PPh$_3$)	1864	1850	4.67 (s)		49,50
Cp$_2$Ti(CO)(PEt$_3$)	1864	1850	4.75 (d)	1.5	49,50
Cp$_2$Ti(CO)(PMe$_3$)	1863				26,61

$$Cp_2Ti(CO)_2 \quad + \quad PhC \equiv CPh \quad \xrightarrow[\substack{25\,°C,\ 3\,h \\ -CO}]{\substack{vacuum, \\ heptane}}$$

(1)

$$Cp_2Ti(CO)(h^2\text{-}PhC \equiv CPh)$$

(47)

Later, the structure of **47** was determined by X-ray diffraction studies (*95*), and its molecular structure appears in Fig. 5. Examination of the structure reveals the η^2- bonding mode of the acetylene ligand. The phenyl rings of the acetylene are bent back from linearity by 41 and 34° while the carbon–carbon bond length (C≡C) is 1.285(10) Å. The Ti–CO distance is 2.050(8) Å, similar to those found in $Cp_2Ti(CO)_2$ (*87,88*) (**1**) and $Cp_2Ti(CO)(PEt_3)$ (*50*) (**41**) [2.030(11) and 2.009(4) Å, respectively].

In a CO atmosphere, $Cp_2Ti(CO)(\eta^2\text{-}PhC≡CPh)$ (**47**) is readily converted back to $Cp_2Ti(CO)_2$ (**1**). When **47** is allowed to stir in aromatic solvents at temperatures greater than 30°C, it decomposes to $Cp_2Ti(CO)_2$

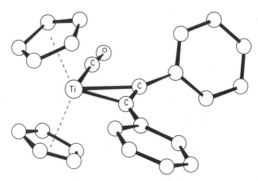

(**1**) and bis(η-cyclopentadienyl)-2,3,4,5-tetraphenyltitanacyclopentadiene (*94,95*). Complex **47** was also found to serve as a very effective hydrogenation catalyst; olefins (e.g., styrene, *trans*-stilbene) could be reduced quantitatively to the corresponding alkanes under 1 atm of H_2 at 10–20°C in only a few minutes.

FIG. 5. Molecular structure of $Cp_2Ti(CO)(\eta^2\text{-}PhC≡CPh)$ (**47**). [Reprinted with permission of the Royal Society of Chemistry (*95*).]

In extending the synthetic utility of $Cp_2Ti(CO)(PEt_3)$ (**41**) as a convenient source of "$Cp_2Ti(CO)$," Rausch and co-workers reacted **41** with various acetylenes and thus obtained the corresponding titanocene monocarbonyl-η^2-acetylene complexes in good yields (*50,96*). These complexes

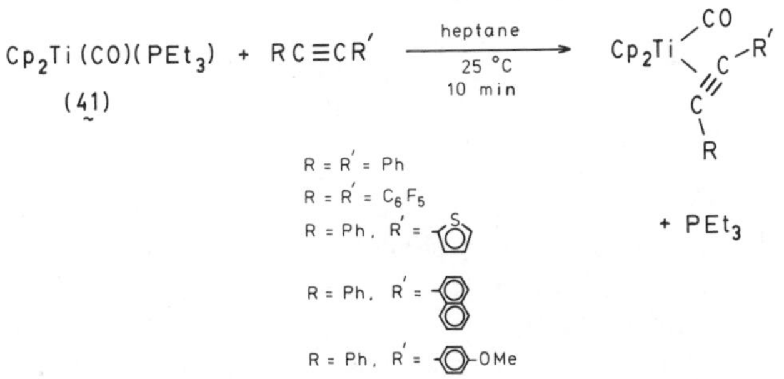

$$Cp_2Ti(CO)(PEt_3) \; + \; RC\equiv CR' \xrightarrow[\substack{25\ ^\circ C \\ 10\ min}]{heptane} Cp_2Ti\overset{CO}{\underset{\substack{C \\ | \\ R}}{\diagdown \overset{|||}{\underset{}{C}}-R'}} \; + \; PEt_3$$

(**41**)

R = R' = Ph
R = R' = C₆F₅
R = Ph, R' = (thienyl)
R = Ph, R' = (naphthyl)
R = Ph, R' = ⟨O⟩-OMe

were characterized by their IR and ^{1}H-NMR spectra and elemental analysis. Noteworthy is the comparison of the metal carbonyl and metal π-acetylene stretching frequencies of the complexes $Cp_2Ti(CO)(\eta^2\text{-}PhC\equiv CPh)$ (**47**) and $Cp_2Ti(CO)(\eta^2\text{-}C_6F_5C\equiv CC_6F_5)$ (**48**). By virtue of the 10 electronegative fluorine atoms in $C_6F_5C\equiv CC_6F_5$, the latter would be expected to be a stronger π-acceptor than $PhC\equiv CPh$. Both $\nu(CO)$ and $\nu(C\equiv C)$ frequencies reflect this effect, appearing at 1995 and 1780 cm^{-1} (Nujol), respectively, for **47**, and at 2025 and 1770 cm^{-1} (Nujol), respectively, for **48**. Complex **48** also appeared to be more thermally robust than **47** since no change in the ^{1}H-NMR spectrum of **48** could be observed when a C_6D_6 solution of it was heated at 58°C for ~2 hours (*51*). Under these conditions, **47** disproportionates to $Cp_2Ti(CO)_2$ (**1**) and $Cp_2Ti(C_4Ph_4)$.

C. Formation of Titanocene Monocarbonyl–η^2-Olefin Complexes

The facile displacement of PEt₃ in $Cp_2Ti(CO)(PEt_3)$ (**41**) has made possible the formation of the first fully characterized monoolefin complexes of titanocene, $Cp_2Ti(CO)(\eta^2\text{-}trans\text{-}RO_2CCH\equiv CHCO_2R)$ [R = Me (**49**), R = Et (**50**)]. Rausch *et al.* discovered that reactions between **41** and either dimethyl fumarate[5] or dimethyl maleate in aromatic solvents at 0°C resulted in dark green solutions, from which the very air-sensitive green monocarbonyl–η^2-olefin complex **49**[6] could be precipitated in 65% yield

---

[5] Treatment of **41** with diethyl fumarate under similar conditions gave **50**, which displayed physical and spectroscopic properties similar to those of **49**.

[6] While **49** is illustrated here as an olefin π-complex, it can alternatively be classified as a titanacyclopropane derivative (*50*).

$$\text{Cp}_2\text{Ti(CO)(PEt}_3) \quad + \quad \text{MeO}_2\text{CCH}=\text{CHCO}_2\text{Me} \quad \xrightarrow[\substack{0\,°\text{C}\,,\,15\,\text{min}\\-\text{PEt}_3}]{\text{toluene}}$$

$$\underset{(\underset{\sim}{41})}{} \qquad\qquad \underset{(\text{cis or trans})}{}$$

(49)

by the addition of cold pentane (50). Complex **49** was characterized by its chemical degradation products, elemental analysis, cryoscopic molecular weight determination, and IR and ^{1}H-NMR spectra.

Degradation of **49** at 0°C in a HCl-saturated toluene solution gave Cp_2TiCl_2 in quantitative yield and dimethyl succinate in 60% yield. Air oxidation at 0°C of a toluene solution of **49** yielded dimethyl fumarate quantitatively.

The IR spectrum of **49** in benzene exhibited strong bands at 2010 and 1675 cm^{-1}, assignable to the metal carbonyl and ligand carbonyl stretching frequencies, respectively. The latter band is shifted only 45 cm^{-1} to lower energy than that of uncomplexed dimethyl fumarate (1720 cm^{-1}), indicating that the ligand carbonyls are not involved in coordination with the metal. Direct interaction of these groups with the metal would result in a much lower shift in wavenumber (97). Furthermore, the diethyl fumarate complex of vanadocene, $\text{Cp}_2\text{V}(\eta^2\text{-}trans\text{-}\text{EtO}_2\text{CCH}=\text{CHCO}_2\text{Et})$, whose X-ray crystallographic determination showed coordination only through the carbon–carbon double bond, displayed the ligand carbonyl band at 1680 cm^{-1} in its IR spectrum (97).

The ^{1}H-NMR spectrum of **49** was consistent with the titanocene monocarbonyl-η^2-olefin assignment and was found to be temperature dependent as determined by variable-temperature ^{1}H-NMR experiments. At −5°C resonances appeared at δ4.95 (s, 5H) and 4.75 (s, 5H) assignable to the protons of nonequivalent cyclopentadienyl rings, at δ 3.58 (s, 3H) and 3.52 (s, 3H) assignable to the protons of nonequivalent methoxy groups, and at δ 2.99 (d, $J_{\text{H–H}} = 11$ Hz) and 2.71 (d, $J_{\text{H–H}} = 11$ Hz) assignable to the nonequivalent vinylic protons. When the temperature was raised to 25°C, coalescence of the respective resonances occurred. Also, when the solution temperature was increased to 25°C, decomposition of **49** began to occur, producing $\text{Cp}_2\text{Ti(CO)}_2$, dimethyl fumarate, and what appeared to be a form of titanocene as evidenced by the appearance of a broad resonance extending from δ 5.9 to 6.5 (50).

Of special significance concerning the reaction of $\text{Cp}_2\text{Ti(CO)(PEt}_3)$ (**41**) with dimethyl maleate was the observation that the 40% molar excess of dimethyl maleate used in this reaction appeared in the ^{1}H-NMR spectrum as uncomplexed dimethyl fumarate rather than uncomplexed dimethyl maleate, suggesting the occurrence of catalytic isomerization. In further studies, when **41** was treated with a sevenfold molar excess of dimethyl maleate at low temperature, complete isomerization to dimethyl fumarate was observed within 20 minutes, as evidenced by the appearance and disappearance of the respective vinylic and methoxy resonances of these isomeric olefins.

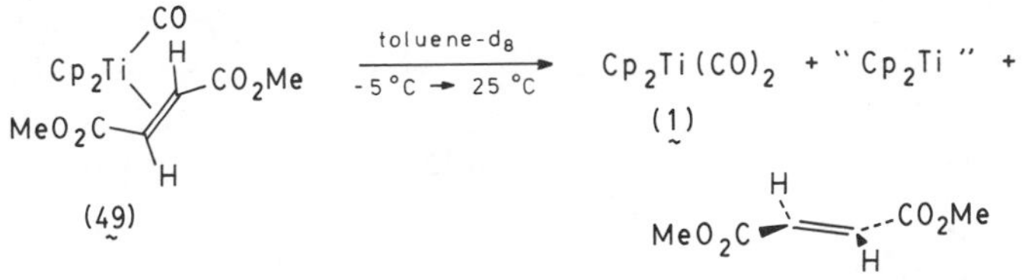

D. Formation of Titanocene Monocarbonyl Complexes Containing a σ-Bonded Ligand

In 1979, Teuben and co-workers found that titanocene complexes, Cp$_2'$TiR (Cp′ = η-C$_5$H$_5$ or η-C$_5$Me$_5$), containing the electron-withdrawing group R would yield carbonyl adducts, Cp$_2'$Ti(R)(CO), on exposure to CO under mild conditions. For example, treatment of a toluene solution of Cp$_2$Ti(C$_6$F$_5$) with CO gave the brown–green Cp$_2$Ti(CO)(C$_6$F$_5$) (**51**) in

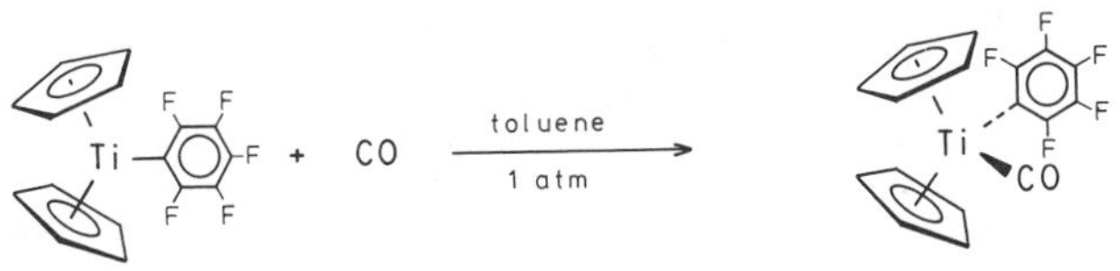

60% yield (98). Solutions of **51** were found to dissociate CO *in vacuo*, giving Cp$_2$Ti(C$_6$F$_5$) and CO. In the solid state under N$_2$ or Ar, **51** slowly decomposed at room temperature producing Cp$_2$Ti(CO)$_2$ (**1**) as one of the products. The IR spectrum of **51** showed a high energy metal carbonyl band at 2060 cm^{-1} as might be expected due to the electron-withdrawing nature of the pentafluorophenyl ligand together with titanium in the 3+ oxidation state. Complex **51** is d^1 paramagnetic. Its EPR spectrum displayed a well-resolved triplet (1/2/1; $g = 1.994$) at −55°C due to the interaction of the lone electron with the *ortho*-fluorine atoms via a through-space mechanism.

Similarly, treatment of a hexane solution of $(\eta\text{-}C_5Me_5)_2TiCl$ with CO at room temperature gave the thermally unstable adduct $(\eta\text{-}C_5Me_5)_2Ti(CO)(Cl)$ **(52)** as evidenced by again a high energy metal carbonyl band at 2000 cm^{-1} (*98*). Complex **52** slowly disproportionated at

$$2 \ (h\text{-}C_5Me_5)_2TiCl \ \xrightarrow{2\ CO} \ 2 \ (h\text{-}C_5Me_5)_2Ti\diagdown\begin{matrix}CO\\Cl\end{matrix}$$

$$(5\underset{\sim}{2})$$

$$2 \ (5\underset{\sim}{2}) \ \longrightarrow \ (h\text{-}C_5Me_5)_2TiCl_2 \ + \ (h\text{-}C_5Me_5)_2Ti(CO)_2$$

$$(2\underset{\sim}{7})$$

room temperature according to the reaction below. The carbonyl band at 2000 cm^{-1} gradually decreased in intensity as the bands for $(\eta\text{-}C_5Me_5)_2Ti(CO)_2$ **(27)** increased in intensity. The EPR spectrum of $(\eta\text{-}C_5Me_5)_2TiCl$ in toluene showed a singlet with $g = 1.956$. On exposure to CO, the singlet shifted to $g = 1.968$ and gradually disappeared over 16 hours on formation of the diamagnetic disproportionation products.

If the R group of $Cp_2'TiR$ was not electron-withdrawing, reaction with CO resulted in the formation of an acyl compound. However, isolation of the initially formed CO adduct, $Cp_2'Ti(CO)(R)$, was unsuccessful in these cases.

E. *Formation of Other Titanocene Monocarbonyl Complexes*

The formation of a Ti(IV) carbonyl complex via the reaction of $Cp_2Ti(CO)_2$ with TCNE (tetracyanoethylene) in benzene at room temperature has been described by Demerseman and co-workers. A green–black, air-sensitive precipitate was formed which liberated 1 equivalent of CO per equivalent of Ti on treatment with iodine and gave an elemental analysis corresponding to $Cp_2Ti(CO)(TCNE)$ (*99*). The precipitate was found to be diamagnetic and exhibited an intense high energy metal carbonyl band at 2055 cm^{-1} in its IR spectrum. The assignment at 2055 cm^{-1} was confirmed by use of ^{13}CO-enriched $Cp_2Ti(CO)_2$. Also present in the IR spectrum were bands at 2181 and 2104 cm^{-1} that were assigned to C≡N vibrations. On the basis of the above data, the cationic titanium carbonyl complex shown below was postulated as the structure for the green–black reaction product.

$$\left[Cp_2Ti \underset{N}{\overset{CO}{\diagdown}} \underset{C}{\overset{}{\diagup}} \underset{NC}{\overset{}{\diagdown}} \underset{}{\overset{}{\diagup}} \underset{CN}{\overset{}{\diagdown}} \underset{N}{\overset{OC}{\diagup}} TiCp_2 \right]^{2+} \quad \left[\overset{-}{IN} \underset{C}{\overset{}{\diagdown}} \underset{NC}{\overset{}{\diagup}} \underset{}{\overset{}{\diagdown}} \underset{CN}{\overset{}{\diagup}} \underset{C}{\overset{}{\diagdown}} \overset{-}{NI} \right]^{2-}$$

The high energy metal carbonyl band would be consistent with a Ti(IV) cationic complex, while the C≡N bands correspond closely to those of the (TCNE)$^{2-}$ dianion at 2160 and 2095 cm^{-1} and also to those in (PPh$_3$)$_2$-(CO)IrN=C=C(CN)C(CN)=C=NIr(CO)(PPh$_3$)$_2$.

Some very meticulous work by Pez and co-workers concerning dinitrogen coordination to titanium metallocenes has resulted in the preparation of a titanium dinitrogen carbonyl complex. When the triply bridging dinitrogen complex, (μ_3-N$_2$)[(η^5:η^5-C$_{10}$H$_8$)Cp$_2$Ti$_2$][(η^1:η^5-C$_5$H$_4$)Cp$_3$Ti$_2$]·[Cp$_2$Ti(diglyme)]·(diglyme), was treated with CO (0.7 atm) in THF for 2.5 hours at room temperature, a compound of approximate composition C$_{20}$H$_{20}$Ti$_2$N$_2$(CO)$_2$ could be isolated. The IR spectrum exhibited a μ_2-bridging dinitrogen band at 1502 cm^{-1} as well as a bridging carbonyl band at 1710 cm^{-1} and terminal carbonyl bands at 1875 and 1965 cm^{-1} (100).

VII

MONOCARBONYL DERIVATIVES OF ZIRCONOCENE

A. Formation of Zirconocene Monocarbonyl–Phosphine Complexes

As with the titanium congener, the first phosphine-substituted derivative of Cp$_2$Zr(CO)$_2$ (2) was also prepared by Demerseman et al. The thermolysis of 2 with excess trimethylphosphine in refluxing hexane gave a maroon product, Cp$_2$Zr(CO)(PMe$_3$), which was identified by a detailed study of its IR spectrum. The low energy metal carbonyl stretching frequency appeared at 1852 cm^{-1} (61).

$$Cp_2Zr(CO)_2 \; + \; PMe_3 \; \xrightarrow[\Delta, \; 30 \; min]{hexane} \; Cp_2Zr(CO)(PMe_3) \; + \; CO$$
$$(2)$$

Schwartz and Gell have reported that a benzene solution of Cp$_2$Zr(PMePh$_2$)$_2$ containing 2 equivalents of PMePPh$_2$ absorbed 1.4 equivalents of CO (based on Zr) over a 4-hour period to give at equilibrium a mixture of $Cp_2Zr(CO)(PMePh_2)$ (25) (60%) and $Cp_2Zr(CO)_2$ (2) (40%) (55). Complexes 25 was characterized by the low energy metal carbonyl band at 1840 cm^{-1} appearing in its IR spectrum and by the cyclopentadienyl doublet at δ 4.90 (J_{H-P} = 1.6 Hz) present in the ^{1}H-NMR spectrum.

Rausch and Sikora have recently described the photochemically induced reaction of $Cp_2Zr(CO)_2$ (2) and triphenylphosphine from which $Cp_2Zr(CO)(PPh_3)$ could be isolated as highly air-sensitive, maroon microcrystals in 28% yield (58). Similarly, when hexane solutions of 2 and PMe_3 or PF_3 were irradiated, the corresponding complexes,

$$Cp_2Zr(CO)_2 \; + \; PR_3 \quad \xrightarrow[\substack{\text{hexane} \\ 15 \text{ °C}}]{h\nu} \quad Cp_2Zr(CO)(PR_3) \; + \; CO$$
$$(2)$$
$$R = Ph, Me, F$$

$Cp_2Zr(CO)(PR_3)$ (R = Me, F), were isolated as oily, air-sensitive solids. The competing photodegradation of $Cp_2Zr(CO)_2$ (2) to oligomeric zirconocene prevented the formation of these monocarbonyl–phosphine complexes in higher yields (51).

While $Cp_2Zr(CO)(PPh_3)$ was found to be more reactive toward acetylenes than $Cp_2Zr(CO)_2$ (2), no monocarbonyl-η^2-acetylene complexes of zirconocene were observed in contrast to the reaction of acetylenes with $Cp_2Ti(CO)(PPh_3)$ (42) (50). Instead the reaction of $Cp_2Zr(CO)(PPh_3)$ with RC≡CR (R = Et, Ph) led directly to the respective zirconacyclopentadienes (58).

The thermolysis of $Cp_2Zr(CO)_2$ (2) and $PF_2N(Me)PF_2$ in refluxing pentane for 4 hours led to the isolation of $Cp_2Zr(CO)[PF_2N(Me)PF_2]$ in 45% yield (51). The 10:3 integration ratio of the cyclopentadienyl protons to the amino methyl protons in the ^{1}H-NMR spectrum established the assignment of $PF_2N(Me)PF_2$ as a monodentate ligand in this complex.

The IR and ^{1}H-NMR spectral data for the various zirconocene monocarbonyl–phosphine complexes are compiled in Table IV.

B. Formation of Other Zirconocene Monocarbonyl Complexes

Bercaw and co-workers reported that treatment of toluene solutions of $[(\eta\text{-}C_5Me_5)_2ZrN_2]_2N_2$ (29) with excess CO at 25°C resulted in the evolution of N_2 (3 equivalents per equivalent of 29) and formation of $(\eta\text{-}C_5Me_5)_2Zr(CO)_2$ (30). However, they also discovered that if the reaction temperature was held at -23°C, only 2 equivalents of N_2 per equivalent of 29 were evolved, giving the metallic green-colored $[(\eta\text{-}$

TABLE IV

IR AND ^{1}H-NMR SPECTRAL PROPERTIES FOR ZIRCONOCENE
MONOCARBONYL–PHOSPHINE COMPLEXES

Complex	$\nu(CO)$ (cm^{-1})	$\delta(\eta^5\text{-}C_5H_5)$ (ppm) (C$_6$D$_6$)	J_{H-P}(Hz) ($\eta^5\text{-}C_5H_5$)	Ref.
Cp$_2$Zr(CO)PF$_3$	1921[a]; 1932[b]	4.92 (d)	2.5	51
Cp$_2$Zr(CO)[PF$_2$N(Me)PF$_2$]	1907[a]; 1925[c]	4.95 (d)	2.5	51
Cp$_2$Zr(CO)(PPh$_3$)	1842[a]	4.93 (d)	1.8	58
Cp$_2$Zr(CO)(PMePh$_2$)	1840[d]	4.90 (d)	1.6	55
Cp$_2$Zr(CO)(PMe$_3$)	1836[a]; 1852[b]	4.98 (d)	2.0	51,61
[η-C$_5$H$_3$(SiMe$_3$)$_2$]$_2$Zr(CO)(PMe$_3$)	1842[b]			47b

[a] THF.
[b] Hexane.
[c] Pentane.
[d] Benzene.

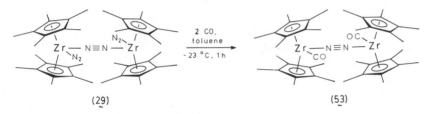

(29)   (53)

C$_5$Me$_5$)$_2$Zr(CO)]$_2$N$_2$ (53). The IR spectrum of 53 showed metal carbonyl bands at 1902 and 1860 cm^{-1} (Nujol), while the ^{1}H-NMR spectrum displayed two singlets of equal intensity at δ 1.80 and 1.82 (101,102).

The carbonyl adduct (η-C$_5$Me$_5$)$_2$Zr(H)$_2$(CO) (54) could be formed by treatment of a toluene solution of the 16-e^- complex, (η-C$_5$Me$_5$)$_2$ZrH$_2$, with CO at $-78°$C. While 54 was not sufficiently stable for its isolation, it was thoroughly characterized in solution at low temperature. The reaction of 54 and excess HCl at $-78°$C gave (η-C$_5$Me$_5$)$_2$ZrCl$_2$, H$_2$, and CO according to the equation below (67).

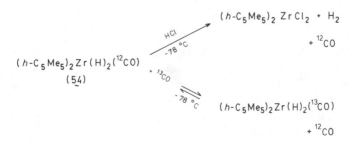

The ^{1}H-NMR spectrum of **54** in toluene-d_8 at $-64°C$ exhibited a singlet at δ 1.84 assignable to the methyl hydrogens of the two equivalent permethylated cyclopentadienyl rings and also a singlet at δ 1.07 due to the two equivalent hydride ligands. The spectrum of $(\eta\text{-}C_5Me_5)_2Zr(H)_2(^{13}CO)$ showed the same singlet at δ 1.84; however, the hydride resonance now appeared as the expected doublet ($J_{H-^{13}C} = 25.1$ Hz) (*67*). The IR spectrum of **54**, recorded at $-65°C$ displayed a very high energy metal carbonyl band at 2044 cm^{-1} as would be expected for a d^0 carbonyl complex. This assignment was corroborated by the observance of the ^{13}CO isotopomer band at 1999 cm^{-1} (*70*). Labeling studies with ^{13}CO have shown that the CO ligand of **54** is labile. At $-78°C$, 0.21 equivalent of ^{13}CO exchanged per equivalent of **54** after 30 minutes (*67*).

In recent mechanistic studies concerning the reduction of coordinated CO by $(\eta\text{-}C_5Me_5)_2ZrH_2$, Bercaw, Grubbs, and co-workers prepared a novel monocarbonyl–η^2-ketene complex of permethylated zirconocene (*103*). The carbonylation of $(\eta\text{-}C_5Me_5)_2Zr(CH_2CMe_3)(Cl)$ gave the haloacyl complex $(\eta\text{-}C_5Me_5)_2Zr(COCH_2CMe_3)(Cl)$ which was then deprotonated with lithium diisopropylamide affording the anionic ketene complex $Li^+[(\eta\text{-}C_5Me_5)_2Zr(COCHCMe_3)(Cl)]^-$ (**55**). On treatment of **55** with CO, the neutral monocarbonyl–η^2-ketene complex **56** was obtained as a green oil and was characterized by its ^{1}H- and ^{13}C-NMR spectra together with its IR spectrum which displayed a metal carbonyl stretching frequency at 1987 cm^{-1} (benzene) (*103*).

$$Li^+ \left[(h\text{-}C_5Me_5)_2 Zr \overset{Cl}{\underset{\underset{C}{\diagdown}\diagup}{\diagup}}O \right]^- \quad \xrightarrow[-\,LiCl]{+\,CO}$$

$$\overset{C}{\diagdown\!\!\diagdown} CHCMe_3$$

(**55**)

$$(h\text{-}C_5Me_5)_2 Zr \overset{CO}{\underset{\underset{C}{\diagdown}\diagup}{\diagup}}O$$

$$\overset{C}{\diagdown\!\!\diagdown} CHCMe_3$$

(**56**)

A monocarbonyl complex of zirconocene has been proposed by Floriani and co-workers as a transient intermediate in the carbonylation of the bridging oxymethylene complex $(Cp_2ZrCl)_2(\mu\text{-}CH_2O)$ (**57**) (*104,105*). The assignment of **58** as the proposed intermediate to the final carbonylation product, $[Cp_2Zr(\mu\text{-}CHO{=}CHO)]_2$ (**59**), was based on the appearance of a metal carbonyl band at 1970 cm^{-1} together with the subsequent CO-induced loss of Cp_2ZrCl_2.

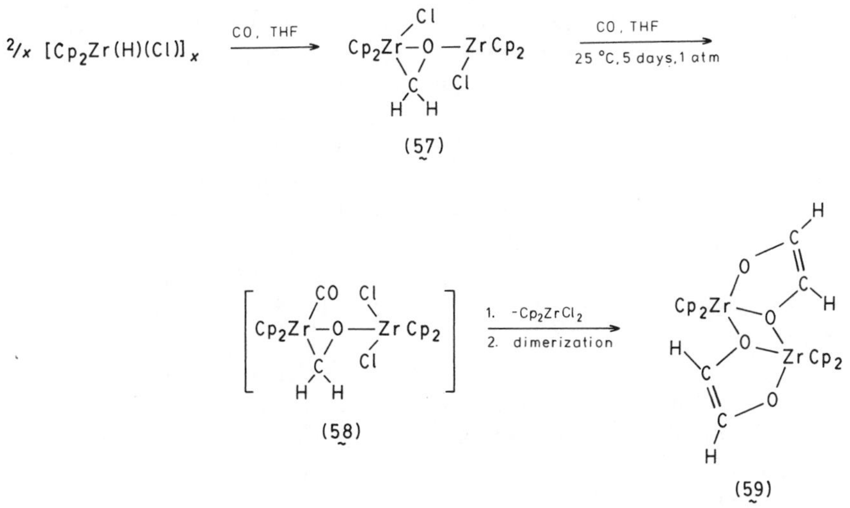

(57)

(58)   (59)

The reaction of CO with the heterobimetallic Zr–Rh complex, $[Cp_2Zr(CH_2PPh_2)(Cl)]_2Rh(CO)Cl$ (60) led to an unexpected zirconocene carbonyl complex 61 (106). Complex 60 was prepared via the reaction of 4 equivalents of $Cp_2Zr(CH_2PPh_2)(Cl)$ with 1 equivalent of $[Rh(CO)_2Cl]_2$. Uncomplexed $Cp_2Zr(CH_2PPh_2)(Cl)$ absorbed CO to give the expected haloacyl complex $Cp_2Zr(COCH_2PPh_2)(Cl)$; however, bimetallic derivative 60 reacted with CO yielding the novel d^0 complex $[Cp_2Zr(CO)-(CH_2PPh_2)(Cl)]_2Rh(CO)Cl$ (61).

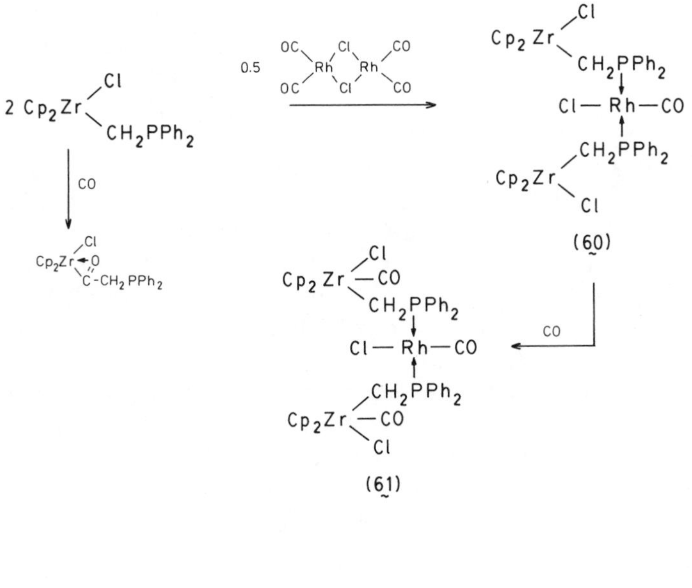

(60)

(61)

Complex **61** was obtained as a relatively unstable, light brown microcrystalline solid, soluble in benzene and THF. Its IR spectrum exhibited a high energy metal carbonyl band at 2040 cm^{-1} due to Zr—CO in addition to ν(Rh—CO) at 1960 cm^{-1} which was also present in **60**. The ^{1}H- and ^{13}C-NMR spectra of **61** were similar to those of **60**; however, when **60** was treated with ^{13}CO, the ^{13}C-NMR spectrum of the resulting complex, **61**—(^{13}CO), showed an intense singlet at δ 188 assignable to Zr—^{13}CO. Neither the ^{13}C resonance for a doublet attributable to Rh—^{13}CO (J_{C-Rh} = ~70 Hz) nor for an acyl carbon was detected. Acyl formation via **61** is postulated as being unfavorable due to steric hindrance and stereorigidity.

While essentially all the metal carbonyl complexes for group 4B contain terminal CO ligands, only recently have some bonafide doubly bridging carbonyl complexes been reported. However, these complexes are heteronuclear, since the carbonyl ligand bridges a zirconium atom with the metal center of a late transition metal.

The treatment of $(\eta\text{-}C_5Me_5)_2ZrH_2$ or $[(\eta\text{-}C_5Me_5)_2ZrN_2]_2N_2$ (**29**) with CpM(CO)$_2$ (M = Co, Rh) and CpRu(CO)$_2$H resulted in the formation of the mixed-metal dimers shown below (*107,108*).

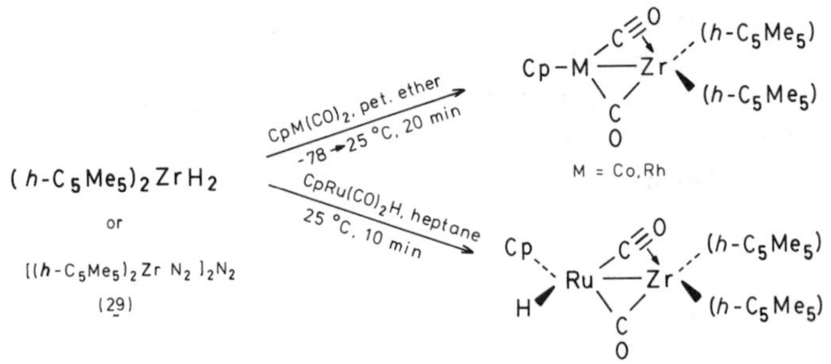

The carbonyl stretching frequency for the conventional bridging CO ligand appeared at 1737 cm^{-1} for the Co complex, 1752 cm^{-1} for the Rh complex, and 1706 cm^{-1} for the Ru complex, while the carbonyl band for the $\mu\text{-}\eta^1$, η^2-bridging CO moiety appeared at 1683, 1696, and 1671 cm^{-1}, respectively. This $\mu\text{-}\eta^1$, η^2-bonding mode has been also found in Cp$_2$Zr(μ-OCMe)(μ-CO)Mo(CO)Cp (*109*), Cp$_2$M(μ-CC$_6$H$_4$Me-4)(μ-CO)W(CO)Cp (M = Ti, Zr) (*110*), and Cp$_2$Ti(μ-CR=CH$_2$)(μ-CO)-W(CO)Cp (R = C$_6$H$_4$Me-4) (*111*).

All three of the mixed-metal dimers react with CO to give the respective dicarbonyl compounds, CpM(CO)$_2$ (M = Co, Rh, RuH), and (η-C$_5$Me$_5$)$_2$Zr(CO)$_2$ (**30**). It is presumed that this reaction is initiated by the breaking of the zirconium–$\mu\text{-}\eta^1,\eta^2$-carbonyl bond.

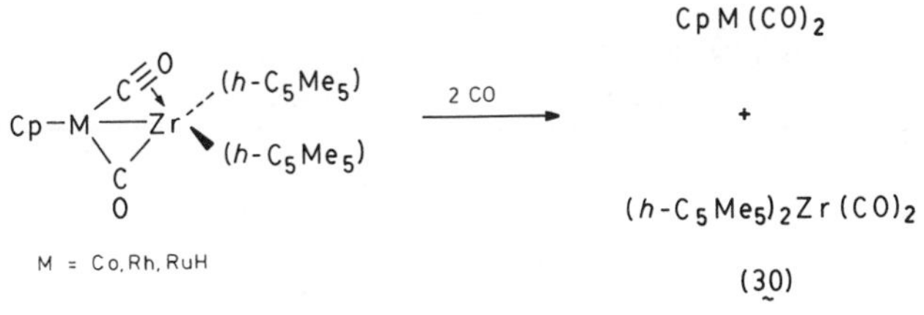

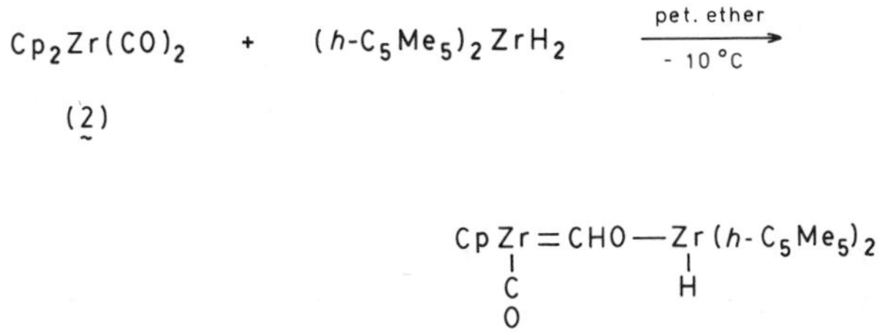

$$Cp-M\overset{\displaystyle C\overset{O}{\diagdown}}{\underset{\displaystyle \underset{O}{C}}{\diagup}}Zr\overset{(h\text{-}C_5Me_5)}{\underset{(h\text{-}C_5Me_5)}{\diagdown}} \xrightarrow{2\ CO} \begin{array}{c} CpM(CO)_2 \\[6pt] + \\[6pt] (h\text{-}C_5Me_5)_2Zr(CO)_2 \end{array} \quad (30)$$

M = Co, Rh, RuH

$$Cp_2Zr(CO)_2 \quad + \quad (h\text{-}C_5Me_5)_2ZrH_2 \xrightarrow[-10\,°C]{pet.\ ether}$$

(2)

$$CpZr=CHO-Zr(h\text{-}C_5Me_5)_2$$
$$\underset{O}{\overset{|}{C}} \qquad \overset{|}{H}$$

Recently Bercaw and co-workers have reacted $Cp_2Zr(CO)_2$ (2) with $(\eta\text{-}C_5Me_5)_2ZrH_2$ in toluene to produce a zirconoxy carbene complex $[\nu(CO) = 1925\ cm^{-1}$ (Nujol)] (112).

VIII

MONOCARBONYL DERIVATIVES OF HAFNOCENE

A. Formation of Hafnocene Monocarbonyl–Phosphine Complexes

The first phosphine-substituted derivatives of $Cp_2Hf(CO)_2$ (3) were reported by Rausch and co-workers in 1979 and were prepared photochemically, similarly to the zirconium congeners (90). The irradiation of $Cp_2Hf(CO)_2$ (3) with either PPh_3 or $Ph_2PCH_2CH_2PPh_2$ (diphos) led to the isolation of $Cp_2Hf(CO)(PPh_3)$ (58) (62) and $Cp_2Hf(CO)(Ph_2PCH_2CH_2PPh_2)$ as very air-sensitive, maroon solids in 31 and 14% yields, respectively. Similarly, the photolysis of heptane solutions of 3 with PMe_3 and PF_3 gave the corresponding complexes, $Cp_2Hf(CO)(PR_3)$ (R = Me, F), as oily, air-sensitive solids. Like the zirconium reactions, the competing photodegradation of $Cp_2Hf(CO)_2$ (3) to oligomeric hafnocene prevented the formation of the hafnocene monocarbonyl–phosphine complexes in higher yields.

While $Cp_2Hf(CO)(PPh_3)$ (62) was found to be more reactive toward acetylenes than $Cp_2Hf(CO)_2$ (3), no monocarbonyl–η^2-acetylene complexes of hafnocene were observed, in contrast to the reaction of acetylenes with $Cp_2Ti(CO)(PPh_3)$ (42) (50). Instead, the reaction of 62 with $RC\equiv CR$ (R = Et, Ph) led to the respective hafnacyclopentadienes (58).

TABLE V

IR AND ^{1}H-NMR SPECTRAL PROPERTIES FOR HAFNOCENE
MONOCARBONYL–PHOSPHINE COMPLEXES

Complex	$\nu(CO)$ (cm^{-1})	$\delta(\eta^5\text{-}C_5H_5)$ (ppm) (C$_6$D$_6$)	J_{H-P}(Hz) (η^5-C$_5$H$_5$)	Ref.
Cp$_2$Hf(CO)(PF$_3$)	1910^a; 1922^b	4.79 (d)	2.5	90
Cp$_2$Hf(CO)(PPh$_3$)	1830^a	4.75 (d)	1.8	58,90
Cp$_2$Hf(CO)[Ph$_2$PCH$_2$CH$_2$PPh$_2$]	1827^a	4.75 (d)	1.8	90
Cp$_2$Hf(CO)(PMe$_3$)	1824^a	4.83 (d)	2.0	90

^{a} THF. ^{b} Hexane.

Characterization of the various monocarbonyl–phosphine complexes of hafnocene was achieved by IR and ^{1}H-NMR spectroscopy. These data appear in Table V.

B. Formation of $(\eta\text{-}C_5Me_5)_2Hf(H)_2(CO)$

Treatment of a hexane solution of the 16-e^- complex, $(\eta\text{-}C_5Me_5)_2HfH_2$, with CO at $-35°C$ resulted in formation of the thermally unstable $(\eta\text{-}C_5Me_5)Hf(H)_2(CO)$ as evidenced by the appearance of a high energy metal carbonyl band at 2036 cm^{-1} in its IR spectrum (70). As the solution warmed, this band decreased in intensity concurrent with the appearance of metal carbonyl bands at 1941 and 1850 cm^{-1} assignable to $(\eta\text{-}C_5Me_5)_2Hf(CO)_2$ (33) (70,72).

$$(h\text{-}C_5Me_5)_2HfH_2 \xrightarrow[-35\ °C]{\substack{CO \\ \text{hexane}}} (h\text{-}C_5Me_5)_2Hf(H)_2(CO)$$

$$\xrightarrow{-5\ °C} (h\text{-}C_5Me_5)_2Hf(CO)_2$$

$$(33)$$

IX

GROUP 4B CARBONYL COMPLEXES CONTAINING π-BONDED ACYCLIC HYDROCARBON LIGANDS

Without question the vast majority of group 4B metal carbonyl complexes contain a metallocene framework. Only two carbonyl complexes of group 4B have been reported that contain π-bonded acyclic hydrocarbon ligands that are not metallocenes.

In 1982 Ernst and Liu prepared the novel complex, bis(η-2,4-dimethylpentadienyl)carbonyltitanium (63) by exposure of a pentane solution of bis(η-2,4-dimethylpentadienyl)titanium (64) ("open titanocene") to CO (113). The yellow, thermally sensitive compound (63) exhibited a metal carbonyl band at 1942 cm^{-1}, while the ^{1}H- and ^{13}C-NMR

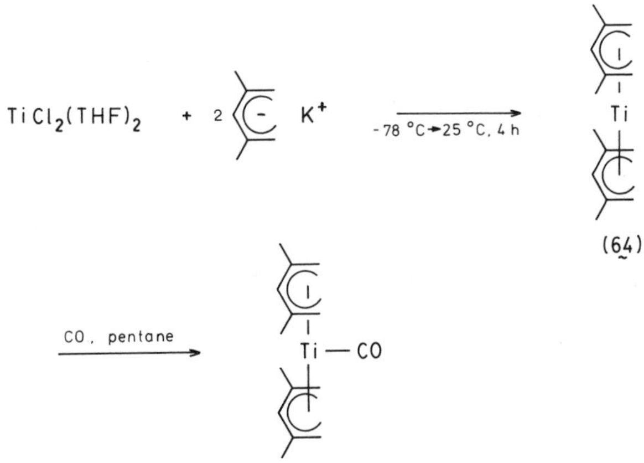

spectra each displayed four resonances denoting the equivalency of the pentadienyl ligands and probably the existence of a mirror plane of symmetry perpendicular to the ligand plane. While a structural determination of 63 by X-ray diffraction has yet to be done, the pentadienyl ligands are thought to be nearly cis-eclipsed. Their central portions are bent toward each other such that the CO is bound near the open portions of the pentadienyl ligands. It is noteworthy that only one CO rather than two coordinates to 64, thus 63 possesses only a 16-e^- configuration.

The only Zr(0) carbonyl complex has been prepared by Wreford and co-workers. When ZrCl$_4$ was treated with 1,2-bis(dimethylphosphino)ethane (dmpe), and then subsequently reduced with sodium amalgam in the presence of 1,3-butadiene, the dmpe bridged dimer, [(η-C$_4$H$_6$)$_2$Zr(dmpe)]$_2$(dmpe) (65), resulted (114). The brown crystalline dimer 65 was found to be in equilibrium with the 16-e^- coordinatively unsaturated complex, (η-C$_4$H$_6$)$_2$Zr(dmpe) (66), and free dmpe. When toluene solutions of 65 were exposed to CO at $-45°$C, 1 equivalent of CO per equivalent of Zr was consumed and the CO adduct (η-C$_4$H$_6$)$_2$Zr-(dmpe)(CO) (67) precipitated as a yellow solid. If these mixtures were allowed to warm above $-22°$C under vacuum, the precipitate dissolved and the consumed CO evolved (114). Complex 67 could be isolated by

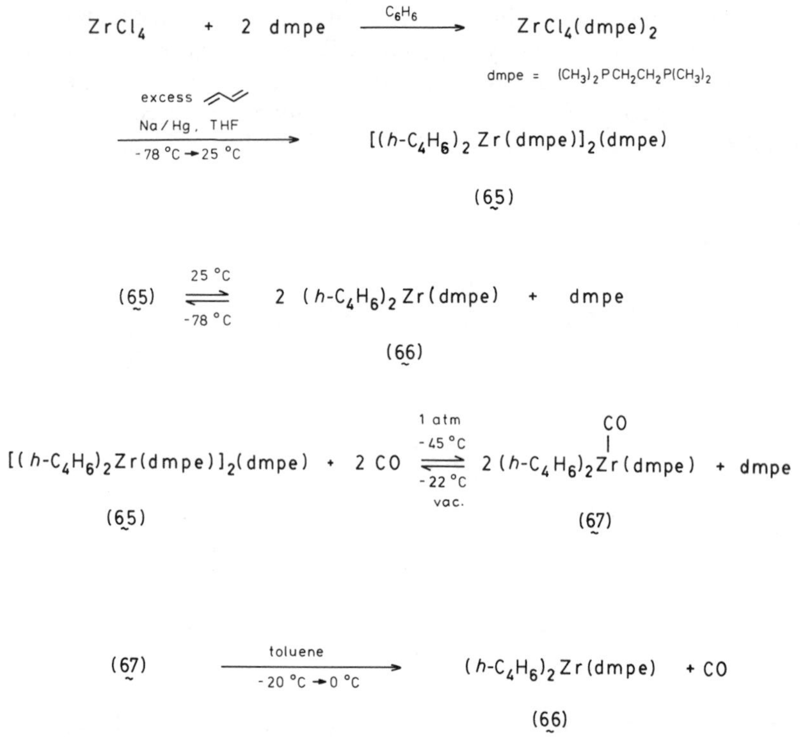

$$ZrCl_4 \quad + \quad 2 \ dmpe \quad \xrightarrow{C_6H_6} \quad ZrCl_4(dmpe)_2$$

$$dmpe = (CH_3)_2PCH_2CH_2P(CH_3)_2$$

$$\xrightarrow[-78\,°C \to 25\,°C]{\substack{excess \ \diagup\!\!\diagdown \\ Na/Hg, \ THF}} \quad [(h-C_4H_6)_2 Zr(dmpe)]_2(dmpe)$$

$$(\underline{65})$$

$$(\underline{65}) \quad \underset{-78\,°C}{\overset{25\,°C}{\rightleftharpoons}} \quad 2 \ (h-C_4H_6)_2 Zr(dmpe) \quad + \quad dmpe$$

$$(\underline{66})$$

$$[(h-C_4H_6)_2Zr(dmpe)]_2(dmpe) \ + \ 2\ CO \quad \underset{-22\,°C \atop vac.}{\overset{1\,atm \atop -45\,°C}{\rightleftharpoons}} \quad 2\ (h-C_4H_6)_2 \overset{\overset{\displaystyle CO}{|}}{Zr}(dmpe) \ + \ dmpe$$

$$(\underline{65}) \qquad\qquad\qquad\qquad\qquad\qquad\qquad (\underline{67})$$

$$(\underline{67}) \qquad \xrightarrow[-20\,°C \to 0\,°C]{toluene} \qquad (h-C_4H_6)_2 Zr(dmpe) \quad + \ CO$$

$$(\underline{66})$$

low-temperature filtration and was characterized by ^{31}P and ^{13}C NMR using 90% enriched ^{13}CO. In the solid state **67** is pyrophoric and is stable at $-20°C$ under an atmosphere of CO. However, while under CO at $25°C$ it violently decomposes in only a few minutes. Complex **66** could be isolated when toluene solutions of **67** were allowed to warm to $0°C$ under vacuum (*114*).

X

GROUP 4B CARBONYL COMPLEXES CONTAINING ONE η^5-C_5H_5 RING

Mono(η-cyclopentadienyl) carbonyl complexes of Ti, Zr, and Hf have been noted only in the patent literature. A U.S patent issued to Ethyl Corporation in 1962 described the use of such complexes as anti-knock agents and lubricant additives. Two examples of complexes purported to have been made are shown below (*115*).

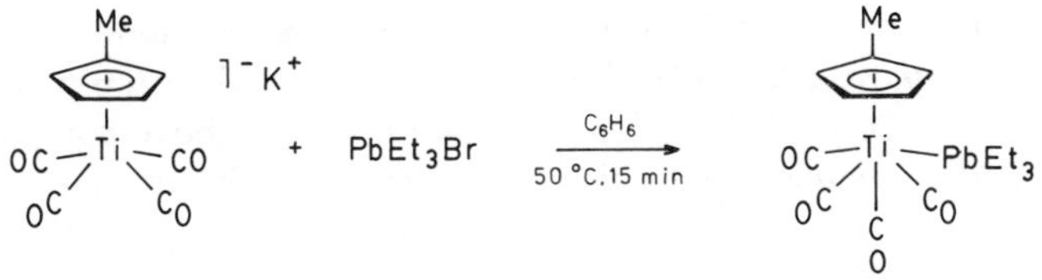

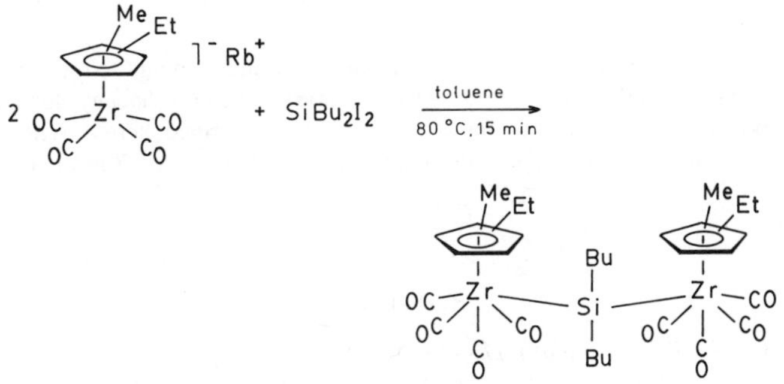

The preparation of the various mono(η-cyclopentadienyl)tetracarbonyl-metallates was not addressed.

Demerseman has briefly noted that the reduction of $CpTiCl_3$ with Al or Mg in a CO atmosphere resulted in the identification of $Cp_2Ti(CO)_2$ (**1**) as the only titanium carbonyl species (27).

XI

GROUP 4B CARBONYL COMPLEXES CONTAINING η-ARENE RINGS

The only description of (η-arene)carbonyl complexes of Ti, Zr, and Hf has again been in the patent literature. Ethyl Corporation reported the preparation of (η-C_6H_6)Ti(CO)$_3$ via the carbonylation of $TiPh_2Cl_2$ (116).

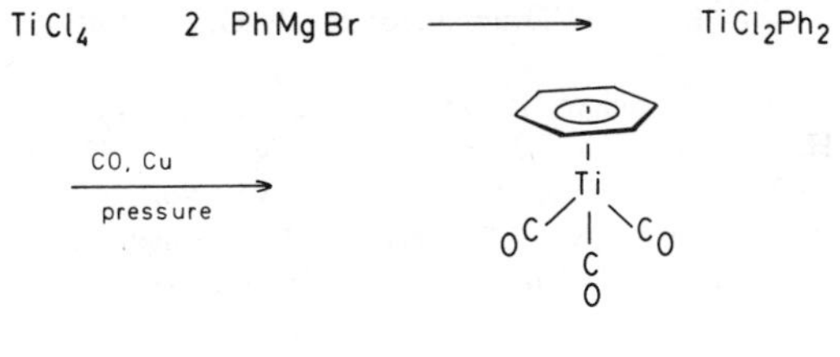

Other such complexes which have been described include $(\eta\text{-}C_6H_5Et)$-Ti(CO)$_4$, $(\eta\text{-}para\text{-}xylene)Zr(CO)_4$, and $(\eta\text{-mesitylene})(\eta\text{-methylcyclo-}$pentadiene)Ti(CO)$_2$ (117). Characterization of the above complexes was not discussed. Their alleged existence seems unlikely without supporting evidence.

ACKNOWLEDGMENTS

The authors are very grateful to the National Science Foundation and to the donors to the Petroleum Research Fund, administered by the American Chemical Society, for grants that have made possible their contributions to the research described herein. They are also grateful to Professor Jerry Atwood of the University of Alabama for fruitful collaborative efforts in this area.

REFERENCES

1. Abel, E. W., *Rev. (London)* **17**, 133 (1963).
2. Basolo, F. and Pearson, R. G., "Mechanisms of Inorganic Reactions," p. 533. Wiley, New York, 1967.
3. Calderazzo, F., Ercoli, R., and Natta, G., "Organic Synthesis via Metal Carbonyls" (Wender, I. and Pino, P., eds.), Vol. 1, p. 1. Wiley(Interscience), New York, 1968.
4. Murray, J. G., *J. Am. Chem. Soc.* **81**, 752 (1959).
5. Murray, J. G., *J. Am. Chem. Soc.* **83**, 1287 (1961).
6. Thomas, J. L. and Brown, K. T., *J. Organomet. Chem.* **111**, 297 (1976).
7. Demerseman, B., Bouquet, G., and Bigorne, M., *J. Organomet. Chem.* **107**, C19 (1976).
8. Fachinetti, G., Fochi, G., and Floriani, C., *J. Chem. Soc. Chem. Commun.* 230 (1976); Fochi, G., Guidi, G., and Floriani, C., *J. Chem. Soc. Dalton Trans.* 1253 (1984).
9. Jazwinski, S. T. and Sisto, J. A., U.S. Pat. 2,793,106 (May 21, 1957); *Chem. Abstr.* **51**, 11978 (1957).
10. Jazwinski, S. T. and Sisto, J. A., U.S. Pat. 2,793,107 (May 21, 1957); *Chem. Abstr.* **51**, 11979 (1957).
11. Busby, R., Klotzbücher, W., and Ozin, G. A., *Inorg. Chem.* **16**, 822 (1977).
12. Wreford, S. S., Fischer, M. B., Lee, J., James, E. J., and Nyberg, S. C., *J. Chem. Soc. Chem. Commun.* 458 (1981).
13. Domaille, P. J., Harlow, R. L., and Wreford, S. S., *Organometallics* **1**, 935 (1982).
14. Ballintine, T. A. and Schmulbach, C. D., *J. Organomet. Chem.* **164**, 381 (1979).
15. Röder, A., Thiele, K., Palyi, G., and Marko, L., *J. Organomet. Chem.* **199**, C31 (1980).
16. Calderazzo, F., Salzmann, J. J., and Mosimann, P., *Inorg. Chim. Acta* **1**, 65 (1967).
17. Clauss, K. and Bestian, H., *Justus Liebigs Ann. Chem.* **654**, 8 (1962).
18. Watt, G. W., Baye, L. J. and Drummond, Jr., F. O., *J. Am. Chem. Soc.* **88**, 1138 (1966).
19. Salzmann, J. J. and Mosimann, P., *Helv. Chim. Acta* **50**, 1831 (1967).
20. Brintzinger, H. H. and Bercaw, J. E., *J. Am. Chem. Soc.* **92**, 6182 (1970).
21. Davison, A. and Wreford, S. S., *J. Am. Chem. Soc.* **96**, 3017 (1974).
22. Pez, G. P. and Armor, J. N., "Advances in Organometallic Chemistry" (Stone, F. G. A. and West, R., eds.), Vol. 19, p. 1. Academic Press, New York, 1981.

23. Van Tamelen, E. E., Cretney, W., Klaentschi, N., and Miller, J. S., *J. Chem. Soc. Chem. Commun.* 481 (1972).
24. Bercaw, J. E., Marvich, R. H., Bell, L. G., and Brintzinger, H. H., *J. Am. Chem. Soc.* **94,** 1219 (1972).
25. El Murr, N. and Chaloyard, A., *J. Organomet. Chem.* **231,** 1 (1982).
26. Demerseman, B., Bouquet, G., and Bigorne, M., *J. Organomet. Chem.* **93,** 199 (1975).
27. Demerseman, B., Bouquet, G., and Bigorne, M., *J. Organomet. Chem.* **101,** C24 (1975).
28. Sobota, P., Jezowska-Trzebiatowska, B., and Janas, Z., *J. Organomet. Chem.* **118,** 253 (1976).
29. Demerseman, B., Bouquet, G., and Bigorne, M., *J. Organomet. Chem.* **145,** 41 (1978).
30. Battaglia, L. P., Nardelli, M., Pellizzi, C., Predieri, G., and Chiusoli, G. P., *J. Organomet. Chem.* **209,** C7 (1981).
31. Battaglia, L. P., Nardelli, M., Pelizzi, C., Predieri, G., and Chiusoli, G. P., *J. Organomet. Chem.* **259,** 301 (1983).
32. (a) Siegert, F. W. and de Liefde Meijer, H. J., *J. Organomet. Chem.* **20,** 141 (1969). (b) McDermott, J. X., Wilson, M. E., and Whitesides, G. M., *J. Am. Chem. Soc.* **98,** 6529 (1976).
33. Fischer, E. O. and Löchner, A., *Z. Naturforsch.* **15b,** 266 (1960).
34. Masai, H., Sonogashira, K., and Hagihara, N., *Mem. Inst. Sci. Ind. Res. Osaka Univ.* **25,** 117 (1968).
35. Bercaw, J. E. and Brintzinger, H. H., *J. Am. Chem. Soc.* **91,** 7301 (1969).
36. Marvich, R. H. and Brintzinger, H. H., *J. Am. Chem. Soc.* **93,** 2046 (1971).
37. Bottomley, F., Lin, I. J. B., and Mukaida, M., *J. Am. Chem. Soc.* **102,** 5238 (1980).
38. Fachinetti, G. and Floriani, C., *J. Chem. Soc. Chem. Commun.* 654 (1972).
39. Fachinetti, G. and Floriani, C., *J. Organomet. Chem.* **71,** C5 (1974).
40. Wailes, P. C., Coutts, R. S. P., and Weigold, H., "Organometallic Chemistry of Titanium, Zirconium, and Hafnium," p. 240. Academic Press, New York, 1974.
41. Alt, H. G. and Rausch, M. D., *J. Am. Chem. Soc.* **96,** 5936 (1974).
42. Rausch, M. D., Boon, W. H., and Alt, H. G., *J. Organomet. Chem.* **141,** 299 (1977).
43. Rausch, M. D., Boon, W. H., and Mintz, E. A., *J. Organomet. Chem.* **160,** 81 (1978).
44. Teuben, J. H., de Boer, E. J. M., Klazinga, A. H., and Klei, B., *J. Mol. Catal.* **13,** 107 (1981).
45. Klei, B., Teuben, J. H., and de Leife Meijer, H. J., *J. Chem. Soc. Chem. Commun.* 342 (1981).
46. Klei, B., Teuben, J. H., de Liefde Meijer, H. J., Kwak, E. J., and Bruins, A. P., *J. Organomet. Chem.* **224,** 327 (1982).
47. (a) Bristow, G. S., Lappert, M. F., Martin, T. R., Atwood, J. L., and Hunter, W. F., *J. Chem. Soc. Dalton Trans.* 399 (1984). (b) Antinolo, A., Lappert, M. F., and Winterborn, D. J. W., *J. Organomet. Chem.* **272,** C37 (1984).
48. Chang, M., Timms, P. L., and King. R. B., *J. Organomet. Chem.* **199,** C3 (1980).
49. Sikora, D. J., Rausch, M. D., Rogers, R. D., and Atwood, J. L., *J. Am. Chem. Soc.* **103,** 982 (1981).
50. Edwards, B. H., Rogers, R. D., Sikora, D. J., Atwood, J. L., and Rausch, M. D., *J. Am. Chem. Soc.* **105,** 416 (1983).
51. Sikora, D. J. Ph.D. dissertation, University of Massachusetts, Amherst, MA, 1982.
52. Gell, K. I. and Schwartz, J., *J. Organomet. Chem.* **162,** C11 (1978).
53. Schwartz, J., *Pure Appl. Chem.* **52,** 733 (1980).
54. Gell, K. I. and Schwartz, J., *J. Chem. Soc. Chem. Commun.* 244 (1979).
55. Gell, K. I. and Schwartz, J., *J. Am. Chem. Soc.* **103,** 2687 (1981).

56. Kropp, K., Skibbe, V., Erker, G., and Krüger, C., *J. Am. Chem. Soc.* **105,** 3353 (1983).
57. Erker, G., Engel, K., Krüger, C., and Müller, G., *Organometallics* **3,** 128 (1984).
58. Sikora, D. J. and Rausch, M. D., *J. Organomet. Chem.* **276,** 21 (1984).
59. Sikora, D. J., Moriarty, K. J., and Rausch, M. D., *Inorg. Syn.* **24** (1986).
60. Bercaw, J. E. and Brintzinger, H. H., *J. Am. Chem. Soc.* **93,** 2045 (1971).
61. Demerseman, B., Bouquet, G., and Bigorne, M., *J. Organomet. Chem.* **132,** 223 (1977).
62. Sikora, D. J., Rausch, M. D., Rogers, R. D., and Atwood, J. L., *J. Am. Chem. Soc.* **103,** 1265 (1981).
63. Tung, H. and Brubaker, Jr., C. H., *Inorg. Chim. Acta* **52,** 197 (1981).
64. Cohen, S. A., Auburn, P. R., and Bercaw, J. E., *J. Am. Chem. Soc.* **105,** 1136 (1983).
65. Manriquez, J. M., McAlister, D. R., Sanner, R. D., and Bercaw, J. E., *J. Am. Chem. Soc.* **98,** 6733 (1976).
66. Manriquez, J. M., and Bercaw, J. E., *J. Am. Chem. Soc.* **96,** 6229 (1974).
67. Manriquez, J. M., McAlister, D. R., Sanner, R. D., and Bercaw, J. E., *J. Am. Chem. Soc.* **100,** 2716 (1978).
68. Bercaw, J. E., *Adv. Chem. Ser.* **167,** 136 (1978).
69. Wolczanski, P. T. and Bercaw, J. E., *Acc. Chem. Res.* **13,** 121 (1980).
70. Marsella, J. A., Curtis, C. J., Bercaw, J. E., and Caulton, K. G., *J. Am. Chem. Soc.* **102,** 7244 (1980).
71. Berry, D. H., Bercaw, J. E., Jircitano, A. J., and Mertes, K. B., *J. Am. Chem. Soc.* **104,** 4712 (1982).
72. Roddick, D. M., Fryzuk, M. D., Seidler, P. F., Hillhouse, G. L., and Bercaw, J. E., *Organometallics* **4,** 97 (1985).
73. Bercaw, J. E., personal communication.
74. Moriarty, K. J. and Rausch, M. D., 11th International Conference on Organometallic Chemistry, Pine Mountain, GA, October 10–14, p. 65, 1983.
75. Alt. H. G. and Rausch, M. D., *Z. Naturforsch.* **30b,** 813 (1975).
76. Smith, J. A. and Brintzinger, H. H., *J. Organomet. Chem.* **218,** 159 (1981).
77. Smith, J. A., Von Seyerl, J., Huttner, G., and Brintzinger, H. H., *J. Organomet. Chem.* **173,** 175 (1979).
78. Curtis, M. D., D'Errico, J. J., Duffy, D. N., Epstein, P. S., and Bell, L. G., *Organometallics* **2,** 1808 (1983).
79. Le Blanc, J. C., Moise, C., Maisonnat, A, Poilblanc, R., Charrier, C., and Mathey, F., *J. Organomet. Chem.* **231,** C43 (1982).
80. Rausch, M. D., Edwards, B. H., Rogers, R. D., and Atwood, J. L., *J. Am. Chem. Soc.* **105,** 3882 (1983).
81. Macomber, D. W., Hart, W. P., and Rausch, M. D., *J. Am. Chem. Soc.* **104,** 884 (1982).
82. Hart, W. P., Ph.D. dissertation, University of Massachusetts, Amherst, MA, 1981.
83. Thomas, J. L. and Brintzinger, H. H., *J. Am. Chem. Soc.* **94,** 1386 (1972).
84. Thomas, J. L., *J. Am. Chem. Soc.* **95,** 1838 (1973).
85. Fragalia, I., Ciliberto, E., and Thomas, J. T., *J. Organomet. Chem.* **175,** C25 (1979).
86. Böhm, M. C., *Inorg. Chim. Acta* **62,** 171 (1982).
87. Atwood, J. L., Stone, K. E., Alt, H. G., Hrncir, D. C., and Rausch, M. D., *J. Organomet. Chem.* **96,** C4 (1976).
88. Atwood, J. L., Stone, K. E., Alt, H. G., Hrncir, D. C., and Rausch, M. D., *J. Organomet. Chem.* **132,** 367 (1977).
89. Atwood, J. L., Rogers, R. D., Hunter, W. F., Floriani, C., Fachinetti, G., and Chiesi-Villa, A., *Inorg. Chem.* **19,** 3812 (1980).

90. Sikora, D. J., Rausch, M. D., Rogers, R. D., and Atwood, J. L., *J. Am. Chem. Soc.* **101**, 5079 (1979).

91. Atwood, J. L., personal communication.

92. King, R. B., *Acc. Chem. Res.* **13**, 243 (1980).

93. Tsumura, R. and Hagihara, N., *Bull. Chem. Soc. Jpn.* **38**, 861 (1965).

94. Fachinetti, G. and Floriani, C., *J. Chem. Soc. Chem. Commun.* 66 (1974).

95. Fachinetti, G., Floriani, C., Marchetti, F., and Mellini, M., *J. Chem. Soc. Dalton Trans.* 1398 (1978).

96. Rausch, M. D. and Xiu-Zhong, Z., unpublished studies..

97. Fachinetti, G., Del Nero, S., and Floriani, C., *J. Chem. Soc. Dalton Trans.* 1046 (1976).

98. de Boer, E. J. M., Ten Cate, L. C., Staring, A. G. J., and Teuben, J. H., *J. Organomet. Chem.* **181**, 61 (1979).

99. Demerseman, B., Pankowski, M., Bouquet, G., and Bigorne, M., *J. Organomet. Chem.* **117**, C10 (1976).

100. Pez, G. P., Apgar, P., and Crissey, R. K., *J. Am. Chem. Soc.* **104**, 482 (1982).

101. Manriquez, J. M., Sanner, R. D., Marsh, R. E., and Bercaw, J. E., *J. Am. Chem. Soc.* **98**, 3042 (1976).

102. Manriquez, J. M., McAlister, D. R., Rosenberg, E., Shiller, A. M., Williamson, K. L., Chan, S. I., and Bercaw, J. E., *J. Am. Chem. Soc.* **100**, 3078 (1978).

103. Moore, E. J., Straus, D. A., Armantrout, J., Santarsiero, B. D., Grubbs, R. H., and Bercaw, J. E., *J. Am. Chem. Soc.* **105**, 2068 (1983).

104. Gambarotta, S., Floriani, C., Chiesi-Villa, A., and Guastini, C., *J. Am. Chem. Soc.* **105**, 1690 (1983).

105. Floriani, C., *Pure Appl. Chem.* **55**, 1 (1983).

106. Choukroun, R. and Gervais, D., *J. Chem. Soc. Chem. Commun.* 1300 (1982).

107. Barger, P. T. and Bercaw, J. E., *J. Organomet. Chem.* **201**, C39 (1980).

108. Barger, P. T. and Bercaw, J. E., *Organometallics* **3**, 278 (1984).

109. Longato, B., Norton, J. R., Huffman, J. C., Marsella, J. A., and Caulton, K. G., *J. Am. Chem. Soc.* **103**, 209 (1981); Marsella, J. A., Huffman, Caulton, K. G., Longato, B., and Norton, J. R., *J. Am. Chem. Soc.* **104**, 6360 (1982).

110. Dawkins, G. M., Green, M., Mead, K. A., Salaün, J.-Y., Stone, F. G. A., and Woodward, P., *J. Chem. Soc. Dalton Trans.* 527 (1983).

111. Barr, R. D., Green, M., Howard, J. A. K., Marder, T. B., Moore, I., and Stone, F. G. A., *J. Chem. Soc. Chem. Commun.* 746 (1983).

112. Barger, P. T., Santarsiero, B. D., Armantrout, J., and Bercaw, J. E., *J. Am. Chem. Soc.* **106**, 5178 (1984).

113. Liu, J. and Ernst, R. D., *J. Am. Chem. Soc.* **104**, 3737 (1982).

114. Beatty, R. P., Datta, S., and Wreford, S. S., *Inorg. Chem.* **18**, 3139 (1979).

115. Gorsich, R. D., U.S. Pat. 3,069,445 (Dec. 18, 1962).

116. Brit. Pat. 896,391 (May 16, 1962); *Chem. Abstr.* **58**, 4597g (1963).

117. Coffield, T. H. and Closson, R. D., U.S. Pat. 3,361,779 (Jan. 2, 1968); *Chem. Abstr.* **68**, 95983 (1968).

Index

Cumulative List of Contributors

Abel, E. W., **5,** 1; **8,** 117
Aguilo, A., **5,** 321
Albano, V. G., **14,** 285
Alper, H., **19,** 183
Anderson, G. K., **20,** 39
Armitage, D. A., **5,** 1
Armor, J. N., **19,** 1
Atwell, W. H., **4,** 1
Baines, K. M., **25,** 1
Behrens, H., **18,** 1
Bennett, M. A., **4,** 353
Birmingham, J., **2,** 365
Blinka, T. A., **23,** 193
Bogdanović, B., **17,** 105
Bradley, J. S., **22,** 1
Brinckman, F. E., **20,** 313
Brook, A. G., **7,** 95; **25,** 1
Brown, H. C., **11,** 1
Brown, T. L., **3,** 365
Bruce, M. I., **6,** 273; **10,** 273; **11,** 447; **12,** 379; **22,** 59
Brunner, H., **18,** 151
Cais, M., **8,** 211
Calderon, N., **17,** 449
Callahan, K. P., **14,** 145
Cartledge, F. K., **4,** 1
Chalk, A. J., **6,** 119
Chatt, J., **12,** 1
Chini, P., **14,** 285
Chiusoli, G. P., **17,** 195
Churchill, M. R., **5,** 93
Coates, G. E., **9,** 195
Collman, J. P., **7,** 53
Connelly, N. G., **23,** 1; **24,** 87
Connolly, J. W., **19,** 123
Corey, J. Y., **13,** 139
Corriu, R. J. P., **20,** 265
Courtney, A., **16,** 241
Coutts, R. S. P., **9,** 135
Coyle, T. D., **10,** 237
Craig, P. J., **11,** 331
Cullen, W. R., **4,** 145
Cundy, C. S., **11,** 253
Curtis, M. D., **19,** 213
Darensbourg, D. J., **21,** 113; **22,** 129
Deacon, G. B., **25,** 237
de Boer, E., **2,** 115
Dessy, R. E., **4,** 267

Dickson, R. S., **12,** 323
Eisch, J. J., **16,** 67
Emerson, G. F., **1,** 1
Epstein, P. S., **19,** 213
Erker, G., **24,** 1
Ernst, C. R., **10,** 79
Evans, J., **16,** 319
Evans, W. J., **24,** 131
Faller, J. W., **16,** 211
Faulks, S. J., **25,** 237
Fehlner, T. P., **21,** 57
Fessenden, J. S., **18,** 275
Fessenden, R. J., **18,** 275
Fischer, E. O., **14,** 1
Forster, D., **17,** 255
Fraser, P. J., **12,** 323
Fritz, H. P., **1,** 239
Furukawa, J., **12,** 83
Fuson, R. C., **1,** 221
Gallop, M. A., **25,** 121
Garrou, P. E., **23,** 95
Geiger, W. E., **23,** 1; **24,** 87
Geoffroy, G. L., **18,** 207; **24,** 249
Gilman, H., **1,** 89; **4,** 1; **7,** 1
Gladfelter, W. L., **18,** 207; **24,** 41
Gladysz, J. A., **20,** 1
Green, M. L. H., **2,** 325
Griffith, W. P., **7,** 211
Grovenstein, Jr., E., **16,** 167
Gubin, S. P., **10,** 347
Guerin, C., **20,** 265
Gysling, H., **9,** 361
Haiduc, I., **15,** 113
Halasa, A. F., **18,** 55
Harrod, J. F., **6,** 119
Hart, W. P., **21,** 1
Hartley, F. H., **15,** 189
Hawthorne, M. F., **14,** 145
Heck, R. F., **4,** 243
Heimbach, P., **8,** 29
Helmer, B. J., **23,** 193
Henry, P. M., **13,** 363
Herberich, G. E., **25,** 199
Herrmann, W. A., **20,** 159
Hieber, W., **8,** 1
Hill, E. A., **16,** 131
Hoff, C., **19,** 123
Horwitz, C. P., **23,** 219

Housecroft, C. E., **21,** 57
Huang, Yaozeng (Huang, Y. Z.), **20,** 115
Ibers, J. A., **14,** 33
Ishikawa, M., **19,** 51
Ittel, S. D., **14,** 33
James, B. R., **17,** 319
Jolly, P. W., **8,** 29; **19,** 257
Jonas, K., **19,** 97
Jones, P. R., **15,** 273
Jukes, A. E., **12,** 215
Kaesz, H. D., **3,** 1
Kaminsky, W., **18,** 99
Katz, T. J., **16,** 283
Kawabata, N., **12,** 83
Kettle, S. F. A., **10,** 199
Kilner, M., **10,** 115
King, R. B., **2,** 157
Kingston, B. M., **11,** 253
Kitching, W., **4,** 267
Köster, R., **2,** 257
Krüger, G., **24,** 1
Kudaroski, R. A., **22,** 129
Kühlein, K., **7,** 241
Kuivila, H. G., **1,** 47
Kumada, M., **6,** 19; **19,** 51
Lappert, M. F., **5,** 225; **9,** 397; **11,** 253; **14,** 345
Lawrence, J. P., **17,** 449
Lednor, P. W., **14,** 345
Longoni, G., **14,** 285
Luijten, J. G. A., **3,** 397
Lukehart, C. M., **25,** 45
Lupin, M. S., **8,** 211
McKillop, A., **11,** 147
Macomber, D. W., **21,** 1; **25,** 317
Maddox, M. L., **3,** 1
Maitlis, P. M., **4,** 95
Mann, B. E., **12,** 135
Manuel, T. A., **3,** 181
Mason, R., **5,** 93
Masters, C., **17,** 61
Matsumura, Y., **14,** 187
Mingos, D. M. P., **15,** 1
Mochel, V. D., **18,** 55
Moedritzer, K., **6,** 171
Morgan, G. L., **9,** 195
Mrowca, J. J., **7,** 157
Müller, G., **24,** 1
Mynott, R., **19,** 257
Nagy, P. L. I., **2,** 325

Nakamura, A., **14,** 245
Nesmeyanov, A. N., **10,** 1
Neumann, W. P., **7,** 241
Ofstead, E. A., **17,** 449
Ohst, H., **25,** 199
Okawara, R., **5,** 137; **14,** 187
Oliver, J. P., **8,** 167; **15,** 235; **16,** 111
Onak, T., **3,** 263
Oosthuizen, H. E., **22,** 209
Otsuka, S., **14,** 245
Pain, G. N., **25,** 237
Parshall, G. W., **7,** 157
Paul, I., **10,** 199
Petrosyan, W. S., **14,** 63
Pettit, R., **1,** 1
Pez, G. P., **19,** 1
Poland, J. S., **9,** 397
Poliakoff, M., **25,** 277
Popa, V., **15,** 113
Pourreau, D. B., **24,** 249
Pratt, J. M., **11,** 331
Prokai, B., **5,** 225
Pruett, R. L., **17,** 1
Rausch, M. D., **21,** 1; **25,** 317
Reetz, M. T., **16,** 33
Reutov, O. A., **14,** 63
Rijkens, F., **3,** 397
Ritter, J. J., **10,** 237
Rochow, E. G., **9,** 1
Roper, W. R., **7,** 53; **25,** 121
Roundhill, D. M., **13,** 273
Rubezhov, A. Z., **10,** 347
Salerno, G., **17,** 195
Satgé, J., **21,** 241
Schmidbaur, H., **9,** 259; **14,** 205
Schrauzer, G. N., **2,** 1
Schulz, D. N., **18,** 55
Schwebke, G. L., **1,** 89
Setzer, W. N., **24,** 353
Seyferth, D., **14,** 97
Shen, Yanchang (Shen, Y. C.), **20,** 115
Shriver, D. F., **23,** 219
Siebert, W., **18,** 301
Sikora, D. J., **25,** 317
Silverthorn, W. E., **13,** 47
Singleton, E., **22,** 209
Sinn, H., **18,** 99
Skinner, H. A., **2,** 49
Slocum, D. W., **10,** 79
Smith, J. D., **13,** 453